Noise and Vibration Mitigation for Rail Transportation Systems

Proceedings of the 10th International Workshop on Railway Noise, Nagahama, Japan, 18–22 October 2010

Tatsuo Maeda, Pierre-Etienne Gautier,
Carl E. Hanson, Brian Hemsworth,
James Tuman Nelson, Burkhard Schulte-Werning,
David Thompson, and Paul de Vos (Eds.)

Springer

Dr. Tatsuo Maeda
Principal Researcher,
Research & Development Promotion Division
Railway Technical Research Institute
2-8-38 Hikari-cho Kokubunji-shi
Tokyo 185-8540 Japan
E-mail: maeda@rtri.or.jp

Dr. Pierre-Etienne Gautier
Scientific Director, INEXIA,
1 place aux Etoiles 93212
Saint Denis La Plaine cedex, France
E-mail:
pierre-etienne.gautier@inexia-ingenierie.com

Carl E. Hanson
Founder
Harris Miller Miller & Hanson Inc.
77 South Bedford Street
Burlington MA 01803 USA
E-mail: chanson@hmmh.com

Brian Hemsworth
Partner
Brian Hemsworth Noise Consultant LLP
16 Whistlestop Close Mickleover
Derby DE3 9DA UK
E-mail: brian.hemsworth@btinternet.com

Dr. James Tuman Nelson
Vice President
Wilson, Ihrig & Associates
6001 Shellmound Suite 400
Emeryville California 94608 USA
E-mail: jnelson@wiai.com

Dr. Burkhard Schulte-Werning
General Manager, Integrated Systems and Infrastructure Technology
Deutsche Bahn AG, DB Systemtechnik
Volckerstr. 5 80939 Munchen Germany
E-mail:
burkhard.schulte-werning@deutschebahn.com

Prof. Dr. David Thompson
Professor of Railway Noise and Vibration
Institute of Sound and Vibration Research,
University of Southampton
Southampton SO17 1BJ UK
E-mail: djt@isvr.soton.ac.uk

Paul de Vos
Principal Consultant,
Environmental Quality Department,
DHV BV
Postbus 1132 NL 3800
BC Amersfoort The Netherlands
E-mail: paul.devos@dhv.com

ISBN 978-4-431-53926-1 e-ISBN 978-4-431-53927-8

DOI 10.1007/978-4-431-53927-8

Notes on Numerical Fluid Mechanics and Multidisciplinary Design ISSN 1612-2909

Typeset & Cover Design: Scientific Publishing Services Pvt. Ltd., Chennai, India.

Printed on acid-free paper

5 4 3 2 1 0

springer.com

NNFM Editor Addresses

Preface

This book contains the presentations given during the 10th International Workshop on Railway Noise (IWRN10), which took place in Nagahama, Japan, October 18–22, 2010. This was the first time that the event had been staged in Asia. The workshop was organized by the Railway Technical Research Institute, Japan (RTRI), and was supported by the Ministry of Land, Infrastructure, Transport and Tourism and the Ministry of the Environment, Japan.

In total, there were 146 participants from 15 countries across the world: 73 from Japan, 11 from the United Kingdom, 11 from China, 11 from Sweden, 7 from Germany, 7 from Australia, 6 from France, 5 from the United States, 5 from Korea, 3 from Belgium, 2 from the Netherlands, 2 from Denmark, 1 from Austria, 1 from Singapore, and 1 from Spain.

IWRN10 covered not only railway noise but also other environmental problems such as vibrations and micro-pressure waves from tunnel portals. Sessions were classified into 7 categories: 1. Prospects, legal regulation, and perception; 2. Wheel and rail noise; 3. Structure-borne noise and squeal noise; 4. Ground-borne vibration; 5. High-speed trains (aerodynamic noise and micro-pressure waves from tunnel portals); 6. Interior noise and sound barriers; and 7. Prediction, measurements, and monitoring. As many as 50 papers were presented in oral sessions and 20 papers were given in poster sessions.

Papers submitted to IWRN10 from many countries showed that the understanding of environmental phenomena and their countermeasures are essential if railway networks are to prosper and high-speed railways are to be developed around the world. IWRN10 presented state-of-the-art technology on environmental phenomena and their countermeasures and demonstrated aspects, theoretical models, and prediction tools on wheel and rail noise, squeal noise, structure-borne noise, aerodynamic noise, micro-pressure waves from tunnel portals, interior noise, and ground-borne vibrations.

Following the tradition of the previous workshops, IWRN10 was held as a single-session event for participants to exchange information on all facets of railway environmental problems.

The International Committee of IWRN supports the chairman of IWRN10 during the preparation process with the experience and expertise of its members. Assistance is given to formulate the scientific program, to release the Call for Papers, to perform the paper selection process, to act as session chairmen at IWRN10, and to act as a peer review group for the IWRN10 proceedings.

Special thanks are due to Masao Uchida, Koichi Goto, and other members of the planning committee and the executive committee of RTRI for all the hard work and care in organizing the conference.

The editors are grateful to Prof. E.H. Hirschel as the general editor of the "Notes on Numerical Fluid Mechanics and Multidisciplinary Design" and also to the staff of Springer Japan for the opportunity to publish the proceedings of IWRN10 in the series.

We look forward to this volume being used as a "state-of-the-art" reference by scientists and engineers involved in solving environmental problems of railways.

November 2010

Tatsuo Maeda
Pierre-Etienne Gautier
Carl E. Hanson
Brian Hemsworth
James Tuman Nelson
Burkhard Schulte-Werning
David Thompson
Paul de Vos

Contents

Session 1: Prospects, Legal Regulation, Perception

Session 2: Wheel and Rail Noise – Part 1

Session 3: Wheel and Rail Noise – Part 2

Poster Session 1: Wheel and Rail Noise

Session 9: Interior Noise, Sound Barrier

Poster Session 3: High Speed Trains, Interior Noise, Sound Barrier, Prediction, Measurements, Monitoring

Session 10: Prediction, Measurements, Monitoring – Part 1

Session 6: Ground-Borne Vibration – Part 2

Poster Session 2: Prospects, Legal Regulation, Perception, Structural-Borne Noise, Squeal Noise, Ground-Borne Vibration

Session 4: Structural-Borne Noise, Squeal Noise

Session 5: Ground-Borne Vibration – Part 1

Session 11: Prediction, Measurements, Monitoring – Part 2

Transportation Noise Annoyance, Cognitive Performance and Sleep Disturbances Related to Temporal Structures and Traffic Modes (Deufrako Project "RAPS")

F. Margiocchi and F. Poisson

SNCF, Innovation and Research Department,
45 rue de Londres, 75739 Paris cedex 8, France
tel: +33153422356; fax: +33153429784
e-mail: florence.margiocchi@sncf.fr

Summary

The aim of this project is to further understand the influence of transport noise on sleep, to apply this knowledge to the development of measures for the reduction in noise. The topic comes within the scope of a DEUFRAKO cooperation (Noise group) and is subsidised by the ADEME, for the French part of the program. The project brings together some major French and German specialists in this field; the SNCF and DB have contributed to the development of scenarios that are the most representative of the current situation. The French part of this project has now been completed, and is presented in this paper. Two subjects were studied. On one hand, the LINC concentrated on sleep studies. The analysis of train noise influence on the sleeping pattern and the cognitive performances, as well as its impact on the cardiovascular system carried out by the LINC is developed in this paper. On the other hand, tests on cognitive performances and short-term discomfort have been carried out by the LMRTE. The experiments linked to the sound quality of the pass-by as well as the results related to the study of the influence of temporal parameters on functional discomfort are presented. The results concerning railway noise will be compared with road and aircraft noise at the end of the project in partnership with the German part.

1 Introduction

The noise generated by road and aircraft is the main source of noise pollution mentioned by the populations surveyed. However, in the context of policies of local and state authorities regarding growth of sustainable development, the train, which is not very polluting, and is expected to be the major means of transport in

T. Maeda et al. (Eds.): Noise and Vibration Mitigation for Rail Trans. Sys., NNFM 118, pp. 1–8.
springerlink.com

the future. Moreover, railway traffic is predicted to grow in the coming years. To avert the inconvenience and noise annoyance caused by more railway traffic or the construction of new lines, new studies have been planned.

The Project RAPS aims at the provision of applicable knowledge that allows the directed development of noise abatement measures. Within the framework of a Deufrako (Noise group) co-operation, this project brings together some major specialists. The French partners involved are SNCF, LINC and LMRTE, with French funding by ADEME. The German partners involved are DB, CUE, DLR and IfADO.

ADEME :	*Agence de l'Environnement et de la Maîtrise de l'Energie*
CUE :	*Catholic University Eichstätt-Ingolstadt*
DB :	*Deutsche Bahn*
DLR :	*Deutsches Zentrum für Luft und Raumfarhrt*
IfADO :	*Leibniz Research Centre for Working Environment and Human Factors at TU Dortmund*
LINC :	*Laboratoire d'imagerie et de neurosciences cognitives*
LMRTE :	*Laboratoire Mobilités Réseaux Territoires et Environnement*
SNCF :	*Société Nationale des Chemins de fer Français*

This project concerns

1) the quantitative difference of sleep disturbances resulting from noise emitted by rail, road and air traffic,
2) the significance of temporal structure of railway noise on annoyance, activity interference and sleep to determine if human beings react generally more to fluctuations than to steady states.

The project focuses on noise emitted from currently operating trains, while including possible future technical advances.

Until now, studies have been carried out concerning the noise of aircraft and road traffic in relation to sleep disturbance, but not much research has been related to railway noise. A previous study with unrealistic noise scenarios led to the result that the railway noise is more disturbing than street traffic. These noise scenarios used nearly the same number of trains, cars and airplanes. In reality, the number of cars is much more than one hundred times greater than the number of trains for the same noise level averaged over a time period.

Realistic noise scenarios were used for sleep laboratory studies at IfADO with real measured train pass-byes during a night period at different sections of the DB network; and for sleep laboratory studies at LINC with real measured train pass-byes of the SNCF network.

The project started in April 2006 on the French side and in January 2007 on the German side.

Here, the French contribution to the study is presented in two parts. In the first part, tests on cognitive performances and short-term discomfort have been carried out by the LMRTE. The experiments linked to the sound quality of the pass-by as well as the results dedicated to the study of the influence of temporal parameters on functional discomfort are presented. In the second part, the LINC concentrated on sleep studies. The analysis of train noise on the sleeping pattern and the cognitive performances, as well as their impact on the cardiovascular system carried out by the LINC is discussed here.

2 Influence of the Temporal Structure of Railway Traffic on Sound Quality and Functional Discomfort

2.1 Sound Quality of the Pass-By

LMRTEs objective was to characterize the perception of the passage of various types of trains with particular attention paid to the temporal dimensions. The sound quality of the pass-by was studied to identify which information was perceived and to what extent it had an influence on the choices of subjects.

Methodology: A collection of 24 stimuli of equal length and each containing only a single passage was built. During the recording of the signals, two variable parameters were retained: the type of train (Corail, Freight, TER, TGV) and the measurement distance from the track (7,5m, 50m and 100m), the spectral and temporal properties being different according to these two parameters. The signals were recorded in free field. The speed of the pass-by was constant and dependent on the type of train (Corail/TGV: 160kph; Freight: 100kph; TER: 140kph). The sites were selected to minimize the environmental effects (ground effect etc.).

Because the level is strongly related to the annoyance [Fastl: 90], the 12 stimuli were standardized in relation to level L_{Aeq}. A low variation of level was then introduced to compare the influence of a given level to those of other perceptive effects. New versions of the standardized recordings were created by increasing the mean level by 2 dB, before being included in the collection.

The stimuli evaluation was carried out with a perceptive test on the comparison method by pairs used in numerous studies (see for example [Parizet: 05]): the subjects had to estimate the dissimilarity between the signals, indicate their preference and explain their choice verbally.

For each test, only part of the complete collection was used. Four sub-groups were dedicated to a single category of train for the three measuring distances. The fifth sub-group made the crossed study possible for all the categories at 50m and 100m. The study for the TGV trains is detailed in [CFA2010/66].

Conclusions: The analysis of the results (INDSCAL, correlation, linguistic study, preference scale, etc.) showed the following: (1) the most important perceptive dimension is related to the fluctuations resulting from the passage of different wagons. It explains the differences heard between the sounds recorded at 7,50 m and the others. (2) The variations in noise level are well perceived (even if strong fluctuations can make them less evident) 5 but do not mask the other temporal or spectral effects. (3) There might be confusion between the variations of level and the temporal variations (length of passage, suddenness of the arrival) or spectral variations, the subjects being able to assimilate both of them. (4) No consensus appears on the question of determining if the length is preferred to the suddenness (and vice-versa). Some subjects qualify a passage as "*rapid*" (positive adjective) when others define it as "*brutal*" (negative). Likewise, some will focus on the "*progressive*" (positive adjective) aspect when others on the "*slow*" (negative) aspect. (5) The noise disruption study shows the spectral aspect remains secondary in relation to the noise level or the temporal aspects.

2.2 Impact of the Temporal Effects on Functional Discomfort

In a second stage, the impact of the repetitions of train pass-by on the functional discomfort in the evening with windows half-open was studied.

Methodology: To be in a realistic noise context, each hour-long scenario is made up of continuous road background noise and various train pass-bys (derived from recordings used for the noise quality tests). The subjects are then in a multi-exposure situation. For the construction of these scenarios, three variable parameters were selected: the number of pass-by, the distance and the state of dominance. For the same reasons as before, the scenarios were standardized in relation to level L_{Aeq}.

A study on the evening French railway traffic gives the average number of pass-by for each category of train over the 18.00 - 22.00 period [Cremezy: 07]. On this basis, two traffic densities were selected for one hour duration: 10 events per hour (6 Corail + 2 Freight + 1 TER + 1 TGV) and 20 events per hour (12 Corail + 4 Freight + 2 TER + 2 TGV).

The recordings at 7.5m had differenct characteristics from the other distances (see section 3.1). Consequently only the pass-bys captured at 50m and 100m were kept: the effect of the suddenness of the arrival can be perceived, whereas the fluctuations due to the pass-by of the wagons are less. The distance will also influence the length of the passage and the distribution of the energy in frequency.

As the "perturbing" aspect of the pass-by can be correlated with the discomfort [Beaumont: 05], the emergence of the pass-by in relation to the background noise was included as a third variable parameter. The dominance of the railway traffic in relation to the road traffic [Champelovier: 03] was therefore considered account. In a dominance situation, we then obtain: L_{Aeq}Trains = L_{Aeq}Background noise + 6dB(A) ; in a non-dominance situation: L_{Aeq}Trains = L_{Aeq} Background noise + 3dB(A).

During the tests, the subjects were placed in a rest situation and invited to carry out a reading activity in a pseudo-anechoic room. At the end of the session, the participants had to evaluate the discomfort, due to the noise environment, to do their reading. This evaluation was carried out according to two scales: a relative scale of evaluation of importance ("To what extent were you disturbed by the noise environment? Twice as much? Three times as much? Etc.") and an absolute scale of categories from 0 (not disturbed at all) to 50 (extremely disturbed). Only the scale of the categories alone made it possible to obtain significant results for the 8 sequences elaborated.

Conclusions: The distribution of the answers, as well as the analyses of variances were carried out to study the subjects' answers. It appears thus that (1) the more pass-bys there are, the less the subjects are disturbed. This can be explained by the fact that the overall level being maintained constant, the peak level values are lower when the traffic is heavier. (2) The non-dominance situations are preferred in general. The subjects reject the trains when they become perturbing elements of the "background" noise. (3) There is an interaction between the dominance and the other variables. The subjects will then adopt distinct strategies as a function of this.

In a dominance situation, the sudden and distinct aspect of the passage will involve an increase in discomfort despite a reduced length of passage. Thus, at a constant overall level, the nearer the participants are to the tracks, the more they are disturbed. In a non-dominance situation, the lengths of the pass-by will influence the discomfort: the emergence being reduced, the trains are more easily associated with the background noise. This noise will thus become more perturbing and the discomfort will increase with the distance from the tracks (at constant overall level).

The suddenness/length opposition appears once more. We rediscover here the various listening strategies already observed during the noise quality evaluation (section 3.1). Some subjects focus their attention on the suddenness of the arrival, whereas others focus on the length of the passage. Putting them in a noise situation with several listening contexts makes it possible to understand better on which temporal aspect(s) the participants base their evaluation of their level of functional discomfort.

2.3 Discussion

In this study, the use of recorded stimuli in situ involves a limitation directly due to the natural correlation between the various characteristics of the signals (distance in relation to the tracks, suddenness of the arrival, length of the passage, spectra etc.). It would be interesting to pursue this work using synthesis stimuli. The various properties could be decorrelated and their impact evaluated independently. We could then associate one single perceptive aspect to one single acoustic magnitude.

During this study, we were confronted by the lack of pertinence of some psychoacoustic tools as regards the characterization of temporally variable phenomena. It is, therefore, necessary to persevere in the development of tools aimed at the characterization of these phenomena to define, for example, a robust and representative indicator of the rhythm associated with the passage of the wagons.

Lastly, to increase the statistical significance of the study, it would be advisable to build tests where each subject would listen to all the noise sequences. We would thus limit the inter-individual variability, but we would also introduce an habituation effect to the test. Moreover, it would be possible to measure the effect of the noise environment on the performance, not in terms of the number of correct responses (which seems to be independent from the noise load), but in terms of response time (or reading time). In fact, a slowing down could appear under the effect of the pass-by, as suggested by some subjects ("The passage of the train stopped my reading and distracted me; "The ear is attracted by the train noise and therefore concentration is impaired", etc.).

3 Sleep Disruption Linked to Transport Noise

In general, this last section carried out by the LINC demonstrated that night train noise administrated sporadically have not much effect on the sleeping pattern and the cognitive performances the next day, but that permanent exposure to night train noise produces adverse effects on the cognitive performances.

3.1 Material and Methods

The study involved two groups of 20 subjects (men and women) divided into age groups: a group of young subjects (age 20-30: 26.2 years ± 3.6 years) called the Junior Group, and a group of middle-aged subjects (age 45-60: 56.2 years ± 4.2 years) called the Senior Group. Each of these groups was in turn divided into two sub-groups, living either near a railway line (noisy environment) or in a quiet area (quiet environment). The recruitment of the subjects in a noisy area was done due to the SNCF selecting portions of line with an average night railway traffic of 25 to 30 freight trains per night.

The study was carried out in two stages: a study on site, including three consecutive nights of actigraphic recording (and subjective questionnaires: discomfort related to noise pollution, subjective perception and somnolence, ...), and a laboratory study consisting of four non-consecutive nights spaced out by about one week each (habituation night, control night, low density night (30 trains/night), high density night (60 trains/night)). The laboratory nights were all followed by a whole day dedicated to the iterative realization of various cognitive tasks with subjective measurements. The same subjects took part in both stages of this study.

Electrophysiological recordings were carried out during the experimental nights (EEG, EOG, EMG, ECG). In addition, then each session of cognitive performance tests included, in addition to the subjective questionnaires (already used in the first phase at home), three very attentional tasks, very sensitive to effects of sleep deprivation, even partial:

- ANT (Attentional Network task) (Fan et al., 2002)
- PVT (Psychomotor Vigilance Task) (Dinges and Powell, 1985)

3.2 Results Analysis

The night train noise produces an increase in arousal and partial arousal proportional to the density of traffic. The subjects describe an impression of having slept badly which translates from the subjective point of view into an increase in complaints about stress and a bad night. It is essentially from the cardiovascular point of view that the effects of noise pollution are felt, with, in particular, an increase in the cardiovascular response percentage and the amplitude of these responses. Nonetheless, it would seem that the important factor is not so much the density of the traffic, and therefore the frequency of train pass-by, but the noise intensity of the trains as is proven by the more adverse effect of the low traffic night (but with a higher noise intensity per train) in comparison to the high traffic night.

This study also showed that age itself translated, as many other studies have demonstrated, into a deterioration in the sleep quality with in particular a lighter sleep pattern marked by an increase in the lightest stages and a diminution of the slow and deep sleep stages, an increase in its fragmentation and its instability in terms of the number of stage changes. On the other hand, contrary to the majority of studies, the cognitive performances the next day were not very poor for the seniors. However, this could be the result, not of better performances in this senior group in particular, but worse performances by the young subjects. In fact, the objective evaluation of the daytime sleep pressure by the wakefulness EEG clearly demonstrated that the young subjects presented higher spectral densities than the seniors, which translates into a higher somnolence, and, therefore, a higher sensitivity of the young subjects to night noise pollution.

However, the most striking result of this study arises from the difference observed between the subjects living near a railway line in comparison to those living in a quiet area. It appears that these subjects, irrespective of their age, present in general, poor performances essentially at the level of reaction times and, therefore, the rapidity of information processing.

This could be the sign of a chronic lack of sleep. However, the examination of their sleeping pattern clearly reveals that when they are laboratory evaluated, the subjects normally living near a railway line have a deeper sleep than those living in a quiet area, which probably translates into a phenomenon of compensation linked to the lack of sleep associated with a behavioural habituation to the noise translating into a form of desensitizing.

Furthermore, these subjects also differ on the cardiovascular level. The cardiovascular reactivity is very clearly reduced in seniors living in a noisy area in comparison to seniors living in a quiet area. On the contrary, among the juniors, those living in a noisy area present a cardiovascular reactivity much higher than those living in a quiet area. Therefore, everything happens as if, in the short term, the chronic exposure to noise produces a hypersensitivity of the vegetative system to noise, whereas with time, and contrary to what was suggested by previous studies, we notice an habituation translating into an extremely reduced reactivity. In fact, even if it is a known fact that the cardiovascular reactivity is considerably reduced with age, probably due to the ageing of the arteries, this would not explain why there is such an obvious difference between the two groups of seniors.

4 Conclusion

The LMRTE studied the sound quality of the train pass-by and their influence on functional discomfort. Their work, within a post-project perspective, would need to be improved, in particular by decorrelating the signal characteristics (use of synthesised signals). The LINC, for its part, analyzed the effect of night train noise on sleeping pattern and cognitive performances. They demonstrated that when they are administered sporadically, they have not much effect on the sleeping pattern and the cognitive performances the next day, as well as their repercussions on the cardiovascular sphere. The SNCF supplied traffic data to define the most representative scenarios of real current and future situations (traffic doubled), brought its railway expertise in the implementation of methodologies and contributed to discussions on the results analysis.

In general, the study made it possible to demonstrate that train noises, while being generally considered as harmless or relatively ecological, present nonetheless all the effects that will have to be taken into consideration if we want to develop railway freight further.

The exchanges with the German part will be presented later and complete and enhance the analyses and conclusions presented here. In fact, the work carried out by the French part only concerns the railway transport mode, contrasting a quiet environment to an environment beside a railway line. The German laboratories have also been worked on all the transport noise: road, aircraft and railway noise. Their work makes it possible to compare and takes into perspective the relative effects of each mode of transport.

Acknowlegments

This project comes within the scope of a DEUFRAKO cooperation (Noise Group) and is subsidised by the ADEME, for the French part.

We would like to thank in particular, all the partners involved in the project [CFA2010/525], and the persons who took part in the experiments and made the realization of these studies possible.

References

[1] Champelovier, P., Cremezy-Charlet, C., Lambert, J.: Evaluation of discomfort due to combined exposure to road and railway noise, INRETS, Collections (2003)

[2] Cremezy-Charlet, C.: DEUFRAKO Project: Transport noise, influence of the temporal pattern on discomfort, cognitive performance and sleep. Night Traffic, SNCF Internal report (2007)

[3] Fastl, H., Widmann, U.: Technical note: Subjective and physical evaluation of aircraft noise. Noise Control Engineering Journal 35(2), 61–63 (1990)

[4] Parizet, E., et al.: Comparison of some listening test methods: a case study. Acta Acustica 91(2), 356–364 (2005)

[5] Lavandier, C., Terroir, J.: Sound quality in of TGV trains: influence of perceptive factors –multidimensional and preferential analysis. In: CFA 2010, Lyon (2010)

[6] Margiocchi, F., Tassi, P., Bonnefond, A., Terroir, J., Lavandier, C.: Bruit des transports, influence de la structure temporelle sur la gêne, les performances cognitives et le sommeil (Projet RAPS - coopération Deufrako). In: CFA 2010, Lyon (2010)

[7] Beaumont, J., Semidor, C.: Event descriptors for qualifying the urban sound environment. EAA European Acoustics Association: Forum Acusticum, Budapest, Hungary (2005)

Advancements in Noise and Vibration Abatement to Support the Noise Reduction Strategy of Deutsche Bahn

B. Schulte-Werning, B. Asmussen, W. Behr, K.G. Degen, and R. Garburg

Deutsche Bahn AG
DB Systemtechnik, Völckerstr. 5, 80939 München, Germany
Tel: +49 89 1308 2581, Fax: +49 89 1308 2491
e-mail: Burkhard.Schulte-Werning@deutschebahn.com

Summary

The Environmental Noise Directive of the European Union requires the member states to produce noise maps and to establish action plans for noise reduction. This directive has been transposed into national law in Germany and directly affects Deutsche Bahn (DB) and its noise reduction strategy. DB has set itself the goal of halving, by the year 2020, the level of rail traffic noise experienced by local residents in 2000.

1 Introduction

The European Parliament and the Council have put into force the Directive on the Assessment and Management of Environmental Noise ("Environmental Noise Directive (END)" [1]), aiming at avoiding, preventing or reducing harmful effects of environmental noise on human health. END requires member states to produce "strategic noise maps" using noise indicators assessing the number of people affected by noise, to inform the public about noise exposure, and to draw up "action plans" to reduce noise where necessary. The directive was transposed into German legislation [2] in 2005 and has tightened the requirements governing traffic noise abatement. In future, local authorities will have to lay down action plans to prevent and reduce environmental noise based on the noise-mapping results. Cost-effectiveness will play a crucial role in the noise reduction strategies.

Deutsche Bahn takes its responsibilities seriously and has set itself the ambitious goal of halving, by the year 2020, the level of rail traffic noise experienced by local residents in 2000 [3]. If this target of 10 dB(A) reduction in the residents noise perception is to be achieved, all available and new abatement techniques need to be used.

T. Maeda et al. (Eds.): Noise and Vibration Mitigation for Rail Trans. Sys., NNFM 118, pp. 9–16.
springerlink.com

2 Noise Reduction Strategy of DB

DB is implementing the voluntary noise rehabilitation programme of the German government. With the introduction of this scheme, it became possible to implement noise protection measures along existing sections of railway tracks. Meanwhile 100 Million € have been made available annually for this programme to install noise protection walls or sound proofed windows. The implementation regulations for this noise rehabilitation programme are set out in the "Guidelines for promoting rail noise abatement measures" published by the Federal Ministry for Transport, Building and Urban Development [4].

Major reductions in noise emission levels of freight wagons are now feasible. The novel composite brake block (known as the "K-block"), which has undergone extensive testing, was recently approved for international use [5]. Up till now, wheel treads were roughened every time the cast-iron brake blocks were applied. With this new development, the treads remain smooth — a fact which will lead to a reduction of 8 to 10 dB(A) in rolling noise. For mixed freight and passenger traffic conditions, the overall effect is 4 to 5 dB(A) noise reduction.

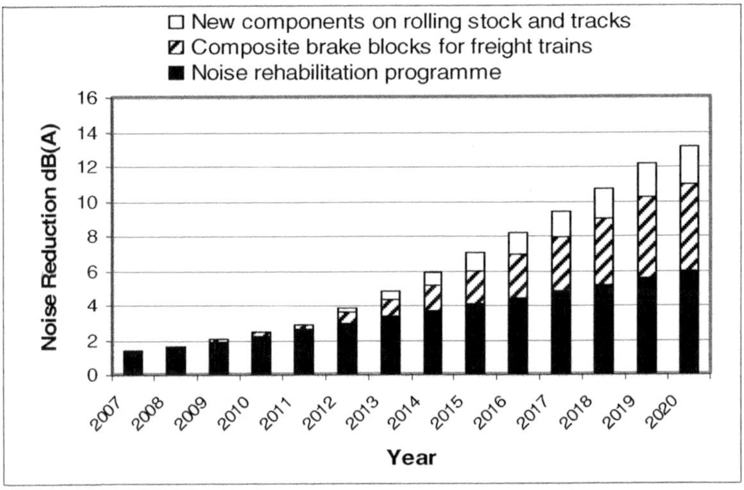

Fig. 1. The three building elements in DB's strategy to reduce railway noise by at least 10 dB(A) by 2020 compared to 2000

To achieve the goal of at least 10 dB(A) noise reduction all over the country, up to 5 dB(A) diminution is necessary in addition. Therefore, the third important building element in DBs national plan to reduce railway noise by at least 10 dB(A) by 2020 is the development of new noise-reducing components for rolling stock and track. The outcome of these efforts will add notably to the effect of the federal noise rehabilitation programme and to noise reduction by retrofitting composite brake-blocks to freight wagons. Fig. 1 shows how the three elements add up to achieve the goal.

3 Development of New Noise-Reducing Components for Rolling Stock and Track

3.1 EU Project SILENCE

Substantial strategies and measures for noise reduction options were under development within the SILENCE project [6], where Deutsche Bahn was involved as a major partner during the project's duration from 2005 to 2008.

SILENCE was an integrated research project funded by the Sixth Framework Programme of the European Commission to develop methodology and technology for improved control of surface transport noise in urban areas. "Integrated" means the combined consideration of city authorities, individual traffic (on road) and mass transport (on rail and road). Within SILENCE, 15 partners from railway undertakings (Trenitalia, STIB, SNCF, DB), the railway industry (ALSTOM, Ansaldobreda, Bombardier, CRF, Lucchini S.A, Corus, D2S), as well as engineering and academic institutions (ISVR, KTH, VTC, Deltarail) formed a railway-related sub-consortium. The railway-related activities were concentrated on two subprojects led by SNCF ("Railway Vehicles") and by DB ("Railway Infrastructure"), the latter focusing on the development and implementation of efficient infrastructure-based noise reduction technologies. DBs major contribution to the project consisted of conducting field tests in combination with extensive measuring campaigns.

In general, there are two options to reduce rolling noise at source: (1) minimization of contact forces by keeping the running surfaces of both rail and wheel smooth and (2) reduction in the intensity of the radiated sound field by increasing the damping of the rail and wheel. The second option was followed within SILENCE. The goal of considerably further reduce the rolling noise of freight wagons equipped with K-blocks (which within SILENCE was considered to be state of the art) by adding dampers to the wheels was particularly challenging. Unlike disk-braked wheels, where dampers are known to be an efficient way to reduce noise emission by about 4 dB(A), wheel dampers for tread-braked wheels are not yet commercially available due to the high temperatures the wheel is subjected to during braking.

It is well-known that increasing the track decay rate by dedicated "rail dampers" attached to the rail web can considerably reduce rolling noise emission. The development of the rail dampers was carried out by Corus Rail and was supplemented by extensive modeling and computer simulation performed by the Institute of Sound and Vibration Research at the University of Southampton. Rail dampers with improved acoustic coupling to the rail web were installed on a test track near Augsburg (see Fig. 2). In an extensive measuring campaign in September 2006, all relevant parameters (sound pressure, rail roughness, rail vibration, track decay rates) were recorded. Noise reductions between 2 dB(A) and 4 dB(A) were observed [7]. The mitigation effect of the dampers depended on the train speed and on track stiffness.

Fig. 2. DBs test track with rail dampers developed by Corus within the SILENCE project (left) and concept of the wheel dampers for freight wagons developed by Lucchini S.A. (right) (color figure online)

Prototypes of the wheel dampers developed by Lucchini S.A. (see Fig. 2) were tested in another measuring campaign in September 2007. A total of 24 wheel dampers were mounted on to the wheels of a test train with sliding wall freight wagons. Their noise emission was measured and compared with that of wheels without dampers. These tests were performed on rails with and without rail dampers. Fig. 3 shows a summary of the measured noise reduction obtained with rail dampers, wheel dampers and a combination of both.

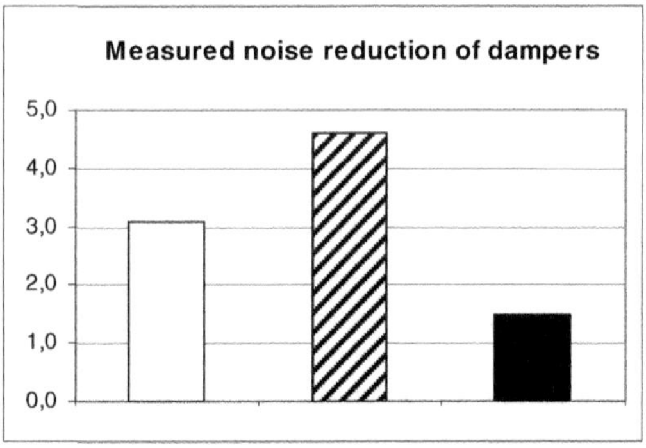

Fig. 3. Measured noise reduction in dB(A) obtained with rail dampers (left column), wheel dampers (right column) and the combination of both (middle column).

3.2 Federal Project "LZarG" - Low-Noise Train on Existing Tracks

To further reduce railway noise, DB started a national research and technical development project in which both academic and industrial partners are involved. Within the projects duration 2007 to 2010 economically satisfying solutions are to be developed which will act in addition to the K-block technology. The project focuses on the different noise emitting components and their interdependencies.

The project comprises the optimization of wheels, bogies and the track system [8]. One subproject covers the wheel/rail contact in detail with the aim of optimizing the bogies of freight trains to find single low-noise components and to reduce the thermal stress on the wheels during the braking process. Minimizing the sound radiation of the wheels of regional trains and freight trains will be effected by developing a new wheel shape design, as well as wheel dampers within a second subproject. In addition, the disks of the braking system connected to the wheelset in the case of regional trains are taken into consideration. A third subproject covers the acoustic optimization of the track system. This is to include damping devices mounted on to the rail, and under sleeper pads. The rail dampers developed by Vossloh have already been mounted on to the rail with hard rail pads near Augsburg (see Fig. 4). First acoustic measurements show a damping effect of 1-2 dB(A) depending on train categories.

Fig. 4. Rail dampers are mounted on an existing track near Augsburg (color figure online)

These three subprojects are combined in the part "reduction of the rolling noise" and cover the technical aspects of the project. At the end of the project, all improvements achieved will be evaluated in field tests with a test train on different track systems. By an adequate combination of measures on rolling stock and the track system, the noise abatement effect will be maximized (color figure online).

3.3 DB Handbook 80025 "Ground-Borne Vibrations"

During the passage of track-bound vehicles, not only noise, but also vibrations are generated at the wheel—rail contact point, transmitted via track and superstructure into the substructure, transferred through the ground (soil) and can generate noticeable vibrations inside adjacent buildings. Germany's national legislation expressly mentions vibrations as potentially annoying or harmful. As a consequence, vibration emissions have to be predicted in the planning process when building new lines or when upgrading existing railway lines and their impact has to be assessed within environmental impact studies. If necessary, vibration mitigation measures have to be foreseen.

Therefore, DB is at present in the process of putting the DB Handbook 80025 "Ground-borne vibrations and secondary air-borne noise" into force as an internal directive [9]. It describes general principles and fundamentals, measurements and prediction, evaluation of ground-borne vibrations caused by rail traffic, mitigation measures and vibrations in connection with construction work.

For the prediction, the guideline describes a spectral forecasting procedure, generally based on third-octave vibration—velocity spectra. In contrast to noise predictions, the reception has to be predicted at a point inside the building, so that all relevant characteristics of the building have to be known. The complete path of the vibrations from the track via the soil to the inside of the building is divided into several sub-systems: the source system (origin of the dynamic excitation from the concurrence of vehicle and track), the transmission system (propagation of the vibration through the soil towards the building) and the reception system (the transfer function describes for example foundation vibrations and the secondary air-borne noise by vibrations of the walls and ceilings inside the building).

Based on the vibration—velocity spectra, which have been determined in the way described, the KB values have to be calculated. KB is the frequency-weighted non-dimensional vibration—velocity (a definition can be found e.g. in [10]). KB is used in several European countries for the assessment of vibrations.
For new lines, these KB values can be evaluated with certain "thumb-rule" values. For existing and upgraded lines, normally for the evaluation of annoyance, the augmentation of vibration exposure level has to be determined. In the past, DB implemented several annoyance studies to find characteristic values which help in describing the disturbance level [11].

In cases where the forecasted values exceed the vibration limit values, some mitigation measures have to be determined. Usually "active" mitigation measures — which mean measures near the track system to prevent the occurrence or transmission of vibrations — are preferred, but in some cases also mitigation measures within the path of transmission or connected to the buildings are the first-choice solutions.

Several established mitigation measures exist on the basis of additional resilient elements in the track. The optimum measure for a particular situation depends on the frequency range of the excitation of the main peaks in the KB spectrum. If the maximum of the vibration emission spectrum is at very low frequencies (below 15 Hz) floating—slab track systems can be used very effectively. Ballast mats are effective when the maximum in the vibration emission or the needed mitigation at the perception point is at medium frequencies (15 to 35 Hz). These mitigation measures are very efficient for railway lines in tunnels and DB has invested a great deal in investigations and specific measures in the last few years (e.g. in connection with the new railway tunnel crossing the centre of Berlin [12] and other projects). For surface lines, no similar effective measures are known. Sometimes, sleepers with elastic supports (under-sleeper pads or USPs) are an alternative with moderate costs (compared to floating slab track systems and ballast mats) which not only increase the track quality, but can also achieve a significant reduction in ground-borne vibrations and structure-borne noise provided the relevant excitation frequencies are above 40 Hz. USPs have the additional advantage that they are a retrofit solution (Fig. 5).

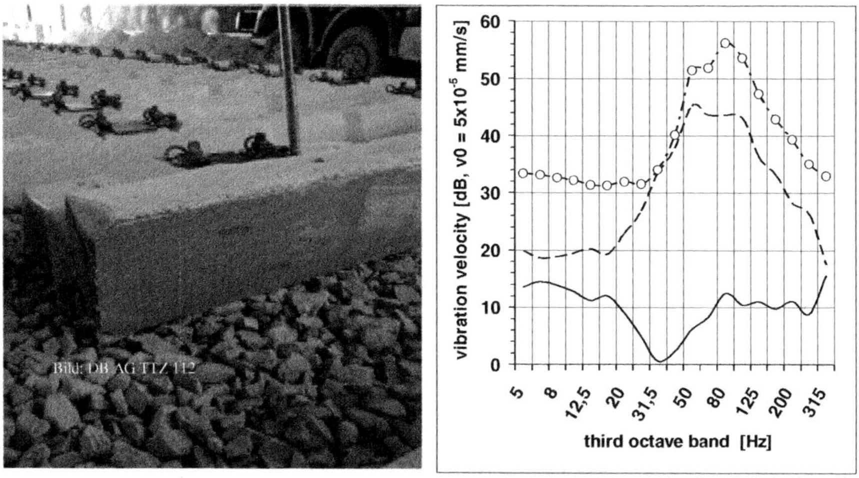

Fig. 5. Retrofit of a tunnel with sleepers with USP and measured vibration level in vertical direction on first floor of an adjacent building before (-0-0-) and after (-----) the installation of under-sleeper pads and the insertion loss as the difference spectrum (______). (color figure online)

DB has been participating in a UIC-funded project (UIC Project No I/05/U/4440 "Under Sleeper Pads") to share information and experience with USPs in different fields of applications. Measurements in conjunction with field tests in tunnels and on open lines have shown insertion losses of about 5-15 dB for frequencies above 50 Hz [13] (see Fig. 5). DB has summarized its specifications for USPs in the standard DBS 918 145 [14]. It should be pointed out that USPs can have a potentially negative effect on air-borne noise emission. Measurements carried out at different locations have not provided a consistent picture yet. Some measurements indicated a negligible effect, while others recorded considerably increased noise emission. Further research is necessary here.

4 Conclusions

This article describes the three building elements of DBs noise reduction strategy to support DBs self-obligation of halving, by 2020, the level of rail traffic noise compared with the amount experienced by local residents in 2000. To achieve this ambitious target, well-coordinated implementation of state-of-the-art noise abatement techniques, as well as new noise reduction techniques covering both vehicles and infrastructure is needed. After completing the projects mentioned, DB will be able to use the end results to halfen the railway noise level. Tools, methodologies and input data for decision support systems are available as well.

Acknowledgements

The SILENCE project receives funding from the European Commission, Research Directorate-General, which is gratefully acknowledged.
The LZarG project receives funding from the German Federal Ministry of Economics and Technology, which is gratefully acknowledged.

References

[1] Directive 2002/49/EC of the European Parliament and of the Council of 25 June 2002 relating to the assessment and management of environmental noise
[2] Gesetz zur Umsetzung der EG-Richtlinie über die Bewertung und Bekäm- pfung von Umgebungslärm, Bundesgesetzblatt 2005, Teil I Nr. 38, Bonn (2005)
[3] Deutsche Bahn AG, Berlin (2009), `http://www.deutschebahn.com/site/nachhaltigkeitsbericht`
[4] Bundesministerium für Verkehr, Bau und Stadtentwicklung, `http://www.bmvbs.de/Klima_-Umwelt-Energie/Mobilitaet-Verkehr/Laermschutz`
[5] de Vos, P., Bergendorff, M., Brennan, M., van der Zijpp, F.: Implementing the Retrofitting Plan for the European Rail Freight Fleet. In: Proceedings of the 8th International Workshop on Railway Noise, September 8-11, pp. 555–562 (2004)
[6] `http://www.silence-ip.org/site/`
[7] Asmussen, B., Stiebel, D., Kitson, P., Farrington, D., Benton, D.: Reducing the noise emission by increasing the damping of the rail: Results of a field test. In: Noise and Vibration Mitigation for Rail Transportation Systems. NNFM, vol. 99. Springer, Heidelberg (2008)
[8] Description of Work Leiser Zug auf realem Gleis - LZarG, BMWi-Verbundprojekt Förderkennzeichen 19 U 7020A (03.08.2007)
[9] Handbuch 80025 Erschütterung und sekundärer Luftschall (2008), `http://www.dbportal.db.de/zrwd`
[10] DIN 45669 Teil 1 Messung von Schwingungsimmissionen; Schwingungsmesser; Anforderungen, Prüfungen (1995)
[11] Degen, K.G., Behr, W., Grütz, H.-P.: Investigations and results concerning railway-induced ground-borne vibrations in Germany. In: 8th IWRN, Buxton, UK (September 2004); Journal of Sound and Vibration 293, 865–872 (2006)
[12] Jaquet, T., Garburg, R.: Measurements and investigations at the floating-track-bed system in the North-South tunnel in Berlin. In: Noise and Vibration Mitigation for Rail Transportation Systems. Notes on Numerical Fluid Mechanics and Multidisciplinary Design(NNFM), vol. 99. Springer, Heidelberg (2008)
[13] Garburg, R.: UIC Project No I/05/U/4440 Under Sleeper Pads, WP 3; Acoustics and Vibration, final report (July 2007)
[14] Deutsche Bahn Standard DBS 918145, Spannbetonschwellen mit elastischer Sohle (2004)

Planning and Controlling Railway Noise in a Metropolis: Our Practical Experience

Benson Lee, W. Chau, John Lam, and Maurice Yeung

Assessment and Noise Group, Environmental Protection Department,
The Government of the Hong Kong Special Administrative Region
People's Republic of China
Tel: +852 2835 1093, Fax: +852 2802 4511
e-mail: bensonlee@epd.gov.hk

Summary

Hong Kong is a metropolis and railway is the key transportation mode in Hong Kong with about 4 million trips made in an average weekday. Currently the railway network in Hong Kong includes 211km of rail and 85 railway stations. In the coming decade, 56km of rail will be added to the railway network. Planning and controlling railway noise in Hong Kong is a great challenge to both the railway operator and the planning / enforcement authorities.

The paper will give an account from planning perspective how the railway noise problems in Hong Kong are prevented and also from the enforcement perspective how the noise problem from existing railway lines are tackled. Details regarding noise standards, abatement programme and the noise mitigation measures adopted for recently planned/completed railway systems will also be discussed.

1 Introduction

Hong Kong is a metropolis with an area of about 1100 square kilometre (85% are of hilly areas) and a population of about 7 million. Railway is the key transportation mode in Hong Kong. Railway service in Hong Kong has a history of 100 years and the first railway line in Hong Kong was in operation in 1910. Currently there are 10 heavy railway lines in Hong Kong. The railway network includes 211km of rail and 85 railway stations, with about 4 million trips made in an average weekday. There are currently at least 5 new railway lines under planning/construction, which will further add 56km of rail to the network in the coming decade.

The existing local commuter trains in Hong Kong run with a maximum speed of 80-130 kph. Most of the stations in urban areas are underground stations located at 12-37m below ground. Most of the stations at the less populated New Territories are at grade or elevated on viaduct. Both air-borne and ground-borne railway noise

T. Maeda et al. (Eds.): Noise and Vibration Mitigation for Rail Trans. Sys., NNFM 118, pp. 17–23.
springerlink.com

would be a great problem if noise issue has not been examined seriously at the planning stage of railway projects. On the other hand, noise impact from railway lines already in service, which were built many decades ago, would also need to be tackled to bring relief to the people living near these railway lines.

2 Environmental Planning to Avoid Excessive Railway Noise in Hong Kong

2.1 Environmental Impact Assessment Ordinance

Since 1997, the noise impact from new railway lines has been put under the scrutiny of the Environmental Impact Assessment Ordinance (EIAO). Under this legal framework, all designated projects shall require Environmental Impact Assessment (EIA) and obtain Environmental Permits prior to proceeding with the works. Railway systems including the tracks, stations, depots, etc are designated projects. The Environmental Permit issued under the ordinance will require the project proponent to implement the EIA specified noise mitigation measures. The provision of such mitigation measures becomes a statutory requirement.

2.2 Noise Standards for Planning New Railways

Under the EIAO all noise sensitive uses including dwellings, hospitals, educational institutions, hotels, places of public worship have to be protected. Noise standards for planning purpose are contained in the Technical Memorandum on Environmental Impact Assessment Process [1] issued under the EIAO. The noise standards depend on the location of the noise sensitive receiver (NSR), the characteristics of its surrounding environment, and the time of day under consideration.

Area-sensitive ratings (ASR) are used to describe the type of area within which the NSR is located and the degree of the effect of particular influencing factors (e.g. major roads or industrial establishments) on the NSR in concern. Normally, a low-density residential area not affected by such influencing factors will be assigned an ASR of A, which represents the quiet surrounding environment. A multi-storey housing estate with its noise climate dominated by a heavily trafficked road will be assigned an ASR of C, which basically represents urban setting.

The acceptable noise levels at the façade of NSR for the respective ASRs at different time periods are as follows:

Table 1. Acceptable Noise Levels in Leq (30 min) dB(A)

Time Period	ASR = A	ASR = B	ASR = C
7 am to 11 pm	60	65	70
11 pm to 7 am	50	55	60

NB The assessment point is normally at a position one metre from the external façade.

For most NSRs in urban areas, the noise limits at the receivers are 65dB(A) in Leq(30min) from 7 am to 11 pm , and 55dB(A) from 11 pm to 7 am. On the other hand, an additional night-time (2300-0700 hours) criterion of 85dB(A) in Lmax is also stipulated in the said Technical Memorandum to prevent sleep disturbance, irrespective of the ASR.

2.3 Noise Mitigation Measures Adopted at Planning Stage of Railway Systems

Following the Hong Kong Government Railway Development Strategy 2000 [2] that gives priority to railways in Government's plans for infrastructure development, a number of new railway systems were commissioned and would be commissioned after the enactment of the EIAO. For example, the West Rail Line and Ma On Shan Line were commissioned in 2003 and 2004, respectively, which adopted the same viaduct and noise mitigation design.

Due to the scarcity of land, the viaduct sections of the West Rail and Ma On Shan Line run between some high-rise residential developments within 30m. While a full enclosure could be a solution solely from noise point of view, it causes concerns in other aspects like adverse visual impact and the need of smoke extraction system. As a result, special design was adopted to comply with the stringent noise standards without the need for a complete cover over the trackway.

Firstly, floating slab trackform (FST) with low stiffness rail fasteners had been installed on the viaduct to reduce the structure-radiated noise from the viaduct structure, especially at low frequencies. Furthermore, a special mitigation concept called "multi-plenum" was developed at the planning of West Rail Line. This noise reduction system consists of three sound absorptive components, namely, undercar plenum, under walkway plena on two sides of the vehicle, and parapet walls (Fig.1). The undercar plenum is bounded by vehicle skirts (Fig. 2) on the sides of the car extended down close to an upstand in the concrete trackform, the bottom of the car body and the floating slab track. The under walkway plenum is bounded by the parapet wall, the deck, safety walkway and the vehicle skirt.

Apart from the multi-plenum design described above, enhanced noise mitigation measures have also been installed at critical locations in view of the feedback from local communities during commission stage. The enhanced measures included a) extended rubber nosing at the walkway to reduce the gap between the walkway and the car body; b) sandwiching the track with non-metallic sound absorption panels (Fig. 3); and c) installation of additional absorption panels under the walkway. All these measures (i.e. multi-plenum and enhanced measures) are able to reduce the wayside noise by a total of about 15dB(A).

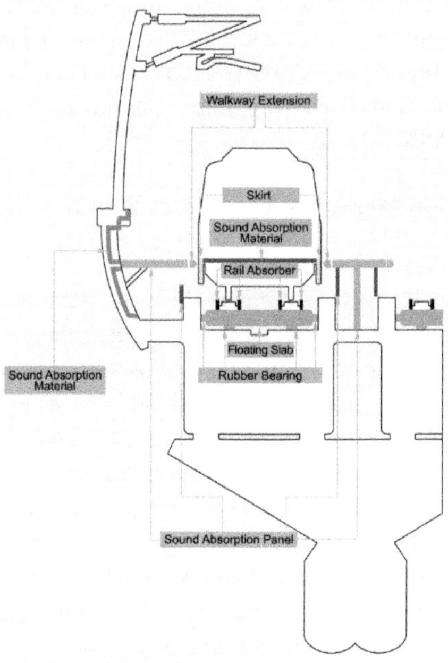

Fig. 1. Typical multiple-plenum viaduct design for Ma On Shan Rail (color figure online)

Fig. 2. Ma On Shan Rail train with noise-reducing skirts (color figure online)

In an underground railway project — the Kowloon Southern Link, the greatest challenge acoustically was having the railway alignment at about 25m from the Hong Kong Cultural Center (HKCC), a world class performing arts center in Hong Kong. The project proponent had proposed a stringent noise standard of Lmax 25dB(A) for the performance theatres and concert hall in the HKCC. Floating slab tracks (FST) with natural frequency of 10Hz have been installed in the vicinity of the HKCC to meet the noise standard.

Fig. 3. Non-metallic Sound Absorption panels on Ma On Shan Rail (color figure online)

2.4 Planning of Noise Sensitive Uses in the Vicinity of Railways

The effort of good environmental planning for new railway systems will be jeopardized if new noise sensitive uses are placed near the railway alignment without giving due consideration and proper control. In Hong Kong, major change in land use (e.g. from industrial to residential) requires rezoning/planning approval from the Town Planning Board under the Town Planning Ordinance. The proponent is required to carry out noise impact assessment to support the application for sites near railway alignment to demonstrate that there would be no adverse noise impact on the planned noise sensitive use. The criteria adopted for noise impact assessment in the town planning application are the same as those adopted in the EIAO mentioned above, though the criteria are referenced to another document, "Hong Kong Planning Standards & Guidelines" [3].

3 Tackling Noise from Railway Systems Already in Place

3.1 Noise Control Ordinance

Noise from trains and railway depots is controlled under the Noise Control Ordinance (NCO). A specific Technical Memorandum [4] was issued under the NCO laying down the acceptable noise levels. In fact, the 30min Leq noise standards of the EIAO mentioned in Section 2.2 are the same set of acceptable noise levels in this Technical Memorandum.

The control on such noise in Hong Kong is by means of Noise Abatement Notices, requiring the noise producer to reduce the noise to specified limits within a prescribed period. Non-compliance with the Notice is an offence, which will lead to prosecution and fines. The control also applies to all railway lines irrespective of whether they were built before or after the enactment of the NCO in 1989. Nonetheless, its control shall apply to the railway operator "only so far as is practicable and compatible with the discharge of any function or the exercise of any power or duty conferred or imposed upon them according to law".

3.2 Abatement Programme to Reduce Noise from Railway Already in Place

Close liaison has been maintained between the railway operator and the noise control authority, which resulted in a HK$900 million (US$115 million) [5] abatement programme to bring relief to some 110,000 people living near the existing railway lines at more than 20 locations. Because of the high-rise nature of the residential buildings in Hong Kong, the mitigation measures required for meeting the noise standards are usually a full or partial enclosure. Fig. 4 and Fig. 5 show an example of a 300-m long semi-enclosure built next to a housing estate benefiting more than 4,000 residents. Another full enclosure of 430m long, which has benefited 19,000 residents, is shown in Fig.6.

Fig. 4. Track before Installation of Enclosure (color figure online)

Fig. 5. Semi-enclosure to Protect the Housing Estate (on the left of the photo) (color figure online)

Fig. 6. 430m Long Full Enclosure Benefiting 19,000 Residents (color figure online)

4 Challenge Ahead

Currently, 2 new railway lines are in their construction stage while 3 railway lines are in the environmental impact assessment stage. Notwithstanding that these 5 railways have their own operating characteristics, they all have one thing in common — all of them have underground sections running in urban areas, though some of them also have viaduct / at grade section as well. Ground-borne noise impact is the major challenge in designing these new railway systems. The standards adopted in assessing ground-borne noise are 10dB below those for air-borne noise presented in Table 1.

One of the projects currently in its early construction stage is the first "high speed" railway line in Hong Kong —— The Guangzhou—Shenzhen—Hong Kong Express Rail Link (XRL). It is a 26-km long (in Hong Kong section only) railway line linking the urban Kowloon area to the Mainland China. The whole alignment in the Hong Kong section will run underground with a maximum operating speed of 200kph (higher speed in the China section). In some sections the high-speed train will even run directly under densely populated residential areas, making the design of the railway system a more challenging task in order to design away the ground-borne noise problem. In the preliminary design, isolated slab track (slab on resilient mat) would be installed at critical locations [6]. The measures are capable of reducing ground-borne railway noise by more than 10dB(A). The design is subject to further review when more ground response data are available during construction phase.

5 Conclusions

Railway is the key transportation mode in Hong Kong. Consideration of noise in planning new railway lines and new sensitive receivers, and the effort in mitigating noise from existing railways are equally important. Planning against and controlling railway noise in Hong Kong is a great challenge to both the railway operator and the planning / enforcement authorities. Innovative design has been developed to make railway transport a sustainable transportation mode in a metropolis.

References

[1] Technical Memorandum on Environmental Impact Assessment Process, Environmental Protection Department, Hong Kong

[2] Hong Kong Government Railway Development Strategy 2000, Transport Bureau, Hong Kong (2000)

[3] Hong Kong Planning Standards & Guidelines, Chapter 9 – Environment, Environmental Protection Department and Planning Department, Hong Kong

[4] Technical Memorandum for the Assessment of Noise from Places other than Domestic Premises, Public Places or Construction Sites, Environmental Protection Department, Hong Kong

[5] Kowloon-Canton Railway Corporation Annual Report (2002)

[6] Environmental Impact Assessment of Hong Kong Section of Guangzhou-Shenzhen-Hong Kong Express Rail Link, Environmental Impact Assessment Report

Human Response to Groundborne Noise and Vibration in Buildings Caused by Rail Transit: Summary of the TCRP D-12 Study

Jeffrey A. Zapfe[1], Hugh Saurenman[2], and Sanford Fidell[3]

[1] Acentech Incorporated
33 Moulton Street, Cambridge, MA, USA 02138
Tel: +1 617 499 8033, Fax: +1 617 499 8074
e-mail: jzapfe@acentech.com
[2] ATS Consulting
801 S. Grand Avenue, Suite 575, Los Angeles, CA, USA 90017
Tel: +1 213 488 7770, Fax: +1 213 488 0270
e-mail: hsaurenman@ATSConsulting.com
[3] Fidell Associates, Inc.
23139 Erwin Street, Woodland Hills, CA, USA 91367
Tel: +1 818 298 5542
e-mail: fidell@pacbell.net

Summary

Ground vibration produced by rail transit systems can be annoying to nearby building occupants when they perceive some combination of feelable vibration, re-radiated sound, and vibration-induced rattling of household paraphernalia. Community response to rail-induced ground vibration has not been extensively researched. While the well-known Schultz dosage—response curve is routinely used to predict the prevalence of annoyance produced by airborne transportation noise, no similar relationship has gained widespread acceptance for groundborne noise and vibration. The principal goal of the present study was to develop a dosage response relationship for predicting community annoyance due to ground vibration produced by rail transit systems. The research was conducted as part of the Transit Cooperative Research Program D-12 project. Telephone interviews were conducted with 1306 individuals in five North American cities. Field measurements were made in each city to estimate vibration and noise exposure at each interview location. The work produced several dosage—response relationships between vibration/noise exposure and annoyance. When compared to the current noise and vibration criteria specified by the United States Federal Transit Administration (FTA), the dosage response analysis predicted a probability of 0.05 to 0.10 that a D-12 respondent would be highly annoyed by vibration and noise at the current FTA criterion levels.

T. Maeda et al. (Eds.): Noise and Vibration Mitigation for Rail Trans. Sys., NNFM 118, pp. 25–32.
springerlink.com

1 Introduction

Ground vibration produced by rail transit systems can be annoying to occupants of nearby buildings when they perceive some combination of feelable vibration, re-radiated sound, and vibration-induced rattling of household items. Community response to rail-induced ground vibration has not been extensively researched. While the well-known Schultz dosage-response curve [1] is routinely used to predict the prevalence of annoyance produced by airborne transportation noise, no similar relationship has gained widespread acceptance for groundborne noise and vibration.

In 2005, the National Academy of Sciences in the United States awarded the Transit Cooperative Research Program (TCRP) D-12 contract to Acentech Incorporated and its subcontractors ATS Consulting and Fidell Associates. The goal of the D-12 study was to develop dosage response relationships that would be useful for predicting community annoyance due to ground vibration produced by rail transit systems.

2 Literature Review

Most countries that have standards related to human response to building vibration use a weighted overall vibration level as the exposure metric. Many of these weighting curves are derived from ISO 2631 [2]. Interestingly, in the frequency region where vibration from rail transit is typically most severe (16 Hz to 100 Hz), the weighted overall vibration level is not significantly different from an overall measure of vibration velocity, the metric that is used to assess vibration exposure in the United States [3].

Other studies that have investigated the impact of rail vibration on human response have been done, or are being planned in Scotland [4], Sweden [5], Norway [6, 7, 8] and the United Kingdom [9]. The methods and results from these studies were used in developing the design for the D-12 project.

3 Transit System Survey

Survey responses were received from 30 of 53 North American transit systems with rail operations. The primary finding of the agency survey was that complaints from rail vibration are not widespread in North America. The agencies that reported a complaint history were primarily older systems with track constructed before environmental impact assessments were required and modern track mitigation measures were available. All systems that received complaints in the previous year reported that at least some of them had come through informal channels. Community annoyance may therefore be underreported due to a lack of formal complaint procedures. The most common resolution to vibration-related issues was the development of, or intensification of, maintenance programs.

4 Field Survey and Measurement Program

Five transit systems were selected for field testing: New York City, Toronto, Boston, Sacramento, and Dallas. Sacramento and Dallas were at-grade light rail systems with ballast and tie track. The other three systems were rapid transit subway systems.

Test areas were selected in each city based on discussions with agency personnel. Ideal test areas had a history of complaints, a large number of potential respondents and a wide range of exposure to rail transit vibration. Fortunately, a range of exposures could easily be accommodated by including residences in the sample out to 100 m from the alignment, where train-induced vibrations would be expected to be minimal.

After a test area was selected, a telephone questionnaire was administered to a simple random sample of residential households within the survey area. Survey questions with standardized wording were prepared along the lines of prior questionnaires addressing the annoyance of transportation noise [10, 7, 4]. The questionnaire was optimized for telephone administration, with predominantly closed response category items. Respondents were constrained to describe degrees of annoyance as "not at all annoying", "slightly annoying", "moderately annoying", "very annoying", or "extremely annoying". The survey was designed so that it could be completed in 5 to 10 minutes.

A total of 1306 respondents completed a telephone interview: 582 in Toronto, 281 in New York, 304 in Boston, 103 in Dallas, and 36 in Sacramento. The final survey question asked whether the respondent would consider allowing measurements to be made inside their home. A portion of respondents who agreed were re-contacted to arrange field measurement appointments. The field measurements consisted of:

- Noise and vibration measurements inside the residence, in a room where the resident thought that the vibration/noise was most noticeable.
- Vibration measurements on the ground immediately outside the residence.
- A simultaneous vibration measurement at a common fixed outdoor reference position near the alignment. The reference measurements were used to adjust for any differences in the fleet vibration characteristics.

Because it was not practical to measure inside all 1306 surveyed homes, measurements were made at selected residential locations. The residence measurements were supplemented with grid measurements at locations distributed throughout each test area. The grid measurements were used to estimate the exposure at respondent locations, where field tests were not conducted.

4.1 Results of Field Measurements

4.1.1 Indoor Vibration Compared to Outdoor Vibration

The average difference between the indoor vibration and the outdoor vibration was close to 0 dB, both for the entire population of test residences and for the residences specific to each transit system. However, substantial variation (±5 dB) was observed in the residential data. No clear pattern to the variation was apparent (at some locations, indoor vibration levels varied by as much as 10 dB between adjacent residences of the same apparent design). For the D-12 data set:

- on average, the indoor vibration level equaled the outdoor level, and
- the indoor vibration exceeded the outdoor vibration by less than 5 dB more than 95% of the time.

4.1.2 Outdoor Vibration Compared to Indoor Sound

Given that the vibrating surfaces in a room are responsible for the radiated sound, it is reasonable to assume that groundborne noise and vibration are related. For plane wave radiation from a vibrating surface ($p = \rho c\, v$, where, ρ is the air density, c is the speed of sound, and v is the vibration velocity of the surface) the sound pressure and vibration velocity are related by Lp (dB re 20μPa) = -5.7 + Lv (VdB re 1μin/s). In rooms with reverberation or multiple radiating surfaces, the sound level could be higher. The Federal Transit Administration (FTA) Guidance Manual [3] recommends that Lp (dB re 20μPa) = Lv (VdB re 1μin/s) is a reasonable way to account for these effects. The D-12 data however suggests that $Lp = Lv$ might be overly conservative. In the frequency range where train-induced groundborne noise was loudest (31.5 Hz to 100 Hz), the average measured difference between vibration level and sound pressure level was about –5 dB. Because of the limited amount of data collected and the uncertainties caused by measuring vibration at just one location in the room, it is not clear what the correct adjustment should be. However, the D-12 data suggest that the method cited in the FTA manual tends to over predict the levels of groundborne noise by approximately 5 dB.

4.1.3 Indoor Vibration Compared to Indoor Sound

The correlation between the indoor vibration and indoor noise was only slightly better than the correlation between outdoor vibration and indoor noise. Again, based on the D-12 data, this implies that reasonable estimates of indoor groundborne noise can be obtained with measurements of exterior vibration, and that the additional expense of indoor vibration measurements offers only a slight improvement in the accuracy of the estimated indoor sound.

4.1.4 Vibration Spectra

Although differences were observed between vibration levels and frequency spectra measured at the five transit systems, the frequency spectra for all of the systems was greatest between 40 Hz and 80 Hz (1/3 octave frequency bands).

4.1.5 Relationships among Noise and Vibration Measures

More than 200 noise and vibration metrics were considered as potential predictor variables for the dosage—response analysis. Most of these metrics proved to be so highly correlated (r = 0.9 or greater) with one another that for practical purposes, they differed only by a scale factor and/or a constant. Such high correlation among measures indicates that most of the derived measures were, in effect, only trivially variant measures of the underlying physical quantities. A factor analysis was unable to identify physically interpretable subsets of vibration metrics that were both closely related to one another and, at the same time, different from other subsets. Since, essentially, any metric was as good as any other in predicting annoyance, a set of commonly used vibration and noise measures were used to develop dosage—response relationships.

5 Dosage Response Analysis Results

The methods employed in the social survey yielded satisfactory interview completion rates and orderly and interpretable questionnaire responses. Only small numbers of respondents spontaneously mentioned rail-related noise or vibration as the least-favored aspect of neighborhood living conditions, or were consequentially affected by train passbys. Less than 6% of respondents living near surface tracks and less than 4% of those living above subway tracks described themselves as being highly annoyed by groundborne noise from trains.

5.1 Logistic Regression Models to Predict Response Due to Passbys

The relationship between vibration exposure and annoyance was obtained using a logistic-regression model [11] whose general form is,

$$p = \frac{e^{A+B\,x}}{1+e^{A+B\,x}}$$

where p is the probability (of annoyance in this case), x is the vibration metric and A and B are constants derived from the logistic-regression analysis.

5.2 Vibration Results and Relationship to Current US Criteria

Fig. 1 shows the dosage—response relationships expressed in terms of the passby maximum vibration level (1-second true RMS average). The solid curves in the figure represent the opinions of people who described themselves as being highly annoyed (questionnaire response of "very" or "extremely" annoyed) by train vibration for any reason (rumble, rattle or shake). The dotted curves represent people who described themselves as moderately or more greatly annoyed by the train vibration. Vibration exposure is defined in two ways: 1) the mean of the passby maximum levels (that is, the energy average vibration level over all observed passbys), and 2) the mean vibration level plus two standard deviations (mean+2σ). For a normal distribution, the probability that a train would exceed the mean+2σ level is 5%, so this level would correspond to the "loudest" trains that passed by the site, and hence, is the more appropriate measure if it is believed that people are more likely to be disturbed by the loudest trains in the fleet, rather than the fleet-average train. The FTA's 72 VdB limit for frequent service (>70 events/day) and residential occupancy is also shown in Fig. 1. The probability of high annoyance due to train vibration at the FTA limit of 72 VdB varied by a factor of two (from about 0.05 to 0.10), depending on whether the prediction is based on the mean level or the mean level plus two standard deviations.

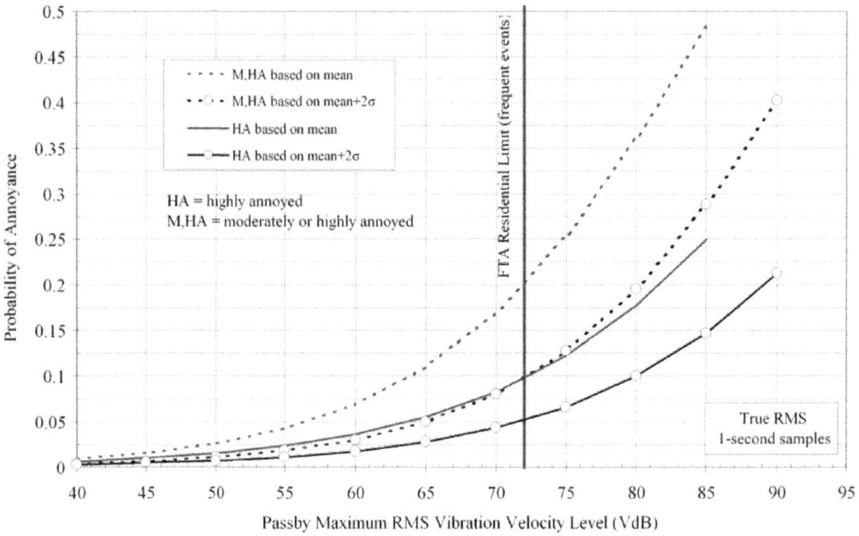

Fig. 1. Probability of Annoyance Based on Passby Maximum Vibration Velocity Level (color figure online)

5.3 Groundborne Noise Results and Relationship to US Criteria

Since the rail-induced sound pressure in a room can be reliably estimated from the floor vibration, it follows that the A-weighted vibration level can reliably predict the A-weighted sound pressure level. The FTA manual states that, for typical residential rooms, the sound pressure level in dB is approximately equal to the floor vibration velocity level in VdB. In the subsequent discussion, A-weighted velocity level (dB) and sound level (in units of A-weighted decibels) are used interchangeably. Fig. 2 shows the D-12 dosage—response relationships based on the passby maximum A-weighted velocity level as the exposure metric. The FTA's 35 dB A-weighted limit for residential occupancy (and frequent events) is also shown on the figure. At 35 dB (A-weighted), the probability that a D-12 respondent would have reported high annoyance from train passbys varied from about 0.03 to 0.06, depending on whether the prediction was based on the mean exposure level or the mean +2σ level.

The US Federal Aviation Administration uses *Ldn* = 65dB as a threshold for noise mitigation which corresponds to a prevalence of high annoyance of 0.12-based FICON's [12] dosage—response relationship. Note that no corresponding policy criteria are used in the US for groundborne noise. However, if 0.12 were to be considered as a reasonable rate of highly annoyed for policy reasons, Fig. 2 shows that the mean A-weighted level at which the probability of high annoyance reaches 0.12 is about 46 dBA. Consequently, if a probability of 0.12 of high annoyance is taken as a reasonable impact threshold, then the present information suggests either that the current FTA criterion of 35 dBA is too stringent, or alternatively, the conversion from velocity to sound is overly conservative. As was mentioned previously, the latter explanation is more strongly supported by the present data. If the curves in Fig. 2 are shifted to the left by 5 dB, then the probability of high annoyance at 35 dBA would increase from 5.5% to 7.8%.

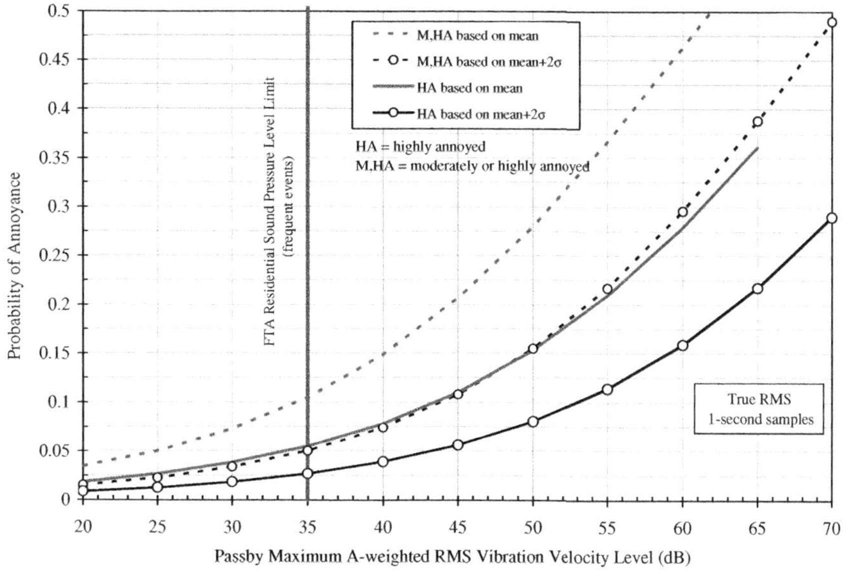

Fig. 2. Prob. of Annoyance Based on Passby Maximum A-weighted Vibration Velocity Level (color figure online)

6 Discussion

6.1 Duration of Residency in the Community

At the time of interviewing, more than 95% of the D-12 respondents had lived in their homes for longer than a year. Hence, their opinions regarding annoyance reflect their long-term familiarity with the exposure conditions, rather than transient reactions to a novel exposure. More adverse reaction might be expected if a transit system were suddenly "turned on" in an otherwise quiet community. The D-12 data were obtained at systems that have been in the community for many years. If one considers the D-12 result as an asymptotic value, then it would be expected that short-term adverse community reaction associated with the start of operations by a new transit system would diminish with time as people acclimatize to the noise and vibration.

6.2 Vibration Measurement Location

Noise and vibration criteria are typically applied at the receiver location inside the building. This approach can be costly and potentially inaccurate if there is ever a need to verify compliance with criteria. In the present study, large variations in indoor vibration level were observed among seemingly identical homes that were exposed to similar levels of ground vibration.

Given appreciable building-to-building variation, a measurement in one residence does not necessarily lead to a reliable prediction of the vibration in the residence next door. Thus, unless one were to measure in every house in the potentially affected community, exterior vibration levels could provide as useful an indication of expected vibration levels as a limited number of interior residential measurements. The present data are not comprehensive enough to conclusively justify this reasoning. Nonetheless, the compelling logistical reasons for exterior-based vibration measurements are apparent; exterior measurements obviate the need for access to people's homes, and reduce the cost and complexity of field measurements. This is one area where future research could yield a large payback in simplifying the analysis of transit-related noise and vibration in communities.

References

[1] Schultz, T.J.: Synthesis of Social Surveys on Noise Annoyance. Journal of the Acoustical Society of America 64(2), 377–405 (1978)
[2] Mechanical vibration and shock - Evaluation of human exposure to whole-body vibration," IS0 2631-1 (1997): Part 1: General requirements, ISO 2631-2 (2003): Part 2: Continuous and shock-induced vibrations in buildings (1 to 80 Hz)
[3] Transit Noise and Vibration Impact Assessment, Federal Transit Administration, FTA-VA-90-1003-06 (2006)
[4] Woodroof, H.J., Griffin, M.J.: A Survey of the Effect of Railway-induced Building Vibration on the Community, University of Southampton, ISVR Tech. Rpt. to British Rail Tech. Ctr., Derby (1986)
[5] Öhrström, E.: Effects of Exposure to Railway Noise – A Comparison Between Areas With and Without Vibration. Journal of Sound and Vibration 205(4), 555–560 (1997)
[6] Turunen-Risem, I.H., Brekke, A., et al.: Vibration in Dwellings from Road and Rail Traffic – Part I: A New Norwegian Measurement Standard and Classification System. Applied Acoustics 64, 71–87 (2003)
[7] Klæboe, R., et al.: Vibration in Dwellings from Road and Rail Traffic – Part II: Exposure-effect Relationships Based on Ordinal Logit and Logistic Regression Models. Applied Acoustics 64, 89–109 (2003)
[8] Klæboe, R., et al.: Vibration in Dwellings from Road and Rail Traffic – Part III: Towards a Common Methodology for Socio-Vibrational Surveys. Applied Acoustics 64, 111–120 (2003)
[9] Human Response to Vibration in Residential Environments, NANR172, Department for Environment, Food and Rural Affairs, London, UK (2007)
[10] Fidell, S., et al.: Relationship between low-frequency aircraft noise and annoyance due to rattle and vibration. Journal of the Acoustical Society of America 111(4), 1743–1750 (2002)
[11] Tabachnick, B., Fidell, L.: Using Multivariate Statistics, 5th edn. Allyn and Bacon, Boston (2007)
[12] Federal Agency Review of Selected Airport Noise Analysis Issues, Federal Interagency Committee on Noise, FICON (1992)

An Investigation of Velocity-Dependent Friction in Wheel-Rail Rolling Contact

B.E. Croft[1], E.A.H. Vollebregt[2,3], and D.J. Thompson[4]

[1] SLR Consulting Australia Pty Ltd, 2 Lincoln St, Lane Cove 2066, Australia
Tel.: +61 2 9427 8100; Fax: +61 2 9427 8200
e-mail: bcroft@slrconsulting.com
[2] Delft University of Technology
[3] VORtech BV
[4] ISVR, University of Southampton

Summary

Friction between sliding surfaces decreases as the velocity of sliding increases ('falling friction'). This paper investigates the velocity-dependent friction relationships in non-Hertzian models of wheel-rail rolling contact. The effect of falling friction on tangential stress distribution and slip is examined. In otherwise steady rolling with constant creep, falling friction introduces a stick-slip oscillation to the trailing edge of the contact. This oscillation is increasingly unstable with increasing creep. The stick-slip behaviour at the trailing edge leads to sudden changes in the tangential stress distribution in the stick zone, including intermittent slip at the leading edge of the contact.

1 Introduction

Over the last 40 years, significant efforts have been made to model and understand the sources of rolling noise [1]. The fundamental source of rolling noise is the roughness of the wheel and rail surfaces. Increasingly, the attention of researchers is being drawn to the mechanisms of roughness development and growth. Initial interest in roughness growth mechanisms was driven by the problem of corrugation. More recently, attention is being paid to broadband roughness development with a view to reducing noise by maintaining relatively smooth tracks and also with the aim of reducing the need for rail grinding.

The theory of contact mechanics and friction is fundamental to studies of wheel-rail wear and roughness development. However, models of wheel and rail wear, roughness and corrugation development often simplify the contact problem. It is common to assume Hertzian contact theory to predict the stress distribution in the contact patch, even though this is valid only for specific geometries. It is also

T. Maeda et al. (Eds.): Noise and Vibration Mitigation for Rail Trans. Sys., NNFM 118, pp. 33–41.
springerlink.com

common to assume a constant coefficient of friction throughout the contact area. These assumptions have been necessary partly because of the high computational cost of alternative theories. However, making these assumptions about the contact conditions when modelling roughness or corrugation development affects the conclusions that are reached [2,3].

Studies have shown that friction between sliding surfaces decreases as the velocity of sliding increases ('falling friction'). Velocity-dependent friction coefficients have been used in models of wheel squeal [4], but only in conjunction with simplified models of the tangential stress distribution relying on local traction-displacement relationships (e.g. FASTSIM). This paper investigates more sophisticated friction laws in non-Hertzian models of wheel-rail rolling contact, using a variational method, which incorporates full linear elastic theory and non-local interactions. The effect of falling friction on tangential stress distribution is examined, for different relationships between friction and slip velocity. Results are presented from two independently programmed models based on Kalker's rolling contact theory [5]: a specially modified version of CONTACT [6] and the time-stepping model of Croft [7].

2 Velocity-Dependent Friction in Rolling Contact

Kalker's theory for the calculation of the stress distribution and slip in rolling contact is documented in [5]. Here, a two-dimensional analysis is proposed of a smooth wheel or cylinder of radius 0.46 m rolling over a smooth rail with velocity 40 m/s and constant 0.1% longitudinal creep. The normal load per unit width has been set to give a length of the contact in the rolling direction of 8 mm. Three different friction relationships have been considered, as shown in Fig. 1.

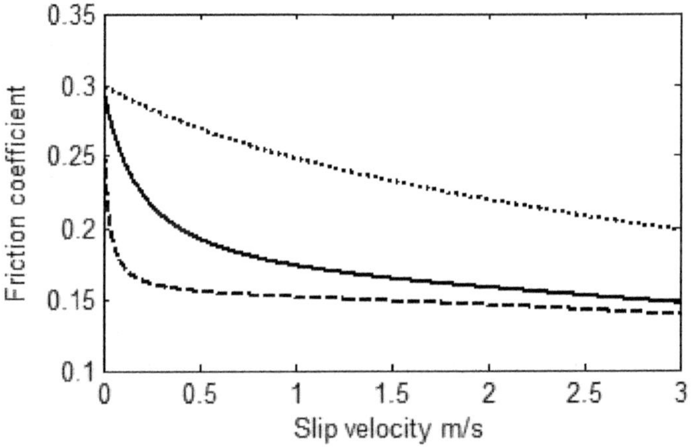

Fig. 1. Friction relationships as in Eq. (2.1): ——— a0=0, a1=0.1, a2=50, b1=0.2, b2=100; ······· a0=0, a1=1, a2=50, b1=2, b2=100; — — — a0=0, a1=0.01, a2=50, b1=0.02, b2=100.

These friction relationships have been selected to demonstrate a range between Amonton's law of friction (constant friction) and Coulomb's theory (different static and dynamic friction coefficients) and are in the form

$$\mu(s) = \mu_0 \left[a_0 + \frac{a_1}{(b_1 + |s|)} + \frac{a_2}{(b_2 + |s|^2)} \right] \tag{2.1}$$

where the friction coefficient μ is a function of slip velocity s, $\mu_0 = 0.3$ is the static coefficient of friction and a_0, a_1, a_2, b_1 and b_2 are constants.

2.1 Determining Stress Distributions and Slip Velocity in Rolling

Results are presented from two models. Croft [7] uses a time domain method, beginning with the wheel at rest (zero initial slip and tangential stress). Tangential stress is applied in the form of constant longitudinal creep and the wheel rolls a distance along the rail until a steady or quasi-steady state is reached. At each time step the normal stress distribution is calculated, and from this the traction bound, tangential stress distribution and slip velocity are obtained. If the friction coefficient is constant and there is no variation in the normal force, Carter's solution [8] develops as the steady state. To include velocity-dependent friction, the slip velocity from the previous time-step is used to give an initial estimate of the traction bound, iterating to refine the slip velocity and traction bound for the current time step as described in [7]. With falling friction, a quasi-steady stick-slip oscillation is observed at the trailing edge of the contact. This results in the typical stick-slip saw-tooth shape of the overall tangential force supported by the contact area, as shown in Fig. 2.

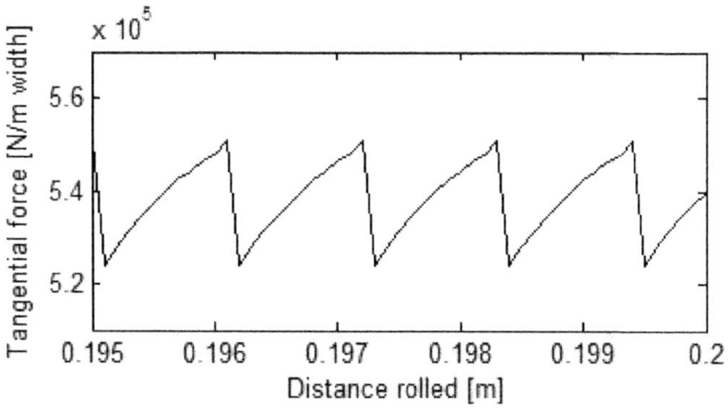

Fig. 2. Detail of stick-slip instability in otherwise steady rolling contact with a velocity-dependent friction coefficient.

2.2 Including Velocity-Dependent Friction in CONTACT

The second model used in this investigation is a version of CONTACT [5,6,9] that has been modified to include velocity-dependent friction in the form of Eq. (2.1). Instead of starting with a wheel at rest, a steady state solution is solved directly to give an initial case for the time-stepping process. With falling friction, the steady state solution is an unstable equilibrium if the slip velocity and velocity-dependence are sufficiently large. Any small perturbation of the system then causes the system to jump to a different state. In subsequent cases, the system settles to a stick-slip pattern as observed using the pure time-stepping model.

The dependence of friction on the slip velocity has been implemented using fixed point iteration for the slip velocity and corresponding traction bound:

1. Set k=0 and make an appropriate initial estimate s^0 for the slip velocity;
2. Set k=k+1 and start a new iteration;
3. Compute the traction bound $g^k = \mu(s^{k-1}) \cdot p_n$;
4. Solve the frictional contact problem with traction bound g^k;
5. Determine the slip velocity s^k and its difference to the estimate s^{k-1}.
6. If the difference is considered too large, go to step 2, else stop.

This extra iteration loop increases the computational work, because the traction distribution has to be calculated multiple times (step 4). A strong point of this approach is that the solution of step 4 is guaranteed to be unique, and robust solvers are available.

Different approaches can be used for the initial estimate in step 1: $s^0=0$ (low estimate), $s^0=V$ (high estimate) and $s^0=s(t-\delta t)$ (solution of previous time step). In certain cases, these choices lead to different results for a time step. This shows that the solution to the overall problem with velocity-dependent friction is not unique. This behavior seems closely related to the findings of Ionescu and Paumier [10].

3 Tangential Stress in the Stick Zone

The stick-slip oscillation in rolling contact with velocity-dependent friction can be seen in both the time history of the slip velocity throughout the contact and in the time history of the tangential stress distribution. An example is shown in Fig. 3 calculated using the model of Croft [7]. The parameters are as described in Sect. 2, for the middle friction relationship (solid line in Fig. 1). The potential contact is 9 mm long, divided into 0.1 mm elements in the rolling direction.

Of particular interest for this work is the variation in the tangential stress distribution in the stick zone, particularly the peaks shown in Fig. 3. This phenomenon was observed in [7] but not investigated in detail as it was initially thought to be an artefact of the calculation method, similar to that reported in [9]. It was expected that the peaks could be removed by incorporating a "leading edge correction" such as the one described in [9]. Further investigation using the extended version of CONTACT described above has shown that the variation in the stick zone is not eliminated by the introduction of the leading edge correction.

The peaks in the tangential stress distribution are initiated at the leading edge of the contact at the point when the maximum slip occurs in the contact as shown in Fig. 3(e). The peaks remain at that location as the wheel rolls along, with a new peak initiated in the next 'gross slipping' event. As the slip increases, the velocity-dependent traction bound decreases until there is a sudden large slip. But when this slip occurs, the tangential stress is relieved, the traction bound is no longer exceeded and the two surfaces stick again.

The sudden large slip event at the trailing edge appears to result in a small slip occurring at the leading edge. This is not expected (in a case with no spin creepage and identical materials), as at the leading edge new material is constantly coming into contact with initially zero relative slip.

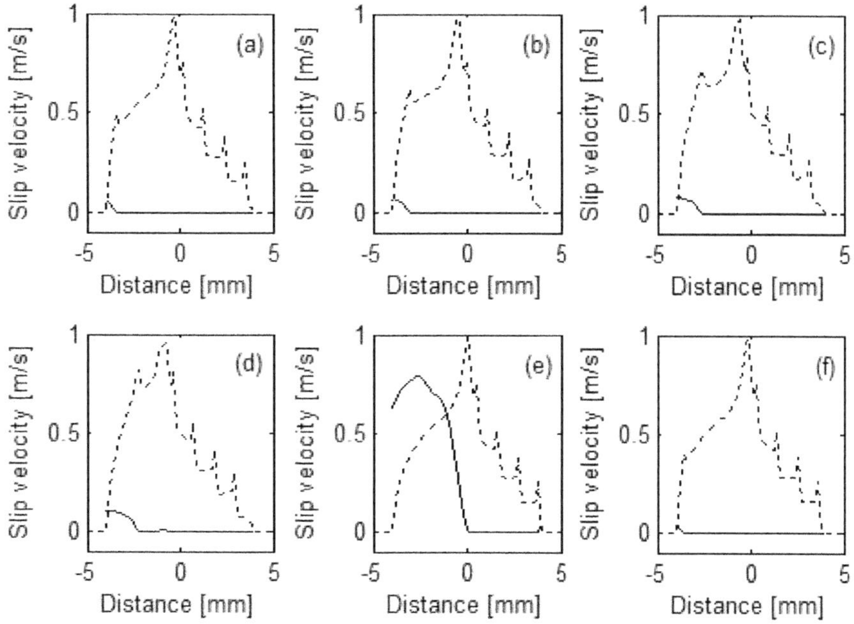

Fig. 3. History of: ——— slip velocity and ······· tangential stress distribution (not to scale) in rolling contact. Rolling increment is 0.2 mm between each figure, calculated in two 0.1 mm steps. (a) Small slip only at trailing edge to left of figure, rolling from left to right; (b)-(d) Increasing slip velocity and length of slip zone; (e) Sudden slip occurs, peak slip velocity, also small slip at leading edge; (f) Cycle resumes with small slip at trailing edge only.

3.1 Comparison of Results from the Two Models

Similar effects are seen in both models, as shown in Fig. 4.

In this figure, the model parameters are the same but the models and calculation methods are different as described in Section 2. However, the frequency of the stick-slip oscillation is different between the two models – with CONTACT, the peaks in the tangential stress in the stick zone are closer together, indicating a higher frequency stick-slip cycle.

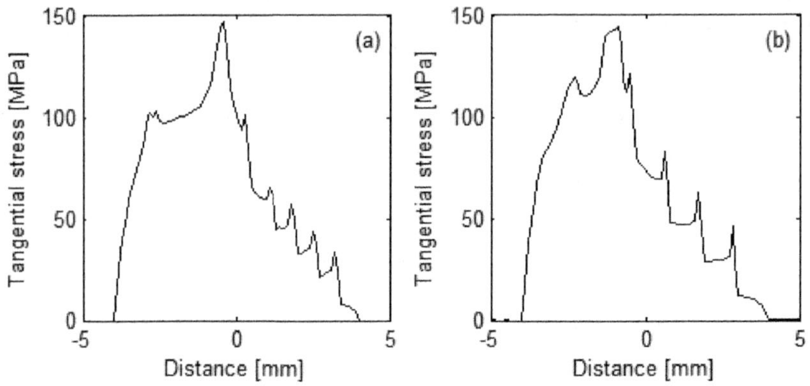

Fig. 4. Tangential stress distribution from (a) CONTACT and (b) the model of Croft. Parameters as in Fig. 3. Results are shown at the same stage of the stick-slip cycle as Fig. 3(d).

3.2 Effect of Element Size

The effect of element size has been examined by calculating the same case but with larger and smaller elements than previously. The distance rolled in each step changes accordingly. The element sizes examined were 0.2 mm, 0.1 mm (as before) and 0.05 mm. Again both models show similar trends. With larger elements, the height of the peaks in the tangential stress are reduced. Reducing the element size increases the height of the peaks, that is the observed effect becomes more pronounced when the elements are refined.

No convergence has been observed with increasing the number of elements and reducing the step size used in rolling. An explanation for this is that the sudden slip and "collapse" of tangential tractions occurs in a single time step that is halved with each grid refinement. This affects the slip velocity, as with the current approach the slip velocity is given by the magnitude of the slip divided by the time. Reducing the element size, therefore, increases the slip velocity in the time step in which the sudden slip occurs.

3.3 Effect of Creep Magnitude

Some experimentation has been done with the different values of constant longitudinal creep. Increasing the magnitude of the creep exaggerates the phenomenon, as might be expected with higher tangential loads and typically larger slip zones. The stick-slip oscillation becomes less stable with the increasing creep. Reducing the magnitude of the creep reduces the observed effect. The resulting tangential stress distributions with low creep show very small peaks in the stick zone, occurring more closely together than with higher creep, as shown in Fig. 5.

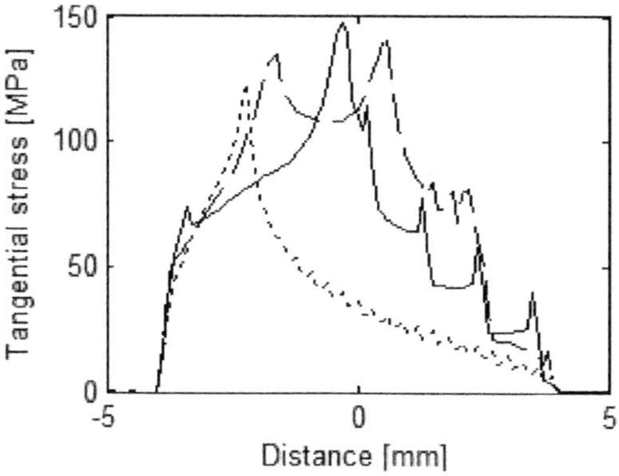

Fig. 5. Tangential stress distribution for different longitudinal creep: ——— 0.1%; ······· 0.05%; — — — 0.15%.

3.4 Effect of Friction Relationship

The friction relationship used in the models has a large effect on the results obtained, especially on the peaks in the tangential stress distribution in the slip zone. Example results are presented in Fig. 6 for the three friction relationships considered. If the reduction in friction is small with increasing slip velocity, the tangential stress distribution resembles that obtained from a constant coefficient of friction, i.e. there is no sudden large slip, no stick-slip oscillation at the trailing edge and no peaks in the tangential stress in the slip zone. The more rapid the fall-off in the falling friction relationship, the more pronounced the stick-slip phenomena described here become.

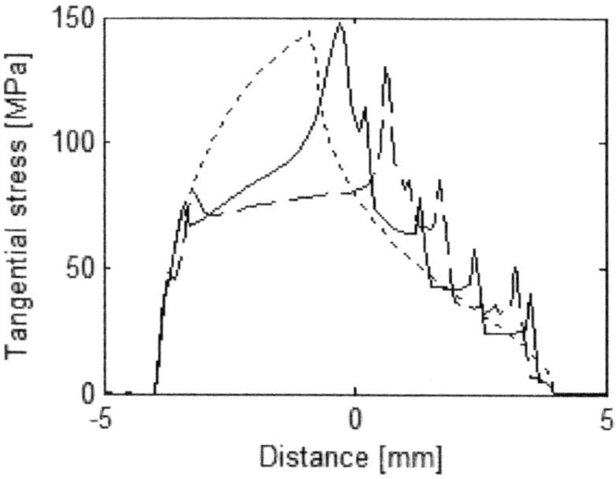

Fig. 6. Tangential stress distribution for different friction laws as shown in Fig. 1: ——— Medium falling friction; ······· Small reduction in friction with increasing slip velocity — — — Sharp reduction in friction with increasing slip velocity.

4 Discussion

The inclusion of velocity-dependent (falling) friction in non-Hertzian models of rolling contact has been investigated. Similar effects have been seen in two independent models. With falling friction, a stick-slip oscillation occurs at the trailing edge of the contact. This stick-slip oscillation also causes intermittent small slip at the leading edge of the contact, which results in peaks in the tangential stress distribution in the stick zone. Since it is thought that the leading edge of the contact should not experience slip, there is some doubt about whether the phenomenon is real or if it is a result of the way that the models are set up.

The two models differ in the distance between the peaks. Further, the grid resolution affects the amount of slip that is found after "collapse". These results are attributed to the difficulties of modeling the velocity-dependent phenomena. It has been found that multiple solutions to the same problem exist, and that the results to the continuous equations may be discontinuous at certain moments in time. This is problematic for the current models that, at this stage, do not include dynamic effects.

The effects described here occur at very high-frequency, and are therefore not likely to affect predictions of rail wear and roughness growth at the wavelengths of interest for roughness. However, the effects raise some interesting questions about modelling variable friction in rolling contact. At this stage, no dynamic effects have been included in the models, which may be the reason for some of the unexpected results. Further, it is not certain what the behaviour of the system might be if the observed stick-slip mechanism was combined with wheel and track dynamics. In the case of curve squeal, it is known that stick-slip occurs at a lightly damped resonance of the wheel [1] so it may be expected that wheel and track dynamics can have a strong effect on the frequency of stick-slip oscillation.

References

[1] Thompson, D.J.: Railway Noise and Vibration: Mechanisms, Modelling and Means of Control. Elsevier, Oxford (2008)

[2] Croft, B.E., et al.: The effect of rail dampers on roughness growth rates with multiple wear mechanisms, non-Hertzian contact and velocity dependent friction. In: Proceedings of the 8th Int. Conf. on Contact Mechanics and Wear of Rail/Wheel Systems 2009, Firenze, Italy (2009)

[3] Xie, G., Iwnicki, S.D.: Simulations of roughness growth on rails - Results from a 2D non-Hertzian, non-steady contact model. Vehicle System Dynamics 46(1-2), 117–128 (2008)

[4] Xie, G., et al.: Introduction of falling friction coefficients into curving calculations for studying curve squeal noise. Vehicle System Dynamics 44, 261–271 (2006)

[5] Kalker, J.J.: Three-Dimensional Elastic Bodies in Rolling Contact. Kluwer Academic Publishers, Dordrecht (1990)

[6] Vollebregt, E.A.H.: User guide for CONTACT, J.J. Kalker's variational contact model. Tech.report TR09-03, VORtech BV, Delft, The Netherlands (2009), http://www.kalkersoftware.org

[7] Croft, B.E.: The Development of Rail-head Acoustic Roughness. Thesis for the degree of Doctor of Philosophy, University of Southampton (2009)
[8] Carter, F.W.: On the action of a locomotive driving wheel. Proceedings, Royal Society A112, 151 (1926)
[9] Vollebregt, E.A.H.: Refinement of Kalker's rolling contact model. In: Proceedings of the 8th Int. Conf. on Contact Mechanics and Wear in Rail/Wheel Systems 2009, Firenze, Italy (2009)
[10] Ionescu, I.R., Paumier, J.C.: On the contact problem with slip displacement dependent friction in elastostatics. Int. J. of Eng. Sci. 34, 471–491 (1996)

High-Frequency Vertical Wheel–Rail Contact Forces at High Vehicle Speeds –The Influence of Wheel Rotation

P.T. Torstensson[1], J.C.O. Nielsen[1], and L. Baeza[2]

[1] CHARMEC/Department of Applied Mechanics,
Chalmers University of Technology, SE-412 96 Gothenburg, Sweden
Tel: +46 31 7721295, Fax: +46 31 7723827
e-mail: peter.torstensson@chalmers.se
[2] Centro de Investigación en Tecnología de Vehículos, Universidad Politécnica de Valencia, Camino de Vera s/n, 46022 Valencia, Spain

Summary

Dynamic vehicle–track interaction at high vehicle speeds is investigated in a frequency range from about 20 Hz to 3 kHz. The inclusion of wheel rotation in the vehicle model is investigated by implementing a structural dynamics model of a rotating wheelset in an existing simulation environment. Calculated wheel–rail contact forces using this comprehensive flexible wheelset model are compared with contact forces based on less detailed, non-rotating wheelset models. Load cases including broad-band wheel–rail excitation due to vertical rail irregularities (rail corrugation) and impact excitation caused by a local deviation from the nominal wheel radius (wheel flat) are considered. Viewed from the excitation point, each wheelset resonance peak of multiplicity two splits into two peaks; the separation of the two peaks increases with increasing rotational speed. If the wheelset model is excited at a frequency where two different mode shapes, due to the wheel rotation, have coinciding resonance frequencies, the contact force calculated for the rotating wheelset model differs significantly from those of the non-rotating models. Further, the use of a flexible rotating wheelset model is recommended for load cases leading to large magnitude contact force components in the high-frequency range (above 1.5 kHz). In particular, the influence of the radial wheel eigenmodes with two or three nodal diameters is significant.

1 Introduction

At frequencies above 20 Hz, the design of the primary and secondary suspensions of a railway vehicle causes the unsprung masses of the vehicle (wheelsets including

T. Maeda et al. (Eds.): Noise and Vibration Mitigation for Rail Trans. Sys., NNFM 118, pp. 43–50.
springerlink.com

bearings, axle-hung traction motor and gearing) to be effectively isolated from the bogie structure and the car body [1]. Often the unsprung mass corresponding to one wheelset is treated as a rigid mass. If the structural flexibility of the wheelset is accounted for the influence of wheel rotation is in most cases neglected [2]. In such cases, a finite element (FE) model of the wheelset with the (pitch) rotation angle constrained to zero is used. The importance of accounting for the inertial effects due to rotation, such as the gyroscopic forces and the centrifugal stiffening, seems to be not fully understood as this matter is rarely discussed in the literature.

In work by Thompson [3], the effect of rotation was considered by modelling a wheel subjected to a force which was rotating around the wheel perimeter with a constant rotational velocity. The inertial effects were neglected. Viewed from the excitation point, it was concluded that the rotation causes resonance peaks to split into two peaks corresponding to two contra-rotating waves. Fayos et al. [4] adopted Eulerian coordinates to develop a model of a rotating wheelset that captures the inertia effects. The mathematical formulation was validated versus the analytical solution for a simply supported rotating Rayleigh beam. For railway wheelset applications, the functionality of the formulation was demonstrated in [5]. Significant differences in the vertical wheel–rail contact force, calculated with a rotating or a non-rotating flexible wheelset model, were shown.

The current work focuses on dynamic vehicle–track interaction at high vehicle speeds. In particular, the effects of wheel rotation are investigated. Several different load cases are considered. The computer program DIFF [6] is applied. It has previously been validated in the frequency range 20 – 2000 Hz [7].

2 Track Model

The dynamic behaviour of a discretely supported railway track is accounted for by applying a FE model of finite length that is clamped at its ends. The simulation model DIFF assumes structural and loading symmetry with respect to an imaginary track centre-line, and hence it is sufficient to only incorporate half of the track in the model. Only vibration in the vertical plane is studied. The rail pads, sleepers and ballast are represented by equi-distant mass-spring-damper supports of the rail with sleeper spacing $L_S = 0.65$ m, see Fig. 1(a). The length of the track model is 70 sleeper bays. This length ensures a negligible influence of the clamped boundary conditions on the system response at the centre portion of the track. The 60E1 rails are modelled by undamped Rayleigh–Timoshenko beam elements with bending stiffness $EI = 6.4$ MNm^2, shear stiffness $kGA = 250$ MN, mass per unit length $m = 60$ kg/m and rotational inertia per unit length $mr^2 = 0.24$ kgm. The rail model is valid up to a frequency of about 2.5 kHz [1]. There are eight rail beam elements in each sleeper bay. The discrete supports under the rail model include half a sleeper treated as rigid with the mass $M_S = 125$ kg. Rail pads and ballast are modelled with linear springs (k) and viscous dampers (c) coupled in parallel. Here, the input data for the rail pads and ballast are $k_p = 120$ MN/m, $c_p = 25$ kNs/m and $k_b = 100$ MN/m, $c_b = 82$ kNs/m, respectively.

The number of degrees-of-freedom (dofs) of the linear and non-proportionally damped track model is reduced by performing a complex-valued modal expansion retaining a truncated set of modal parameters. In the current study, the number of complex-conjugated pairs of eigenmodes accounted for in DIFF is 260 (corresponding to 4 kHz for the real part of the maximum eigenfrequency). The wheel-rail contact is modelled by a nonlinear spring with stiffness according to Hertz theory.

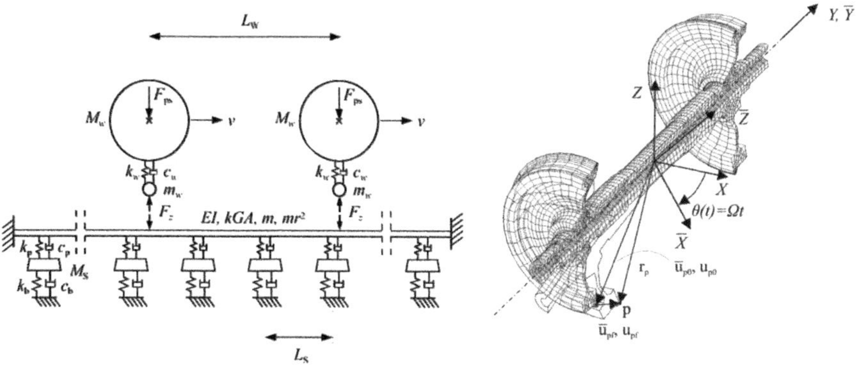

(a) (b)

Fig. 1. (a) Sketch of the dynamic vehicle–track interaction model with wheelset model D. (b) Cross-section of wheelset models A and B. Rotating ($\overline{XYZ}$) and non-rotating (XYZ) coordinate systems are outlined.

3 Vehicle Model

The current work considers dynamic vehicle–track interaction in the frequency range 20 – 3000 Hz. Four vehicle models are compared. Each vehicle model contains two wheelsets moving along the track model at constant speed v. The load from the primary suspension with magnitude F_{ps}=66.3 kN is assumed constant in the frequency range of interest. The axle load is 15.3 tonnes. The two wheelsets in one bogie are modelled without any coupling via the primary suspension and bogie frame.

The non-rotating (A) and rotating (B) flexible wheelset models are based on a FE model, see Fig. 1(b). Solid linear hexahedron elements (8-nodes) were applied and the model comprised of 14 040 elements and 56 052 dofs. Rigid attachments between the wheel axle and the wheels were assumed. The masses of axle box, primary suspension and gear box were included by increasing the nodal masses symmetrically around the circumference of the axle at their respective locations. Wheelset model C includes a single rigid body with mass 1779/2 kg corresponding to the weight of half a wheelset. In wheelset model D, a two-dof linear mass-spring-damper system is applied, see Fig. 1(a) and [7,8].

The mathematical formulation applied to establish the equations of motion for wheelset model B is described in Section 3.1. Properties of the wheelset models are compared in Section 3.2.

3.1 Mathematical Formulation for a Flexible Rotating Wheelset

To establish the equations of motion for the rotating wheelset, the methodology by Fayos et al. [4] is applied. By adopting Eulerian coordinates, the difficulties in modelling interaction forces between rotating and non-rotating structures are avoided. A detailed description is found in [4].

The wheelset is assigned two coordinate systems: one inertial coordinate system (XYZ) and one floating coordinate system ($\overline{XYZ}$) that rotates with the wheelset at a constant angular velocity Ω, see Fig. 1(b). Using time-variant Lagrangian modal coordinates $\mathbf{p}(t)$, the location of grid point p in Fig. 1(b) can be expressed in the inertial coordinate system as

$$\mathbf{r}_{\mathrm{p}} = \mathbf{A}(\theta)(\overline{\mathbf{u}}_{\mathrm{p0}} + \overline{\mathbf{u}}_{\mathrm{pf}}) = \mathbf{u}_{\mathrm{p0}} + \mathbf{A}(\theta)\overline{\mathbf{\Phi}}(\overline{\mathbf{u}}_{\mathrm{p0}})\mathbf{p}(t) \tag{3.1}$$

where $\mathbf{A}(\theta)$ is the transformation matrix from the floating coordinate system to the inertial coordinate system, $\overline{\mathbf{u}}_{\mathrm{p0}}$ and $\overline{\mathbf{u}}_{\mathrm{pf}}$ are the undeformed location vector and the deformation vector of grid point p with respect to the floating coordinate system and $\overline{\mathbf{\Phi}}(\overline{\mathbf{u}}_{\mathrm{p0}})$ is the partition of the mass-normalised eigenmode matrix associated with grid point p in the floating coordinate system.

For structures with axial symmetry, the location of grid point p using time-dependent Eulerian modal coordinates $\mathbf{q}(t)$ can be written as

$$\mathbf{r}_{\mathrm{p}} = \mathbf{u}_{\mathrm{p0}} + \overline{\mathbf{\Phi}}(\mathbf{u}_{\mathrm{p0}})\mathbf{q}(t) \tag{3.2}$$

where $\mathbf{u}_{\mathrm{p0}}$ is the undeformed location vector of grid point p with respect to the inertial coordinate system, while $\overline{\mathbf{\Phi}}(\mathbf{u}_{\mathrm{p0}})$ is the partition of the mass-normalised eigenmode matrix associated with grid point p in the inertial coordinate system.

The equations of motion for a rotating solid of arbitrary geometry in Lagrangian coordinates can be expressed as, see [4],

$$\ddot{\mathbf{p}} + 2\Omega\tilde{\mathbf{J}}\dot{\mathbf{p}} + (\tilde{\mathbf{K}} - \Omega^2\tilde{\mathbf{E}})\mathbf{p} = \mathbf{Q}_{\mathrm{p}} + \Omega^2\tilde{\mathbf{L}} \tag{3.4}$$

where $\tilde{\mathbf{K}}$ is the diagonal matrix that contains the square of the undamped natural frequencies associated with the eigenmodes in $\overline{\mathbf{\Phi}}(\mathbf{u})$. The matrices $\tilde{\mathbf{J}}$ and $\tilde{\mathbf{E}}$ are associated with the gyroscopic and centrifugal effects, respectively. The vector $\Omega^2\tilde{\mathbf{L}}$ contains the constant radial centrifugal force associated with the displacement of eigenmodes with multiplicity one and $\mathbf{Q}_{\mathrm{p}}$ is the external force vector. A force applied at a constant spatial location, such as the wheel–rail contact force, will require $\mathbf{Q}_{\mathrm{p}}$ to be calculated in each step of the ODE time integration procedure.

The equations of motion in Eulerian coordinates are obtained as, see [4],

$$\ddot{\mathbf{q}} + 2\Omega\tilde{\mathbf{G}}^{\mathrm{T}}\dot{\mathbf{q}} + (\tilde{\mathbf{K}} + \Omega^2(\tilde{\mathbf{G}}\tilde{\mathbf{G}} + \tilde{\mathbf{G}}\tilde{\mathbf{J}} - \tilde{\mathbf{J}}\tilde{\mathbf{G}} - \tilde{\mathbf{J}}\tilde{\mathbf{J}} - \tilde{\mathbf{E}}))\mathbf{q} = \mathbf{Q}_{\mathrm{q}} + \Omega^2\tilde{\mathbf{L}} \tag{3.5}$$

where $\tilde{\mathbf{G}}$ is the gyroscopic matrix. To establish the external force vector $\mathbf{Q}_q$, a computationally expensive integration over the volume of the solid is required. However, in Eulerian coordinates, for a force applied at a constant spatial location, it is sufficient to perform this integration only once in a pre-step to the simulations of vehicle–track interaction.

3.2 Comparison of Wheelset Models

Given a symmetric pair of radial, harmonic and synchronous force excitations at the two wheel–rail contact positions (one position on each wheel), calculated receptances (displacement over force) at the excitation point are shown for all wheelset models in Fig. 2(a). The possibility to tune the parameters of wheelset model D to mimic the anti-resonance at about 300 Hz results in a receptance in the high-frequency range that on average is in good agreement with the more complex models. The tuned parameters for model D are: $M_w = 1779/2$ kg, $k_w = 2900$ MN/m, $c_w = 68$ kNs/m and $m_w = 3$ kg.

Due to the wheel rotation, eigenmodes of multiplicity two split into so-called backward and forward whirl modes which diverge in frequency with increasing rotational speed. For some of the eigenmodes, this feature is observed in Fig. 2(b).

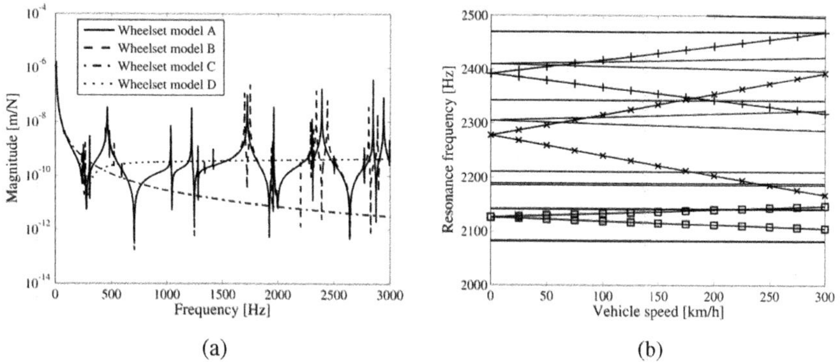

Fig. 2. (a) Calculated direct receptance for four different wheelset models. Symmetric excitation at one wheel–rail contact position per wheel. Vehicle speed 300 km/h is used in wheelset model B. All wheelset models are hinged at their primary suspensions. (b) Campbell diagram for wheelset model B illustrating the resonance frequencies in frequency range 2 kHz – 2.5 kHz. Corresponding eigenmodes marked with symbols are: (□) 4th anti-symmetric mode in bending, (x) axial wheel mode with four nodal diameters, (+) radial wheel mode with three nodal diameters.

4 Results

Calculated wheel–rail contact forces using the four different wheelset models are compared. An impact excitation is achieved by introducing a wheel flat (a local deviation from the nominal rolling radius caused by unintentional sliding) of length

0.06 m and depth 0.45 mm on the wheel running surface. The time history of the vertical contact force after impact, and the corresponding Fourier spectrum, are shown in Fig. 3(a) and Fig. 3(b), respectively. The largest magnitude of the vertical contact force is calculated for wheelset models C and D, and it occurs shortly after the wheel recovers contact with the rail. The contact force calculated for the rigid and flexible wheelset models show significant differences in the high-frequency range, see Fig. 3(b). At frequencies 1725 Hz and 2400 Hz, corresponding to the radial wheel modes with two or three nodal diameters, the results for wheelset model B are influenced by the forward and backward whirl modes. The largest components of the contact force are calculated at frequencies corresponding to the P2 resonance (wheelset, rail and sleepers moving in phase on the sleeper support stiffness) at around 80 Hz and the third symmetric mode in bending of the wheelset at about 1 kHz.

The vertical contact forces calculated for the different wheelset models when subjected to excitation by a sinusoidal irregularity are also investigated. Two cases are considered: in Fig. 4(a), the excitation frequency is corresponding to the radial mode with three nodal diameters of wheelset model A; in Fig. 4(b), frequency leading to a simultaneous excitation of the radial mode with three nodal diameters and the axial wheel mode with four nodal diameters of wheelset model B (see intersection point in Fig. 2(b)). Fig. 4 shows that excitation at a resonance of a wheelset reduces the magnitude of the vertical contact force. Further, for excitation at a frequency corresponding to an intersection point between two eigenmodes of the rotating wheelset model B, a phase-shift is observed in the vertical contact force calculated for model B, see Fig. 4(b).

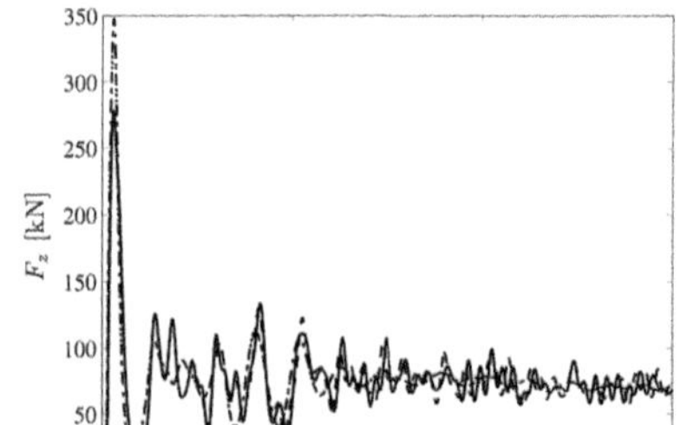

(a)

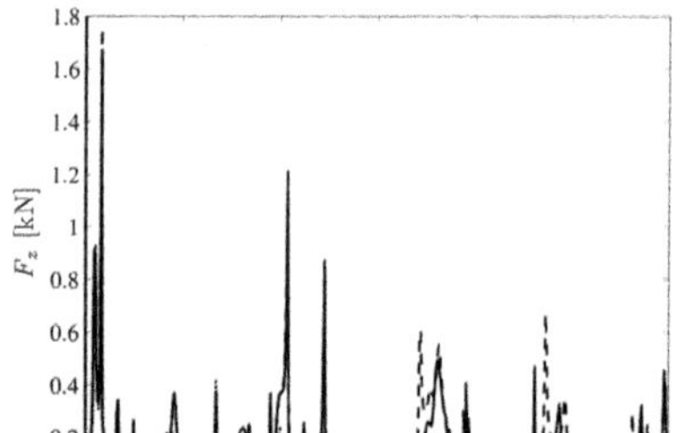

(b)

Fig. 3. (a) Time history and (b) Fourier spectrum of vertical wheel–rail contact force of the trailing wheelset. Excitation by a wheel flat of length 0.06 m and depth 0.45 mm, vehicle speed 197 km/h. —: wheelset model A, - -: wheelset model B, ·-·: wheelset model C,····: wheelset model D.

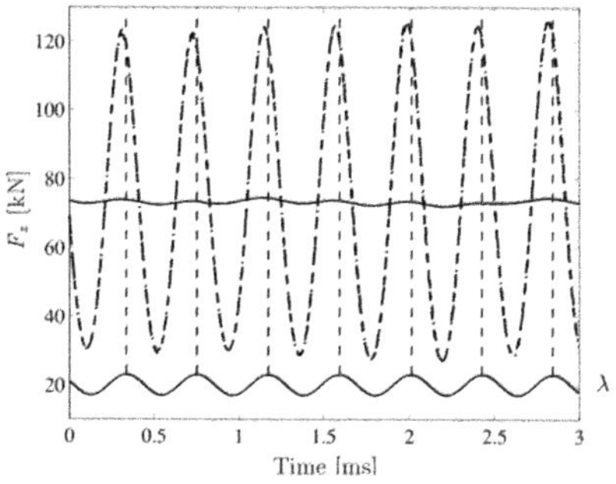

(a)

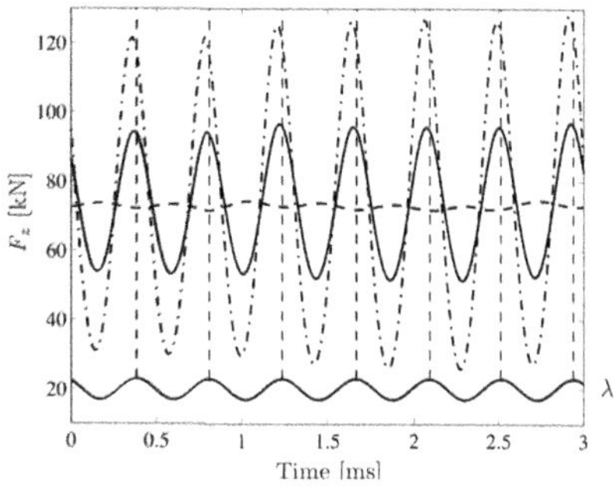

(b)

Fig. 4. Time history of vertical wheel–rail contact force due to excitation by a sinusoidal irregularity on the running surface of the rail. The shape of the irregularity (λ) is outlined. Vehicle speed 182 km/h: (a) λ = 0.0211 m (2396 Hz), (b) λ = 0.0215 m (2351 Hz).—: wheelset model A, - -: wheelset model B, ·-·: wheelset model D. Results for wheelset models B and D are overlapping in (a).

Finally, the influence of vehicle speed on the vertical contact force is investigated. Roughness wavelengths shorter than 2.25 cm were set to zero and hence, at vehicle speed 200 km/h, no excitation of contact force occurs above a frequency of about 2.5 kHz. In Fig. 5, the quotient of the power spectral densities (PSD) of the vertical contact force calculated for wheelset models D and B is shown. The quotient shows large deviations from 1 at frequencies corresponding to resonances of wheelset model B. Significantly higher magnitudes of the vertical contact force are calculated for wheelset model D at frequencies corresponding to the radial wheel modes with two or three nodal diameters at around 1725 Hz and 2400 Hz, respectively. The influence of the second symmetric mode in bending of the wheelset at about 460 Hz is also observed. Deviations from 1 of the PSD quotient are particularly significant for frequencies above 1.5 kHz.

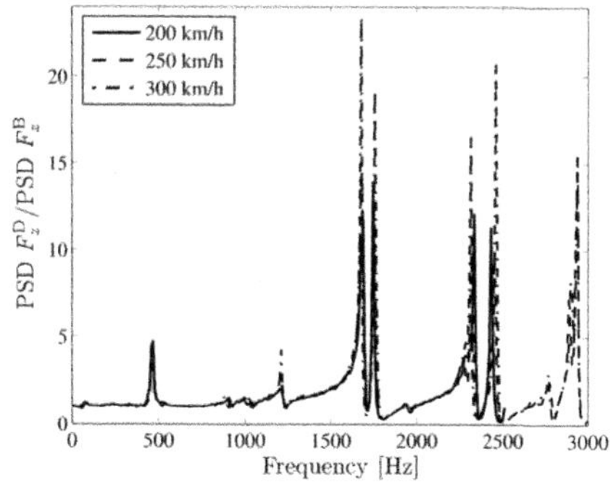

Fig. 5. Quotient between power spectral densities (PSD) of vertical wheel–rail contact force calculated for wheelset models D and B. Results calculated for a single wheelset model travelling at different speeds on a corrugated rail. Level of rail corrugation according to measured data.

5 Concluding Remarks

The in-house computer program DIFF has been applied to investigate dynamic vehicle–track interaction at high vehicle speeds. The importance of considering the inertial effects due to wheel rotation in the modelling of the wheelset has been assessed for different load cases.

When comparing results based on rotating and non-rotating flexible wheelset models, this study has shown that if the wheelset is excited at a frequency where two different mode shapes, due to the wheel rotation, have coinciding resonance frequencies, significant differences in the amplitude and phase of the calculated vertical wheel–rail contact forces may appear. This implies that it is important to consider the rotational effects in order to accurately capture the wheelset dynamics at high vehicle speeds and high excitation frequencies.

For an impact excitation caused by a wheel flat, the Fourier spectrum of the vertical contact force calculated using a flexible wheelset model showed large magnitudes in the mid-frequency range below about 1.5 kHz. Peaks were found at frequencies corresponding to the P2 resonance at about 80 Hz and the third symmetric mode in bending of the wheelset at 1 kHz. At and around these frequencies, the spectral content of the contact force calculated for the rotating and non-rotating flexible wheelset models were similar. However, large differences were found if instead a broad-band excitation with significant components in the high-frequency range (above 1.5 kHz) was introduced. The influence of structural flexibility was observed for frequencies corresponding to the radial wheel modes with two or three nodal diameters at about 1725 Hz and 2400 Hz.

References

[1] Knothe, K., Grassie, S.L.: Modelling of railway track and vehicle/track interaction at high frequencies. Vehicle System Dynamics 22, 209–262 (1993)
[2] Andersson, C.: Modelling and simulation of train–track interaction including wear prediction, PhD Dissertation, Department of Solid Mechanics, Chalmers University of Technology, Göteborg, Sweden (2003)
[3] Thompson, D.J.: Wheel–rail noise generation, Part V: Inclusion of wheel rotation. Journal of Sound and Vibration 161, 467–482 (1993)
[4] Fayos, J., Baeza, L., Denia, F.D., Tarancon, J.E.: An Eulerian coordinate-based method for analyzing the structural vibrations of a solid of revolution rotating about its main axis. Journal of Sound and Vibration 306, 618–635 (2007)
[5] Baeza, L., Fayos, J., Roda, A., Insa, R.: High frequency railway vehicle–track dynamics through flexible rotating wheelsets. Vehicle System Dynamics 46, 647–662 (2008)
[6] Nielsen, J.C.O., Igeland, A.: Vertical dynamic interaction between wheel and track – Influence of wheel and track imperfections. Journal of Sound and Vibration 187, 825–839 (1995)
[7] Nielsen, J.C.O.: High-frequency vertical wheel–rail contact forces – validation of a prediction model by field testing. Wear 265, 1465–1471 (2008)
[8] Thompson, D.J.: Railway Noise and Vibration. Elsevier, Oxford (2009)

Optimization of New Noise Shielded and Vibration Damped Assemblies for High Speed Train Wheels Using FEM-BEM Simulation

C. Wang, X.B. Xiao, and X.S. Jin

State Key Laboratory of Traction Power,
Southwest Jiaotong University, Chengdu 610031, China
Tel: +86 28 87634355, Fax: +86 28 87600868
e-mail: xsjin@swjtu.cn

Summary

This paper presents the design and optimization of a new noise screen for the noise control of a high speed train wheel. This apparatus consists of multiple circular metal sheets, which are alternately anchored to the inner edge of the wheel rim and the hub area, respectively. Viscoelastic damping layers are inserted between the metal sheets. It acts as a noise shield, which can obstruct the sound radiation from the wheel web region. Meanwhile, it also works as a vibration absorber and can mitigate the vibration of the wheel through energy dissipation in the intermediate viscoelastic damping layers. A combined numerical procedure of FEM and BEM is used to analyze the harmonic response and sound radiation of a wheel of the CRH3 high speed train with this assembly installed.

1 Introduction

As wheel is one of the main contributors of rolling noise, various types of noise control devices for it have been studied for many years. Tuned resonance dampers, ring dampers, friction dampers and resilient wheels are intended to reduce the vibration of the wheel and hence its radiated noise. Meanwhile, there are alternative ways to minimize the sound that is radiated due to a particular vibration level. One means for reducing the radiation is to mount a shield on the wheel so that sound radiation from the wheel web region is obstructed. In the RONA project, one type of web screen with 1mm thick steel plates was mounted resiliently onto the wheel and reduced the wheel component of noise by about 6 dB [1]. Wheels with similar shields were tested in the Silent Freight project and were estimated to reduce the wheel component of noise by 8 dB [2]. Another type of cover-like wheel absorber was reported to reduce noise by 8-9 dB with respect to an 'untreated' wheelset with a relatively high roughness, during the field test of the ETR500 [3].

T. Maeda et al. (Eds.): Noise and Vibration Mitigation for Rail Trans. Sys., NNFM 118, pp. 51–59.
springerlink.com

The present work develops an innovative web-fitted assembly. It can shield noise radiated from the wheel web region and reduce the vibration level of the wheel effectively. A combined numerical procedure of a finite element method (FEM) and a boundary element method (BEM) is used to analyze the harmonic response and sound radiation of a wheel of the CRH3 high speed trains with this assembly.

2 Noise Shielded and Vibration-Damped Assembly

This cover assembly comprises multiple circular metal sheets. These sheets are alternately anchored to the inner edge of the wheel rim and the hub area, respectively. It acts substantially in a twofold manner. On the one hand, it operates as a shield which prevents the sound in the wheel web region transmitting into the surrounding air. On the other hand, it acts as a vibration absorber, which not only damps the cover plates itself, but also mitigates the vibration of the wheel. When the contact force induces large flexural deformation of the wheel rim due to its resonance, the large relative displacement occurs between the alternately anchored plates. In such a situation, this cover assembly effectively induces energy dissipation during high cyclic shear strains in the intermediate viscoelastic layers.

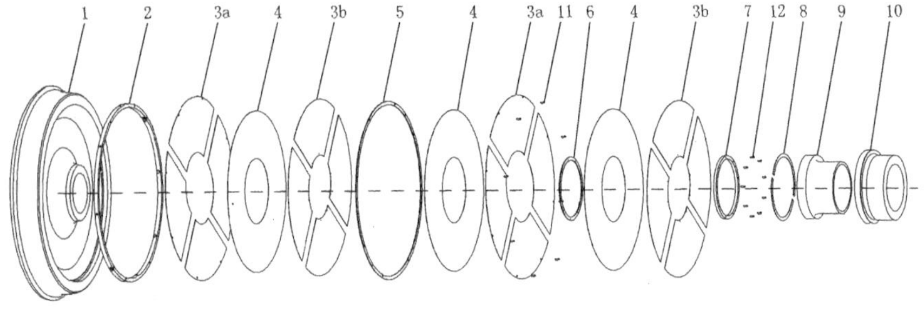

Fig. 1. Assembly diagram of the prototype

To describe the cover assembly more specifically, a prototype assembly diagram is shown in Fig. 1. The anchored portions of the metal sheets (4) are separated by the spacers (5, 6). Each metal sheet has a free end. The middle area between the metal plates is adhered together using the layers of adhesive viscoelastic material (3a, 3b). Then, as an integral anchored sandwich, it is fixed to mounting rings (2, 8). The larger ring (2) is installed at the inner side of the rim; the smaller ring (8) is installed at the hub. To install these rings, there are grooves milled into the wheel body in which the rings are clamped. In addition, a thermal insulating layer is placed between the large ring and the integral damping layers to protect the viscoelastic damping material from the high temperature of the rim during braking. The straight-web wheel of CRH3 high speed train is used for the investigation of the absorber device.

3 The Finite Element and Boundary Element Model

The whole structure is meshed exclusively with hexahedral solid elements, as shown in Fig. 2. It can be sorted into three groups: the wheel body (green), the anchored metal plates (blue) and the viscoelastic damping layers (yellow). The average element size is 18mm.

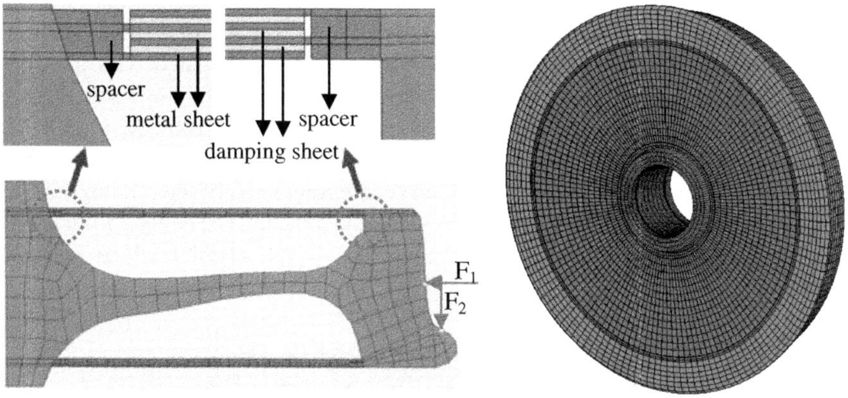

Fig. 2 Mesh of the wheel with shielding assembly (cross section view, iso view of the model) (color figure online)

Table 1. Material parameters

material	Avg. E (N/m^2)	density (kg/m^3)	Poisson's ratio	Avg. loss factor
steel	2.1×10^{11}	7800	0.3	
aluminium	7.2×10^{10}	2700	0.3	
damping mat 1	3.4×10^{6}	1000	0.49	0.5
damping mat 2	3.4×10^{7}	1000	0.49	0.5
damping mat 3	3.4×10^{6}	1000	0.49	1.0
damping mat 4	3.4×10^{7}	1000	0.49	1.0
ZN03		1000	0.49	

The study considers several different choices of material, including choices of the metal plates (steel and aluminium) and the damping layers. Several combinations of average shear modulus and average loss factor of damping material are studied for the effectiveness of the damping layer in terms of noise reduction. The material parameters used in the analysis are listed in Table 1.

The China-made ZN03 material is used to study the dependence of the elastic and viscous properties of the intermediate damping layer material on frequency and temperature. The analysis considers frequency-temperature-dependent shear modulus and loss factor in each 1/3 octave frequency band. The chosen material has a large loss factor over a relatively wide temperature and frequency range as shown in its nomogram in Fig. 3. Moreover, it can accommodate relatively well to the harsh railway environment, such as oil pollution and aging, and can meet self-extinguishing demands.

The commercial code NASTRAN is used to calculate the mode shapes with the Lanzos method and the harmonic response of the untreated wheel with the modal frequency response method. A nominal value of 0.0002 is selected for the damping ratio. The harmonic response of the shielded wheel is calculated with the direct frequency response method. In the dynamic response calculation, three types of excitation force are considered: vertical unit force from 100 Hz to 5000Hz, horizontal unit force in the same frequency range and a vertical irregularity force. The vertical irregularity force reflects the average roughness of the running surfaces of the wheel and rail, which is filtered by the contact patch. The contact filter used is for the CRH3 wheel (920mm wheel diameter, 60kN wheel load) at speed of 300km/h.

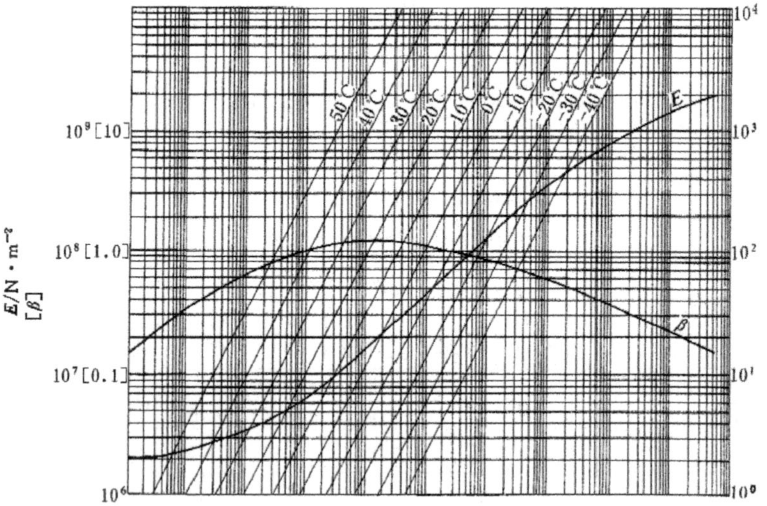

Fig. 3. Damping Polymer Property Specification of ZN03

With the FEM results as the boundary conditions, the sound radiation characteristics of the shielded and damped wheel are investigated using the BEM software SYSNOISE. The radiation boundary of the wheel is meshed mainly with quadratic elements. The average size of the elements is 18mm. To prevent sound leakage through the hole at the hub, additional elements are placed at this area with zero velocity boundary conditions.

4 Numerical Results of Vibration and Sound Radiation

4.1 Dynamic Response of Wheel

Fig. 4 shows the vibration results (velocities) at typical points on the surface of the wheel under roughness excitation at the nominal contact point. The velocities are normal to the surface at these points. Other results under unit force excitation are not described in detail in this paper. With the proposed assembly, the vibration levels of the wheel at the concerned points of the wheel tread, rim and web are all mitigated considerably under different excitations, when compared with the wheel without damping treatment. More specifically, this damping assembly itself has a lower vibration level, when compared with that of the corresponding position at the

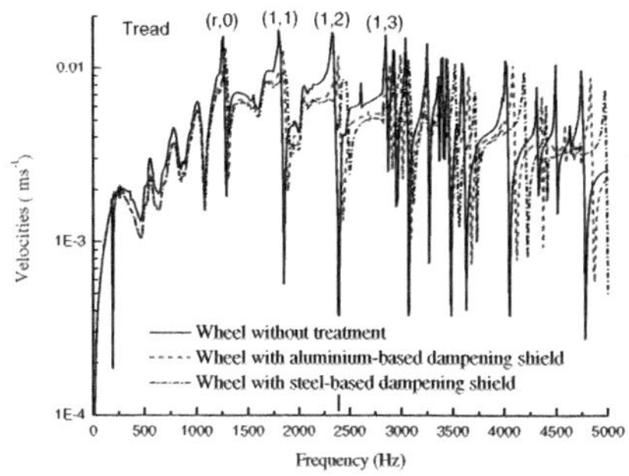

(a) Tread

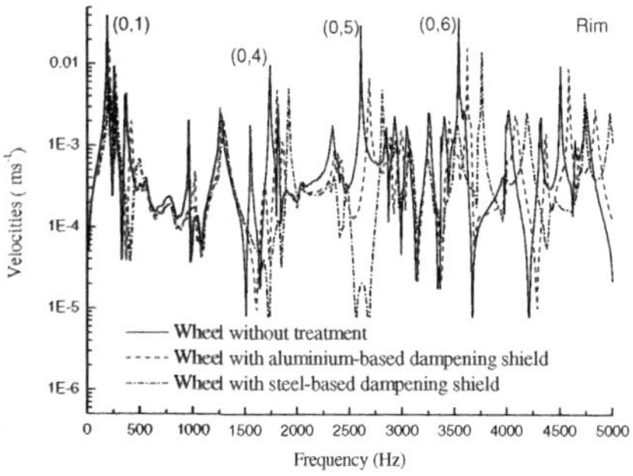

(b) Rim

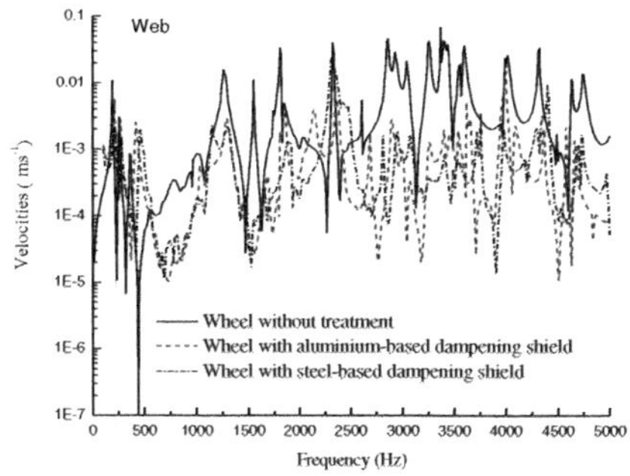

(c) Web

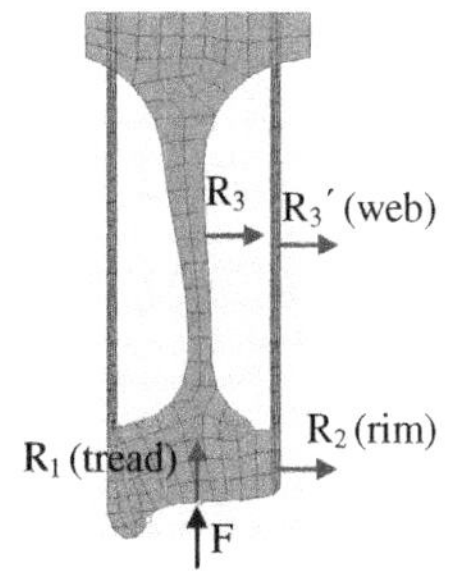

(d) excitation point and specified response points

Fig. 4. Vibration (velocities) of the wheel under roughness excitation (color figure online)

*For a wheel, similarly to a flat disc, out-of-plane (axial) modes correspond to various numbers of nodal diameters, n, possibly in combination with a number of nodal circles, m. The mode with m nodal circles and n nodal diameters will be denoted mode (m, n).

web, as shown in Fig.4 (c). Also, it can effectively reduce the vibration level of the wheel rim, especially in vibration modes showing the rim tilting. These modes are, respectively, modes (0, 1), (0, 6), (0, 5) and (0, 4)*, as shown in Fig.4 (b). The vibration of the tread is dominated by one-nodal-circle modes and radial modes (r, 0), and this assembly can also reduce these vibration peaks, to some extent, as shown in Fig.4 (a).

4.2 Sound Radiation Power of Wheel

Figure 5 compares the sound radiation power spectrum of the untreated wheel, the wheels with an aluminium-based damping shield and a steel-based damping shield under vertical equivalent force excitation. The results show that the damping shields effectively reduce the sound radiation power level of the wheel above 500Hz. The damping shields can sufficiently reduce the vibration level (above 500Hz) of all main parts of the wheel (web, rim and tread), see Fig.4. From previous studies,, the wheel contribution to railway noise is mainly concentrated between 1000Hz and 5000Hz. So this damping shields show satisfactory results in the relevant frequency range. The ZN03 material parameters used in this study are for 30°C and 0°C. The Young's modulus and loss factor are frequency-dependent in the direct frequency response analysis. Other damping shield cases with different damping materials are analyzed in an attempt to find suitable material properties for the intermediate damping layers. The overall sound radiation power levels are listed in Table 2.

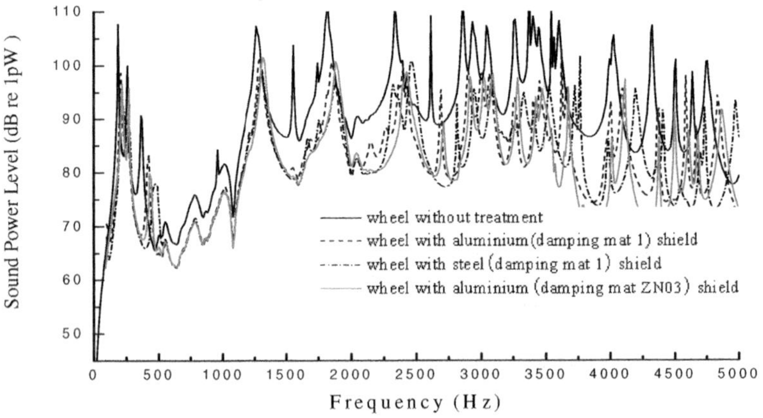

Fig. 5. Sound radiation power level of the wheel under roughness excitation (color figure online)

Table 2. Overall sound radiation power level of the wheel

	vertical unit force	lateral unit force	roughness equivalent force
untreated wheel	66.68 dB(A)	72.40 dB(A)	111.44 dB(A)
alum-damping-mat1	55.68 dB(A)	66.69 dB(A)	102.88 dB(A)
alum-damping-mat2	52.11 dB(A)		101.05 dB(A)
alum-damping-mat3	49.39 dB(A)		97.77 dB(A)
alum-damping-mat4	49.36 dB(A)		98.76 dB(A)
steel-damping-mat1	52.68 dB(A)	65.23 dB(A)	101.58 dB(A)
steel-damping-mat2	47.57 dB(A)		100.08 dB(A)
steel-damping-mat3	48.27 dB(A)		98.52 dB(A)
steel-damping-mat4	44.66 dB(A)		98.38 dB(A)
alum-damping-mat ZN03 (30°C)	52.76 dB(A)		102.23 dB(A)
alum-damping-mat ZN03 (0°C)	56.65 dB(A)		103.21 dB(A)

4.3 Sound Pressure Contours of Field Points

Fig. 6 shows the results of the field point planes from the front and back side of the wheel web. The reduction of the sound pressure level is about 16 dB under the vertical roughness equivalent force excitation. The sound pressure levels calculated at the standard points [4] are listed in Table 3. And sound radiation at the standard point 1 (Distance: 7.5 m, Height: 1.2 m), is typically used to indicate the wheel-rail rolling noise contribution, as shown in detail in Fig.7.

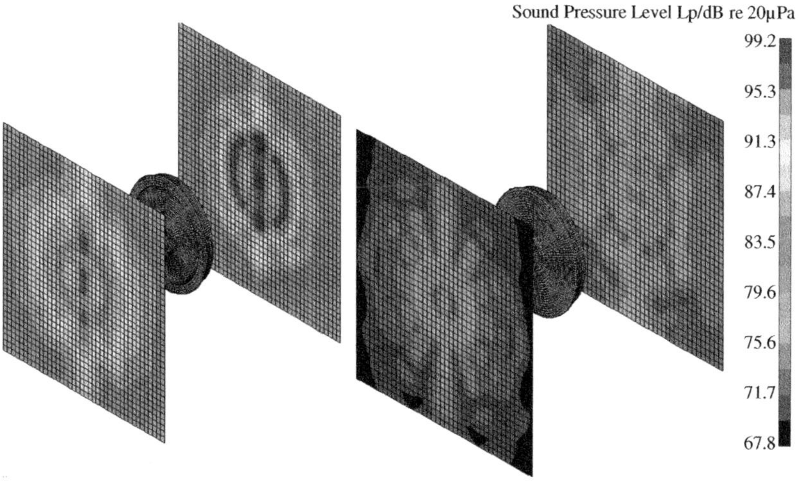

Fig. 6. Sound pressure contour at field point planes (dB, 100~5000Hz, full range integration) (color figure online)

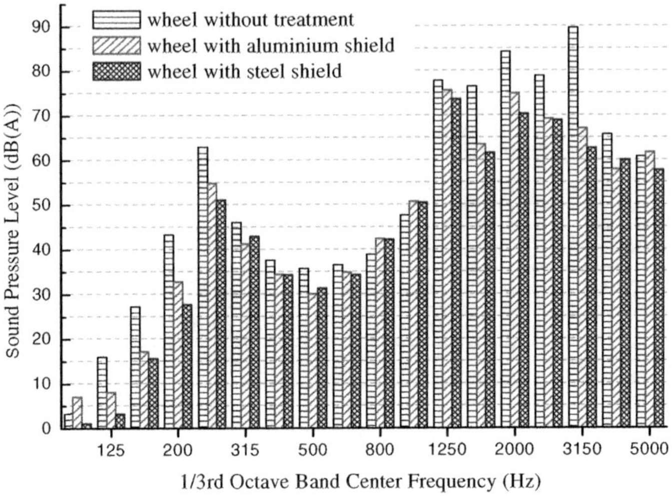

Fig. 7. Sound pressure level at standard point 1 (Distance: 7.5 m, Height: 1.2 m) (color figure online)

Table 3. Sound pressure level at standard field points

	Position 1	Position 2	Position 3
	L: 7.5m, H: 1.2m	L: 7.5m, H: 3.5m	L: 25m, H: 3.5m
untreated wheel	91.4 dB(A)	85.7 dB(A)	80.3 dB(A)
alum-damping-mat1	79.3 dB(A)	77.0 dB(A)	69.6 dB(A)
steel-damping-mat1	76.7 dB(A)	74.9 dB(A)	67.1 dB(A)

5 Conclusion and Outlook

The numerical results show good effectiveness in reducing the noise of CRH3 high speed wheel by around 10dB when applying the noise shield and vibration damping assembly to the wheel. Further laboratory and field measurements on the shield assemblies of the wheel noise are under way.

Acknowledgements

The present work has been supported by the National Natural Science Foundation of China (50821063) and the Basic Research Program of Railway Ministry (2008J001-A).

References

[1] Thompson, D.J., Gautier, P.E.: A review of research into wheel/rail rolling noise reduction. Proceedings of the Institution of Mechanical Engineers, Part F. Journal of Rail and Rapid Transit 220F, 385–408 (2006)
[2] Hemsworth, B., Gautier, P.E., Jones, R.: Silent Freight and Silent Track projects. In: Proceedings of Internoise 2000, Nice, France (2000)
[3] Bracciali, A., Ciuffi, L., Ciuffi, R.: Calibration of an On-Board Noise Measuring Device by Simultaneous Measurement of Trackside Noise of Three Different Wheelsets for the ETR500 F.S. Train. Journal of Rail and Rapid Transit. 211, 41–49 (1997)
[4] ISO 3095-2005 Railway applications – Acoustics – Measurement of noise emitted by railbound vehicle

Rolling Noise Simulation: A New Approach in Time Domain

V. Delavaud[1], A. Chaigne[2], and F. Poisson[1]

[1] SNCF, Innovation and Research Departement, Batiment Lumière, 40 avenue des terroirs de France, 75611 Paris cedex 12
e-mail: virginie_delavaud@hotmail.com
[2] ENSTA ParisTech, Unité de Mécanique, Chemin de la Hunière 91761 Palaiseau cedex

Summary

Rolling noise is the main source of railway noise for a wide speed range. It is then essential for a railway company to understand the generation mechanisms to control it. Until now, SNCF, the French railway company, has used TWINS, the Track Wheel Interaction Noise Software, to predict the rolling noise. Based on the frequency—domain approach, this software is very efficient for the case of straight standard ballasted tracks. However, as it runs in the frequency domain, this software is limited to cases of wheel and rail without discrete irregularities. A time—domain approach has been then considered to extend the rolling noise modelling to the case of impact noise due to discontinuities, such as wheel flats or rail joints.

The main challenge in the development of a model running in the time domain is to simulate an infinite structure with a finite numerical model. Here, absorbing boundary conditions, using an impedance technique, have been developed for the case of a finite Euler—Bernoulli beam model representing the rail. The full problem has been solved using a finite difference scheme. The performance of the absorbing boundary conditions has been studied by comparing the results with the ones obtained for the case of free-boundary conditions.

1 Introduction

The modelling of rolling noise has been widely investigated in the frequency domain. Complete presentations have been given by Knothe and Grassie [1] and by Thompson [2]. Thompson's model has been implemented in TWINS (Track Wheel Interaction Noise Software) [2, 3]. This software is very efficient in the case of typical rolling noise. However, the loss of contact between wheel and rail can occur in the presence of severe roughness or discrete irregularities, such as wheel flats or rail joints. This kind of feature leads to nonlinearities in the contact model, which can be only solved with a time—domain model [1, 2].

T. Maeda et al. (Eds.): Noise and Vibration Mitigation for Rail Trans. Sys., NNFM 118, pp. 61–69.
springerlink.com

The main difficulty of a time—domain model is to simulate an infinite track with a finite numerical model [1]. A good approach based on the time domain Green's functions has been used by Pieringer [4]. Another alternative is to develop artificial absorbing boundary conditions. The method proposed by Svensson et al. [5] uses an impedance technique to cancel boundary reflections of an Euler Bernoulli beam in the frequency domain. This technique has been used to build absorbing boundary conditions in the time domain for a finite track model.

The technique of impedance matching has already been investigated in sound propagation (see for example [6]). In this situation, the impedance is a scalar quantity relating the pressure with the acoustic particle velocity. For the case of an Euler—Bernoulli beam, the impedance is a 2x2 matrix relating the shear force F and the bending moment M to the normal velocity v_R and the rotational velocity w_R.

The procedure used is to calculate an absorbing impedance matrix which cancels the reflection matrix. The reflection matrix is defined using four impedance matrices; two of these describe the Euler—Bernoulli beam, the third defines the arbitrary boundary and the last is an external active impedance load which is defined to cancel the reflection matrix [5].

2 Finite Track Model with Absorbing Boundary Conditions

2.1 Track Model

The track model (Fig.1) is composed of a finite Euler—Bernoulli beam of length L, to represent the rail and spring-damper systems to represent the periodic supports. The equation of the vertical motion of the rail is given by

$$\rho S\partial_t^2 z_R + EI\partial_x^4 z_R + \sum_{n=1}^{n_A}(C_A\partial_t z_R + K_A z_R)\delta(x-nl) = f(x,t) \tag{2.1}$$

where z_R is the rail vertical displacement. E is the steel Young's modulus, I is the second moment of area, ρ is the beam density and S is the beam transverse

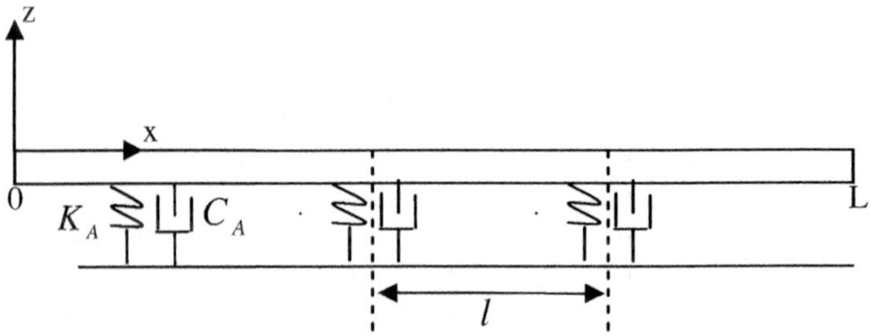

Fig. 1. Track model with discrete supports.

section area. C_A and K_A are the support damping and is the support stiffness, respectively. n_A represents the number of discrete supports and l is the distance between each support. $f(x,t)$ is the external load per unit length. The beam is assumed to have free ends.

2.2 Absorbing Impedance in Frequency Domain

To avoid boundary reflections, some special absorbing boundary conditions have been specified using an impedance technique [5]. Although this method is defined in the frequency domain, it has been used to develop absorbing boundary conditions in the time domain. The impedance matrices are defined for all beam points except those above discrete supports. Thus, it can be considered that the impedance technique is applied on boundary spans of the beam.

If the beam has free ends, the arbitrary boundary impedance matrix is zero [5]. Then, the shear force F and the bending moment M can be described by the following relation:

$$\begin{pmatrix} F(x,\omega) \\ M(x,\omega) \end{pmatrix} = Z^{abs} \begin{pmatrix} v_R(x,\omega) \\ w_R(x,\omega) \end{pmatrix} \tag{2.2}$$

where Z^{abs} is the 2x2 absorbing impedance matrix. v_R and w_R are the normal and the rotational velocities, respectively. ω is the circular frequency.

The reflection matrix R links the incident propagating wave and near-field amplitudes to the reflected ones by the relation:

$$\begin{pmatrix} a_p^- \\ a_n^- \end{pmatrix} = R \begin{pmatrix} a_p^+ \\ a_n^+ \end{pmatrix} \tag{2.3}$$

where the exponents + and - denote the direction of wave propagation and the subscripts *p* and *n* denote the wave nature (propagating or near-field). The reflection matrix can also be obtained are to characteristic impedances of the Euler—Bernoulli beam and the boundary impedance matrix. The objective is to calculate the absorbing boundary impedance matrix Z^{abs} which cancels the reflection matrix R . It has been decided to cancel the reflected propagating wave amplitudes only (i.e. the first column of the reflection matrix).

The absorbing impedance has been arbitrarily defined by:

$$Z^{abs} = \begin{bmatrix} z_{11} & 0 \\ 0 & z_{22} \end{bmatrix} \tag{2.4}$$

to link the shear force with only the normal velocity, and the bending moment with only the rotational velocity. The problem is thus to find the impedances z_{11} and z_{22} which cancel the first column of the reflection matrix. The results, in the frequency domain, for incident propagative waves travelling in the positive x direction are given by:

$$z_{11}(\omega) = \frac{EI}{a}\sqrt{\frac{\omega}{a}} \tag{2.5}$$

and

$$z_{22}(\omega) = \frac{EI}{\omega}\sqrt{\frac{\omega}{a}} \tag{2.6}$$

whe re E , I are defined in section 2.1 and a is such that $a^2 = \sqrt{EI/\rho S}$.

2.3 Interpolation of Absorbing Impedance in the Frequency Domain

The absorbing boundary conditions in the frequency domain are obtained from z_{11} and z_{22} in equations (2.5) and (2.6). However, the presence of the square root in each equation makes the conversion from the frequency domain to the time domain difficult. Therefore, an interpolation of z_{11} and z_{22} has been carried out using rational fractions such as

$$z_{11}(\omega) = \frac{a_1 + a_2(j\omega)}{a_3 + a_4(j\omega)} \tag{2.7}$$

and

$$z_{22}(\omega) = \frac{b_1 + b_2(j\omega)}{b_3 + b_4(j\omega)} \tag{2.8}$$

where j is the imaginary unit. This form of rational fraction makes the conversion into the time domain easier as the $j\omega$ terms correspond to the first time derivative. However, the continuous expressions of z_{11} and z_{22} are real functions. Only the real part of equations (2.7) and (2.8) have been considered in the interpolation, which has been achieved for frequencies in the range [1;4000] Hz. The results of this interpolation are shown in Fig.2. They are considered valid up to 2000 Hz, with a relative error less than 20%.

The results presented in this part are carried out for waves travelling in the positive x direction. In this case, the parameters $a_i, b_i\ \ i = \{1,\ldots,4\}$ are positive. In the case of waves travelling in the opposite direction, the expressions of z_{11} and z_{22} are the exact opposite of Eqs (2.5) and (2.6) and $a_i, b_i\ \ i = \{2,3\}$ are negative, whereas $a_i, b_i\ \ i = \{1,4\}$ are positive.

2.4 Absorbing Boundary Conditions in Time Domain

Substituting equations (2.7) and (2.8) into (2.2) leads to a relation which can be easily converted into the time domain. Writing equation (2.2) in terms of displacements rather than velocities, the expressions obtained for the shear force and the bending moment are given by

$$\begin{cases} F(x,\omega) = \dfrac{a_1(j\omega)+a_2(j\omega)^2}{a_3+a_4(j\omega)} z_R(x,\omega) \\ M(x,\omega) = \dfrac{b_1(j\omega)+b_2(j\omega)^2}{b_3+b_4(j\omega)} \partial_x z_R(x,\omega) \end{cases} \tag{2.9}$$

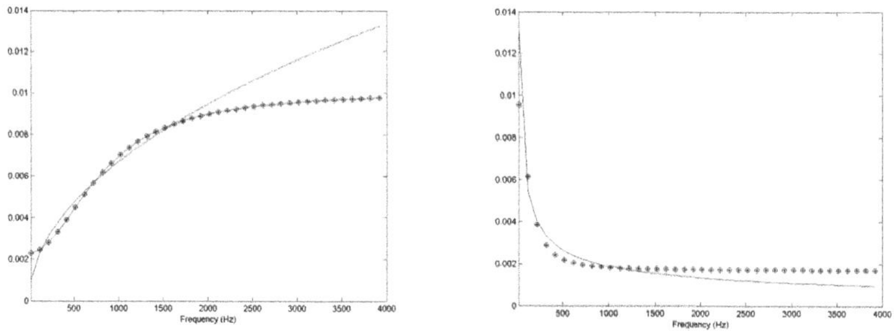

(a) Interpolation of Z_{11} (2.7) (b) Interpolation of Z_{22} (2.8)

Fig. 2. Interpolation of the absorbing impedance matrix coefficient in the frequency domain using rational fractions. (—) Real part of the continuous impedance, (***) interpolated impedance.

Then, the expressions of shear force and bending moment in the time domain are given by

$$\begin{cases} F(x,t) = \dfrac{a_1}{a_3}\partial_t z_R(x,t) + \dfrac{a_2}{a_3}\partial_t^2 z_R(x,t) - \dfrac{a_4}{a_3}\partial_t F(x,t) \\ M(x,t) = \dfrac{b_1}{b_3}\partial_t\partial_x z_R(x,t) + \dfrac{b_2}{b_3}\partial_t^2\partial_x z_R(x,t) - \dfrac{b_4}{b_3}\partial_t M(x,t) \end{cases} \tag{2.10}$$

The shear force and bending moment of the Euler—Bernoulli beam model can also be written in the form

$$\begin{cases} [F(x,t)] = [EI\partial_x^3 z_R(x,t)] \\ M(x,t) = -EI\partial_x^2 z_R(x,t) \end{cases} \tag{2.11}$$

where [.] denotes the shear force difference. By substituting Eq. (2.11) into (2.10), the absorbing boundary conditions in the time domain are finally derived.

A simple observation of Eqs. (2.10) shows that the first term on the right hand side is equivalent to a damping and that the second term is equivalent to a mass.

3 Implementation of the Track Model in the Time Domain

3.1 Simplification of the Absorbing Boundary Conditions

An energy analysis of the track model described above has been done for the case of a harmonic displacement. This analysis shows that a physical instability is introduced in the system due to the second and third terms on the right hand side of Eqs. (2.10), as the energy of the system increases in local time intervals. If these terms are considered in the boundary conditions, the problem is consequently not well-posed.

To avoid this physical instability, the parameters a_2, a_4, b_2, b_4 are set to zero. This results in the assumption of a constant impedance in Eqs (2.7) and (2.8). The absorbing boundary conditions are now equivalent to damping terms and are expressed by

$$\begin{cases} [F(x,t)] = [EI\partial_x^3 z_R(x,t)] = R_0 \partial_t z_R(x,t) \\ M(x,t) = -EI\partial_x^2 z_R(x,t) = \eta_0 \partial_t \partial_x z_R(x,t) \end{cases} \tag{3.1}$$

where $R_0 = a_1/a_3$ and $\eta_0 = b_1/b_3$. When considering the remark at the end of section 2.3, R_0 and η_0 are positive at $x = L$ and negative at $x = 0$.

3.2 Numerical Procedure of Equation Solving

Equation (2.1) is solved together with the boundary conditions (3.1) using a finite difference scheme. The fourth, order spatial derivative in the left side of eq. (2.1) has been discretized with a fourth, order scheme which has been optimized to reduce its numerical dispersion. The time derivatives have been discretized with first, and second, order centre schemes, depending on the order of the derivative. The Courant—Friedrichs—Levy condition (numerical stability condition) of the full scheme is such that

$$a\frac{dt}{dx^2} \leq 0.189 \tag{3.2}$$

where a is defined in section 2.1, dt is the time step and dx is the spatial step.

4 Comparison between Free and Absorbing Boundary Conditions

Simulations have been carried out for the case of a point force of magnitude F_0 applied at the middle of the beam at $t = 0$. Then, the external load in Eq. (2.1) is expressed as

$$f(x,t) = -F_0\delta(x - L/2)\delta(t) \tag{4.1}$$

The parameter values of the track are $L = 6.38$ m, $E = 2\times10^{11}$ Pa, $I = 20.18\times10^{-6}$ m^4, $\rho = 7850$ kg.m^{-3}, $S = 6.48\times10^{-3}$ m^2. The distance between the periodic supports is $l = 0.58$ m, then there are $n_A = 11$ supports. The parameter values of the periodic supports are $K_A = 964\times10^6$ Nm^{-1}, $C_A = 51066$ Nsm^{-1}. The force magnitude is $F_0 = 2$ kN. The total observation time of the beam displacement is $T = 0.3$ s.

4.1 Results in the Time Domain

The displacement of the beam at $x = L - l/2$ (middle point of the last span), for an excitation of the form (4.1) at $x = L/2$, is represented in Fig.3 for the cases of free and absorbing boundary conditions for a time observation equal to two, thirds of T, that is $t_{obs} = 0.2$ s.

It can be seen that the wave amplitude is almost two times smaller for the case of absorbing boundary conditions. It can be seen also that the energy of the system decreases with the time in both cases. This is due to the supports damping.

The total energy of the system is absorbed after $t = 0.15$ s for the case of absorbing boundary conditions, whereas there are small oscillations around the equilibrium of the beam with free ends during all the time observation. Despite the energy absorption of the supports, the free end beam keeps energy during the time observation, while the absorbing end beam dissipates it.

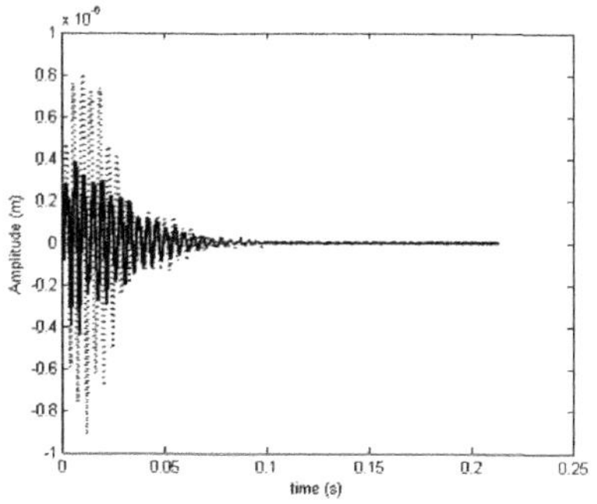

Fig. 3. Vertical displacement of the beam at $x = L\text{-}l/2$ in time domain, $t \in [0;0.1]$ s. (….) free boundary conditions, (—) absorbing boundary conditions.

4.2 Spectrograms

The spectrograms of the two displacements in Fig. 3 are shown in Fig.4(a) for free boundary conditions and in Fig.4(b) for absorbing boundary conditions for frequencies in the range $[50{,}2000]$ Hz. To achieve them, the total observation time T has been divided in to 6 windows. The time step for each window is then $dt_{win} = 0.0532$ s.

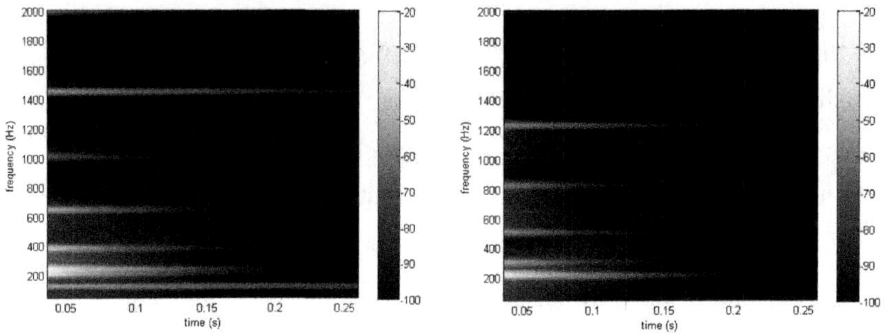

(a) Free boundary conditions. (b) Absorbing boundary conditions.

Fig. 4. Spectrograms of the vertical displacements (dB ref. 1 mm) at $x = L - l/2$ (color figure online)

In the case of absorbing boundary condition, Fig.4(b), the major part of the energy is dissipated after $t = 0.15$ s for frequencies in the range $[50{,}2000]$ Hz. If the beam has free ends, Fig.4(a), two resonances frequency at about 150 and 1400 Hz are present during all the time observation. These observations confirm the one made in section 4.1.

5 Conclusions

The construction of absorbing boundary conditions for an Euler—Bernoulli beam has been achieved in the time domain using the method proposed by Svensson et al. [5]. Some simulation results have been presented for the case of a simple point force excitation. A good efficiency of the absorbing boundary conditions has been shown as compared to a free boundary. However, the impact of some parameters, such as the length of the beam must be studied.

Another way to validate the absorbing boundary conditions would be to compare the Fourier transform of the beam response at an arbitrary point, given by the model described above, with the frequency response of an infinite beam at the same point.

In future work, the track model developed in this paper will be used in the presence of surface roughness and a moving load to predict the track behaviour in the context of the rolling noise modelling. The discrete supports will be then modified to take into consideration the sleeper displacement.

References

[1] Knothe, K.L., Grassie, S.L.: Modelling of railway track and vehicle track interaction at high frequencies. Vehicle System Dynamics 22, 209–262 (1993)

[2] Thompson, D.J.: Railway Noise and Vibration: Mechanisms, Modelling and means of control. Elsevier Ltd., Amsterdam (2009)

[3] Thompson, D.J., Janssens, M.H.A., De Beer, F.G.: TWINS Theoretical Manual, 3rd edn. (1999)

[4] Pieringer, A.: Modelling of wheel/rail interaction considering roughness and discrete irregularities, Licenciate Thesis, Gotenburg, Sweden (2008) ISSN 1652-9146

[5] Svensson, J.L., Andersson, P.B.U., Schueren, J., Kropp, W.: Feedforward control of bending waves in frequency domain at structural junctions using an impedance technique formulation. Journal of Sound and Vibration 323, 555–573 (2009)

[6] Doutaut, V., Matignon, D., Chaigne, A.: Numerical simulations of xylophones. II. Time domain modelling of the resonator and of the radiated sound pressure. Journal of Acoustical Society of America 104(3), 1633–1647 (1998)

Wave Propagation in an Infinite Track Having an Irregular Region

K. Abe[1], A. Kikuchi[2], and K. Koro[1]

[1] Niigata University,
8050 Igarashi 2-Nocho, Nishi-ku, Niigata, 950-2181, Japan
Tel: +81 25 262 7028, Fax: +81 25 262 7021
e-mail: abe@eng.niigata-u.ac.jp
[2] Toa Corporation

Summary

This paper presents a numerical method for wave transmission through a periodic track having a disturbed region. The infinite track analysis is realized by attaching the transmitting boundaries at both ends of the finite sub-domain. To achieve this, an impedance matrix representing the transmitting boundary is constructed by utilizing eigenmodes for the transfer matrix of a unit cell. The developed method is applied to infinite tracks subjected to incident harmonic waves. Through numerical analyses, the influence of the stiffness of the pads on the wave transmission is examined. The relation between the transmittance and the band structure of track is also discussed.

1 Introduction

The dynamic behavior of railway tracks is closely related to the passenger comfort, ballast settlement and ground vibration. The understanding of it is accordingly very important in the field of railway engineering, particularly, since continuous welded rails have a very long dimension along the track axis, waves propagating in the rails will dominate the dynamic response of the track. Consideration of a damped system allows us to use a finite track model due to the dissipation of waves reflected at the truncated ends. Therefore, the quantitative assessment of the dynamic response can be achieved satisfactorily by a finite track model with damping.

Although, needless to say, a practical simulation of track behavior with a realistic model provides worthwhile knowledge, an idealized mathematical model without damping can also serve as a useful hypothesis to understand fundamental features, such as wave dispersion and dominant modes [1, 2]. Since in such a structure waves may propagate without any dissipation, the existence of the truncation must affect the dynamic response [3]. Therefore, the capture of the infinite extent is essential in the analysis of undamped track model.

T. Maeda et al. (Eds.): Noise and Vibration Mitigation for Rail Trans. Sys., NNFM 118, pp. 71–79.
springerlink.com

When a track model consists of infinite rails and equidistant sleepers, its periodicity can be represented by an irreducible sub-structure called a unit cell. A dynamic problem of a periodic structure is reduced to the motion of the unit cell by virtue of the Floquet transform [4]. However, the method becomes invalid for a track having an irregular region in which the periodic structure is disturbed, such as a rail joint. Abe et al. [3] have developed a numerical method for such structures. In the method, the transmitting boundaries representing semi-infinite periodic structures are derived by way of a Floquet transform. The dynamic problem of an infinite system is then reduced to that of the finite irregular domain by attaching the transmitting boundaries at both ends.

In Ref.[3] railway tracks subjected to harmonic loads acting only on the non-periodic region were considered. However, from the viewpoint of the ground vibration and settlement of ballast, it is also important to examine the transmission of waves emitted from running trains, especially, the existence of a disturbance of the periodicity in some region may insulate the transmitting waves [5, 6, 7], and contribute to the mitigation of track deterioration.

In this paper, to analyze these problem, a numerical method is developed with the aid of the transmitting boundaries as in Ref.[3]. As mentioned above, in Ref.[3], the impedance matrices are derived by way of a Floquet transform. Because it is achieved by integration with respect to the Floquet wavenumber, calculation of the transmitting boundary needs finite element analysis for each integral point and thus consumes computation time. To save the computational cost, in this study, the impedance matrices are derived by means of the spectral analysis for the transfer matrices of the unit cells. Through numerical examples, the transmittance of propagating waves is analyzed. Frequencies at which the transmitting energy is attenuated are deduced with regard to the band structure of sub-regions. Influence of the track structure on the transmittance is also discussed.

2 Outline of the Method

2.1 Equation of Motion of Irregular Region

Let us consider an infinite railway track subjected to an incident wave u^I propagating to rightwards as shown in Fig. 1. The track consists of three sub-structures, i.e. the left and right semi-infinite periodic tracks Ω_L and Ω_R, and the finite irregular region Ω_H which has a different structure from Ω_L and Ω_R. Ω_L and Ω_R are not necessarily the same structure. Interfaces between these sub-regions are denoted by Γ_L and Γ_R. Notice that an external force P may act on Ω_H. Each domain is composed of a rail, sleepers, pads and a rigid foundation. In the following, the rail is discretized by Bernoulli—Euler beam elements. The sleeper and pad are represented by a rigid body and an elastic spring, respectively.

To achieve the dynamic analysis, the track is divided into three sub-structures as illustrated in Fig. 2. In the figure, u^R is the reflected wave from the interface Γ_L. $\mathbf{F}_L$, $\mathbf{U}_L$ and $\mathbf{F}_R$, $\mathbf{U}_R$ are internal force and displacement vectors at end nodes of Ω_L and

Ω_R. $\mathbf{F}_1$, $\mathbf{U}_1$ and $\mathbf{F}_3$, $\mathbf{U}_3$ are nodal vectors at both ends of Ω_H. The equation of motion of Ω_H can be written as

$$\begin{bmatrix} \hat{\mathbf{K}}_{11} & \hat{\mathbf{K}}_{12} & \hat{\mathbf{K}}_{13} \\ \hat{\mathbf{K}}_{21} & \hat{\mathbf{K}}_{22} & \hat{\mathbf{K}}_{23} \\ \hat{\mathbf{K}}_{31} & \hat{\mathbf{K}}_{32} & \hat{\mathbf{K}}_{33} \end{bmatrix} \begin{Bmatrix} \mathbf{U}_1 \\ \mathbf{U}_2 \\ \mathbf{U}_3 \end{Bmatrix} = \begin{Bmatrix} \mathbf{F}_1 \\ \mathbf{F}_2 \\ \mathbf{F}_3 \end{Bmatrix}, \tag{2.1}$$

where $(\ )_2$ is a sub-vector associated with nodes in Ω_H except on Γ_L and Γ_R. $\hat{\mathbf{K}}_{ij} = \mathbf{K}_{ij} - \omega^2\mathbf{M}_{ij}$, $\mathbf{K}_{ij}$ and $\mathbf{M}_{ij}$ are stiffness and mass sub-matrices, respectively. ω is the circular frequency. In eq. (2.1) $\mathbf{F}_1$ and $\mathbf{F}_3$ are unknowns.

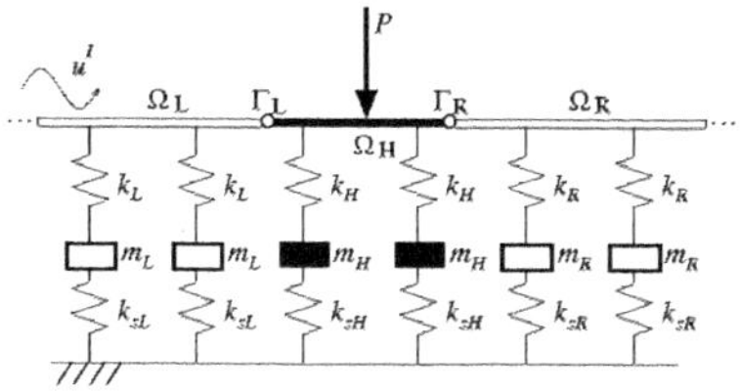

Fig. 1. Infinite Railway Track

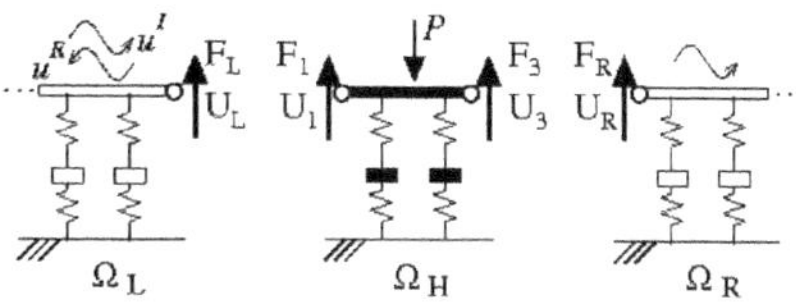

Fig. 2. Sub-structures of Infinite Track

2.2 Derivation of {F$_1$} and {F$_3$}

The compatibility and equilibrium conditions hold at the interfaces as

$$\{\mathbf{U}_L\}=\{\mathbf{U}_1\},\ \{\mathbf{U}_R\}=\{\mathbf{U}_3\},\ \ \{\mathbf{F}_L\}+\{\mathbf{F}_1\}=\{\mathbf{0}\},\ \{\mathbf{F}_R\}+\{\mathbf{F}_3\}=\{\mathbf{0}\}. \tag{2.2}$$

Therefore, in eq. (2.1), the unknown force vectors $\{\mathbf{F}_1\}$ and $\{\mathbf{F}_3\}$ can be replaced by $-\{\mathbf{F}_L\}$ and $-\{\mathbf{F}_R\}$, respectively. The present problem is then reduced to the evaluation of the internal force vectors $\{\mathbf{F}_L\}$ and $\{\mathbf{F}_R\}$. The dynamic response in Ω_L is composed of the incident and reflected waves. Hence, $\{\mathbf{U}_L\}$ and $\{\mathbf{F}_L\}$ can be decomposed into two terms as

$$\{\mathbf{U}_L\} = \{\mathbf{U}_{LL}^I\} + \{\mathbf{U}_{LL}^R\},\ \ \{\mathbf{F}_L\} = \{\mathbf{F}_{LL}^I\} + \{\mathbf{F}_{LL}^R\}, \tag{2.3}$$

where $(\ )^I$ and $(\ )^R$ denote components related to the incident and reflected waves, respectively.

We introduce an impedance matrix $[\mathbf{K}_{LL}]$ for Ω_L, which gives the following relation between the external force $\{\mathbf{F}_{LL}\}$ and its response $\{\mathbf{U}_{LL}\}$ at Γ_L,

$$[\mathbf{K}_{LL}]\{\mathbf{U}_{LL}\}=\{\mathbf{F}_{LL}\}. \tag{2.4}$$

Notice that $\{\mathbf{U}_{LL}\}$ and $\{\mathbf{F}_{LL}\}$ represent waves radiated from Γ_L to the left. In a similar manner, we define an impedance matrix $[\mathbf{K}_{LR}]$ for $\{\mathbf{U}_{LR}\}$ and $\{\mathbf{F}_{LR}\}$ at the left-end node of the semi-infinite complement to Ω_L (Fig. 3) as

$$[\mathbf{K}_{LR}]\{\mathbf{U}_{LR}\}=\{\mathbf{F}_{LR}\}. \tag{2.5}$$

Let us now consider propagation of the incident wave in an infinite track having a periodicity identical to Ω_L. At the position corresponding to Γ_L the internal force coincides with $\{\mathbf{F}_{LL}^I\}$. Since u^I is assumed to be a wave traveling to the right, $\{\mathbf{F}_{LL}^I\}$ can be evaluated by the nodal force resulting from the excitation $\{\mathbf{U}_{LL}^I\}$ at the left-end node of the complement of Ω_L. From eq. (2.5) this is given by

$$\{\mathbf{F}_{LL}^I\} = -[\mathbf{K}_{LR}]\{\mathbf{U}_{LL}^I\}. \tag{2.6}$$

In contrast, $\{\mathbf{F}_{LL}^R\}$ is related to the reflected wave propagating to the left. Therefore, it can be reproduced by an excitation $\{\mathbf{U}_{LL}^R\}$ at the right end of Ω_L, i.e.

$$\{\mathbf{F}_{LL}^R\} = [\mathbf{K}_{LL}]\{\mathbf{U}_{LL}^R\}. \tag{2.7}$$

Substituting eqs (2.6) and (2.7) into eq. (2.3) and recalling the relation $\{\mathbf{U}_{LL}^R\} = \{\mathbf{U}_L\} - \{\mathbf{U}_{LL}^I\}$, we reach the expression:

$$\{\mathbf{F}_1\} = -\{\mathbf{F}_L\} = [\mathbf{K}_{LR} + \mathbf{K}_{LL}]\{\mathbf{U}_{LL}^I\} - [\mathbf{K}_{LL}]\{\mathbf{U}_L\}. \tag{2.8}$$

The response in Ω_R is given only by transmitted waves propagating to the right. Therefore, the relation between the displacement $\{\mathbf{U}_R\}$ and force $\{\mathbf{F}_R\}$ can be described by the impedance matrix $[\mathbf{K}_{RR}]$ defined at the left-end node of Ω_R as

$$\{\mathbf{F}_3\} = -\{\mathbf{F}_R\} = -[\mathbf{K}_{RR}]\{\mathbf{U}_R\}. \tag{2.9}$$

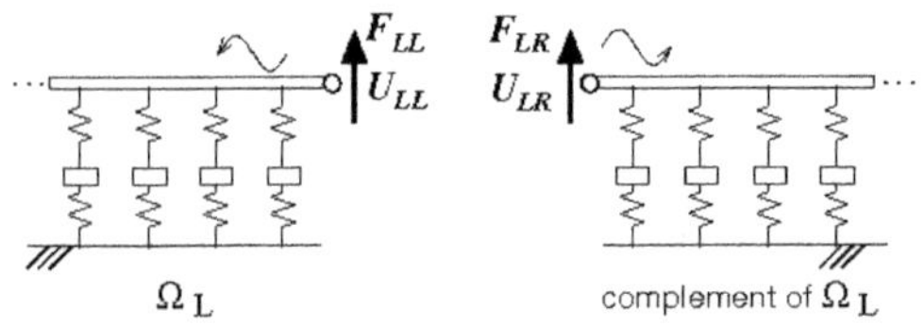

Fig. 3. Half-Infinite Track Ω_L and its Complement

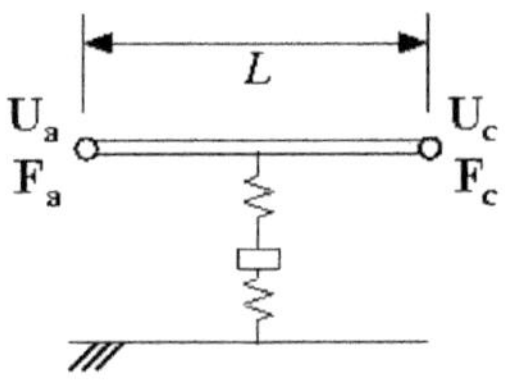

Fig. 4. Unit Cell of Periodic Track

2.3 Solving Equation

Substituting eqs (2.8) and (2.9) into eq. (2.1), since from eq. (2.2) $\{\mathbf{U}_L\}$ and $\{\mathbf{U}_R\}$ can be replaced with $\{\mathbf{U}_1\}$ and $\{\mathbf{U}_3\}$, we can eliminate the unknown force vectors. Arranging with respect to the displacement vector, we obtain the equation:

$$\begin{bmatrix} \hat{\mathbf{K}}_{11} + \mathbf{K}_{LL} & \hat{\mathbf{K}}_{12} & \hat{\mathbf{K}}_{13} \\ \hat{\mathbf{K}}_{21} & \hat{\mathbf{K}}_{22} & \hat{\mathbf{K}}_{23} \\ \hat{\mathbf{K}}_{31} & \hat{\mathbf{K}}_{32} & \hat{\mathbf{K}}_{33} + \mathbf{K}_{RR} \end{bmatrix} \begin{Bmatrix} \mathbf{U}_1 \\ \mathbf{U}_2 \\ \mathbf{U}_3 \end{Bmatrix} = \begin{Bmatrix} [\mathbf{K}_{LL} + \mathbf{K}_{LR}]\{\mathbf{U}_{LL}^I\} \\ \mathbf{F}_2 \\ \mathbf{0} \end{Bmatrix}. \tag{2.10}$$

Note that there are no unknowns on the right-hand side. We can thus accomplish the dynamic response analysis of the infinite track by solving eq. (2.10).

3 Impedance Matrix

From eq. (2.10), the present problem is reduced to the calculation of the impedance matrices $[\mathbf{K}_{LL}]$, $[\mathbf{K}_{LR}]$ and $[\mathbf{K}_{RR}]$. This section is devoted to the derivation of these matrices.

3.1 Derivation of Transfer Matrix

As an example, we will describe the derivation procedure of $[\mathbf{K}_{LR}]$. Consider an infinite track having the same periodicity as Ω_L. The equation of motion in a unit cell is given by

$$[\hat{\mathbf{K}}]\{\mathbf{U}\} = \{\mathbf{F}\}. \tag{3.1}$$

$\{\mathbf{U}\}$ and $\{\mathbf{F}\}$ are composed of sub-vectors as

$$[\mathbf{U}]^T=[\mathbf{U}_a\,\mathbf{U}_b\,\mathbf{U}_c]^T,\ \ [\mathbf{F}]^T=[\mathbf{F}_a\,\mathbf{F}_b\,\mathbf{F}_c]^T, \tag{3.2}$$

where $(\ )_a$ and $(\ )_c$ stand for nodal vectors at both ends of the unit cell characterized by the sleeper spacing L as in Fig. 4. $(\ )_b$ is a sub-vector given by the remainder of the nodal vector.

To derive the transfer matrix relevant to $\{\mathbf{U}_a\}$, $\{\mathbf{F}_a\}$ and $\{\mathbf{U}_c\}$, $\{\mathbf{F}_c\}$, we introduce the following eigenvalue problem,

$$[\mathbf{K}]\{\boldsymbol{\varphi}_i\} = \omega_i^2[\mathbf{M}]\{\boldsymbol{\varphi}_i\},\ \ (i = 1,\cdots,N), \tag{3.3}$$

where ω_i is the ith natural circular frequency and $\{\boldsymbol{\varphi}_i\}$ is the ith eigenmode. N is the degrees of freedom. In general, the solution $\{\mathbf{U}\}$ in eq. (3.1) can be represented by a linear combination of the eigenvectors as

$$\{\mathbf{U}\} = \sum_i^N \alpha_i\{\boldsymbol{\varphi}_i\}, \tag{3.4}$$

where α_i is a constant. Owing to the orthogonality of $\{\boldsymbol{\varphi}_i\}$, α_i is given by

$$\alpha_i = \frac{1}{m_i(\omega_i^2 - \omega^2)}[\overline{\boldsymbol{\varphi}}_i]^T\{\mathbf{F}\},\ \ m_i = [\overline{\boldsymbol{\varphi}}_i]^T[\mathbf{M}]\{\boldsymbol{\varphi}_i\}, \tag{3.5}$$

where $(\ \bar{}\ )$ stands for the conjugate of a complex vector.

Providing that the structure is subjected to no external loads, i.e. $\{\mathbf{F}_b\}=\{\mathbf{0}\}$, eqs (3.4) and (3.5) lead to the relation between nodal displacement and force vectors at both ends as

$$\begin{Bmatrix}\mathbf{U}_a\\ \mathbf{U}_c\end{Bmatrix} = \sum_i^N \frac{1}{m_i(\omega_i^2-\omega^2)}\{\boldsymbol{\varphi}'_i\}[\overline{\boldsymbol{\varphi}}'_i]^T\begin{Bmatrix}\mathbf{F}_a\\ \mathbf{F}_c\end{Bmatrix} = \begin{bmatrix}\mathbf{H}_{aa} & \mathbf{H}_{ac}\\ \mathbf{H}_{ca} & \mathbf{H}_{cc}\end{bmatrix}\begin{Bmatrix}\mathbf{F}_a\\ \mathbf{F}_c\end{Bmatrix}, \tag{3.6}$$

where $\{\boldsymbol{\varphi}'_i\}$ consists of nodal components of both ends.

Rearranging eq. (3.6) with respect to the left- and right-end nodes, we obtain the transfer matrix $[\mathbf{G}]$:

$$[\mathbf{G}]\begin{Bmatrix}\mathbf{U}_a\\ \mathbf{F}_a\end{Bmatrix}=\begin{Bmatrix}\mathbf{U}_c\\ -\mathbf{F}_c\end{Bmatrix},\ [\mathbf{G}]=\begin{bmatrix}\mathbf{H}_{cc}\mathbf{H}_{ac}^{-1} & -\mathbf{H}_{cc}\mathbf{H}_{ac}^{-1}\mathbf{H}_{aa}+\mathbf{H}_{ca}\\ -\mathbf{H}_{ac}^{-1} & \mathbf{H}_{ac}^{-1}\mathbf{H}_{aa}\end{bmatrix}. \tag{3.7}$$

3.2 Derivation of Impedance Matrix

By virtue of Floquet's theorem [8], a steady state wave f propagating in a periodic structure satisfies the following relation,

$$f_{\mathrm{c}}=e^{-i\kappa L}f_{\mathrm{a}}, \tag{3.8}$$

where κ is the Floquet wavenumber.

Due to eq. (3.8), (3.7) results in the eigenvalue problem:

$$[\mathbf{G}]\begin{Bmatrix}\mathbf{U}_a\\ \mathbf{F}_a\end{Bmatrix}=\lambda\begin{Bmatrix}\mathbf{U}_a\\ \mathbf{F}_a\end{Bmatrix},\ (\lambda=e^{-i\kappa L}). \tag{3.9}$$

Half of the eigenmodes travel to the right, while the other half travel to the left. Arranging the former modes $\{\mathbf{U}_i\}$, $\{\mathbf{F}_i\}$, (i=1,2), the 2×2 impedance matrix $[\mathbf{K}_{LR}]$ can be obtained by

$$[\mathbf{K}_{LR}]=[\mathbf{F}_1\ \mathbf{F}_2]\ [\mathbf{U}_1\ \mathbf{U}_2]^{-1}. \tag{3.10}$$

$[\mathbf{K}_{LL}]$ and $[\mathbf{K}_{RR}]$ can be derived in the same way as $[\mathbf{K}_{LR}]$.

4 Examples

4.1 Harmonic Excitation of Periodically Supported Rail

To verify the present approach for the impedance matrices, the developed method is applied to harmonic loading of a periodic track. The infinite rail is given by JIS 50kgN rail and supported evenly with L=0.6m by sleepers of m=80kg. The stiffness of the pads is set to be $k=k_s$=60MN/m. The rail in the unit cell is discretized into 16 beam elements. The transmitting boundaries are attached at both ends of a finite track consisting of two units as shown in Fig. 5. A unit harmonic load acting on the mid-span is considered.

The real part of the deflection at the loading point is shown in Fig. 6. In the figure, the numerical solution for which the impedance matrices are calculated by means of the Floquet transform [3] is also shown by dotted curves. The validity of the present approach can be evidenced by good agreement.

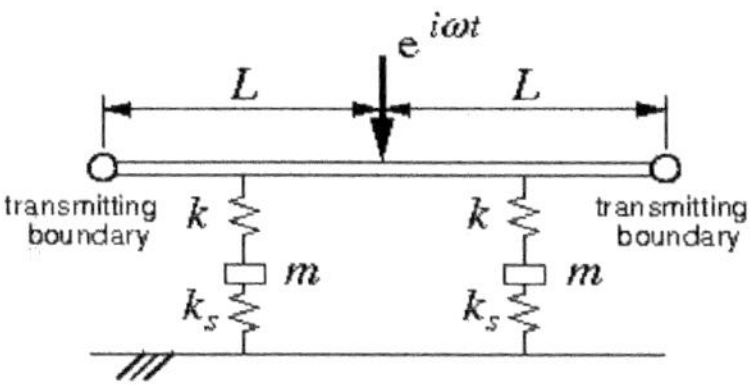

Fig. 5. Finite Track with Transmitting Boundaries

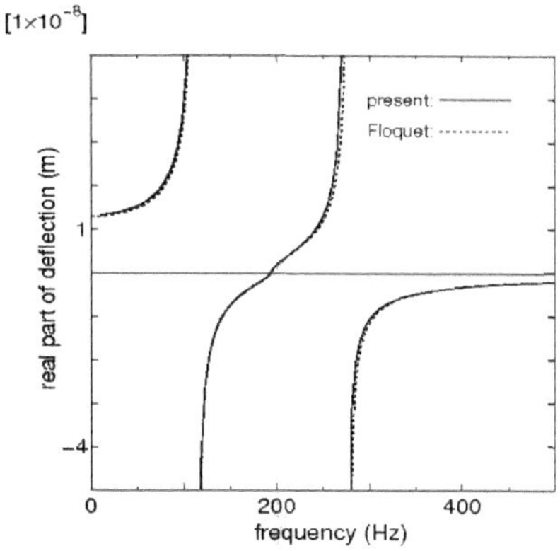

Fig. 6. Deflection due to Harmonic Load

4.2 Transmission of Waves through Irregular Region

An infinite track having a different structure in a finite region, as illustrated in Fig. 1, is considered in the context of wave propagation analysis. Ω_L and Ω_R have the same structure as that of the previous example. The irregular region Ω_H has the identical rail and evenly spaced sleepers as Ω_L and Ω_R, but the pads k_H and k_{sH} are different from those of both sides. Ω_H consists of two spans as in Fig. 1. In the following analyses, the periodic structure discussed in Sect. 4.1 is also considered as a reference track (Case 1). Two types of modification are employed, i.e. k_H=60MN/m, k_{sH}=5MN/m (Case 2) and k_H=100MN/m, k_{sH} =60MN/m (Case 3).

In the analysis, the transmittance E_r which is defined as the energy ratio of the transmitted wave to the incident wave is evaluated. E_r is expressed by

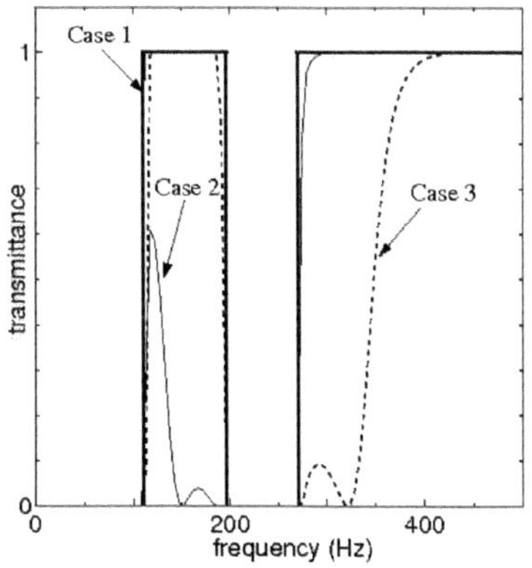

Fig. 7. Transmittance of Wave Modes

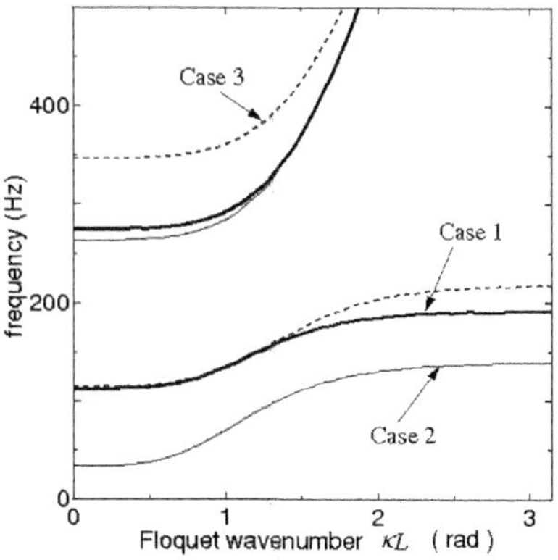

Fig. 8. Dispersion Curves

$$E_r = \frac{\bar{E}_t}{\bar{E}_i},$$
$$\bar{E}_i = \frac{\omega}{2}\mathrm{Im}([\bar{\mathbf{U}}_{LL}^I]^T[\mathbf{K}_{LR}]\{\mathbf{U}_{LL}^I\}),\ \bar{E}_t = \frac{\omega}{2}\mathrm{Im}([\bar{\mathbf{U}}_3]^T[\mathbf{K}_{RR}]\{\mathbf{U}_3\}), \tag{4.1}$$

where $\bar{E}_i$ and $\bar{E}_t$ are time averages of the energy of incident and transmitting waves, respectively. Im( ) stands for the imaginary part of a complex number.

Numerical results are shown in Fig. 7. Incident waves are given by wave modes propagating in the periodic track of Ω_L. These can be obtained by the eigenmodes $\{\mathbf{U}_i\}$ of eq. (3.10). Therefore, E_r can be calculated only for frequencies at which the propagating modes exist. Notice that the transmittance of the periodic track (Case 1) is E_r=1 at these frequencies, otherwise E_r=0. The former frequency band is called a pass band, while the latter is called a stop band.

The band structure of a periodic system can be obtained by dispersion analysis [1]. Fig. 8 shows dispersion curves for Case 1 and periodic tracks given by repetition of Ω_H in Case2 and Case3. It can be realized that pass bands of Case 1 track are composed of frequencies between 112 and 192Hz and higher than 274Hz. The stop bands for the periodic track given by Ω_H of Case 2 are distributed in frequencies lower than 34Hz and in the range of 140-263Hz. Consequently, the first pass bands of Case 1 and Case 2 overlap with each other only in a narrow band of 112-140Hz. That is, in Case 2, incident waves in the first pass band of Case 1 track scarcely propagate through Ω_H as shown in Fig. 7, despite of the short disturbance of two spans. Contrary to this, the first pass band of the infinite track given by Ω_H of Case 3 covers that of the Case 1 track. Therefore, not much reduction of transmittance is observed in the first pass band. However, Case 3 track causes the increase in the second pass band as shown in Fig. 8. Thus, from Fig. 7, we can realize that this change of the band structure results in the reduction in energy transmittance in the frequency range of 274-350Hz.

Although in this analysis, the sleeper support is given by a rigid foundation, in general, the dynamic energy can be transmitted to the substrate. Therefore, the propagating waves may partially reach the opposite side of the irregular region through the ground.

5 Conclusion

In this paper, a numerical method has been developed for wave transmission through an infinite track having a disturbed region. To achieve this, the solving equation was constructed by attaching the transmitting boundaries at both ends of the finite irregular region. The impedance matrix was derived by matrices composed of eigenmodes for the transfer matrix of a unit cell. Infinite tracks subjected to incident waves were employed as examples. It was found that the replacement of rubber pads supporting the sleepers with softer ones will contribute to the reduction in wave transmission at lower frequencies. In contrast with this, the stiffening of rail pads will be effective to control the higher bands. In the present examples, the irregular region composed of only two spans exhibited the capability in filtering incident waves. The dependence of transmittance on the frequency can well be explained based on the band structure of each sub-track.

The numerical results shown in this paper suggest that the rail vibrations at a weak portion, such as a rail joint can be insulated by tuning the track structure around the site.

References

[1] Mead, D.J.: Free wave propagation in periodically supported infinite beams. Journal of Sound and Vibration 11(2), 181–197 (1970)

[2] Thompson, D.J.: Wheel-rail noise generation, Part III: Rail vibration. Journal of Sound and Vibration 161(3), 421–446 (1993)

[3] Abe, K., Fujishita, S., Koro, K.: Development of transmitting boundary for periodically supported rails. Journal of Mechanical Systems for Transportation and Logistics (JSME) 3(1), 44–52 (2010)

[4] Clouteau, D., Arnst, M., Al-Hussaini, T.M., Degrande, G.: Freefield vibrations due to dynamic loading on a tunnel embedded in a stratified medium. Journal of Sound and Vibration 283, 173–199 (2005)

[5] Caballero, D., Sánchez-Dehesa, J., Rubio, C., Mártinez-Sala, R., Sánchez-Pérez, J.V., Meseguer, F., Llinares, J.: Large two-dimensional sonic band gaps. Physical Review. E 60(6), 6316–6319 (1999)

[6] Chen, Y.-Y., Ye, Z.: Acoustic attenuation by two-dimensional arrays of rigid cylinders. Physical Review Letters 87(18), 184301 (2001)

[7] Duclos, G., Clément, A.H.: Wave propagation through arrays of unevenly spaced vertical piles. Ocean Engineering 31, 1655–1668 (2004)

[8] Delph, T.J., Herrmann, G., Kaul, R.K.: Harmonic wave propagation in a periodically layered, infinite elastic body: Antiplane strain. Journal of Applied Mechanics 45, 343–349 (1978)

Slab Track Vibration and Wayside Noise in a High-Speed Rail Tunnel

A. Wang[1], Y. Xi[2], S. Wang[2], S. Tao[2], and J. Sun[2]

[1] Pandrol Limited
63 Station Road, Addlestone, KT15 2AR, England
Tel.: +44 1932 834500; Fax: +44 1932 850858
e-mail: a.wang@pandrol.com
[2] China Ship Scientific Research Center, Wuxi 214082, P.R. of China

Summary

This paper describes track vibration measurements on slab track in a tunnel under normal traffic. The track frequency response functions had also been measured to validate the track dynamic parameters. Comparison of the dynamic characteristics, the track vibrations and wayside noise, of a slab track with traditional ballasted track were made using available data.

1 Introduction

Wayside noise and in-car noise are important considerations in the planning and operation of a high-speed train line. Both rolling noise [1] and aerodynamic noise [2] tend to increase with increasing train speed [3, 4]. The wheels and rails are the dominant sources of the noise received at the wayside of slab tracks, with most of the energy content found in the frequency range from about 500Hz to 5kHz. In addition, there may also be a contribution made by the track slab at low frequencies. For in-car noise, tonal components are important in the frequency range from 70Hz ~ 300Hz [5]. Loaded track resonance and high-speed rail excitation frequencies within this range have also been investigated for a slab track and are affected by the rail fastening stiffness directly [6]. Depending on the design of the different components, and the speed of the train, the relative importance of the different contributions to the overall noise level will change. Investigations into the vibration levels and characteristics on each track component in a high-speed rail tunnel have become important for track design optimisation.

The rail fastening system, which fits between the rail and the slab, is a component which can be modified relatively easily, and its stiffness is known to have a significant influence on the dynamic behaviour of the system. Noise emission models generally show that as the fastening stiffness is decreased, the slab

T. Maeda et al. (Eds.): Noise and Vibration Mitigation for Rail Trans. Sys., NNFM 118, pp. 81–88.
springerlink.com

is increasingly well isolated from the rail, and it becomes less significant as a source of noise [7]. However, another effect of a lower fastening stiffness is to reduce the rate at which vibrations originating at a point source decay with distance along the track. This means that a larger length of the rail tends to radiate sound, thereby increasing the component of noise radiated from it.

By attempting to balance the increased level from the slab against the reduced contributions from the rail as fastening stiffness is increased, it is possible to propose a fastening stiffness which is optimised from the perspective of noise emission.

This paper describes the track vibration measurements in a tunnel on Hefei—Wuhan [9] Passenger Dedicated Line (PDL) slab track under normal traffic. The track frequency response functions were also measured to validate the track dynamic parameters. To compare the dynamic characteristics of a slab track with traditional ballasted track, the track vibrations were also measured on a ballasted track site under the same traffic conditions.

2 Track, Traffic and Measurements

Track measurements had been carried out at a main site in Dabieshan tunnel (length of the double track tunnel is 13.3km) on the high-speed line linking Hefei to Wuhan in China in June 2009, two months after the line was opened to the public. Track in the tunnel is on slab (Figure 1) but outside it is ballasted (Figure 2). The tracks at the main test site in the tunnel are on a 6988-m radius curve, consisting of 60kg Chinese rail fixed every 0.60m with Pandrol SFC baseplate on Rheda slab. The dynamic stiffness of SFC fastening system on the slab track is about 39kN/mm [10] under the traffic condition with a CRH2 high-speed train.

The test was also repeated outside the tunnel on the ballasted track which consists of a pre-stressed monobloc concrete sleeper fitted with the Pandrol FC fastening system. The dynamic stiffness of the FC fastening system is about 86kN/mm [11] under the traffic condition with a CRH2 high-speed train.

The Hefei—Wuhan PDL has been designed for an initial speed of 250km/h, which will later be raised to 300km/h. Unusually for a PDL, some freight will run on the line, which can accept a 25 tonne axle load. The train running during the test recording is a CRH2 with an average axle load of 12 tonnes. The train speed was 250km/h at both test sites: in the tunnel and on the ballasted track outside.

Measurements of rail deflection relative to the slab, track vibration and wayside noise have been made under traffic. Measurements of the track frequency response functions of unloaded tracks had also been made to validate the track dynamic parameters. Wayside noise measurement was also conducted [12] which can be compared with previous results on ballasted tracks [3, 8].

Fig. 1. Slab track in Dabieshan tunnel (color figure online)

Fig. 2. Hefei—Wuhan PDL (color figure online)

3 Track Frequency Response Function Measurements

Track characteristics that cannot easily be calculated or measured in the laboratory include track receptance. Field testing with excitation forces applied to the rail-head using an instrumented hammer is usually used to obtain the track frequency response functions.

Figure 3 gives the rail vertical direct receptance over the sleeper on the ballasted track and over the baseplate on the slab track. The peaks in the response at about 270Hz on the ballasted track and 190Hz on the slab track indicate resonance due to the resilient rail fastening for each track. These indicate that the dynamic stiffness of the rail pad on the ballasted track is about twice that of the baseplate on the slab track. The dynamic stiffness for each track can be estimated using a beam-on-elastic-foundation model from the track receptance measured as 45kN/mm for the baseplate and 100kN/mm for the rail pad. These values of fastener dynamic stiffness have been derived from the model when the predicted receptance has been approximately matched with the measured data in the vicinity of the resonance frequencies at medium frequencies between 100Hz and 300Hz. For rail fastening systems, dynamic stiffness normally increases as the frequency increases. This indicates a good agreement with the estimation in the low frequency range from the track dynamic deflection measured under traffic [10, 11].

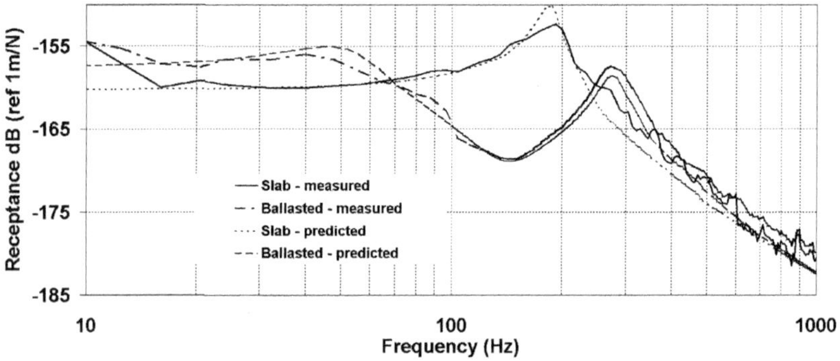

Fig. 3. Rail vertical receptance

4 Track Vibration Measurement

Rail, slab and tunnel wall accelerations are of interest because vibration generates noise. Slab acceleration is also of interest because it indicates the level of dynamic force transmitted into the slab and tunnel structures. In order to investigate the noise sources and their contributions to the noise, vibrations are represented here as velocity level and are also A-weighted. The speed of all trains was similar during the test; therefore the averaged results over a pass-by have been calculated across 8 trains for recordings on one day.

1/3 octave spectra of rail vibration velocity in the vertical and lateral directions on the slab track in the tunnel are shown in Figure 4. Rail vibration levels in the vertical direction are about 10dB(A) higher than those in the lateral direction, predominantly in the frequency range between 500Hz and 4kHz. Rail vibrations on the ballasted track are also plotted on the same graph for comparison. Generally speaking, rail vibration levels on the slab track are 5dB(A) higher than those on the ballasted track, particularly for frequencies over 1kHz. This may be affected by the rail vibration decay rates on different tracks.

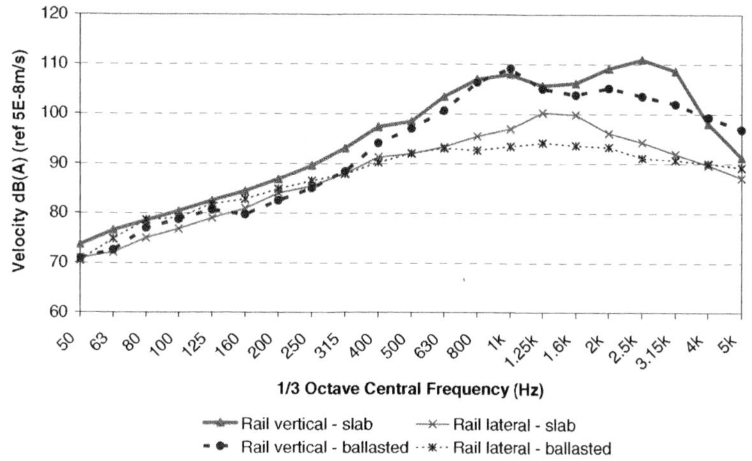

Fig. 4. Rail vibration velocity spectra (color figure online)

1/3 octave velocity spectra of slab vibration in the vertical and lateral directions, and tunnel wall vibration in the lateral direction (normal to the surface) are shown in Figure 5. Sleeper vibration on the ballasted track is also plotted on the same graph for comparison. These indicate that the sleeper vibration levels in the vertical direction on the ballasted track are much higher in all frequency bands than those on the slab. Sleeper vibration in the lateral direction at the low frequencies is higher than the slab but they tend to be similar at high frequencies. Ground vibration velocity at free field has a peak at 80Hz band, as also shown in Figure 5.

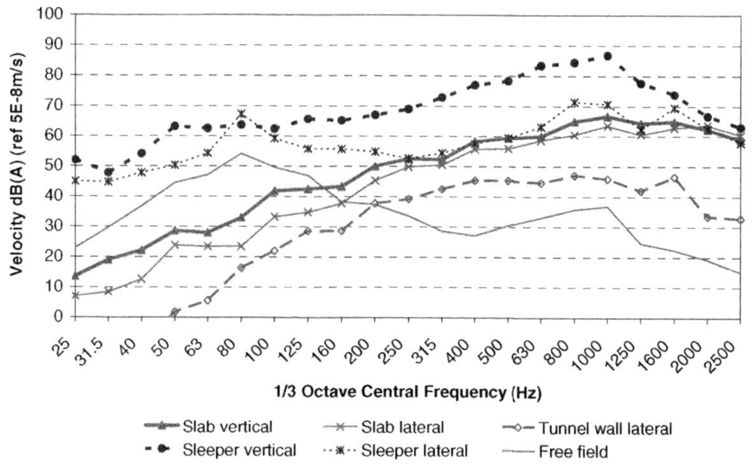

Fig. 5. Slab/sleeper vibration velocity spectra (color figure online)

Total A-weighted rail vertical, rail web lateral and slab/sleeper velocity levels over the frequency range up to 5 kHz for the rail and 2.5 kHz for the slab and sleeper are shown in Figure 6.

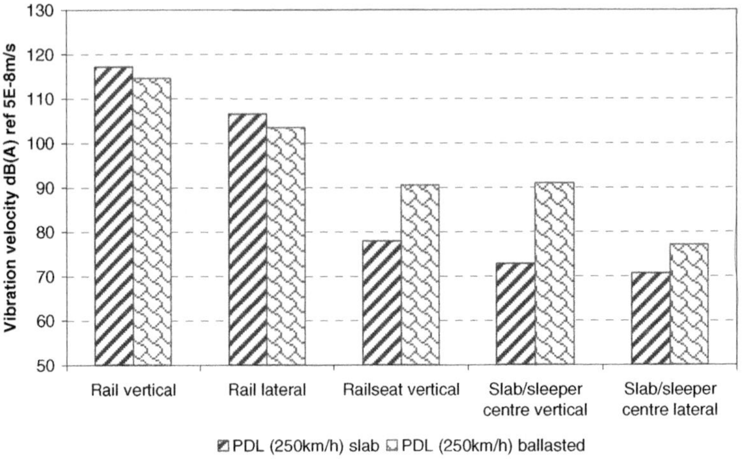

Fig. 6. Track vibration velocity (color figure online)

5 Wayside Noise under Traffic

Wayside noise levels in the tunnel were measured at 5.75m from track centre and reported to be about 114dB(A) [12] with a CRH2 train at a speed of 250km/h. In order to compare the wayside noise level of slab track in the tunnel [12] with ballasted track, measured noise results [3] on a Belgian high speed line with a Thalys train between Brussels and Paris are used here. Wayside noise levels at different positions (3.75m, 7.5m and 20m from the track centre) on different ballasted tracks (fitted with hard rail pad — dynamic stiffness of 275kN/mm; and soft rail pad — dynamic stiffness of 78kN/mm) at different train speeds (between 160km/h and 350km/h) had been reported in reference [3] as represented here in Figure 7. They show the expected increase in noise level with train speed, and decrease in noise level with distance. Figures 7(a) and 7(b) show a large influence of the pad stiffness on near field noise levels. This may indicate that the noise contribution from sleepers with hard pads is higher than that with soft pads. The noise level at 5.75m from the track at train speed of 250km/h can then be estimated at about 103dB(A) from a ballasted track fitted with soft rail pads. This is 11dB(A) lower than that in a tunnel with slab track, for the same distance from the track and the same train speed. It is expected that this difference is due to noise reflection at the tunnel walls, the difference between the two track forms and possibly also the effect of the tunnel walls on aerodynamic noise.

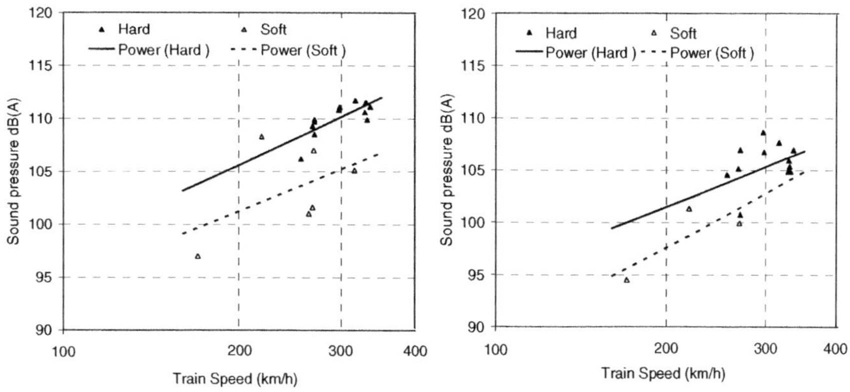

(a) 3.75m from the track centre (b) 7.5m from the track centre

Fig. 7. Wayside noise levels for ballasted track (Thalys in Belgium)

6 Conclusions

Measurements have been made on two different tracks: slab track with resilient baseplates in a tunnel, and ballasted track with concrete sleepers on open ground. Measurements of track frequency response functions indicate that the dynamic stiffness of the rail fastening system is about 45kN/mm for the baseplate on the slab track and 100kN/mm for the rail pad on the ballasted track. These figures agree well with the design and laboratory tests.

Rail vibration levels in the vertical direction are about 10dB(A) higher than those in the lateral direction, predominantly in the frequency range between 500Hz and 4kHz. Rail vibration levels on the slab track are 5dB(A) higher than those on the ballasted track, particularly for frequencies above 1kHz. This may be affected by the rail vibration decay rates on the different tracks. Sleeper vibration levels in the vertical direction on the ballasted track are much higher in all frequency bands than those on the slab track. Sleeper vibration in the lateral direction at the low frequencies is higher than the slab but tends to be similar in the high frequency range. Ground vibration velocity at free field has a peak at 80Hz band.

Wayside noise on slab track in the tunnel is higher than that on the ballasted track, due to noise reflection at the tunnel walls, the difference between the two track forms and possibly also the effect of the tunnel walls on aerodynamic noise.

References

[1] Thompson, D.J., Jones, C.J.C.: A review of the modelling of wheel/rail noise generation. Journal of Sound and Vibration 231, 519–536 (2000)

[2] Talotte, C.: Aerodynamic noise, a critical survey. Journal of Sound and Vibration 231, 549–562 (2000)

[3] Wang, A., Cox, S.J.: Noise characteristics of high speed track with railpads of different stiffness. In: Proceedings of EuroNoise 1998, Munich, Germany (1998)

[4] Poisson, F., Gautier, P.E., Letourneaux, F.: Noise sources for high speed trains: a view of results in the TGV Case. In: 9th International Workshop on Railway Noise, Munich, Germany (2007)

[5] Koh, H.I., Kwon, H.B., You, W.H., Park, J.H.: A study on source mechanism in the interior noise problem of high speed trains. In: 9th International Workshop on Railway Noise, Munich (2007)

[6] Wang, A., Cox, S.J.: High-speed rail: excitation frequencies and track stiffness. In: 10th International Workshop on Railway Noise, Nagahama, Japan (2010)

[7] Greer, R., Allett, T., Manning, C.: Channel Tunnel Rail Link Section 2 – A Review of Innovative Noise and Vibration Mitigation. In: 8th International Workshop on Railway Noise, Buxton, UK (2004)

[8] Wang, A., Cox, S.J.: Effects of railpad stiffness on rail roughness growth and wayside noise levels on high speed track. In: 6th World Congress on Railway Research, Edinburgh, UK (2003)

[9] Zhao, Z., Whitmore, D., Wu, Z., Wang, J.: Pandrol and the Railway in China. Track Report. The Journal of Pandrol Rail Fastenings, 3–7 (2009)

[10] Xi, Y.: Inspection report of deflection and vibration measurements in Dabieshan tunnel on He-Wu PDL. Report No. VS2009-101-1, China Ship Scientific Research Center (August 2009)

[11] Xi, Y.: Inspection report of deflection and vibration measurements on ballasted track of He-Wu PDL. Report No. VS2009-101-2, China Ship Scientific Research Center (September 2009)

[12] Wu, J.: Noise measurement and analysis on high speed rail in Dabieshan tunnel. Dissertation for Beijing Professional Qualification, Tsinghua University Test Centre for Building Environment (July 2009)

Tuned Mass Damper for Rail Noise Control

Wilson Ho[1], Banting Wong[1], and David England[2]

[1] Wilson Acoustics Limited
Unit 616, Technology Park, 18 On Lai Street, Shatin, Hong Kong, China
Tel: (852) 31881170, Fax: (852) 34228117
e-mail: who@wal.hk
[2] MTR Corporation
MTR Headquarters Building, Telford Plaza, Kowloon Bay, Hong Kong

Summary

A new type of tuned mass damper has been developed to reduce rail vibration and noise radiation. The damper comprises multiple masses oscillating along the shear direction of resilient layers forming a multiple mass—spring system. It provides effective broadband damping for a frequency range between 300Hz and 2500Hz. The resilient layers are aligned perpendicularly to the rail alignment such that each mass is allowed to oscillate in both vertical and lateral directions. The natural frequencies of the oscillating masses are individually tuned to match the rail vibration at multiple frequencies to cover both vertical and lateral pinned—pinned resonance frequencies of the rail.

The damper is being installed on a curved track section in Hong Kong to study its noise reduction performance and influence on the corrugation growth rate. The vertical track decay rate is generally increased to 6-11dB/m at frequencies above 500Hz, while the lateral track decay rate is generally increased to 3-6dB/m at frequencies above 300Hz. The dampers reduce rail vibration by about 10dB(A) and noise level by 3.5dB(A). After installation of rail dampers, the sleeper blocks become the dominant noise sources.

Further investigation on the damper influence on corrugation growth is being conducted, and results will be published after completion of the 1-year trial test.

1 Introduction

In the last two decades, various types of rail dampers have been investigated and tested for railway noise control. Among various types, tuned mass dampers are found to be relatively effective. Noise reduction performance of rail dampers strongly depends on the track characteristics. Existing tuned mass dampers generally provide overall noise reduction in the range of 1 to 3dB(A) [1, 2, 3], while in a particular field test, an overall noise reduction of approximately 6dB(A) was achieved using a vehicle fitted with noise-reducing wheels [4].

T. Maeda et al. (Eds.): Noise and Vibration Mitigation for Rail Trans. Sys., NNFM 118, pp. 89–96.
springerlink.com

Rail displacement amplitude at the noise radiation frequency is in the order of microns or sub-microns. Coupling between the damper and rail is a critical item affecting damper performance [2]. With traditional mounting method, rail dampers are glued or clamped to the rail without specific technique to enhance coupling effect. Sub-micron movement gaps at the mounting interface are difficult to avoid after severe vibrations induced by repeated train passages, thus hindering vibration absorption in long-term performance.

2 Damper Design

In view of the above, a new type of rail damper has been developed. It comprises multiple oscillating masses sandwiched by resilient layers as shown in Fig 1. The total mass for one damper pair is approximately 20-30kg subject to specific site requirements.

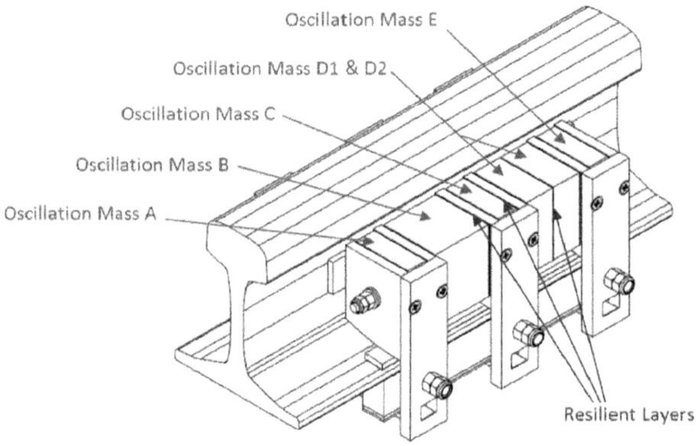

Fig. 1. Rail Damper Isometric View

For rail vibration in the frequency range of 300-1000Hz, a tuned mass damping mechanism is employed. The shear stiffness and mechanical loss factors of the resilient layers are selected such that each oscillating mass covers a resonance bandwidth of 20-40% depending on frequency. As bending waves propagate along the rail, most of the vibration energy is transferred to the damper and dissipated by hysteresis in the resilient layers. For high-frequency rail vibration above 1000Hz, a viscous damping layer is employed for broadband vibration absorption.

The tuned mass damper incorporates three key features to enhance its performance,

- The mass oscillates along the shear directions of the resilient layers. Each oscillating mass provides effective vibration absorption along the directions perpendicular to the rail alignment, including both vertical and lateral rail vibration. It provides clear oscillation modes for vibration absorption at designated frequencies.

- Gap filler is placed at the mounting interfaces. The gap filler behaves differently under different strain rates. Under static compressive mounting force, it behaves as a flexible solid, which deforms to fill up movement gaps at the mounting interfaces. Under high-frequency dynamic train excitation force, it behaves as a stiff solid, which allows vibration energy to be effectively transferred to the dampers.
- Resilient buffer layers are inserted at the mounting components to provide persistent mounting forces. They are pre-compressed to provide a rigid mounting force for typical rail vibration. In events of extremely high vibration, the resilient layer acts as a spring buffer to prevent structural damage of mounting components. After the extreme event, the mounting components would be automatically restored to equilibrium position. Therefore, the design provides a persistent mounting method with minimal interfacing gaps even under repeated severe vibrations.

3 Testing Date and Locations

The selected test site is a curved operational track inside a tunnel with 300 m radius, where high levels of rail vibrations in both vertical and lateral directions are excited. Track properties and train parameters are shown in Table 2. The trial test will last for 1 year to study the long-term noise reduction performance and influence on corrugation growth. Three sets of measurement have been conducted as shown in Table 1.

Fig. 2. Rail Damper installed at the testing location

Table 1. Testing Program

	Before Installation	1st day after Installation	1st month	6th month	12th month
Tunnel N&V	22/07/2010	25/07/2010	05/08/2010	T.B.A	T.B.A
Saloon Noise	22/07/2010	26/07/2010	11/08/2010	T.B.A	T.B.A

Table 2. Track Properties and Train Parameters for Damper Performance Test

	Test Location
Support Type	Low-Vibration Trackform (Resiliently Booted Block)
Static Stiffness of LVT Support Pad	26-35kN/mm
Static Stiffness of Rail Pad	75-90kN/mm
Curvature	300m
Train Speed	60-70 km/h

4 Measurement Results

4.1 Track Decay Rate

The track decay rate [5] is a crucial parameter determining rail noise radiation. For rails having a low decay rate, train-induced vibration could propagate for a long distance along the rail and radiate noise, which is often referred to as "singing rail". The damper increases the track decay rate thus reducing the effective noise radiation length. For low decay rates, doubling the decay rate reduces the sound power level of the rail by about 3dB in the corresponding frequency band [6, 7].

Fig 3 shows the frequency response at the rail head and underneath the rail foot using an impact force hammer. The first resonance peak at 100 Hz in the vertical direction and 70 Hz in the lateral direction corresponds to the natural frequency of rail and booted supports rebounding on resilient support. The sharp peaks at 825 Hz in the vertical direction and 385 Hz in the lateral direction correspond to the vertical and lateral pinned—pinned resonances. These resonances are associated with low decay rates as rail bending waves can effectively propagate through the periodic rail fastening [8]. The decay rate at the pinned—pinned resonances could be enhanced by placing the oscillation mass at the anti-nodes, i.e. mid-section between two supports.

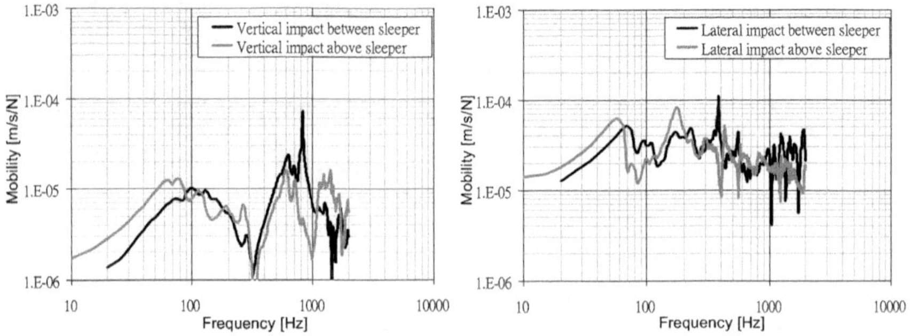

Fig. 3. Frequency Response of the RailLeft Chart –Vertical Frequency Response underneath Rail Foot Right Chart –Lateral Frequency Response at the Rail Head

Fig 4 shows the track decay rate before and after damper installation. A force hammer was used to introduce 5 kN peak force with broadband excitation up to 2000 Hz. At each impact location, the frequency response was calculated from a synchronized time average of 5-10 impact signals. The vertical decay rate was significantly increased to 6-11dB/m above 400Hz, while the lateral decay rate was significantly increased to about 3-6dB/m above 300Hz.

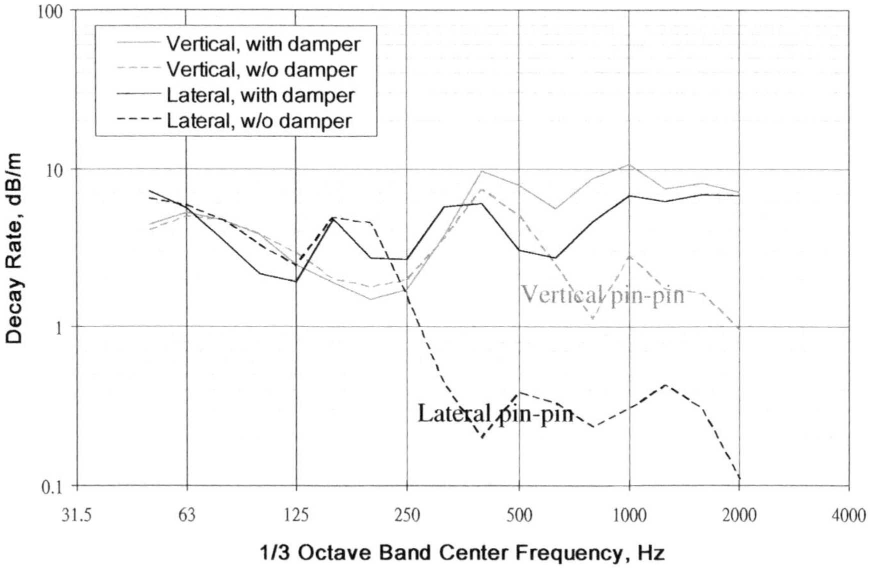

Fig. 4. Measured Track Decay Rate before and after installation of rail damper

4.2 Track Vibration

The rail vibration spectrum is shown in Fig 5. Each set of measurement data includes about 20 train passages. Overall vertical and lateral vibration levels were reduced by 7dB(A) and 10dB(A), respectively. Sleeper vibration spectrum is shown in Fig 6. The sleeper block has a resonance peak at 300 Hz which contributes to significant noise radiation at that frequency. The rail damper has little effect on the sleeper resonance.

The vertical vibration time history at 800 Hz, the vertical pinned—pinned resonance, is shown in Fig 7. Due to the high track decay rate after damper installation, the rail vibration level drops by 40-50 dB between two bogies. The rail becomes localized point sources at the wheel contact points.

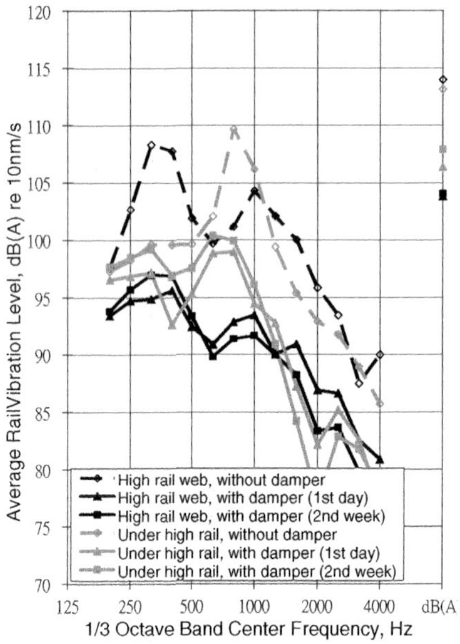

Fig. 5. Average vibration level high rail

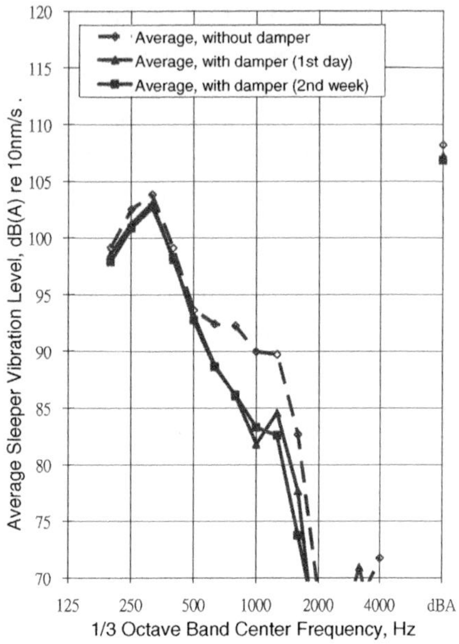

Fig. 6. Vibration level at sleeper block

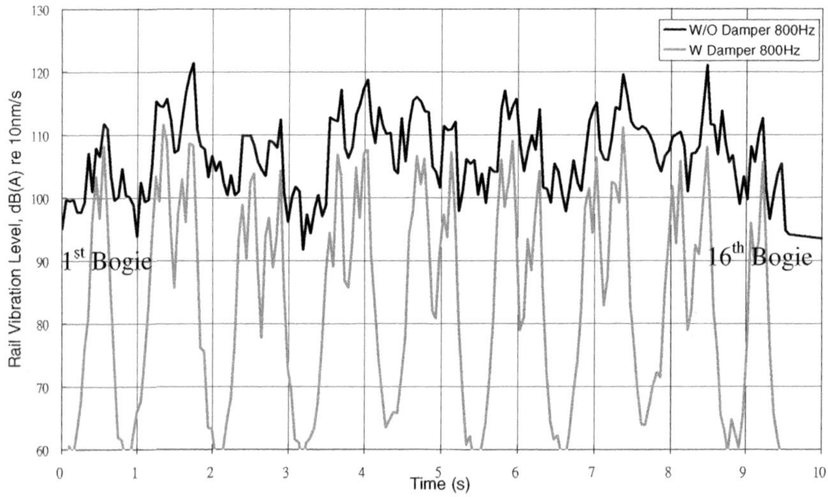

Fig. 7. Vertical Vibration Time History at 800 Hz

4.3 Tunnel Noise Measurement

In Fig 8, reverberant field noise level at the tunnel walkway was reduced by 3.4dB(A), while the near field noise level at 120 mm from the high rail was reduced by 3.5 dB(A). Although the rail web vibration level at the 300 Hz peak was reduced by more than 10 dB(A), the noise level shows little reduction at that frequency. This is because the rail is not the dominant noise source at 300 Hz.

Fig 9 shows the sound power level of each track component as projected from the vibration level. After installation of rail dampers, the sound power level from rail was significantly reduced. The residual sound power level is dominated by the sleeper blocks having a resonance peak at 300Hz, which thus limits further noise reduction by rail dampers.

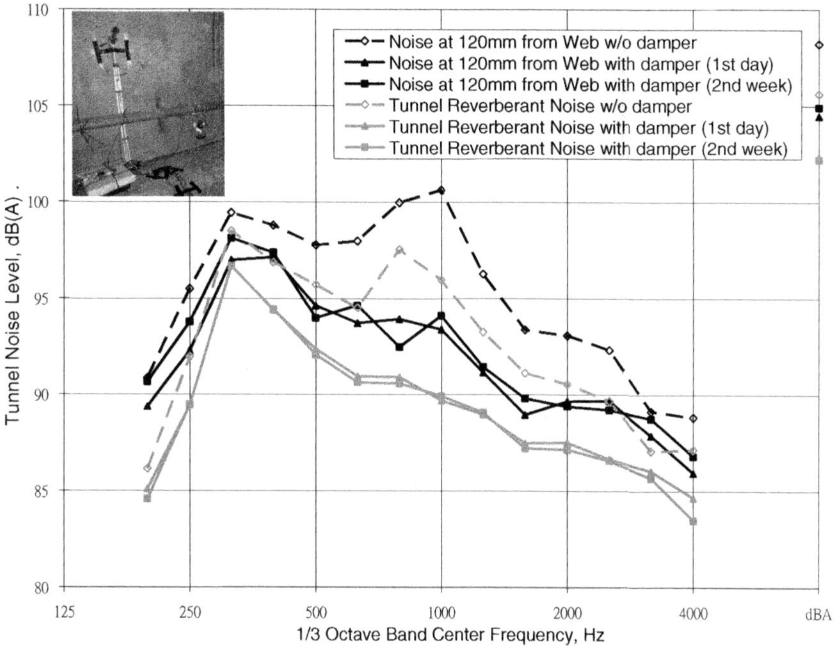

Fig. 8. Noise level Inside Tunnel

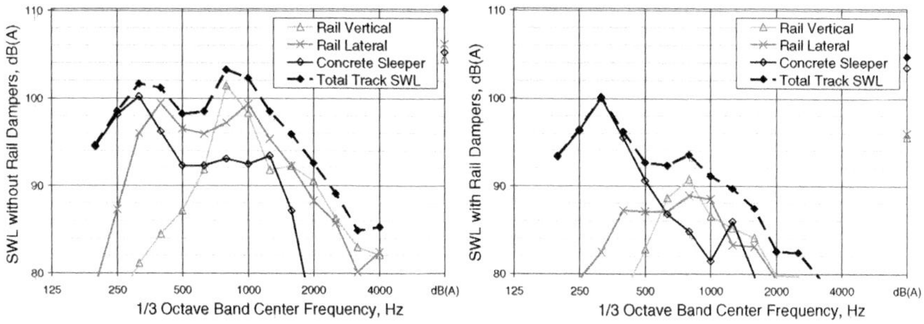

Fig. 9. Projected Sound Power Level from Each Track Component

5 Further Studies

Further investigation lasting for 1 year is being conducted to study the influence on corrugation growth by rail dampers. Preliminary observations in the first 3 months indicate that corrugation growth has been slowing down. Further investigation results will be published after completion of the trial test.

6 Conclusion

A new type of tuned mass damper has been developed to reduce rail vibration and noise radiation. The damper comprises multiple oscillation masses sandwiched between resilient layers forming a multiple mass—spring system. The resilient layers are mounted perpendicular to the rail alignment such that each mass is allowed to oscillate in both vertical and lateral directions.

The damper has been installed at a curved track section in Hong Kong to study its noise reduction performance and influence on the corrugation growth rate. The vertical track decay rate is generally increased to 9-10dB/m at frequencies above 500Hz, while the lateral track decay rate is generally increased to 7dB(A) at frequencies above 300Hz. The dampers reduce rail vibration by about 10dB(A) and noise level by 3.5dB(A). After installation of rail dampers, the sleeper blocks become the dominant noise sources.

Acknowledgement

The author wishes to express sincere thanks to MTR Corporation for their provision of testing sites and approval for publishing of the paper content.

References

[1] Létourneaux, F., et al.: High Speed Railway Noise: Assessment of Mitigation Measures. Notes on Numerical Fluid Mechanics and Multidisciplinary Design 99, 56–62

[2] Asmussen, B., et al.: Reducing the Noise Emission by Increasing the Damping of the Rail: Results of a Field Test. Notes on Numerical Fluid Mechanics and Multidisciplinary Design 99, 229–235

[3] Nelson, T.: TCRP Report 67 – Wheel and Rail Vibration Absorber Testing and Demonstration, Transportation Research Board, ch. 9, pp. 42–44 (2001)

[4] Thompson, D.J., et al.: A Tuned Damping Device for Reducing Noise from Railway Track. Applied Acoustics 68, 43–57 (2007)

[5] BS EN 15461: Railway Applications – Noise Emission – Characterization of the Dynamic Properties of Track Selections for Pass by Noise Measurement (2008)

[6] Thompson, D.J., et al.: Experimental Validation of the Twins Prediction Program For Rolling Noise, Part 1: Description of the Model and Method. Journal of Sound and Vibration 193(1), 123–135 (1996)

[7] Thompson, D.J., et al.: Experimental Validation of the Twins Prediction Program For Rolling Noise, Part 2: Results. Journal of Sound and Vibration 193(1), 123–135 (1996)

[8] Nelson, T.: TCRP Report 23 – Wheel/Rail Noise Control Manual, Transportation Research Board, ch. 4, pp. 41–42 (1997)

Development of Rail Noise Isolating Material (RNIM)

Masanori Hansaka[1], Shogo Mamada[1], Daigo Sato[1], Kiyoshi Sato[1], and Fumiaki Kishino[2]

[1] Railway Technical Research Institute of Japan
Hikari-Cho 2-8-38, Kokubunji-shi, Tokyo, Japan
Tel: +81 042 573 7339, Fax: +81 042 573 7354
e-mail: hansaka@rtri.or.jp
[2] NICHIAS Corporation

Summary

Recently, there have been increasing requirements to control the rolling noise generated by the wheels and rails, but the conventional counter measures such as noise barriers have a big disadvantage in that their installations are very labor-intensive. Therefore, the authors have developed a new type of material, called rail noise isolating material (RNIM), which consists of two laminated layers, the inner one a foamed ethylene—propylene rubber and the outer one vibration-damped steel. An impact test executed in a laboratory and some noise measurements executed on commercial lines have verified that RNIM is effective in reducing rolling noise.

1 Introduction

Recently, as running speeds of railway cars have increased and demands for noise abatement in residential areas have been intensified, there have been increasing requirements to control the rolling noise generated by the wheels and rails, which has a great contribution to the trackside noise. However, the conventional counter measures such as noise barriers have a big disadvantage, in that their installations are very labor-intensive; it is, therefore, essential to develop a new, easily installed measure1),2). For these reasons, the authors have developed a new type of material, called rail noise isolating material (RNIM), which consists of two laminated layers, the inner one a foamed ethylene-propylene rubber and the outer one vibration-damped steel3). RNIM covers approximately the whole area of the web and foot of the rail. It is expected to reduce the noise radiated by the rail, because the vibration of the outer layer of RNIM is significantly reduced by the vibration isolating property of its inner layer made of soft foamed rubber. To

T. Maeda et al. (Eds.): Noise and Vibration Mitigation for Rail Trans. Sys., NNFM 118, pp. 97–105.
springerlink.com

estimate the vibration and noise reduction property of RNIM, an impact test was executed in a laboratory and some noise measurements were executed on commercial lines. In this paper, the outline of RNIM and the results of estimation tests are described.

2 Structure of Rail Noise Isolating Material (RNIM)

Figure 1 illustrates the structure of RNIM. RNIM consists of two laminated layers, the inner one a soft visco-elastic rubber, and the outer one a thin, stiff plate. Two pieces of RNIM cover approximately the whole area of the web and foot of the rail. RNIM is attached to the rail by means of a bracket without being bonded. Thereby it is easy to install and remove so that the rail can be inspected after installation of RNIM. RNIM is expected, to reduce the noise radiated by the rail because with RNIM the noise depends on the vibration of the outer layer of RNIM. This is due to the characteristic that the noise radiated by a vibrating body generally depends on its outer vibration[4] and this is significantly reduced by vibration isolating property of the inner layer of RNIM made of soft-foamed rubber.

A foamed ethylene—propylene rubber, which has a good durability among visco-elastic materials, was selected for the material of the inner layer. As a result of preliminary examination, the thickness of the inner layer was decided to be about 30 mm. Vibration-damped steel was selected for the outer layer considering that it should have high vibration damping ability, because its vibration will generate noise radiation directly as mentioned above.

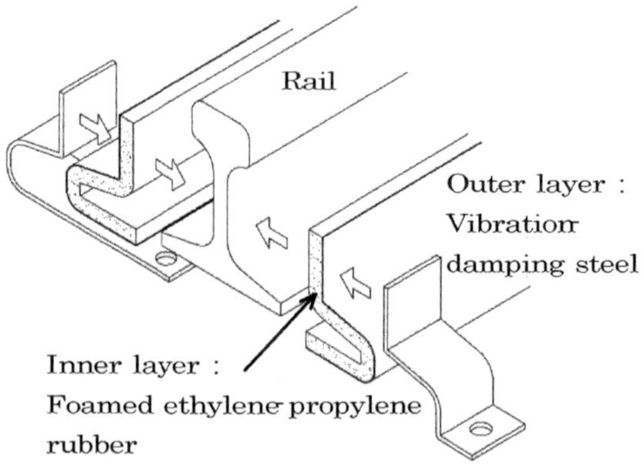

Fig. 1. Structure of Rail Noise Isolating Material (RNIM)

3 Impact Test Executed in Laboratory

This section describes the impact test executed in a laboratory to evaluate the basic vibration and noise reduction properties of RNIM.

3.1 Sample and Test Method

Figure 2 shows a view of the experimental set up of the impact test. A 50-kg/m N-type rail cut to about 800 mm in length was used as a sample. Both its ends were supported simply on two concrete sleepers with rail pads being placed between the rail and the sleepers. A central point on the railhead surface was hit vertically by an impact hammer for both the cases with and without RNIM. The excitation force was measured by the instrumented hammer, the vibration accelerations by piezo-electric accelerometers, and the noise (sound pressure) radiated from the rail by using standard microphones. The vibration acceleration was integrated once with respect to time to obtain vibration velocity, and was analyzed and transformed into mobility, which is the frequency response function (FRF) of vibration velocity per unit force. Noise was also analyzed and transformed into a FRF. In this paper, mobility and noise FRFs are expressed in decibels, where 0 dB corresponds to 1 m/s/N for mobility, and to 2×10^{-5} Pa/N for the noise FRF. The frequency range for analysis was from 10 Hz to 10 kHz.

This paper discusses the results for two points marked as V1 and V2 for vibration and one point marked as S1 for noise. V1 was located on the railhead surface near the impact point, and V2 was located on the rail foot in the case without RNIM, and on the outer layer of RNIM when it was attached. V2 was located just beneath the impact point in both cases.

S1 was positioned about 300 mm away from the surface of the rail web opposite the center of the rail, and at the same height as the center of the rail web.

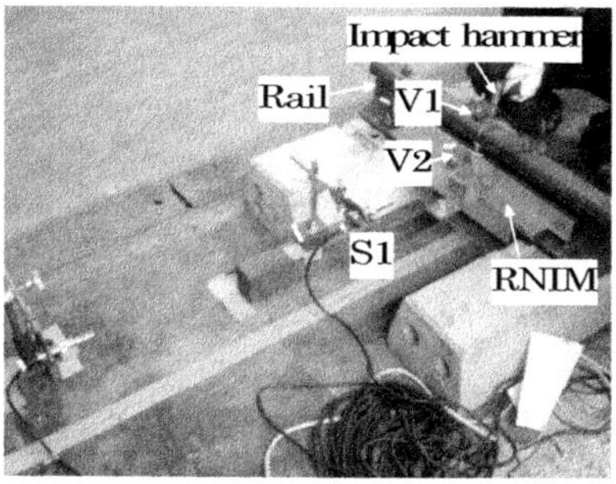

Fig. 2. Experimental setup of the impact test (color figure online)

3.2 Test Results

Figure 3 shows time waveforms of vibration acceleration measured at V1 and V2. Both show the value of vibration acceleration divided by the maximum value of the excitation force. From fig. 3, it is found out that while the vibration acceleration of the rail head (V1) changes little due to the attachment of RNIM that of the outer layer of RNIM (V2) which is much smaller than that of the rail without RNIM. These results mean that, on the one hand RNIM has only a slight effect in reducing the vibration of the rail itself, but on the other hand it has a significant effect in preventing the vibration of the rail from being transmitted to the outer layer because of the vibration isolating performance of the inner layer of RNIM.

Figure 4 shows the mobility measured at V1 and V2. These results make clear in the frequency domain that while the vibration of the railhead surface has little change due to the attaching of RNIM, the vibration of the outer layer of RNIM is significantly reduced. Fig. 4(b) shows that both the vibration of the rail foot (without RNIM) and the outer layer of RNIM have many peaks. Some of them are presumed to be attributable to the vibration modes of the outer layer of RNIM itself, but it is difficult to identify which peaks belong to it. Fig. 4(b) shows also that the reduction of vibration of the outer layer of RNIM becomes obvious above about 200 Hz and it increases as the frequency increases.

Figure 5 shows the time waveform and FRF of noise measured at S1. In this time waveform, the noise was divided by the maximum value of the excitation force just like the vibration. From fig. 5(a) it is found out that the noise radiated from the rail is reduced by attaching RNIM soon after the rail is impacted. Fig. 5(b) shows that while the noise radiated from the rail increased at some peaks, such as around 120 Hz after attaching RNIM, it is reduced obviously and entirely in the high frequency range over 500 Hz. The tendency is similar to the vibration of the outer layer of RNIM. Thus, at some region of low frequency below 200 Hz the noise becomes rather larger after attaching RNIM but as is shown in later, the component of noise at such low frequencies is so small in commercial line that the contribution of this region is negligible.

Table 1 shows the values of summed squared mobility of V1 and V2 and FRF of noise of S1 from 10 Hz to 10 kHz. From this, it is recognized that while the reduction of that value of mobility of V1 is no more than about 2.2 dB, that of V2 is as much as about 12.9 dB. Moreover, the reduction of that value of FRF of noise at S1 by attaching RNIM amounts to about 7.8 dB. However, as shown in fig. 5(b), main components of the decrease are attributable to those in the high frequency region over 2 kHz, in which the noise generated on the trackside is relatively low.

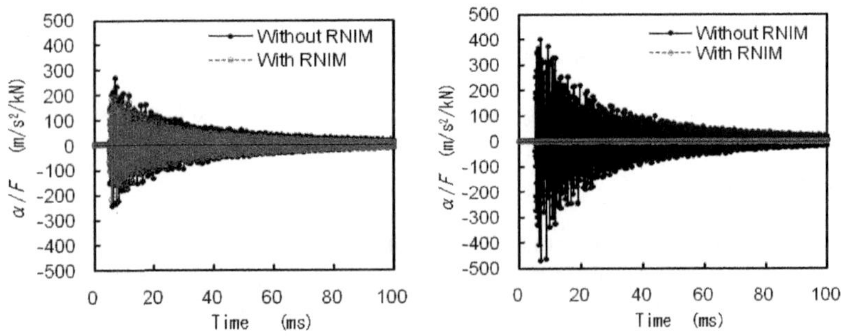

(a) V1 (b) V2

Fig. 3. Time waveform measured at V1 and V2 by the impact test (color figure online)
*) α : Vibration acceleration, F : Excitation force

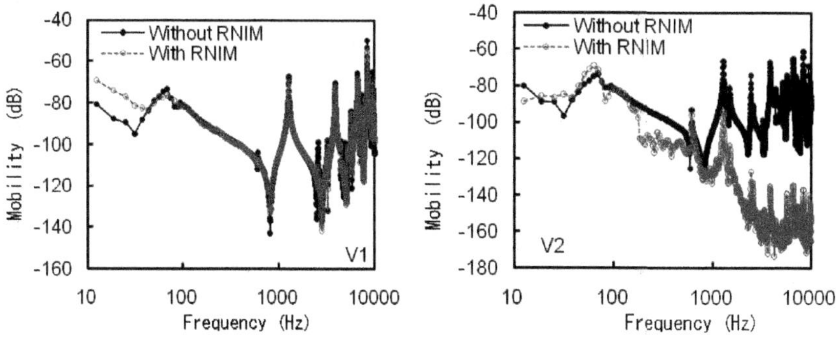

(a) V1 (b) V2

Fig. 4. Mobility measured at V1 and V2 by the impact test (color figure online)

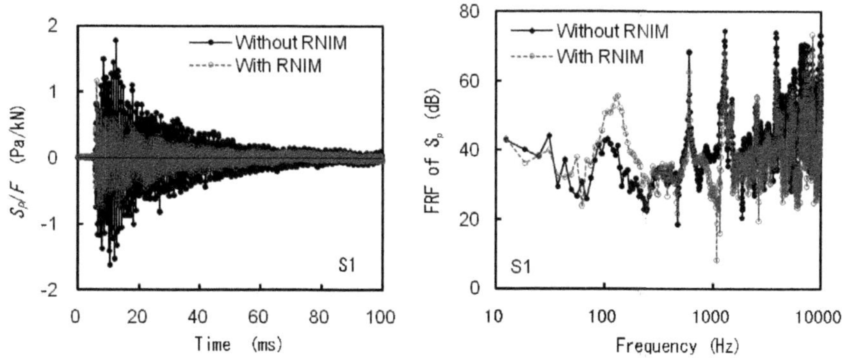

(a) Time waveform (b) FRF

Fig. 5. Time waveform and FRF of noise measured at S1 by the impact test (color figure online)

*) *SP* : Sound pressure, *F* : Exciting force

Table 1. Summed squared mobility of V1 and V2 and FRF of S1 (dB)

Sample	Mobility		FRF of noise
	V1	V2	S1
Without RNIM	-44.8	-51.1	91.1
With RNIM	-47.0	-64.0	83.3
Reduction*	2.2	12.9	7.8

*) The reduction from the value of sample without RNIM to that with RNIM.

4 Noise Measurement Executed on Commercial Line

This section describes the result of a test executed on a conventional electrified line with ballasted track as one of tests executed on commercial lines to estimate the practical effects in noise reduction.

4.1 Installation of RNIM and Test Method

RNIM was installed over a length of approximately 25 m on a conventional electrified line with ballasted track (straight and flat line). The rail pad installed in the track is the standard type (nominal stiffness is about 110 MN/m). Figure 6 illustrates a view of the track installed with RNIM. The installation work was simple enough to be done in 2–3 hours by several workers.

In this paper the results measured at microphone position S7, which was set about 6.25 m away from the near side track center and about 1.2 m above the ground level, are reported. The measuring points are shown in Figure 7. Noise was measured by a standard or directional microphone when trains passed the measuring point. Results are reported that are A-weighted with the time constant fast. Moreover, a one-third-octave band analysis of noise was done during one passage of the test cars.

Fig. 6. View of track installed with RNIM (color figure online)

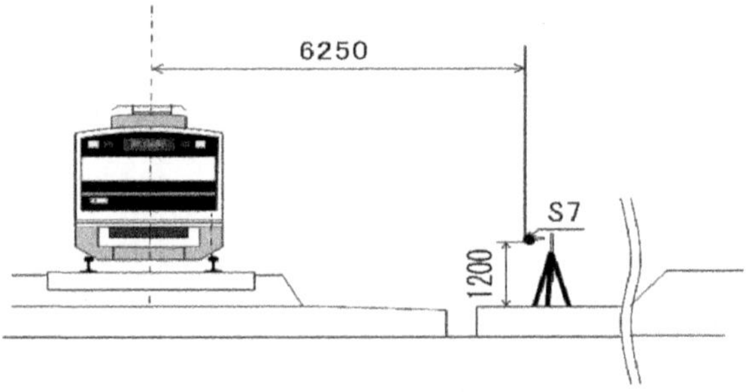

Fig. 7. Measuring points of noise on the conventional line (color figure online)

4.2 The Results of Measurement

Figure 8 shows a time history of noise level measured by a directional microphone whose directivity in the direction right to the rail is intensified by arranging multi-array microphones at point S7 during one passage of the test cars. This figure reveals that peaks and dips of noise level appear repeatedly in the time history of noise. The peaks of noise correspond to the wheelset passages of train cars and the dips of noise are located between them. Fig. 8 reveals also that the noise is entirely reduced throughout test cars passage by installation of RNIM. In addition, the reduction of noise becomes great at the dips of noise and small at the peaks. The reason why the noise reduction becomes small at the peaks, which agrees with the time of the wheelsets passages, is considered as follows. At the time of the wheelsets passage, noise is generated by many sources including wheels, driving devices in motor cars and so on. RNIM has the noise reduction capability only for the rail, thereby, the noise reduction effect by RNIM is weaker at the time of the wheelsets passage since at such times the trackside noise is mixed with the components which are generated by sources other than the rail and suffer little influence of RNIM.

Figure 9 shows the peak values of noise level during one set of train cars passage measured by a standard microphone (actually corresponding to time of motor car passage). From this, it is revealed that a 2-3 dB reduction of the noise level is brought about by the installation of RNIM in a wide range of train velocity from 60 to 100 km/h.

Figure 10 shows the results of a one-third-octave band analysis of data shown in fig. 8. From fig. 10, it is revealed that the noise level is reduced by the installation of RNIM over a wide range of frequency from 500 Hz to 2 kHz. The reduction of noise is, especially, apparent around 1 kHz. According to the previous studies for railway vibration and noise[1),2)], it is presumed that the contribution of noise radiated from the rail is dominant in the trackside noise around 1 kHz. Taking these factors into consideration, it is judged that this result verifies that RNIM has an effect in reducing the noise radiated from rail.

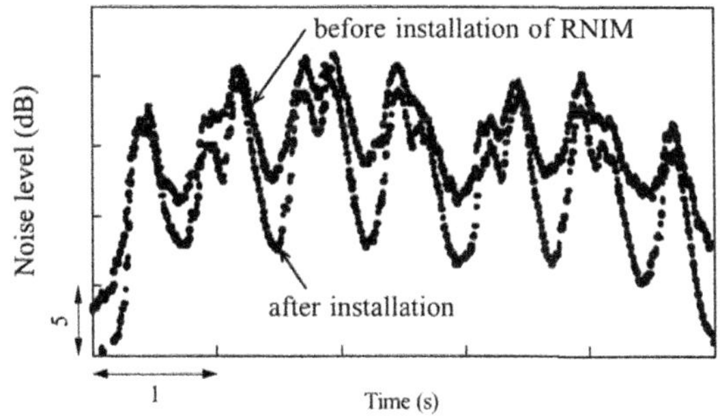

Fig. 8. Time history of noise level during test cars passage

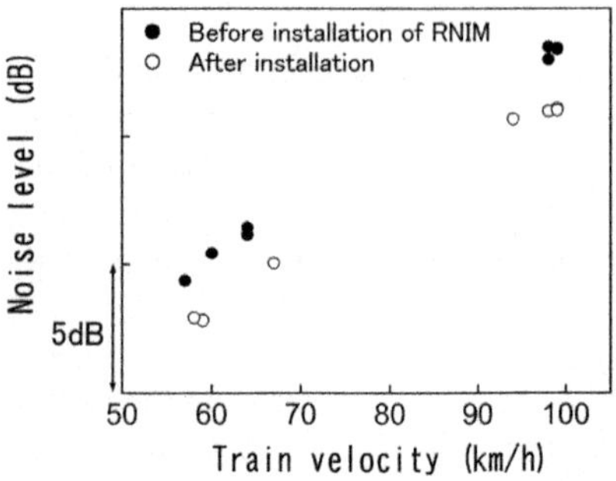

Fig. 9. Peak values of noise level during train cars passage

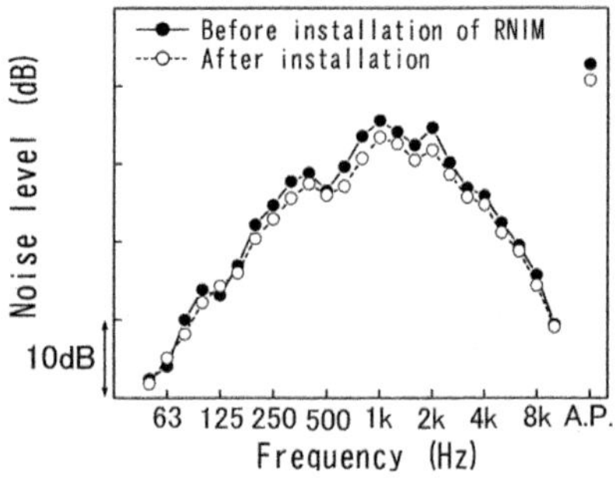

Fig. 10. Result of one-third-octave band analysis of noise

5 Conclusions

To achieve a counter measure against the rolling noise which is easy to install and effective, the authors have developed a new type of material, called rail noise isolating material (RNIM), which consists of two laminated layers, the inner one a foamed ethylene—propylene rubber and the outer one vibration-damped steel. To estimate the vibration and noise reduction properties of RNIM, an impact test was executed in a laboratory, and some measurements of noise on commercial lines. As the results of these tests, the following conclusions were obtained:

1) As a result of the impact test, it was revealed that the vibration of the outer layer of RNIM was significantly reduced when compared with the rail because RNIM has an excellent vibration isolating property.
2) As a result of the impact test, about 7.8 dB reduce in FRF of noise was measured at a near point of the rail (about 300 mm apart from rail web) by attaching of RNIM. However, main components of the decrease are attributable to those in the high frequency region over 2 kHz, in which the noise generated on the trackside is relatively low.

3) As a result of installation of RNIM on a commercial line, it was verified that RNIM was easy to install.
4) As a result of the measurement of the noise on a commercial line, a 2-3 dB reduction in noise level was measured at a point near the rail (about 6.25 m away from the nearside track center) by installation of RNIM.
5) As a result of one-third-octave band analysis of the noise measured on a commercial line, it was revealed that the noise was reduced, especially, around 1 kHz. This result verifies that RNIM is effective in reducing the noise radiated from rail.

Acknowledgment

The authors appreciate the cooperation for the development of RNIM received by Odakyu Electric Railway Co., Ltd.

References

[1] Zenda, Y., Nagakura, K., Moritoh, Y., et al.: Noise Control of High-Speed SHINKANSEN, IWRN, S5.2-1 (1995)
[2] Zenda, Y., Moritoh, Y.: REVIEW: A Review of Methods to Control Railway Noise Source. QR of RTRI 37(4), 197–198 (2007)
[3] Hansaka, M., Mamada, S., Sato, K.: Development of Rail Noise Isolating Material. QR of RTRI 48(4), 215–220 (2007)
[4] Cremer, L., Heckl, M.: Structure-Borne Noise, pp. 491–564. Springer, Heidelberg (1972)

Bridging the Gap from Multibody Simulations to Acoustic Analysis

A. Heckmann, I. Kaiser, and A. Carrarini

DLR, German Aerospace Center, Institute of Robotics and Mechatronics,
Oberpfaffenhofen, 82234 Wessling, Germany
Tel.: +49 8153 281699, Fax: +49 8153 281850
e-mail: andreas.heckmann@dlr.de

Summary

Multibody simulation plays an essential role in the initial design phase of railway vehicles in which conceptual issues are investigated. It is as well the state-of-the-art method in order to study and optimize the running dynamics and comfort of railway vehicles in details. However, acoustic issues could not be adressed so far by multibody analysis. This is due to the fact that the validity of the used multibody models is restricted to the very low frequency range. In addition, multibody methodology is not prepared to evaluate acoustic measures in order to assess a railway vehicle under consideration.

Provided that these two short-comings are resolved, multibody analysis is capable of supplementing and enhancing the tools currently available for assessing railway acoustics such as TWINS, FEM or BEM. The high computational efficiency, the capability to perform time domain analysis, the reuse of model-components and scenarios originating from comfort evaluations are advantages that motivate the extension of the multibody methodology towards acoustic analysis.

Therefore, this paper presents model enhancements of the contact behavior of wheel and rail that are intended to be used in simulations in the frequency range up to 2.5 kHz and higher. Then the concept of an acoustic pre-processing tailored for multibody analysis is proposed based on the evaluation of the sound power on the surface of vibrating elastic bodies. Finally, an example is given in order to demonstrate the application of the proposed methodology.

1 Introduction

The German Aerospace Center has launched a comprehensive long-term research program on high-speed trains. The project is entitled "Next Generation Train"; a name which also refers to a virtual demonstrator which is supposed to present the results of the program in a coherent way.

T. Maeda et al. (Eds.): Noise and Vibration Mitigation for Rail Trans. Sys., NNFM 118, pp. 107–115.
springerlink.com

One particular aim of the project concerns improvements of the virtual design process of railway vehicles. Modern engineering development flows rely on the capability of predicting properties of a specific product lay-out in an early stage and in a reliable and comprehensive manner. Due to recent activities of legislation authorities in Europe and higher public attention on environmental issues, the acoustic characteristics of railway vehicles will gain importance in the future, so that methodological enhancements in this field meet an emerging demand.

The currently available analysis tools in railway acoustics such as TWINS or FEM and BEM tools have reached a high level of maturity and are, therefore, capable of investigating the relevant phenomena of wheel/rail noise generation, see [1] and its extensive list of references. However, they are typically restricted to the frequency domain. In addition, they are not tailored to consider many different railway-specific operation conditions such as hunting motion, buffer and bump stop impacts, approaching curves and switch crossings, etc.

Regarding vehicle dynamics, multibody simulations are already the back-bone in the development of the current generation of high-speed railway vehicles. However, the potential of flexible multibody systems has not been exploited completely. Provided that the validity range of the used models could be extended into the region up to 2.5 to 5 kHz, time domain results of multibody simulations can also deliver information on the acoustic performance of the considered vehicle.

With this background the paper mainly presents two general topics. Firstly, enhancements in the modelling of wheel/rail contact partners, in particular, concerning the wheelset and the track are introduced. Secondly, the acoustic post-processing of multibody simulations in order to predict structure-borne noise levels and their relation to design parameters are described.

In principle the proposed analysis process can be extended to a subsequent radiation analysis using, e.g., BEM in order to evaluate the radiated sound power. However, for conceptual surveys, intermediate optimization steps, and for cases in which the radiation efficiency can be deduced from experience or empiric formulas, the structure borne noise level is already sufficient to assess the acoustic properties of the reviewed design.

2 Modelling Refinement

2.1 Structural Dynamics of the Rotating Wheelset

The introduction of flexible bodies as components of a multibody system has become a standard approach. The deformation field may very efficiently be represented by a selected set of eigenmodes that originates from a modal analysis of a corresponding finite element model [2].

For acoustic analysis it is in particular necessary to consider the structural dynamics of the wheelset, since it is strongly excited by fluctuating wheel/rail forces resulting e.g. from geometric disturbances and it is one of the main sources of railway noise, see [3].

Fig. 1. Elastic wheelset integrated into a SIMPACK model (color figure online)

However, the modelling as an elastic body within a multibody system is a more difficult task for the wheelset than for other flexible components, because the wheelset is rotating and the wheel-rail forces are thereby moving around the wheelset. By taking advantage of the cyclic structure of the wheelset, a formulation of the elastic rotor is possible that corresponds to the so-called Arbitrary-Lagrangian- Eulerian-description (ALE), i.e., the deformations are described as "viewed/observed from the outside". In other words: the deformations of contour of the wheelset which does not perform the overturning motion are considered, see [4] for a detailed description of the ALE-approach.

In a modified form, this formulation has been successfully integrated into a prototype version of the multibody dynamics program SIMPACK, see Fig. 1.

In Fig. 2, two examples of modeshapes of the wheelset are presented, that have been evaluated in advance of the actual multibody analysis.

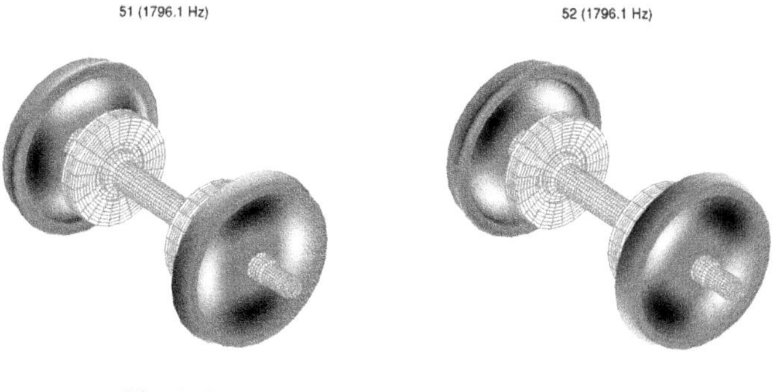

Fig. 2. Eigenforms no. 51 and 52 at 1796 Hz.

2.2 Track Model

Besides the wheelset, the rail is the second element of the vehicle-track system which is directly excited by the wheel-rail forces. The rail is strongly coupled to the sleepers. Therefore, for the analysis of the rail's acoustics, it seems not to be sufficient to consider the rail alone, but analyzing the entire track structure to be removed. To this end, a track model is developed which is displayed in Fig. 3. It is based on the track model by Ripke [5], but uses an enhanced model of the rail.

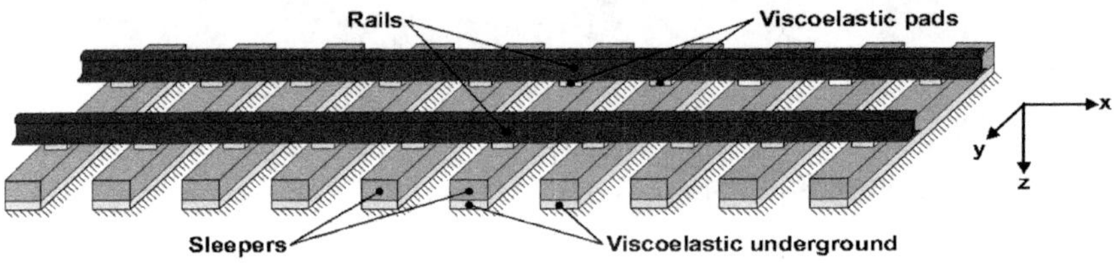

Fig. 3. Model of the track containing elastic rails

The model consists of two elastic rails, which are supported by discrete sleepers. The sleepers are considered as rigid bodies, which can perform all six motions. They are connected to the rails by visco-elastic layers representing the pads and are supported by visco-elastic layers representing the foundation. The inclination of the rails towards the middle of the track is taken into account in this model. For the rails, the profile UIC60 is chosen, which is widely used in Europe.

Since the model is designed for calculations in the time domain, it has to be finite. Of course, this finite model has to describe the properties of a very long track. Therefore, a track with 128 sleepers is used. At the end of the rails, the same boundary conditions are applied so that the track forms a ring, whereas the curvature is neglected. Comparisons show that this is a quite good approximation of the dynamic behaviour over a wide frequency range.

The foot of the rail is connected to the sleepers, lateral wheel-rail forces act at the rail head, and the web of the rail is comparatively thin. Therefore, deformations of the cross-section should be taken into account. This cannot be done by applying a common beam theory, since these theories assume that the cross-section remains undeformed. Therefore, the rail is modelled by using three-dimensional finite elements. As a result, modes describing cross-sectional deformations can be obtained beneath the usual bending modes, as displayed in Fig. 4.

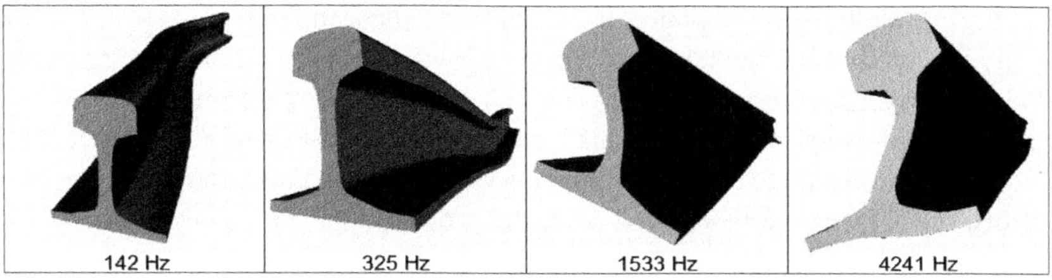

Fig. 4. Eigenmodes of the rail, wavelength λ=2.4 m (color figure online)

3 Concept of Acoustic Post-processing

Since a modal approach is applied, the normal velocity $\boldsymbol{v}$ at the surface of the vibrating body under consideration may be evaluated as:

$$\boldsymbol{v}(\boldsymbol{x}_S, t) = \boldsymbol{\Phi}_\perp(\boldsymbol{x}_S)\, \dot{\boldsymbol{q}}(t)\ , \tag{1}$$

where $\dot{\boldsymbol{q}}(t)$ is the vector of generalized modal velocities and $\boldsymbol{\Phi}_\perp(\boldsymbol{x}_S)$ the modal displacement normal to the surface at the position $\boldsymbol{x}_S$. These modal displacements may also be taken from a modal analysis of the underlying FE-model of the body.

As a consequence of (1), the squared value of the RMS velocity $\tilde{v}$ over a time interval T can then be expressed as:

$$\tilde{v}^2(\boldsymbol{x}_S, t) = \frac{1}{T}\int_0^T \dot{\boldsymbol{q}}^T \boldsymbol{\Phi}_\perp^T \boldsymbol{\Phi}_\perp\, \dot{\boldsymbol{q}}\ \mathrm{d}t\ . \tag{2}$$

In addition the underlining operator is introduced now, which is to be applied to $\boldsymbol{v}$ specifying the corresponding mean value over the surface S, *i.e.*

$$\underline{v}^2 = \frac{1}{S}\int_S v^2 \mathrm{d}s = \dot{\boldsymbol{q}}^T \frac{1}{S}\int_S \boldsymbol{\Phi}_\perp^T \boldsymbol{\Phi}_\perp\ \mathrm{d}s\ \dot{\boldsymbol{q}}\ . \tag{3}$$

The radiated sound power $P$for a vibrating body is then given by [6]:

$$P = \sigma\, \varrho\, c\, S\, \underline{\tilde{v}}^2\ , \tag{4}$$

where σ is the radiation efficiency, ϱ the density of the air, c the corresponding sound velocity. The power is usually expressed as power level, i.e., as logarithm of the ratio of the power and a fixed reference quantity. The reader is referred to the corresponding norms for details (e.g. EN 21683 in Europe).

A central point is that the modal decomposition of the velocity in (1) leads to a separation of the problems in time and space. Consequently the integrations in time and spatial domain in (2) and (3) may be performed independently. In particular, the computationally expensive surface integral in (3), now denoted by $\boldsymbol{A}$, can be evaluated in advance and only once for a given flexible body:

$$\boldsymbol{A} := \int_S \boldsymbol{\Phi}_\perp^T \boldsymbol{\Phi}_\perp\ \mathrm{d}s\ \ \Rightarrow\ \ \underline{v}^2 = \dot{\boldsymbol{q}}^T \boldsymbol{A}\, \dot{\boldsymbol{q}}\ . \tag{5}$$

Once a multibody simulation has been performed, the dynamics of the flexible body included in the model is known. It follows that the motion of every point on the surface of the vibrating bodies can be computed from the time histories of the modal coordinates: see (1). On this basis, the sound power on the body's surface (structure borne sound) can be evaluated with very little additional computational effort, see [7] for a more detailed description of the acoustic post-processing.

4 Application Example

The example scenario deals with a railway vehicle running through a switch. The vehicle is driving at 20 m/s and the leading wheelset, the one modelled as flexible, encounters the rail gap in the centre part of a switch (so called "frog", see Fig. 5) at $t = 0.55$ s. The rail gap, modelled as a discontinuity in longitidunal direction, causes an impact of the wheel rolling over it; the induced vibration of the wheelset originates impulsive noise.

The right hand side picture in Fig. 5 presents an exemplary distribution of the structure borne sound intensity over the wheel surface at one specific point in time. It is not intended for a quantitative analysis, but offers a valuable insight into the physical behavior of the vibrating body.

Fig. 6 shows the frequency content of the sound power level as a function of time. A sliding time window of T=0.125 s (denoted as "fast" in acoustic measurement engineering) has been used. It can be clearly seen that the noise due to the impact of the wheel over the rail gap is of very short duration and decays very rapidly. Note that the peak sound power level is very high (>90 dB). It must be stressed again that this value does not refer to a virtual receiver located somewhere near the wheel but to the power on the surface. Assuming a unitary radiation efficiency, $\sigma = 1$ in (4), as known to be the case for railway wheels above 1000 Hz, the decay due to the propagation should be considered to eventually obtain, for example, the intensity at the standard railway measurement position at a distance of 7.5 m.

The sound power level in Fig. 6 is dominated by a peak between 2000 and 2200 Hz. The proposed post-processing scheme also allows the modal contribution to be studied in order to determine those eigenmodes which are mainly responsible for this peak: see [7].

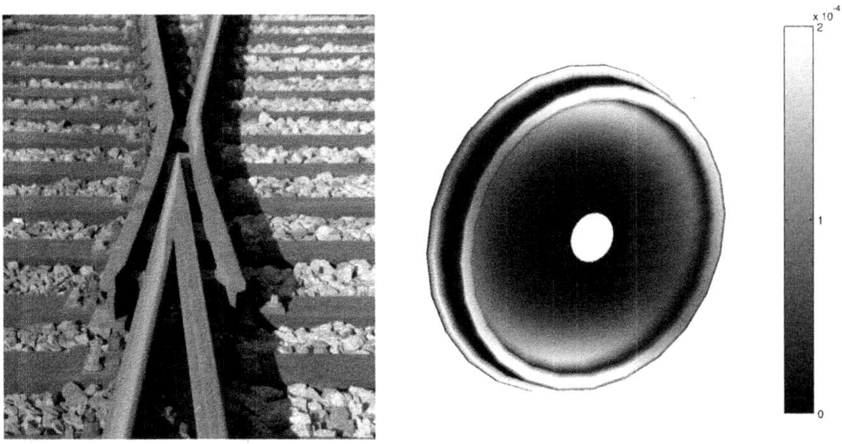

Fig. 5. "Frog" and an example of the distribution of the structure borne sound intensity over the wheel surface (frame from a time simulation) (color figure online)

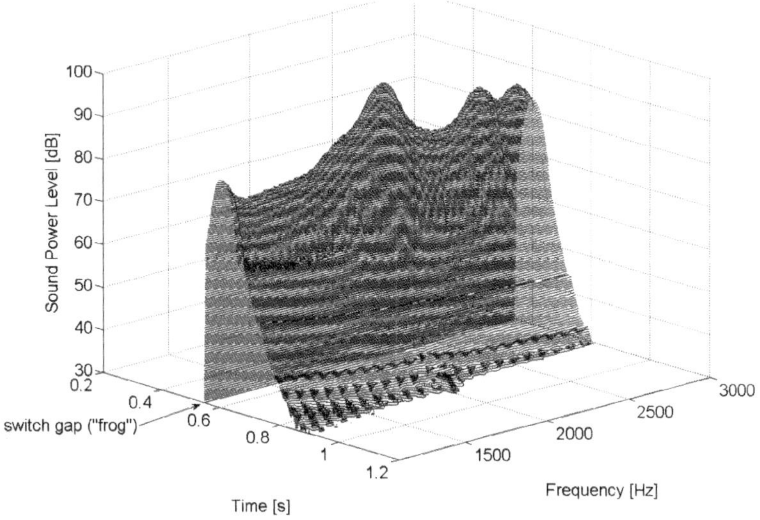

Fig. 6. Frequency content of the structure borne sound power level as a function of time

Fig. 7 presents the modes nos. 51 and 52, see Fig. 2, to strongly dominate the structure-borne noise level and the intensity of the contribution does not change during the short time the wheel vibrates. Since the phenomenon here is an externally excited but not a free vibration the dominating frequencies differ from the eigenfrequency of these modes.

The main computational effort to simulate this application has been spent for the pre-processing, the FE modal analysis of the wheel-set which took less than one-hour cpu-time on a modern 32 bit PC. The mutibody simulation and the post-processing did not require more than one-minute cpu-time.

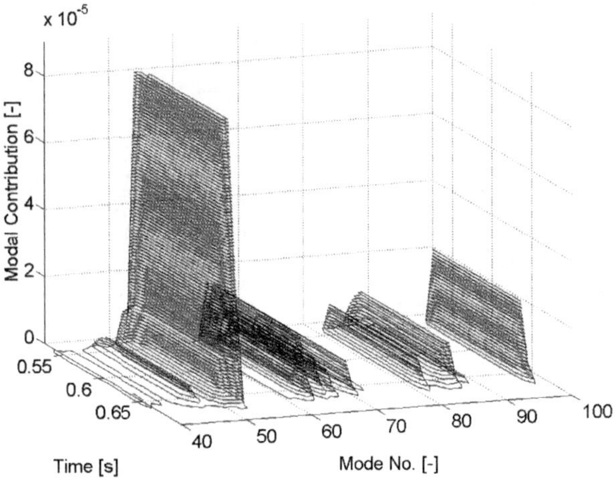

Fig. 7. Switch crossing scenario: main modal contributions as a function of time

5 Conclusions

Due to its environmental and thus social relevance, railway noise is gaining importance and has been object of large legislative and normative intervention on different levels. As a consequence, the acoustic behavior of railway vehicles is becoming a central topic for manufacturers and transportation companies, which need to develop noise optimal configurations and noise abatement concepts. This finally leads to a need for analysis tools which allow the acoustic properties of vehicles and tracks to be analyzed or predicted in an efficient and economic way. From this point of view it makes sense to extend an established tool used to design railway vehicles, namely multibody simulation, to be capable of carrying out acoustic assessment of these vehicles This will supplement tools currently used for this purpose.

The present paper proposes model enhancements and a post-processing scheme in order to meet this emerging demand. The benefits of the methodology has been demonstrated in an simulation study that refers to an application from railway engineering: a vehicle is running through a switch. This transient scenario is well known for creating impulsive noise and the analysis is capable to identify those eigenvectors that dominate the resulting structure borne sound power level.

References

[1] Thompson, D.J., Jones, C.: A review of the modeling of wheel/rail noise generation. Journal of Sound and Vibration 213(3), 519–536 (2000)
[2] Wallrapp, O.: Standardization of flexible body modeling in multibody system codes, Part 1: Definition of standard input data. Mech. Struct. Mach. 22(3), 283–304 (1994)
[3] Möser, M., Müller, G.: Handbook of Engineering Acoustics. Springer, Heidelberg (2009)
[4] Kaiser, I., Heckmann, A., van der Linden, F.: On an ALE-approach for rotating elastic structures. In: Proc. of Multibody Dynamics 2007, ECCOMAS Thematic Conference, Milano (2007)
[5] Ripke, B.: Hochfrequente Gleismodellierung und Simulation der Fahrzeug-Gleis-Dynamik unter Verwendung einer nichtlinearen Kontaktmechanik. VDI-Verlag, Düsseldorf (1995)
[6] Pierce, A.: Acoustics. McGraw-Hill, New York (1981)
[7] Carrarini, A.: Acoustic postprocessing for multibody simulations including flexible bodies. In: Proc. of Multibody Dynamics 2009, ECCOMAS Thematic Conference, Warsaw (2009)

Influence of Subway LIM Train Wheel Polygonization on the Vibration and Sound Radiation Characteristics of It

J.Y. Fang, X.B. Xiao, L. Wu, and X.S. Jin

State Key Laboratory of Traction Power, Southwest Jiaotong University,
Chengdu 610031, China
Tel.: +86-28-87634355, Fax: +86-28-87600552
e-mail: xsjin@home.swjtu.edu.cn

Summary

This paper presents a combined FEM-BEM acoustic prediction model to calculate the effect of polygonal wheels on the vibration and sound radiation of subway LIM train wheel. In the analysis procedure a three-dimensional model for the wheel is firstly put forward and the "Block Landzos" method is used to calculate the natural frequencies of the wheel. Secondly the method of modal superposition is used to determine the dynamical response of the wheel to the vertical excitation of the running surface roughness. Finally the velocity response of the wheel boundary is treated as the boundary input of the boundary model of the wheel sound radiation which calculates the sound radiation of the wheel. The numerical analysis considers the effect of the excitation of the polygonal wheels at the different wear stages on the vibration and acoustic radiation characteristics of the wheels, in which the data regarding the polygonal wear of the wheels are from the site measurements. The numerical results obtained show that the polygonal wheels have a significant influence on the wheel noise radiation. They can provide the important references in the design of the measures against rolling noise at the source by maintaining the surface conditions of the wheel and rail.

1 Introduction

With the increase of train running speeds, especially the continuous rapid growth of busy traffic on modern complex traffic systems such as elevated railroads, underground railways and surface light rails, the problem of noise generated from the traveling train and the track structure has been a matter of increasing importance in the recent years. The noise produced by railway traffic can be classified into different types according to their different sources, mechanisms, and effects. Provided that the train speed is less than about 300 km/h, wheel/rail noise is an

T. Maeda et al. (Eds.): Noise and Vibration Mitigation for Rail Trans. Sys., NNFM 118, pp. 117–124.
springerlink.com

important source that mainly contributes to the overall noise of the railway traffic. The wheel/rail noise is usually classified into three types: rolling noise caused by small amplitude irregularities on the running surfaces of the wheel/rail, squeal noise due to the wheel curving, and impact noise due to the wheel passing the track irregularities with large amplitudes, such as rail joints turnouts, and defects on the running surfaces of the wheel/rail [1].

The rolling noise source of wheel and rail, which generally is the main contributor of the overall wheel-rail noise on tangent track, is attributed to the geometry irregularities of wheel and rail, such as running surface roughness, wheel burns, irregular welds and joints, and rolling contact fatigue defects ("squats" and "shells"), wheel out-of-roundness, etc[2].

The polygonal phenomenon of railway vehicle wheels is often called wheel corrugation, or wheel harmonic wear, or wheel periodic out-of-roundness (OOR). The mechanism of this phenomenon has not been explained perfectly, and further investigation on this subject is still underway. Such surface defect growth is responsible for an evident increase of the noise induced by the wheel/rail interaction, especially at certain train speeds[3]. A large number of polygonal wheels were recently detected on linear induction motor (LIM) trains on one of the subway lines in China. Roughness measurements, using three measuring probes to register nine main peaks and troughs around the wheel circumference, have been performed on many relevant subway wheels. As a part of an in-situ test from August 2008 to July 2009, highly accurate OOR measurements were performed on both worn and re-profiled subway wheels. The test results showed that an initial irregularity, dominating with a fourth-order polygonal form, present on wheels. This irregularity was believed to be induced by the resonance of the machine tool during re-profiling. After four months operation after the re-profiling, these wheels began to present the rudiment of a nine-order polygonal wear on their treads.

In this paper, the vibration and noise radiation characteristics of the polygonal wheel at different harmonic wear stages are analyzed. A new combined numerical procedure based on a combination of finite element method and boundary element method is used to analyze the dynamic response and noise radiation of the wheel at the different stages of harmonic wear. This new method allows for the real 3D geometry size of the wheel, and also takes into account the three-dimensional structure vibration and sound field. In the procedure, several commercial softwares, such as Hypermesh, MSC.Pantran/Nastran and LMS.Sysnoise, are combined for the calculation. In the dynamic response calculation, the wheel is excited at the nominal contact point by the combination of wheel and rail roughnesses. The velocity response is used as a velocity boundary condition in the acoustic boundary element model (BEM) to predict the sound radiation. Other results under unit force excitation are not described in detail in this paper. In the model the vertical interaction of wheel and rail is considered, and the effect of the contact patch filtering on the roughness of the running surfaces of the wheel and rail is taken into account. The wheel/rail interaction is based on the TWINS model, and the rail is also modeled by the FEM-BEM sound radiation model.

2 Vibro-Acoustic Model of Railway Wheel

2.1 FE and BE Models for Railway Wheel

This paper focuses only on the wheel noise, without consideration of the mutual influence of other factors. The cross-section of a standard passenger wheel with straight web is shown in Fig.2. It has a nominal diameter of 730 mm. The wheel is meshed using the commercial software Hypermesh. The FEM model for the wheel uses a dense mesh of quadratic hexahedral elements; see Fig. 1(a).

The mesh sizes of the boundary element model affect the accuracy of the numerical results. The largest element size should be less than or equal to one-sixth of the wavelength corresponding to the maximum analysis frequency. Too small mesh sizes in a local region cannot improve the analysis accuracy, since the analysis accuracy using a fluid model is, in general, controlled by a majority of elements in the mesh [4]. So the acoustic mesh needs to be of reasonable sizes. The boundary element mesh of the wheel is shown in Fig.1(b). The hole of the wheel hub needs to be closed using additional boundary elements with zero vibration amplitude. By this means, acoustic leakage through this hole, which would not be present in reality due to the presence of the axle, is prevented [5].

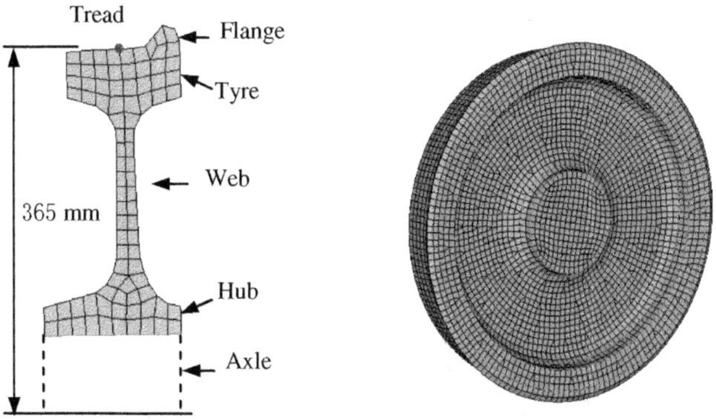

(a) FE mesh of wheel (b) BE mesh of wheel

Fig. 1. The FE model and the BE model of wheel (color figure online)

3 Numerical Results and Discussions

3.1 Measured Data

Using the combined vibration-acoustic FEM-BEM model of wheel, wheel noise radiation is determined using measured data regarding the polygonal wheels.(vehicle 04A024)

In Fig. 2, the wheel profiles of the left wheel of the leading wheelset at different times are shown. Curve1 indicates the harmonic wear before the wheel re-profiling.

On August 21st, 2008 the wheels were re-profiled. Curves 2 to 6 illustrate the harmonic wear at different times after re-profiling. From the measured data of the wheel harmonic wear, it is found that the re-profiling makes the fourth-order polygonal wheel from which the ninth polygonal wheel evolves in about a month after the re-profiling. The wheel reprofiling can efficiently remove the harmonic wear on the wheel tread. A month later after the profiling, the fourth-order polygonal still existed on the treads of the wheels. But, four months later it disappeared and the wheels presented the rudiment of the ninth-order polygonal wear on their treads, as shown in Fig.2(b). 7 months later after the re-profiling, namely on July 28, 2009, the measured data show that the maximum wear depth reached 0.75 mm and occurred on wheel 2, as shown in Fig.2(a).

In the numerical calculation of the combination of the wheel and rail roughness, it is modified by the contact filter and rail roughnesses are from the site measurements.

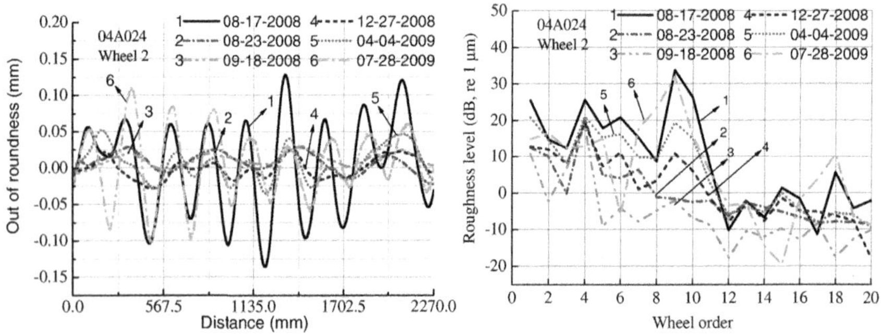

(a) The harmonic wear of the wheel (b) Wheel order analysis

Fig. 2. Measured data of left wheel of the leading wheelset (color figure online)

3.2 Vibration of Railway Wheels

In the numerical calculation, the inner edge of the wheel hub is fixed. This means that the influence of the wheel axle on noise radiation is neglected. Although this approximation cannot accurately predict 0 and 1-nodal-diameter modes, which are strongly coupled with the flexure of the axle, these modes have higher modal damping and are less significant in the vibration response and hence noise radiation. Nevertheless, it is important in wheel/rail squeal noise generation that the rigid-body modes of the whole wheelset are included in these modes [6].

Fig. 3 shows the vibration results (velocities) at the typical points on the surface of the wheel at the different stages of wheel harmonic wear when the vehicle runs on the tangent track at the speed of 90 km/h under roughness excitation at the nominal contact point. The velocities are normal to the surface at these points.

Comparing curve 1 and curve 2, it is shown that the wheel re-profiling can reduce wheel vibration by 4~7 dB. From Curve 2 to Curve 4(from 08-23-2008 to 12-27-2008) after the wheel re-profiling, the dynamic response of the wheel for 12-27-2008 is a little fiercer than that for 08-23-2008. This is because the wheels are in the transition from fourth-order wheel polygonal to ninth-order polygonal for the period. The figures also show the increase of ninth-order harmonic wear increases the vibration response amplitude of the wheel quickly from 12-17-2009 to 07-28-2009, as shown in Fig. 3.

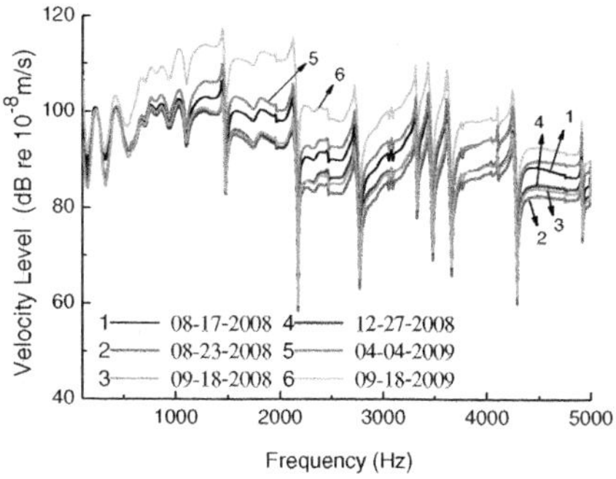

(a) Response at wheel tread

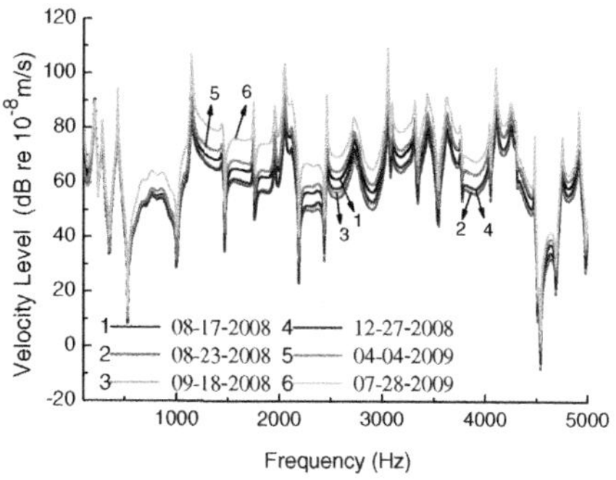

(b) Response at wheel rim

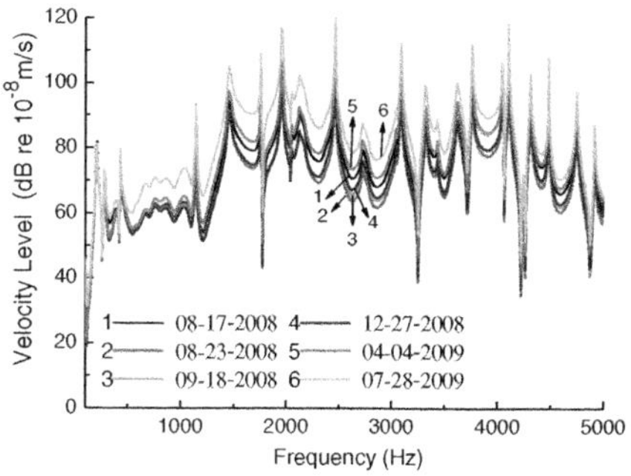

(c) Response at wheel web

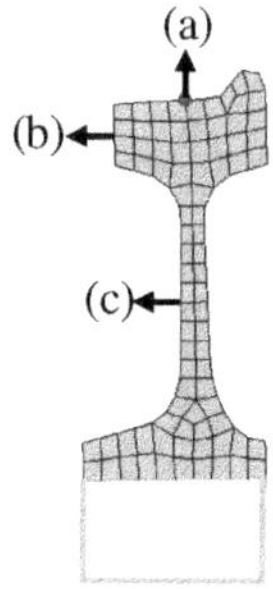

(d) Response location

Fig. 3. Wheel Velocity Level (color figure online)

3.3 Sound Radiation Power of the Wheel

Fig 4 shows sound radiation power level of the wheel at the different stages of wheel harmonic wear when the vehicle runs on the tangent track at the speed of 90 km/h. Curve 1 indicates the sound power level of the wheel before re-profiling. Curves 2~6 illustrate the sound radiation power level of the harmonic wear wheel at different stages after its re-profiling. We can see that the distribution of the sound power level resembles that of the vibration response.

From 08-17-2008 to 07-28-2009, the overall sound radiation power level of the wheel at different stages are 88.5 dB (A), 82.5 dB (A), 83.9 dB (A), 84.4 dB (A), 90.7 dB (A) and 97.6 dB (A), respectively.

We can see that the polygonal wheels play a key role in the wheel noise radiation. The re-profiling of wheel can efficiently mitigate noise radiation of the wheel by 6.0 dB(A). The overall wheel noise level increases with the increasing level of roughness on the wheel, as shown in Fig 4, from curve 4 to curve 6.

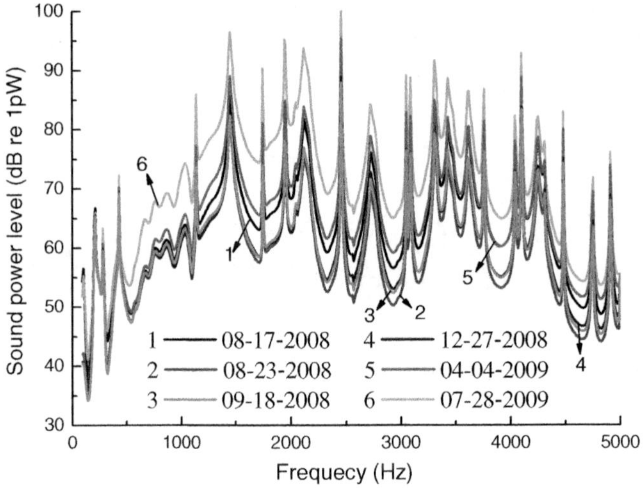

Fig. 4. Sound radiation power level of the wheel at the different stages of wheel harmonic wear (color figure online)

3.4 Sound Pressure Level at Standard Field Points

According to the international railway noise test criterion ISO 3095:2005[7], the sound pressure levels at the standard points are listed in Table 1. The results show that the overall wheel noise level increases with the increasing level of roughness on the wheel. The distribution of the sound pressure level at standard field points resembles that of the vibration response. From 08-23-2008 to 12-27-2008 after the wheel re-profiling, the sound pressure level at standard field points for 12-27-2008 is a little higher than that for 08-23-2008. This is because the wheels are in the transition from fourth-order wheel polygonal to ninth-order polygonal for the period. The depth of the harmonic wear of the wheel grows slowly in this stage.

Table 1. Sound pressure level at standard field points

	Position 1	Position 2	Position 3
	D: 7.5m, H: 1.2m	D: 7.5m, H: 3.5m	D: 25m, H: 3.5m
08-17-2008	65.6 dB(A)	56.8 dB(A)	54.5 dB(A)
08-23-2008	59.2 dB(A)	51.5 dB(A)	48.1 dB(A)
09-18-2008	60.7 dB(A)	52.6 dB(A)	49.6 dB(A)
12-27-2008	61.0 dB(A)	52.8 dB(A)	50.0 dB(A)
04-04-2009	67.1 dB(A)	59.1 dB(A)	56.0 dB(A)
07-28-2009	74.3 dB(A)	66.4 dB(A)	63.1 dB(A)

The table also shows that the wear increase of ninth-order harmonic wear increases the sound pressure level of the wheel quickly from 12-17-2008 to 07-28-2009. All the pressure results are predicted.

The wheel sound pressure level at the standard point 1 (Distance: 7.5 m, Height: 1.2 m), which is typically used to characterise the wheel-rail rolling noise contribution, is shown in detail in Fig.5.

At low frequencies (<500Hz), the harmonic wear of the wheel has little effect on the acoustic radiation of the wheel. However, in high-frequency bands(>500Hz), particularly between 1500 and 5000 Hz when the wheel/rail noise is dominated by wheel noise radiation, the polygonal wheels have a major influence on the wheel noise radiation.

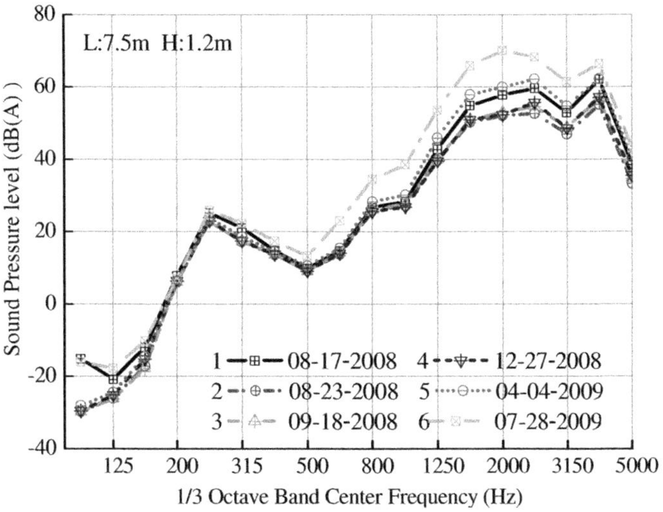

Fig. 5. Sound pressure level at standard point 1 (color figure online)

4 Conclusion

The paper presents a combined FEM-BEM acoustic prediction model to calculate the effect of polygonal wheel on the vibration and sound radiation of subway LIM train wheels.

The re-profiling of wheel can efficiently remove the harmonic wear on the wheel tread. From the measured data of the wheel harmonic wear, it is found that the re-profiling makes the fourth-order polygonal wheel from which the ninth polygonal wheel evolves in about a month after the re-profiling.

The numerical results obtained show that the polygonal wheels have a major influence on the wheel noise radiation. The vibration and acoustic radiation of the wheel change with the depth of the harmonic wear of the wheel, and the wheel velocity level, sound power level and sound pressure level depend mainly on the wavelength and depth of the harmonic wear of the wheel on the wheel circumference. After the re-profiling, the wheel noise radiation mitigates greatly. They can provide the important references for the reduction of rolling noise at the source by maintenance of the surface conditions of the wheel and rail. The numerical results obtained need to be examined by field measurements in the future work.

References

[1] Thompson, D.J., Jones, C.J.C.: A review of the modeling of wheel/rail noise generation. Journal of Sound and Vibration 231(3), 519–536 (2000)
[2] Thompson, D.J.: Railway Noise and Vibration, 1st edn. Elsevier, Amsterdam (2009)
[3] Barke, D.W., Chiu, W.K.: A review of the effects of out-of-round wheels on track and vehicle components. Part F: Rail and Rapid Transit. 219, 151–175 (2000)
[4] Li, Z.: Detailed SYSNOISE Rev 5.6. National Defense Industry Press, Beijing (2005) (in Chinese)
[5] Thompson, D.J., Jones, C.J.C.: Sound radiation from a vibrating railway wheel. Journal of Sound and Vibration 253(2), 401–419 (2002)
[6] Thompson, D.J.: Wheel-rail noise generation, Part II: wheel vibration. Journal of Sound and Vibration 161(3), 401–420 (1993)
[7] International Standard. ISO 3095: 2005 Railway Applications. Acoustics Measurement of noise emitted by rail bound vehicles. International Organization for Standardization (2005)

Dynamic Models of Railway Track Taking Account of Cross-Section Deformation at High Frequency

Jiannan Yang[1], David Thompson[1], and Atul Bhaskar[2]

[1] Institute of Sound and Vibration Research, University of Southampton, Southampton SO17 1BJ, UK
Tel.: +44 (0) 23 8059 2294, Fax: +44 (0) 23 8059 3190
e-mail: jny1e08@soton.ac.uk
[2] School of Engineering Sciences, University of Southampton, Southampton, SO17 1BJ, UK

Summary

Track dynamic behaviour is important for the prediction of the rolling noise, corrugation growth and track damage. Various models have been developed in the literature but problems still remain. On the one hand, analytical models become insufficient because of the need to include cross-section deformation at high frequencies. On the other hand, FE models are straightforward but the truncation of the infinite length is unavoidable. A new tapered plate rail model is developed in this paper. This model takes into account all the main motions required for frequencies below 7 kHz. The rail head is represented by a rectangular beam, the web by a plate of constant thickness and the foot by a plate of variable thickness. The out-of-plane and in-plane motions of the plates are approximated using cubic and linear functions, respectively, based on the relevant wave speeds. Freely propagating waves in the rail are studied by means of Hamilton's principle. Comparing the results in terms of the dispersion relations, the tapered plate rail model shows good agreement with an FE model. Comparison with simpler beam models confirms the improvements at high frequencies due to the taper of the foot.

1 Introduction

Railway rolling noise induced by the roughness of the rail and wheel surfaces forms the principal source of noise from railway operation [1]. These surface irregularities produce dynamic interaction forces between the wheel and rail. The resulting high frequency vibrations are transmitted into both wheel and track and then the sound is radiated from the vibration of the structures.

Dynamic models are required to predict the noise emitted from the track and frequencies up to at least 5 kHz should be considered [2]. Below about 1500 Hz, a

T. Maeda et al. (Eds.): Noise and Vibration Mitigation for Rail Trans. Sys., NNFM 118, pp. 125–133.
springerlink.com

single Timoshenko beam is sufficient to represent the rail dynamic behaviour [3]. However, it is more difficult to study the dynamic behaviour of railway track at high frequencies due to cross-section deformation of the rail as found by Thompson [4]. The vertical vibration is dominated by the rail foot flapping and the lateral motion is even more complicated.

Most of the models taking account of cross-section deformation are based on finite element methods (FEM). Thompson [4] developed an FE model of a finite length of rail using beam elements for the head and plates for the remaining parts. Since such a model only allows frequencies to be predicted for a given wavenumber, this is not enough to study the frequency response of the track. To improve the application of FE methods to an infinite rail, the finite strip method (FSM) and some other derivatives of FEM have been developed, e.g. [5]. The deformations of regularly discretized elements in the infinite rail are connected, by considering harmonic waves in the longitudinal direction. Gavric [6] and Gry [7] used an alternative approach based on factorization of the function describing the displacement field. A general shape function is used to describe the deformation of the cross-section, while along the rail axis wave propagation is assumed.

To avoid the large number of degrees of freedom of FE methods, simplified beam models were developed by Wu and Thompson [8-9], where the rail head and foot were represented by infinite Timoshenko beams. The web was modelled as a spring in the vertical direction or by an array of beams for the lateral direction. The vertical and lateral vibrations of the rail were studied separately based on the assumption of symmetry of the cross-section.

Foot deformation at high frequency is significant and an analytical model which took account of the variable thickness of the rail foot was developed by Bhaskar *et al.* [10]. However, little information is given for this model. In addition, in-plane motions of the plates were approximated by simple beam bending. This requires improvement because they are quite deep beams. Moreover, the stretching of the web in the direction normal to the wave propagation was neglected to simplify the model, which could result in considerable discrepancies.

Therefore, a tapered plate rail model will be implemented in this paper. It will be compared with both a 3D FE model and the simple beam models from [8-9].

2 Tapered Plate Model

The cross-section of the model (UIC60 rail) is shown in Fig. 1. This model takes into account all the main motions required for frequencies below 7 kHz. The rail head is represented by a rectangular beam, the web by a plate of constant thickness and the foot by plates of variable thickness. The motion of the rail head can be characterized by beam bending in two directions, torsion and compression in the longitudinal direction. The deformation of the web and foot can be described by plate out-of-plane bending and twisting, and in-plane stretching and shear. In the rail axis (z direction), harmonic waves are assumed of the form e^{-jkz} at frequency ω (assumed time dependence $e^{j\omega t}$).

The variational principle is employed to determine unknown deflections in order to avoid solving the differential equations of plate theory. The out-of-plane and in-plane motions of the plates are approximated using cubic and linear functions, respectively, in the x–y plane. A total of 17 degrees of freedom are used to represent the cross-section. The components are assembled with the stiffness matrix formulated in terms of the wavenumber in the axial direction. The resulting eigenvalue problem is solved to find the dispersion characteristics and the propagation modes for the freely propagating waves in the rail.

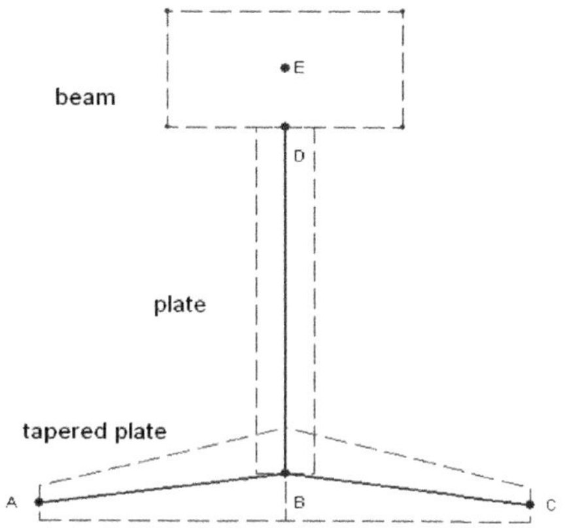

Fig. 1. Mechanical idealization of the theoretical rail model. A, B, C and D are the four nodal points which are located in the middle plane of each plate. Point E represents the centre of the beam and the dashed lines represent the actual shape of the beam and plates. (color figure online)

2.1 Foot Modelling Using a Tapered Plate and Web with Constant Thickness

The foot element AB is shown in Fig. 2 where the cross-section is defined in the x–y plane. The local coordinate ξ_1 denotes the distance from point B. The deflections in the x, y and z directions are written as u, v and w, respectively. The variable thickness can be expressed as $t_f(\xi_1) = S_f\xi_1 + t_f^B$, where S_f is the slope and t_f^B is the thickness at point B. The thickness at the point A is thus $t_f^A = S_f h_f + t_f^B$ where h_f is half of the total foot length.

The vibration energies of plate AB consist of both out-of-plane and in-plane motion, and can be written as

$$U_{ABi} = \frac{1}{2}\iint \left\{ \frac{Et_f^3(\xi_1)}{(1-\nu^2)} \left(\varepsilon_x^*\varepsilon_x + \varepsilon_z^*\varepsilon_z + \nu(\varepsilon_x^*\varepsilon_z + \varepsilon_x\varepsilon_z^*) \right) + G\gamma_{xz}^*\gamma_{xz} \right\} d\xi_1 dz, \tag{2.1}$$

$$T_{ABi} = \frac{1}{2}\iint \rho t \left(\dot{u}^*\dot{u} + \dot{w}^*\dot{w} \right) d\xi_1 dz, \tag{2.2}$$

$$U_{ABo} = \frac{1}{2}\iint D\left\{v^*_{,\xi_1\xi_1}v_{,\xi_1\xi_1} + v^*_{,zz}v_{,zz} + 2(1-\nu)v^*_{,\xi_1 z}v_{,\xi_1 z} + \nu(v^*_{,\xi_1\xi_1}v_{,zz} + v_{,\xi_1\xi_1}v^*_{,zz})\right\}d\xi_1 dz, \quad (2.3)$$

$$T_{ABo} = \frac{1}{2}\iint \rho t\dot{v}^*\dot{v}\,d\xi_1 dz, \quad (2.4)$$

with U for potential energy, T for kinetic energy and subscripts i, o represent in-plane and out-of-plane, respectively. Here E is the Young's modulus, G is the shear modulus, ν is Poisson's ratio, $\varepsilon_x = \frac{\partial u}{\partial \xi_1}$ and $\varepsilon_z = \frac{\partial w}{\partial z}$ are normal strains and $\gamma_{xz} = \frac{\partial u}{\partial z} + \frac{\partial w}{\partial \xi_1}$ is the shear strain. $D = \frac{Et_f^3(\xi_1)}{12(1-\nu^2)}$ is the plate bending stiffness, and the subscripts z and ξ_1 denote the derivatives with respect to z and ξ_1.

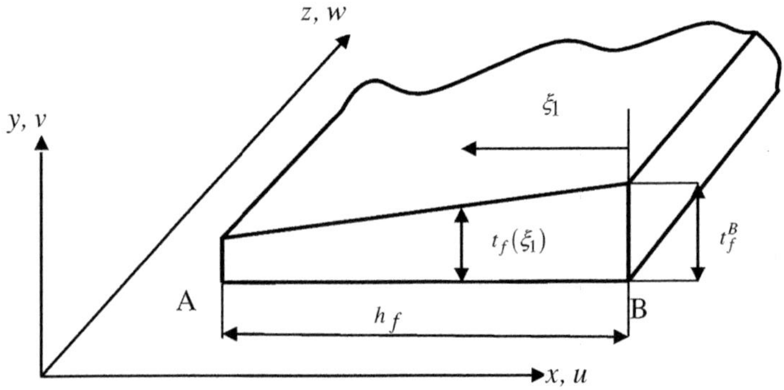

Fig. 2. Schematic diagram of foot AB showing the relevant dimensions and the co-ordinate axes.

The highest derivatives appearing in the energy expressions are second and first, respectively, for plate out-of-plane and in-plane motion. At each node, it is thus necessary to take displacement and rotation as degrees of freedom for the bending motion. Since only the first cantilever mode is of interest, the plate out-of-plane bending deformation can be approximated by a polynomial having four constants, that is, a cubic function. For the in-plane motion, the minimum requirement is a linear function. The shape functions are therefore

$$v_{AB}(\xi_1, z, t) = \left(\sum_{i=1}^{4} a_i \xi_1^{i-1}\right) e^{j(\omega t - kz)}, \quad (2.5)$$

$$u_{AB}(\xi_1, z, t) = (d_1 + d_2\xi_1)\, e^{j(\omega t - kz)}, \quad (2.6)$$

$$w_{AB}(\xi_1, z, t) = (e_1 + e_2\xi_1)\, e^{j(\omega t - kz)}. \quad (2.7)$$

For the rail web, it is quite convenient to put the plate slope equal to zero. Although the web is quite long compared with the foot, for a steel plate with constant thickness, at 5000 Hz the wavelengths of longitudinal and shear waves are about 1.1 m and 0.63 m, respectively. Therefore, one element of length 0.11 m is still sufficient to satisfy the requirement of at least 6 elements per wavelength.

2.2 Rail Head Represented by a Beam

The rail head can be modelled with a rigid cross-section because of its high stiffness. Its dimensions are small compared with the wavelength of shear or longitudinal waves in the frequency range of interest. At around 5000 Hz, the minimum vertical and lateral bending wavelengths are about 0.35 m and 0.27 m from an FE analysis (see [11]), which are about 9 and 4 times the corresponding dimension of rail head. Although the rail behaviour actually becomes mainly dominated by foot vibration, shear deformation and rotational inertia effects become important for the lateral motion at high frequency. Therefore, it is sufficient to use the elementary beam theory for the head for the vertical motion but Timoshenko beam theory has to be employed for the lateral motion.

$$U_h = \frac{1}{2}\int EI_y u^*_{E,zz} u_{E,zz} dz + \frac{1}{2}\int EI_x \phi^*_{E,z}\phi_{E,z} dz + \frac{1}{2}\int \kappa AG\left(v_{E,Z}-\phi_E\right)^*\left(v_{E,Z}-\phi_E\right)dz$$
$$+\frac{1}{2}\int GJ\theta^*_{E,z}\theta_{E,z}dz + \frac{1}{2}\int EAw^*_{E,z}w_{E,z}dz\,, \tag{2.8}$$

$$T_h = \frac{1}{2}\int \rho A\dot{u}^*_E \dot{u}_E dz + \frac{1}{2}\int \rho A\dot{v}_E^{\,*}\dot{v}_E dz + \frac{1}{2}\int \rho I_z \dot{\phi}^*_E \phi_E dz + \frac{1}{2}\int \rho A\dot{w}^*_E \dot{w}_E dz$$
$$+\frac{1}{2}\int \rho J\dot{\theta}_E^{\,*}\dot{\theta}_E dz\,, \tag{2.9}$$

where A is the area of the cross-section of rail head, J is the torsion constant of the cross-section, θ_E is the rotation of the rail head, I_x is the second moments of area of the cross-section about x-axis of the rail, but I_y is the second moments of area of the head base axis (the longitudinal axis where the point D is located). κ is called the Timoshenko shear coefficient and κ=5/6 is used here. ϕ_E describes the rotation of the head cross-section about the vertical.

2.3 Freely Propagating Waves in the Rail

By applying the compatibility at points B and D, the total number of degrees of freedom is 17. The following equation is obtained by applying Hamilton's principle based on a unit length of rail in the integral:

$$\left[(-jk)^4\mathbf{K}_4+(-jk)^2\mathbf{K}_2+(-jk)\mathbf{K}_1+\mathbf{K}_0-\omega^2\mathbf{M}\right]\{\mathbf{q}\} = \{\mathbf{0}\} \tag{2.10}$$

The solutions to this equation represent waves in the free rail without support. To solve this generalized polynomial eigenvalue problem in k for a given frequency, it is convenient to convert it into state space form

$$[\mathbf{A} - jk\mathbf{B}]\begin{Bmatrix}\mathbf{q}_0 \\ \mathbf{q}_1 \\ \mathbf{q}_2 \\ \mathbf{q}_3\end{Bmatrix} = \{\mathbf{0}\}, \tag{2.11}$$

where $\mathbf{A}=\begin{bmatrix}\mathbf{0} & \mathbf{I} & \mathbf{0} & \mathbf{0} \\ \mathbf{0} & \mathbf{0} & \mathbf{I} & \mathbf{0} \\ \mathbf{0} & \mathbf{0} & \mathbf{0} & \mathbf{I} \\ -(\mathbf{K}_0-\omega^2\mathbf{M}) & -\mathbf{K}_1 & -\mathbf{K}_2 & \mathbf{0}\end{bmatrix}$, $\mathbf{B}=\begin{bmatrix}\mathbf{I} & \mathbf{0} & \mathbf{0} & \mathbf{0} \\ \mathbf{0} & \mathbf{I} & \mathbf{0} & \mathbf{0} \\ \mathbf{0} & \mathbf{0} & \mathbf{I} & \mathbf{0} \\ \mathbf{0} & \mathbf{0} & \mathbf{0} & \mathbf{K}_4\end{bmatrix}$ are both 68×68 matrices (**I** is the identity matrix and **0** is a matrix of zeros), and $\{\mathbf{q}_i\}=(-jk)^i\{\mathbf{q}\}$, i=0, 1, 2, 3.

3 Results

The symmetric and anti-symmetric waves of a free rail are shown separately in Figs 3 and 4 since they do not couple with each other due to the symmetry of the cross-section. A short length of free UIC60 rail has been modelled using ANSYS finite element package and the results are used as a reference. The dispersion relations from simplified beam models [8-9] are also presented as comparison.

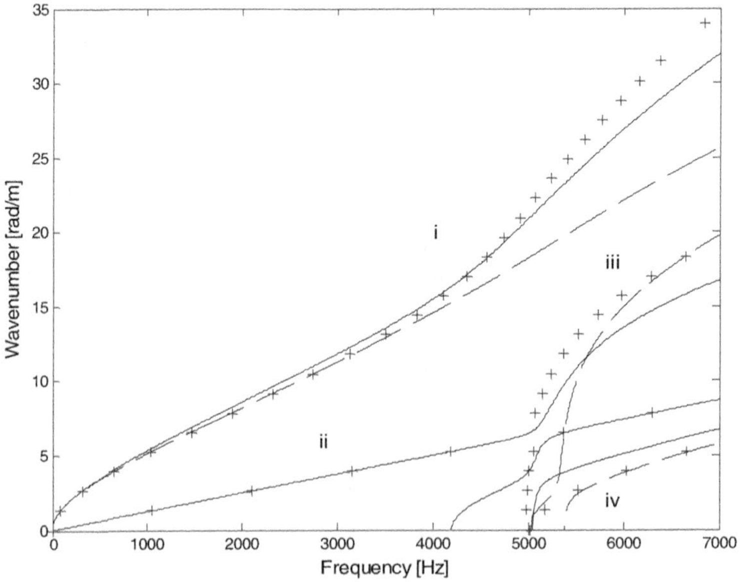

Fig. 3. Dispersion relation for vertical/longitudinal motion of free rail, —— , tapered plate model; - - - - , simplified beam model; + + +, the original FE model.

From these two figures, it can be seen that both analytical models give good results compared to FE analysis. However, the improvements brought by the tapered plate model are still significant especially at high frequencies.

For the vertical motion, the foot flapping mode (iii) which cuts on about 5 kHz is well represented by both plate and beam models, but the tapered plate model gives much better representation for the vertical bending wave (i) for frequency higher than 4 kHz. It is found from the mode shapes (not shown) that the vertical bending wave is dominated by the deformation of the rail foot at high frequency. Although the beam model takes into account the foot flapping by modelling the foot as a separate beam, it cannot represent the deformation of the foot itself. In addition, the double beam model does not give the longitudinal waves (ii) at all because there is no degree of freedom in the rail direction assigned to this model. It should be noted that the wave that cuts on at about 5 kHz from simple beam model is actually the warping behaviour of one of the beams, but not the second-order longitudinal wave (iv) where the head and the foot move in anti-phase.

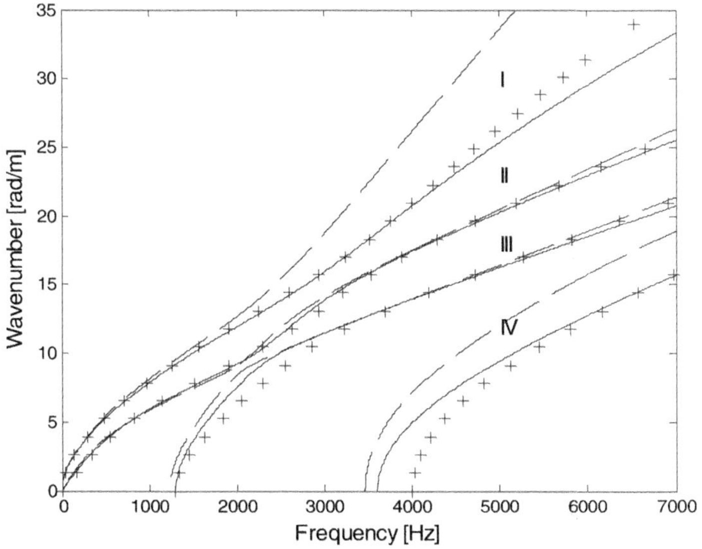

Fig. 4. Dispersion relation for lateral motion of free rail.——, tapered plate model; - - - - , simplified beam model; + + +, the original FE model.

For the lateral motion, the torsion (II) and simple web bending (III) modes are well represented by both analytical models but for the lateral bending wave (I), these two models deviate from about 2 kHz. It is found from the mode shapes from the FE analysis that above 2 kHz the lateral bending wave is dominated by the rocking of the foot. In the beam model, a constant thickness beam is used for the rail foot. As found by Yang [11] this assumption makes the lateral bending motion much less stiff than the real structure at high frequency. However, the tapered plate model makes a big improvement on this by including the taper of the rail foot. The remaining small difference from the FE model is due to the fact that the real structure has a more complicated tapered foot. The discrepancy for the double web bending wave (IV) is also due to the foot modelling.

4 Conclusions

A new analytical model of a rail with a tapered foot is presented. The rail head, web and foot are constructed using a beam, a plate with constant thickness and a tapered plate, respectively. Cubic shape functions are used to approximate the plate out-of-plane motions and linear shape functions are used for the in-plane motions. The head beam is represented by an Euler beam vertically and a Timoshenko beam laterally. The potential and kinetic energy of each part are found first using stress-strain relations. Then the whole structure is assembled and the dispersion relations of the free rail are found using Hamilton's principle. It can be seen that this tapered plate model gives good agreement with an FE model.

In addition, it is clear from the comparison with the simplified beam models that big improvement on this is achieved at high frequency by including the taper of the rail foot. First, it gives more complete modelling of the rail by considering all possible motions up to 7 kHz. Second, at high frequency, the simplified beam models tend to underestimate the vertical response and to overestimate the lateral response. As the tapered plate model considers more details of the rail, more parameters can be varied to optimize the rail design for noise reduction purposes.

To use the model to predict noise, it is necessary to include the support structure (ballast, sleepers and railpads) and to couple the model to a prediction of acoustic radiation. This is beyond the scope of the paper. However, it can be expected that the present model will give improved results, particularly for the decay rate of vibration along the rail.

References

[1] Thompson, D.J.: Railway Noise and Vibration. In: Mechanisms, Modelling and Means of Control, p. 506. Elsevier Science, Oxford (2008)

[2] Thompson, D.J.: Wheel-rail noise: theoretical modelling of the generation of vibrations, PhD thesis, University of Southampton (1990)

[3] Grassie, S.L., Gregory, R.W., Harrison, D., Johnson, K.L.: Dynamic response of railway track to high frequency vertical excitation. Journal of Mechanical Engineering Science 24(2), 77–90 (1982)

[4] Thompson, D.J.: Wheel-rail noise generation, Part III: Rail vibration. Journal of Sound and Vibration 161(3), 421–446 (1993)

[5] Knothe, K., Strzyzakowski, Z., Willner, K.: Rail vibrations in the high frequency range. Journal of Sound and Vibration 169(1), 111–123 (1994)

[6] Gavric, L.: Computation of propagative waves in free rail using a finite element technique. Journal of Sound and Vibration 185(3), 531–543 (1995)

[7] Gry, L.: Dynamic modelling of railway track based on wave propagation. Journal of Sound and Vibration 195(3), 477–505 (1996)

[8] Wu, T.X., Thompson, D.J.: A double Timoshenko beam for vertical vibration analysis of railway track at high frequencies. Journal of Sound and Vibration 224(2), 329–348 (1999)
[9] Wu, T.X., Thompson, D.J.: Analysis of lateral vibration behavior of railway track at high frequencies using a continuously supported multiple beam model. Journal of the Acoustical Society of America 106(31), 1369–1376 (1999)
[10] Bhaskar, A., Johnson, K.L., Wood, G.D., Woodhouse, J.: Wheel-rail dynamics with closely conformal contact. Part 1: Dynamic modelling and stability analysis. Proceedings of the Institution of Mechanical Engineers, Part F: Journal of Rail and Rapid Transit 211(1), 11–24 (1997)
[11] Yang, J.: Dynamic models of railway track at high frequency. MSc Thesis, ISVR, University of Southampton (2009)

Modelling and Performance Analysis of Rail Vibration Absorber by FE and BE Methods

H.P. Liu and T.X. Wu

State Key Laboratory of Mechanical System and Vibration,
Shanghai Jiao Tong University,
800 Dong Chuan Road, Shanghai 200240, P R China
Tel.: +86 21 3420 6332 Ext 819, Fax: +86 21 3420 5875
e-mail: txwu@sjtu.edu.cn

Summary

A finite element model of the railway track with a rail absorber is proposed for investigating the effects of the absorber's multiple motions on the dynamic behavior of the track. The rail vibration and radiation due to a roughness excitation are calculated with the use of the absorber. The calculation results are compared with those of the beam-spring model. The results show that the radiation efficiency of the rail with the absorber is significantly reduced in the frequency range 300-1000 Hz. The noise reduction calculated by the finite element model is less than that of the beam-spring model in the frequency region of 400—1000 Hz due to the transversal rotation of the absorber. The rail-radiated noise can be reduced by about 7.1 dB(A) using the rail absorber.

1 Introduction

A two-frequency tuned rail absorber can increase the rail vibration decay rate along the track, so it can effectively reduce the railway rolling noise. Field tests [1] showed that the rail noise could be decreased by up to 6 dB(A) using a rail absorber on a track with the rail pads of medium stiffness. Researchers have developed some theoretical models [1-4] for investigating the dynamic behavior of the rail with absorber. Compound track models with continuous or discrete rail absorber were developed in Ref. [2] to study the mechanism of rail absorber, where the absorber was represented by a damped two-degree-of-freedom mass-spring system and the rail was modeled as an infinite Timoshenko beam. With the influences of the rail absorber's bending modes taken into consideration, Ref. [3] proposed a combined model of the railway track with rail absorber, which was treated as a beam-spring system.

In practice, the vibration modes of the absorber are composed of both elastic deformation and rigid body motion. The mass-spring model only considers the

T. Maeda et al. (Eds.): Noise and Vibration Mitigation for Rail Trans. Sys., NNFM 118, pp. 135–142.
springerlink.com

absorber's translational motion in the vertical direction, and the beam-spring model considers both the bending modes and the translational motion of the absorber in the vertical direction. However, the vibration motion of an absorber also includes the transversal rotation of the mass bar, which was not taken into consideration in the above models. In addition, the structural shape of the rail changes due to the absorbers glued to the rail. Consequently, when the rail vibrates, the vibration phase difference between the rail and the mass-bar of the absorber causes changes in the sound radiation behavior.

In this study, first, a finite element model of the railway track with a rail absorber is developed to investigate the effects of the absorber's multiple motions on the dynamic behavior of the track. Then, a boundary element model is developed for the rail with the absorber to calculate its radiation efficiency. Finally, the wheel—rail interaction excited by a roughness input is calculated, in addition to the sound radiation power from the rail according to its vibration energy. The results are compared between the finite element and the beam-spring models.

2 Finite Element Model and Vibration Behavior

A quarter of the railway track with rail absorber is modeled by the element SOLID 95 in ANSYS codes due to symmetry, as shown in Fig. 1. The track model considered is 78 and 31-m-long for the untreated rail and treated rail, respectively, and only vertical vibration is taken into account. The length of the absorber, which is attached one by one along the rail, is 0.6 m. The first and second resonance frequencies of rigid body motion of the absorber are designed to be about 400 Hz and 700 Hz, respectively, as shown in Fig. 2. In the first mode of rigid body motion the upper mass-bar oscillates up and down, while in the second one the bottom mass-bar oscillates. The rail supports are discrete and consist of the rail pad, the sleeper, and the ballast. The rail pad and the ballast are modelled by element COMBIN 14, and the sleeper by element MASS 21.

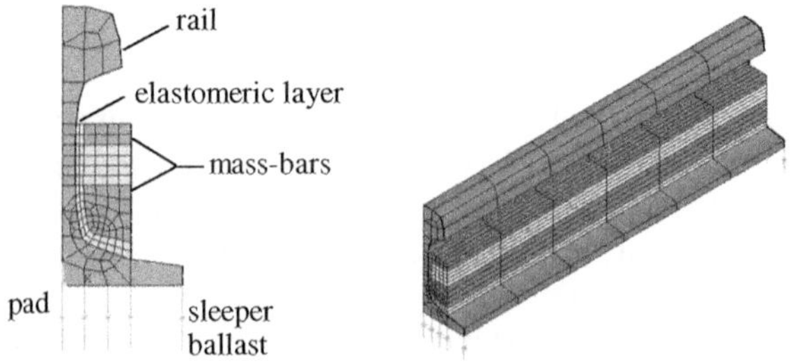

Fig. 1. Mesh of the finite element model for a rail section with absorber.

The parameters of the track absorber used in this paper are listed in Table 1, and they are used to represent a track with UIC 60 rails, stiff rail pads, and concrete monoblock sleepers. The relating active mass and damping parameters of the rail absorber are from Ref. [3].

The calculated dynamic properties of the discretely supported track with and without the absorber are shown in Fig. 3 for a unit harmonic force at mid-span, in terms of the point receptance and decay rate of rail vibration and the vibration energy level of the rail with absorber, upper mass-bar of the absorber. The vibration energy for the finite element model is evaluated by sum $v^2\Delta S$, where v is the magnitude of surface normal vibration velocity of the rail with absorber and ΔS is the area of a surface element. The corresponding result for the beam-spring model is given by sum $hv^2\Delta z$, where h is the total projected width of the section of the rail with absorber in the vertical direction, allowing for the top and bottom of the head and foot, and Δz is the length of a small track section.

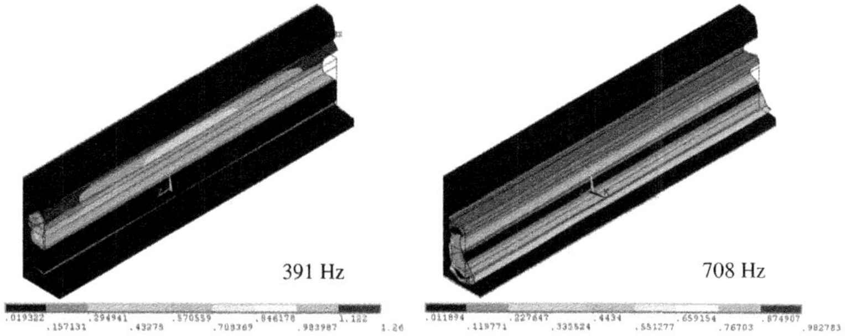

Fig. 2. Selected vibration mode-shapes of 0.6 m long rail with absorber.

Table 1. Parameters used for the railway track including absorber

Rail	Mass per unit length	$\rho_r A_r$	60 kg/m
	Young's modulus	E_r	2.1×10^{11} N/m^2
	Poisson's ratio	v_r	0.3
Support	Pad stiffness	K_p	350 MN/m
	Loss factor	η_p	0.2
	Ballast stiffness	K_b	50 MN/m
	Loss factor	η_b	1.0
	Sleeper mass	M_s	162 kg
	Sleeper spacing	d	0.6 m
Absorber (one side of the rail web)			
Bottom mass-bar	Mass per unit length	$\rho_m A_1$	7 kg/m
Upper mass-bar	Mass per unit length	$\rho_m A_2$	3 kg/m
	Poisson's ratio	υ_m	0.3
Elastomeric layer	Density	ρ_d	930 kg/m^3
	Young's modulus	E_d	5.5×10^6 N/m^2
	Poisson's ratio	v_d	0.499
	Loss factor	η_d	0.25

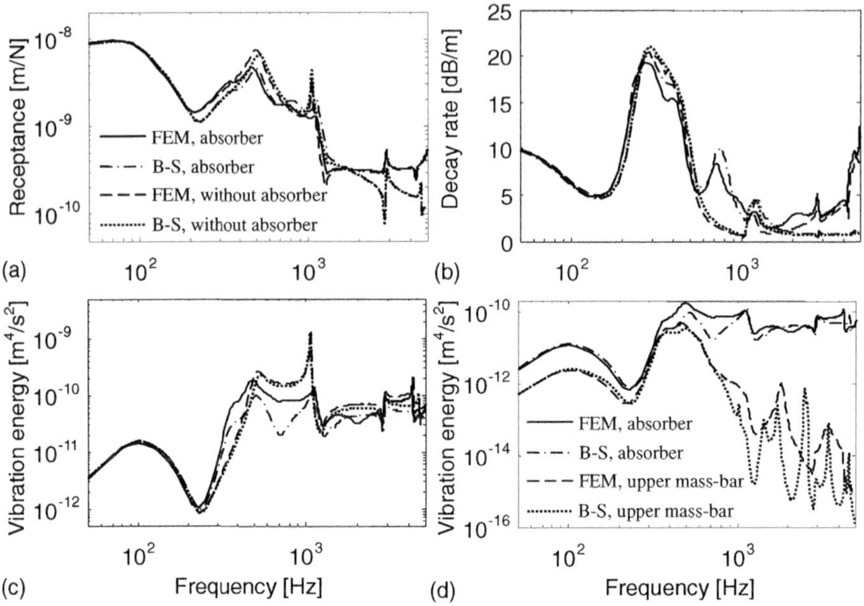

Fig. 3. Vibration receptance, decay rate and vibration energy of the rail with absorber due to a unit force at mid-span, FEM and B-S represent finite element model and beam-spring model, respectively.

Comparisons are made in Fig. 3 between the finite element model and the beam-spring model. It can be seen from Fig. 3(a) the pinned-pinned resonance appears at about 1050 Hz for the track without the absorber, and it is effectively suppressed by the use of rail absorber. The decay rate of rail vibration is estimated over five sleepers from the excitation point, as shown in Fig. 3(b), and can be seen to be moderate at low frequencies and to reach a peak of 18 dB/m around 230 Hz. At high frequencies above 800 Hz, the decay rate is generally lower than 1 dB/m without rail absorber. When the rail absorber is used, the decay rate of rail vibration is significantly increased in the frequency range 500-1200 Hz, with a peak of 8 dB/m appearing at about 700 Hz. However, the decay rate calculated by the beam-spring model is about 2 dB/m higher in this frequency due to ignoring the effects of the transversal rotation of the absorber. At higher frequencies above 2000 Hz, it can be seen that the receptance and decay rate from the finite element model are higher than that from the beam-spring model. The rail vibration energy of the 20-m-long rail section is substantially reduced in the frequency range 400—1200 Hz due to the vibration decay rate enhanced by the rail absorber in the corresponding frequency region, as shown in Fig. 3(c). The vibration energy of the rail with absorber from the finite element model is higher than that from the beam-spring model in the frequency range of 300—900 Hz due to the effects of transversal rotation of the absorber. The energy level of the upper mass-bar of the absorber can be seen from Fig. 3(d) to be higher in frequency region of 300—500 Hz, but it can be seen to be lower at other frequencies.

3 Boundary Element Model and Radiation Efficiency

The radiation efficiency of the rail with the absorber is calculated using ANSYS and SYSNOISE codes. The vibration response of the rail with the absorber from ANSYS is inputted to SYSNOISE to determine the rail radiation efficiency. To reduce the computation time, only a 12-meter-long rail section and its vibration response are taken from ANSYS and used in SYSNOISE. The meshing size of the boundary element model for rail with the absorber, as shown in Fig. 4, is 0.1 m and 0.025 m, in order to meet the requirement of six elements per wavelength of the sound in the air in the frequency regions of 63–1000 Hz and 1000–4000 Hz.

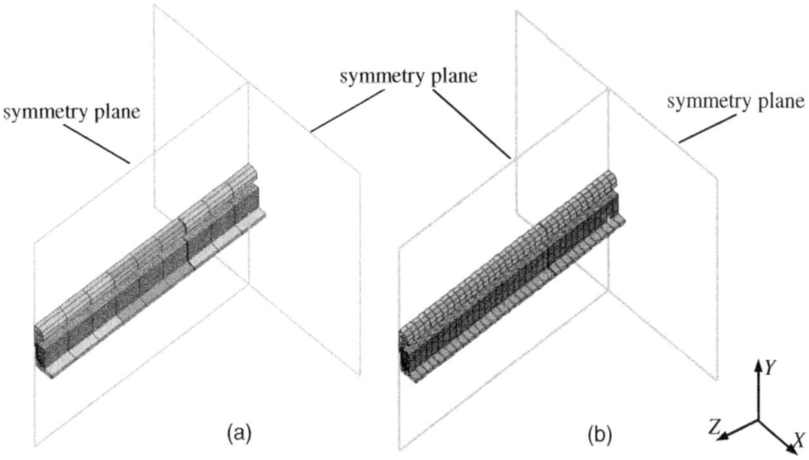

Fig. 4. Mesh of the boundary element models for a rail section with absorber: (a) meshing size 0.1 m, (b) meshing size 0.025 m.

Radiation efficiency is calculated for the rail with and without the absorber. Fig. 5 shows the calculated radiation efficiency for the rail with and without the absorber due to a unit force at mid-span in one-third octave bands. The radiation efficiency of vertical vibration of the rail calculated by Thompson *et al.*[5] is also included. It can be seen that the rail radiation efficiency increases with frequency up to about 1000 Hz and is remarkably lower for the rail with the absorber than for the untreated rail from 300 Hz to 1000 Hz. The reason for this is that the radiated surface of the rail is changed due to the absorbers glued to the rail, and consequently, the phase difference between the rail and mass-bar of the absorber leads to changes in the radiation efficiency of rail vibration. The calculated radiation efficiency of vertical vibration of the untreated rail can be seen to be similar to that obtained by Thompson.

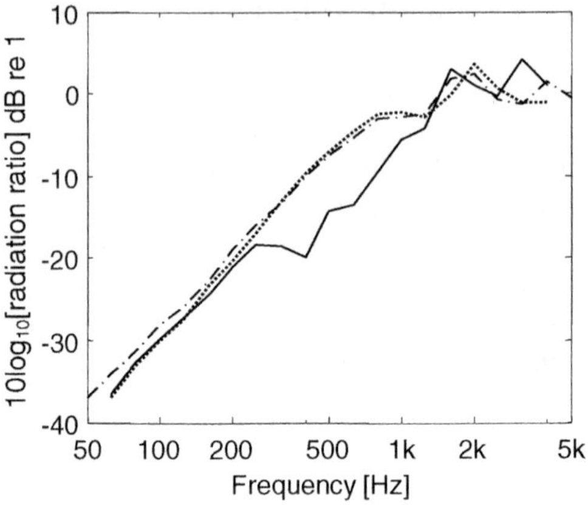

Fig. 5. Calculated radiation efficiency σ of UIC 60 rail with absorber for vertical vibration: — with absorber, ······ without absorber, –·– rail radiation efficiency of vertical motion, from [5].

4 Noise Radiation Due to Roughness Excitation

For a random roughness excitation $R(\omega)$, considering the interaction in the vertical direction only, the wheel-rail contact force $F(\omega)$ is given by

$$F(\omega) = -\frac{R(\omega)}{\alpha^{\mathrm{W}}(\omega) + \alpha^{\mathrm{C}}(\omega) + \alpha^{\mathrm{R}}(\omega)} \tag{4.1}$$

$$\alpha^{\mathrm{C}} = \frac{1}{k_{\mathrm{C}}}, \alpha^{\mathrm{W}} = \frac{1}{K_{\mathrm{M}}} - \frac{1}{(M_{\mathrm{W}}\omega^2)} \qquad (4.2, 4.3)$$

where α^{W}, α^{C}, and α^{R} are the receptance of the wheel, contact spring, and rail at the contact point, respectively. The contact stiffness k_{C} is chosen to be 1.14×10^9 N/m. The parameters of the wheel are chosen from a UIC 920 mm standard freight wheel with the un-sprung mass, M_{W} = 600 kg, and the stiffness of high-frequency modes, $K_{\mathrm{M}} = 4.39\times10^9$ N/m. A typical roughness, as shown in Fig. 6, for a cast-iron block-braked wheel on a smooth rail, and the frequencies correspond to a train speed of 100 km/h.

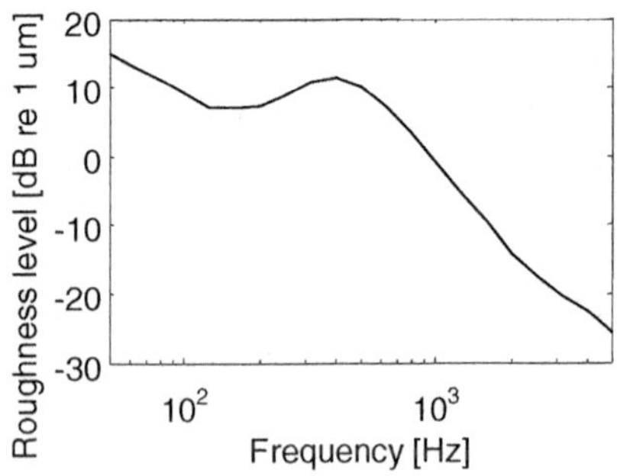

Fig. 6. Roughness level.

For the beam-spring model, the sound power radiated by the rail with the absorber is given by

$$W = \rho c\sigma[(h - w_{\mathrm{a}})\sum v_{\mathrm{r}}^2 + w_{\mathrm{a}}\sum v_{\mathrm{a}}^2]\Delta z \tag{4.4}$$

where ρ is the density of air, c is the speed of sound in the air, σ is the radiation efficiency of the rail with the absorber, v_r^2 and v_a^2 are the mean-square vibration velocity of the rail and upper-mass of the absorber, respectively, w_a is the total width of the absorber, and h is the total projected width of the rail section in the vertical direction.

For the finite element model, the sound power radiated is given by

$$W = \rho c \sigma \sum_{n=1}^{N} v_n^{\ 2} \Delta S_n \tag{4.5}$$

where $v_n^{\ 2}$ is the surface-averaged mean-square normal velocity of the nth surface element of the rail with the absorber, ΔS_n is the nth area of the nth surface element, and N is the total number of the surface element of the finite element model.

The predicted sound power levels from rail radiation due to the roughness excitation in Fig. 6 are given in Fig. 7 in one-third octave bands for the wheel at mid-span.

Fig. 7 shows the sound power level of the rail from the finite element model and beam-spring model. It can be seen that the rail-radiated sound power is similar in different cases up to about 400 Hz. In the middle frequency range of 400—1200 Hz, where the rail component of rolling noise dominates and reaches a maximum, the rail absorber works effectively as the energy consumer and the sound power from the rail radiation is thus substantially reduced. The sound power level from rail radiation is reduced by about 7.1 dB(A). The noise reduction calculated by the finite element model is less than that of the beam-spring model in the frequency range 400—1000 Hz. The reason is that the transversal rotation of the absorber is ignored in the beam-spring model.

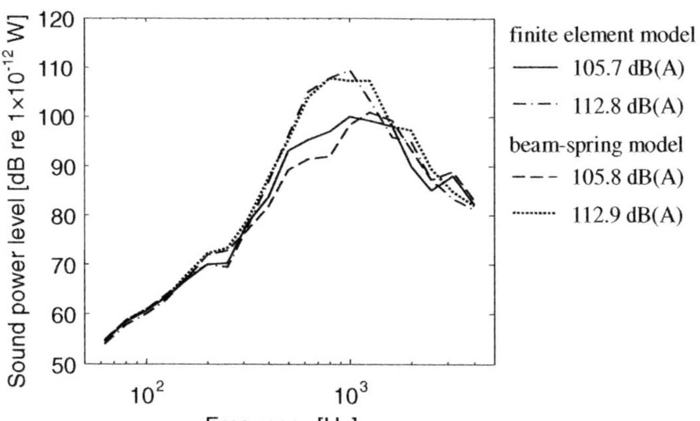

Fig. 7. Sound power level from different rail model due to roughness excitation in Fig. 5, wheel at mid-span: with absorber, — finite element model, – – beam-spring model, without absorber, –·– finite element model, ······ beam-spring model.

5 Conclusions

A finite element model of a railway track with rail absorber is proposed for investigating the performance of rail absorber in the reduction of rail vibration and noise. The effects of the transversal rotation of the absorber are studied. The calculation results are compared with those of the beam-spring model. The results show that the radiation efficiency of the rail with the absorber is significantly reduced in the frequency range 300—1000 Hz for the reason that the radiated surface of the rail is changed due to the absorbers glued to the rail, and the phase difference between the rail and mass-bar of the absorber consequently leads to changes in the radiation efficiency of rail vibration. The noise reduction calculated by the finite element model is less than that of the beam-spring model in frequency region of 400—1000 Hz because of the transversal rotation of the absorber, which is ignored in the latter. However, the total radiated noise from finite element model and beam-spring model are similar. The sound power level from rail radiation is reduced by about 7.1 dB(A) using the rail absorber.

References

[1] Thompson, D.J., Jones, C.J.C., Waters, T.P., et al.: A tuned damping device for reducing noise from railway track. Applied Acoustics 68, 43–57 (2007)

[2] Wu, T.X.: On the railway track dynamics with rail vibration absorber for noise reduction. Journal of Sound and Vibration 309, 739–755 (2008)

[3] Liu, H.P., Wu, T.X., Li, Z.G.: Theoretical modelling and effectiveness study of rail vibration absorber for noise control. Journal of Sound and Vibration 323, 594–608 (2009)

[4] Wu, T.X., Liu, H.P.: Reducing the rail component of rolling noise by vibration absorber: theoretical prediction. Journal of Rail and Rapid Transit 223, 473–483 (2009)

[5] Thompson, D.J., Jones, C.J.C., Turner, N.: Investigation into the validity of two-dimensional models for sound radiation from wave in rails. Journal of Acoustical Society of America 113, 1965–1974 (2003)

Modeling the Directivity of Wheel/Rail Radiation Using a Circular/Straight Line of Perpendicular Dipole Pairs

Xuetao Zhang

SP Technical Research Institute of Sweden,
Box 857, SE-501 15 Borås, Sweden
Tel: 46 10 516 5021, Fax: +46 33 138 381
e-mail: xuetao.zhang@sp.se

Summary

Former measurement investigation on the directivity of wheel/rail radiation has specified that (1) rail radiation is of dipole directivity characteristic in the horizontal direction, whilst it is only about 4 dB more directional than a monopole source in a vertical plane perpendicular to the rail; (2) the directivity of wheel radiation is close to the vertical directivity of rail radiation.

The work presented in this paper intends to interpret the phenomenon. It is found that a model of a perpendicular dipole pair can explain these directivity characteristics specified by the measurement. This model naturally explains why rail radiation has different horizontal and vertical directivity characteristics and why wheel radiation is not a dipole source (at least for wheels with a curved web). The study also emphasizes that, when considering the directivity effect of a dipole source, the orientation of the dipole axis needs to be specified. Moreover when more than one dipole is concerned, a special disposition of the dipoles together with a selection of difference in their sound powers can result in a change of directivity pattern from that of a monopole to that of a dipole.

Since rail radiation dominates at low speed whilst wheel radiation becomes more important at high speed, the horizontal directivity of rolling noise varies with train speed. Therefore, this work on the directivity can help with to construct a proper directivity description of rolling noise, which is important for an accurate wayside noise prediction at different train speeds.

1 Introduction

For a sound source, directivity is an important parameter to specify. In fact, this parameter also reflects the physical feature of the sound generation mechanism. Thus, to reach a proper directivity description is a process to understand better the

T. Maeda et al. (Eds.): Noise and Vibration Mitigation for Rail Trans. Sys., NNFM 118, pp. 143–150.
springerlink.com

sound source. However, in practice, this is often not a simple procedure. As for railway noise engineering, several noise types of different directivity characteristics are often mixed together, such as wheel and rail radiation, engine and cooling fan noise, turbulence sound around bogies, inter-coach areas and pantograph(s). Moreover it is a question if the horizontal directivity of a line source can be measured directly. All these factors increase the difficulty in achieving a proper directivity description that may explain why modeling directivities of railway noise sources is so far behind modeling their sound powers.

In this paper, first the directivity of wheel/rail radiation specified by measurement will be summarized, second a model will be proposed to interpret these directivity characteristics and finally concluding remarks will be made.

2 The Measurement Specified Directivity of Wheel/Rail Radiation

In 1974, Peters measured the noise profile at 200 m from the track, from a 287 m long passenger train travelling at 144 km/h [1]. By fitting the noise profile data, Peters found that rolling noise has a dipole horizontal directivity. Peters also concluded that wheel radiation dominates in rolling noise, based on the comparison between the measured noise data and the calculated track radiation. Putting these two conclusions together, he proposed that wheel radiation is of dipole directivity characteristic. Unfortunately, this understanding turned out not to be correct. It seems that rail radiation is of dipole directivity characteristic in the horizontal direction, because rail radiation dominates at this train speed [2].

Remington reported his measurements of the directivity of wheel radiation and the vertical directivity of rail radiation [3]. After analyzing his data, Remington concluded that these directivities are negligible within 5 dB. Remington did not mention the horizontal directivity of rail radiation, which is probably due to the fact that this quantity cannot be measured directly.

SP Technical Research Institute of Sweden (SP) also made a measurement investigation on the directivity of wheel radiation [4,5]. The SP's directivity data of wheel radiation support Remington's estimation: wheel radiation in one-third octave bands, at least for wheels with a curved web, is of a directivity characteristic about 4 dB different from that of a monopole source. Based on the analysis of the wheel radiation data, a more or less frequency independent directivity function is proposed [6–8]

$$\Delta L_{\text{wheel}}(\varphi) = 10\lg[0.4 + 0.6 * \cos(\varphi)], \tag{1}$$

where the angle is relative to the wheel axle and $\lg$ is for $\log_{10}$.

Since rail radiation is of dipole directivity characteristic in the horizontal direction, which differs a lot from that given by Eq. (1), the relative strength in sound power between rail and wheel radiation plays a role in determining the horizontal directivity of rolling noise. As was illustrated in [2], for a freight vehicle travelling at 100 km/h on a track with a relatively soft rail pad (200 MN/m), the rail radiation contributed about 98.3 dB(A) compared with 94.4 dB(A) by the wheel radiation. However, this example should not be generalized, because the relative strength also depends on train speed. In general, the rail contributes more at low speed whilst the wheel becomes more important at high speed [2].

It then seems acceptable to assume that rail radiation at 150 km/h is about 3 dB stronger than the wheels'. Since rail radiation has dipole directivity in the horizontal direction whilst wheel radiation has the directivity given by Eq. (1), the horizontal directivity of total rolling noise around this speed is found to be [5–7]

$$10\lg\left[0.15+0.85*\cos^2(\varphi)\right], \tag{2}$$

where the angle is defined as shown in Fig. 1. This function differs slightly from the dipole directivity that Peters concluded for the total rolling noise.

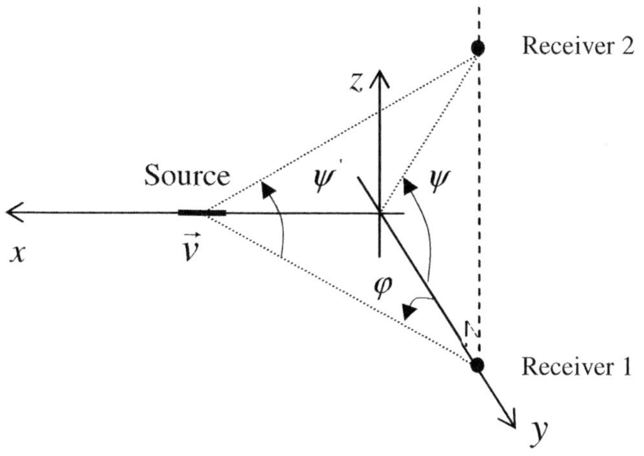

Fig. 1. The definition of angles: φ is a horizontal angle in the x-y plane and relative to the y-axis; ψ is a vertical angle in the y-z plane; ψ' is a vertical angle in a vertical plane containing the receiver and the source (or the centre of the source line). The direction of travel along the track is parallel to x-axis.

Moreover to compare prediction models used in Europe, the horizontal directivity functions for rolling noise used in a number of national models are summarized in Table 1 [8]. As can be seen, the directivity function given by Eq. (2) has already been used in Austria, Germany, Switzerland and the Nordic countries. Now it can be derived. (Note: A directivity description may or may not include a constant, which is the normalization factor.)

SP Acoustics has also measured the vertical directivity of rail/track radiation. The vertical directivity data can be simulated by the following function [4–7]:

$$\Delta L_V^r(\psi)=10\lg\left[0.4+0.6*\cos^2(\psi)\right], \tag{3}$$

where $\Delta L_{\mathrm{V}}^{r}$ denotes the vertical directivity of rail radiation and ψ the vertical angle, defined as shown in Fig. 1. This directivity function is close to that given by Eq. (1).

It once was a puzzle to the author why the vertical directivity of rail radiation differs significantly from the horizontal one and why wheel radiation is not a dipole source. In next section, a model will be made to interpret these directivity characteristics.

Table 1. Horizontal Directivity Functions used in European Prediction Models [8].

Country	**The horizontal directivity function**
Austria	$10\lg[0.15+0.85\cos^2(\phi)]$
Germany	$10\lg[0.22+1.27\cos^2(\phi)]$ $\cong 10\lg[0.15+0.866\cos^2(\phi)]+1.66$
Netherlands	$10\lg[\cos^2(\phi)]$
Switzerland	$10\lg[0.15+0.85\cos^2(\phi)]+10\lg(2.19)$
Nordic countries	$10\lg[0.15+0.85\cos^2(\phi)]+2$

3 To Model the Directivity of Wheel/Rail Radiation

In an ideal fluid, a small rigid sphere under small amplitude oscillations turns out to be a dipole source [9]. By similarity, it seems that a vibrating rail section can be modelled as a line of dipoles. However, this model can only explain the measured horizontal directivity of dipole characteristic, not the measured vertical directivity given by Eq. (3), which shows that, in a vertical plane perpendicular to the rail, a vibrating rail section is close to a monopole source.

As a dipole source is the concern, let us make a close inspection of its directivity characteristic. In Fig. 2, the directivity pattern of a free dipole ($\cos^2\varphi$) is plotted. As can be seen, a free dipole presents the dipole directivity characteristic in a plane containing the dipole axis. However, in a plane perpendicular to the dipole axis, the free dipole presents monopole directivity characteristic! This fact reminds us that, when considering the directivity effect of a dipole source, the orientation of its axis needs to be specified.

With this fact being considered, a trial model of a straight line of perpendicular dipole pairs (PDPs) is made for the directivity of rail radiation. As is shown in Fig. 3, a straight line of dipoles with a vertical orientation simulates the rail head and foot vibration, whilst a straight line of dipoles with a lateral orientation simulates the rail web vibration. And, the latter is assumed to be about 4 dB stronger in sound power than the former.

Fig. 2. 3D directivity character of a free dipole. (color figure online)

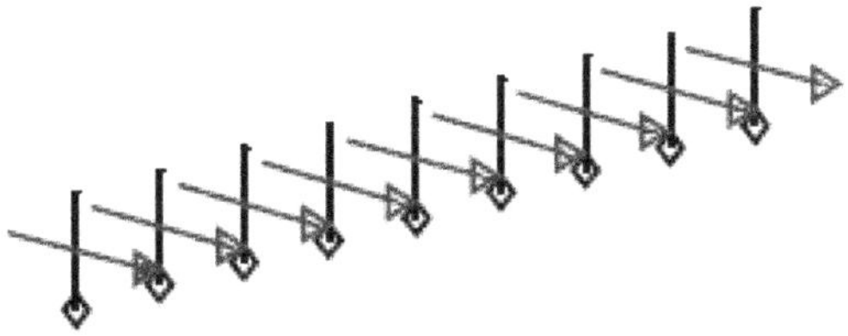

Fig. 3. A straight line of perpendicualr dipole pairs for modeling the directivity of rail radiation. (color figure online)

In a horizontal plane located at about half the rail height, the line of dipoles with the vertical orientation presents monopole directivity characteristic (and is of negligible sound power), while the line of dipoles with the lateral orientation presents dipole directivity characteristic. The combined horizontal directivity of the dipole pair is of dipole directivity characteristic. However, in a vertical plane perpendicular to the rail, they both present dipole directivity characteristic. Thus, the combined vertical directivity of the dipole pair depends on the relative strength in their sound powers. If they were the same in strength, the combined vertical directivity would be of monopole directivity characteristic, because $\cos^2(\psi)+\cos^2(\psi+\pi/2)=1$. Since the line of dipoles with the lateral orientation is assumed to be 4 dB stronger in sound power, the combined vertical directivity is then determined by the following calculation procedure:

$$L_{\text{p, dipole 1}} = L_0 + 10\lg[0.001 + 0.999 * \cos^2(\psi + \pi/2)], \quad (4)$$

$$L_{\text{p, dipole 2}} = L_0 + 4 + 10\lg[0.001 + 0.999 * \cos^2(\psi)], \quad (5)$$

$$L_{\text{p, combined}} = 10\lg\left(10^{L_{\text{p, dipole 1}}/10} + 10^{L_{\text{p, dipole 2}}/10}\right), \quad (6)$$

where L_0 is a constant, dipole 1 is of the vertical orientation and dipole 2 of the lateral orientation. The numerical result shown in Fig. 4 indicates that Eqs. (4)–(6)

produce a vertical directivity identical to the one given by Eq. (3). Thus, it has proved that the model of a straight line of PDPs can properly simulate both the vertical and horizontal directivity characteristics of rail radiation.

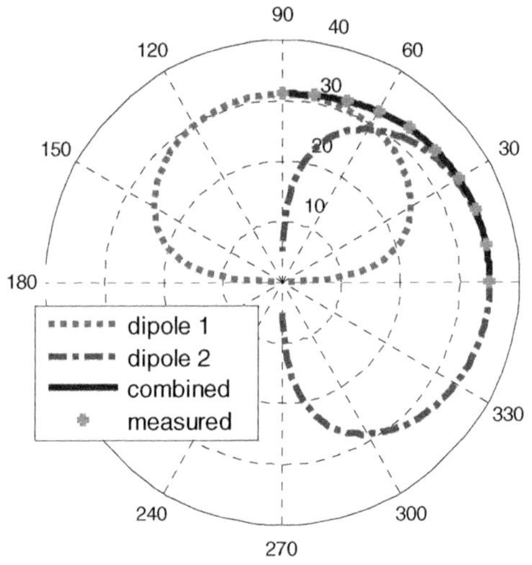

Fig. 4. Comparison between measured vertical directivity of rail radiation and the one predicted by the model of perpendicular dipole pair, "combined", where the directivities of dipole 1, dipole 2 and "combined" are given by Eqs. (4)–(6), respectively. (color figure online)

Next, the directivity of wheel radiation is to be modelled in a similar way. A circular line of dipoles with a radial orientation simulates the wheel tread vibration and a circular line of dipoles with an axial orientation simulates the wheel web vibration, as shown in Fig. 5. Again, the latter is assumed to be about 4 dB stronger in sound power than the former. As has already been learnt, in a plane which contains the wheel axle, the combined directivity should be determined by Eqs. (4)–(6), which will result in a directivity pattern the same as that given by Eq. (3). This directivity is slightly different from the measured one simulated by Eq. (1). However, the difference between these two directivity functions is quite small and can be ignored. (In fact, it has been tried out that a combined directivity closer to that given by Eq. (1) can be achieved, if taking a level difference of 3.5 dB and an orientation difference of 80° between the components of the dipole pair. However, this small "correction" is not used here, considering that the main concept, rather than the details, should be focused on.)

These good results reflect the value of the model: it is promising for understanding wheel/rail radiation. This model naturally explains why rail radiation is of different horizontal and vertical directivity characteristics and why wheel radiation is not a dipole source (at least for wheels with a curved web). Moreover it also emphasizes that, when considering the directivity effect of a dipole source, the orientation of its axis needs to be specified.

It is easy to construct the vertical directivity of wheel-rail vibration noise, because the directivity of wheel radiation given by Eq. (1) is close to the vertical directivity of rail radiation given by Eq. (3). One can use either of them to describe the vertical directivity of wheel-rail vibration noise.

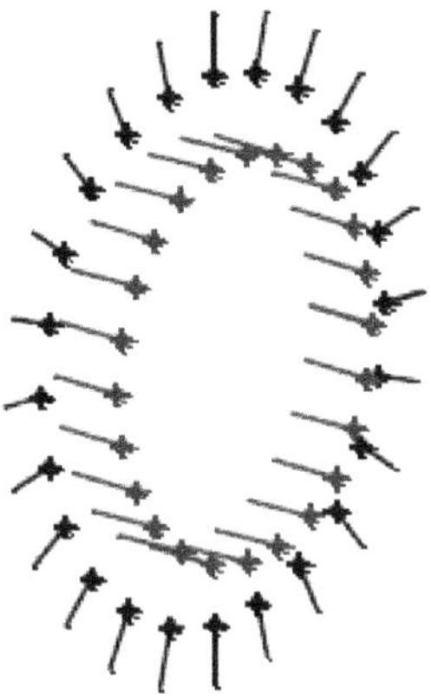

Fig. 5. A circular line of perpendicular dipole pairs for modeling the directivity of wheel radiation (color figure online)

4 Concluding Remarks

The proposed model, a circular/straight line of *perpendicular dipole pairs*, can naturally interpret the measurement specified directivity characteristics of wheel/rail radiation. This model explains why rail radiation is of different horizontal and vertical directivity characteristics and why wheel radiation is not a dipole source. The discussion also emphasizes that, when considering the directivity effect of a dipole source, the orientation of the dipole axis needs to be specified. Especially, if more than one dipole is concerned, a special disposition of the dipoles together with a selection of difference in their sound powers can result in a change of directivity pattern from that of a monopole to that of a dipole.

As for the vertical directivity, Eq. (1) assumes that lateral radiation from the axial wheel web vibration is 4 dB higher than the vertical radiation from the radial wheel tread vibration, which is based on the directivity measurement for wheels with a curved web. For wheels with a straight web and under a vertical/radial excitation, lateral radiation can become closer to the vertical one or even less; this implies that a vibrating wheel of this type could be closer to a monopole source than Eq. (1) predicts. Similarly, Eq. (3) assumes that lateral radiation from the rail web is 4 dB stronger than the vertical radiation from the head and foot combined. This result is based on the directivity measurement on a ballasted track equipped with UIC60 rails and concrete mono-block sleepers. For other different track types, especially for en embedded rail slab track, lateral radiation could become much less, resulting in a vertical directivity pattern different from what Eq. (3) predicts. Therefore, for slab track types, a further measurement investigation on the vertical directivity of rail radiation is needed.

As for the horizontal directivity of rolling noise, since rail radiation (which has dipole directivity) dominates at low speed whilst wheel radiation (which has the directivity given by Eq. (1)) becomes more important at high speed, the horizontal directivity of rolling noise varies with train speed, from that of dipole characteristic to that of nearly monopole characteristic.

With all these understandings, this directivity model can help with to construct a proper directivity description of rolling noise. As we know, an advanced model like the TWINS [10] for the source strength of the wheel and track sources is the most important factor in predicting rolling noise. However, with a proper directivity description also being provided, one can well predict not only the total noise exposure level of a train passage but also the (horizontally and/or vertically) angle-distributed noise impact levels which often become a practical relevance.

References

[1] Peters, S.: The Prediction of Railway Noise Profiles. Journal of Sound and Vibration. Journal of Sound and Vibration 32(1), 87–99 (1974)

[2] Thompson, D.: Railway Noise and Vibration: Mechanisms, Modelling and Means of Control. Elsevier, Amsterdam (2009)

[3] Remington, P.J.: Wheel/rail noise, I: Characterization of the wheel/rail dynamic system. Journal of Sound and Vibration 46, 359–379 (1976)

[4] Zhang, X.: Measurements of Directivity on Test Rig, HAR12TR-020910 -SP04 HAR12TR-020910 -SP04 (Harmonoise technical report) (April 30, 2003)

[5] Zhang, X., Jonasson, H.: Directivity of Railway Noise Sources. Journal of Sound and Vibration 293, 995–1006 (2006)

[6] Zhang, X.: To determine the horizontal directivity of a train pass-by, in03_627. In: The Proceedings for Inter. Noise 2003, Jeju, Korea, August 25-28 (2003)

[7] Zhang, X.: Directivity of Railway Rolling Noise. In: Zhang, X. (ed.) Noise and Vibration Mitigation for Rail Transportation Systems. Notes on Numerical Fluid Mechanics and Multidisciplinary Design, vol. 99, Springer, Heidelberg (2008)

[8] van Leeuwen, J.J.A., Ouwerkerk, M.A.: Comparison of some prediction models for railway noise used in Europe, Report L.94.0387.A, DGMR consulting engineers bv, p. 128, The Hague, The Netherlands (1997)

[9] Howe, M.S.: Theory of Vortex Sound. Cambridge University Press, Cambridge (2003)

[10] Mahé, H., Thompson, D.J., Zach, A., Hölzl, G.: Experimental validation of the prediction model TWINS for rolling noise. In: Proceedings of Inter. Noise 1993, Leuven, Belgium, August 24-26, pp. 1459–1462.

High-Speed Rail: Excitation Frequencies and Track Stiffness

A. Wang and S.J. Cox

Pandrol Limited
63 Station Road, Addlestone, Surrey KT15 2AR, UK
Tel: +44 1932 834500, Fax: +44 1932 850858
e-mail: a.wang@pandrol.com

Summary

This paper considers track vibrations induced by high-speed trains, both in a tunnel on slab track and at grade on ballasted track. In particular, some of the characteristic frequencies present are investigated. These include speed-dependent parametric excitation frequencies related to the geometry of the system, for instance the rate with which bogies and axles pass above a given position in the track, or with which a wheel passes over the fasteners, also speed-independent system aspects such as the loaded track vibration resonance are considered. The circumstances under which these characteristic frequencies are most likely to coincide on high-speed rail systems are considered. Some conclusions are drawn on optimal range for the loaded track resonance and the optimal track stiffness range under different train speeds required to provide this.

1 Introduction

The possibility of coincidence of characteristic frequencies on metro systems was considered in a previous paper [1]. Here, similar principles have been applied to high-speed rail systems. The dynamics of high-speed rail systems have been studied in both experimental and theoretical investigations by many researchers [2, 3, 4]. High-speed trains often produce significant noise and vibration over a wide range of frequencies [5, 6, 7]. But the relatively low frequency range between 10 Hz and 300 Hz seems to be of particular significance in a number of measurements made of different aspects of high-speed systems by other authors. For example, Figure 1 shows high levels in this frequency range in the ⅓ octave spectra for ground-borne vibration and air-borne noise in measurements.

Residents in buildings above and alongside existing railways [8] are becoming less tolerant of vibration from the track and the noise that results. In-car noise is particularly relevant for the passengers on high-speed rail systems [9]. So in this paper, the emphasis has been placed on the possibility of coincidence in characteristic frequencies in this key frequency range of 10 Hz to 300 Hz.

T. Maeda et al. (Eds.): Noise and Vibration Mitigation for Rail Trans. Sys., NNFM 118, pp. 151–158.
springerlink.com

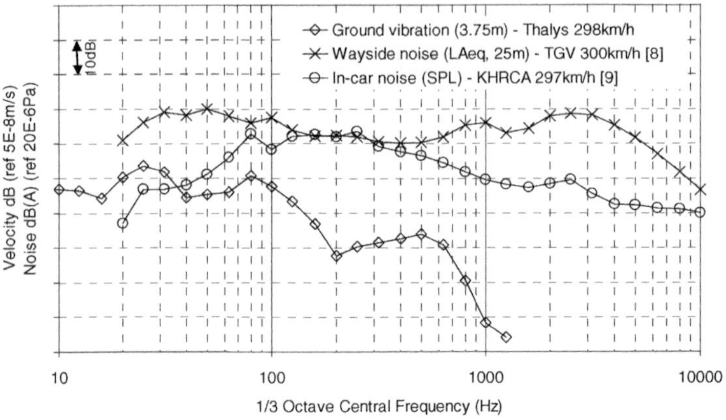

Fig. 1. 1/3 Octave spectra of velocity and noise

One of the most important aspects of behaviour in this frequency range is the transmission of vibrations into the ground, as is illustrated in Figure 1. A common means of controlling ground vibration [10] is through the design of the track structure. Reducing track stiffness [11] is the most widely employed method of mitigating vibration from both new and existing tracks. Track stiffness can be reduced by decreasing the stiffness of individual fasteners, or by spacing them further apart. In both cases, the limitation is usually provided by the requirement to maintain track gauge and to limit rail roll under traffic. A reduction in track stiffness also reduces the frequency of the loaded track resonance, in which the unsprung mass of the train moves on the stiffness of the track. This is an example of a resonant frequency of the train-track system that is not affected by train speed. There are others, such as solid body bounce modes of the vehicle and pinned—pinned resonances of the rail.

But there are other characteristic frequencies that are affected by train speed. Parametric excitation occurs at particular frequencies that are related to the geometry of the system and speed. These include the frequency with which axles, bogies and vehicles pass above a given position; the rotation frequency of a wheel; the frequency with which a wheel passes over the fasteners; and so on. The subject of this paper is to demonstrate how the vehicle-track system resonant frequencies on high-speed rail systems relate to the frequencies of parametrically excited vibration and whether it is possible to avoid coincidence of frequencies by appropriate design.

2 Characteristic Frequencies of High-Speed Rail

2.1 Types of Characteristic Frequency

A number of characteristic frequencies are excited on a high-speed rail system. Three of those considered here are determined by train speed and the fourth is independent of train speed:

i) The frequency with which bogie centres pass a given position on the track. This depends on train speed.
ii) The frequency with which axles pass a given position on the track. This depends on train speed. The spacing and frequency considered here that between adjacent axles on one bogie – that between the last axle on one vehicle and the first on the next may differ slightly, so that two similar frequencies will be excited.
iii) The frequency with which a given wheel on the train passes above regularly spaced rail fastenings. This depends on train speed.
iv) The loaded track resonance. This frequency does not depend on the train speed, but on the unsprung mass of the vehicle and the stiffness of the track. On directly fastened track on a rigid base slab, the track stiffness depends in turn on the stiffness and spacing of the fasteners. On ballasted track it is also dependent on the ballast stiffness.

Other frequencies are also present on high-speed rail systems, but these are not considered here. For example, 'pinned—pinned' frequencies at which the rail vibrates with nodes at the fastening positions - these are usually higher frequencies (from 400 Hz to 1200 Hz) than the range considered here to be of most interest.

2.2 Parametrically Excited Frequencies for Train Pass-by

Two different trains are considered here as typical of high-speed rolling stock: the Thalys [3] in Belgium and CRH2 [12] in China. Important characteristics of both of these train types are listed in Table 1.

Table 1. Geometrical characteristics of the Thalys HST and CRH2

Train Type	Carriage type	Distance between Bogies (m)	Axle distance (m)	Total Axle mass (kg)	Unsprung axle mass (kN)	Sprung axle mass (kg)	No. of axl e	Carriage length (m)
Thalys	Loco	14.0	3.0	17000	1733	15267	4	22.15
	Coach	18.7	3.0	14500	1830	12674	3	21.84
	Coach	18.7	3.0	17000	1826	15170	2	18.70
CRH2	Coach	17.5	2.5	11700	1550	10150	4	25.7
	Coach	17.5	2.5	13100	1550	11550	4	25.0

High-speed rail is often defined as operating at above 250 km/h, with common average revenue service speeds up to 350 km/h. Table 2 gives the first few parametric excitation frequencies of both high-speed train types on typical track.

Table 2. Characteristic frequencies (Hz)

			250 km/h	300 km/h	350 km/h
Thalys	Bogie passage	18.7 m	3.7	4.5	5.2
		14.0 m	5.0	6.0	6.9
	Axle passage	3.0 m	23.1	27.8	32.4
	Sleeper spacing	0.6 m	115.7	138.9	162.0
CRH2	Bogie passage	17.5 m	4.0	4.8	5.6
	Axle passage	2.5 m	27.8	33.3	38.9
	Sleeper spacing	0.6 m	115.7	138.9	162.0

2.3 Loaded Track Resonant Frequency

The Belgian track between Brussels and Paris used by the Thalys train is ballasted, with UIC60 rails fixed every 0.60 m with Pandrol fasteners on pre-stressed concrete monoblock sleepers. The dynamic stiffness of the rail pad can be estimated as 154 kN/mm using a beam-on-elastic-foundation model from the track dynamic deflection measured under traffic [4]. The dynamic stiffness of ballast can be estimated as 460 kN/mm [4] per railseat. The sleeper has a mass of 300kg.

In China between Hefei and Wuhan, Rheda slab track is installed in tunnels used by the CRH2 train. 60kg Chinese rails are fixed every 0.60 m with Pandrol SFC baseplates on cast-in concrete bi-block sleepers. The dynamic stiffness of the fastening system can be estimated as 39 kN/mm under traffic [13].

The loaded track resonance f_p has been calculated from the estimated dynamic stiffness of the track and the unsprung vehicle mass using a simple model [14].

$$f_p \approx (1/2\pi).(1/\, m_w.\alpha_r)^{1/2} \tag{2.1}$$

where m_w is vehicle unsprung mass and α_r is dynamic track receptance. For the ballasted track under Thalys train, the first loaded track resonant frequency is about 74 Hz, and an additional resonance [15] occurs at about 395 Hz where the two elastic layers (rail pad and ballast) are separated by an intermediate mass element – the sleeper. There are also bending resonances of sleeper, which usually occur at high frequencies. For the slab track under CRH2 train, the loaded track resonant frequency is 53 Hz.

3 Vibration Responses on Typical High-Speed Rail

Comparison can now be made between the calculated frequencies given in the previous section and those found in measurements made in track under the Thalys train in Belgium running at a speed of 300 km/hr [3, 16] and under the CRH2 train in China at a speed of 250 km/hr [14]. The measurements made included dynamic deflections of the rail and vibrations of the rail and sleeper or slab. Free-field vibration was also measured for the ballasted track. The speed of the trains at the measurement site was calculated from the time history of the data recordings.

Calculated frequencies are identified in Figure 2 on vibration spectra obtained from measurements made on the sleeper on the Belgian ballasted track and the slab in the Chinese track. The vibration levels on the slab are much lower because it is less mobile. Calculated frequencies are also identified in Figure 3 on the measured free field and slab centre vibration spectra. In both cases, the frequency of bogie passage lies below 10Hz and is not identified in Figures 2 and 3. There are a number of peaks in the spectrum, and as would be expected, several of these correspond quite well with the characteristic frequencies calculated from the train speed, track and train parameters. The figures confirm that resonances in the vehicle-track system lead to high levels of track vibration and that the frequencies at which a number of these occur can readily be calculated from a number of simple parameters.

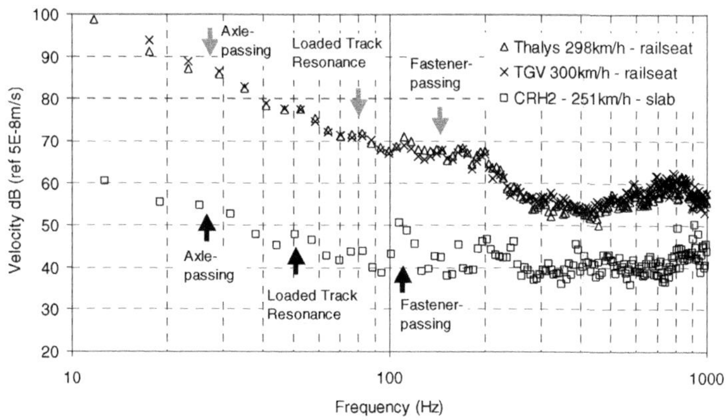

Fig. 2. Spectra of vibration on sleeper or slab

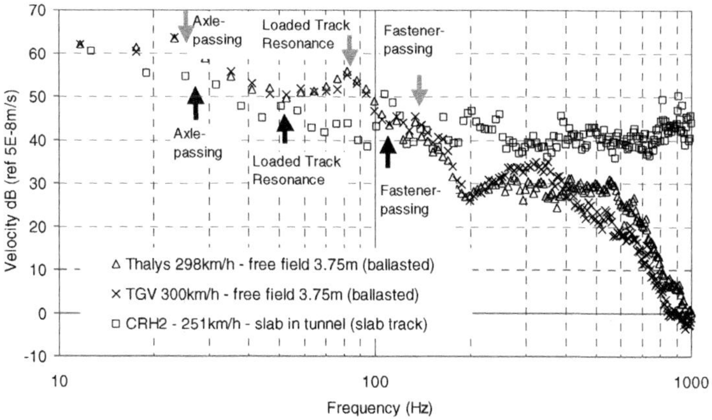

Fig. 3. Spectra of vibration on free field of ballasted track and slab centre of slab track

4 Track Stiffness Design on High-Speed Rail Systems

Track stiffness is a key parameter that directly affects the track resonances and the ground-borne vibration, wayside noise and in-car noise level. It should be optimized in modern high-speed rail designs. It is suggested that one of the considerations in optimizing the stiffness should be avoidance of coincidence between the loaded track resonance and the axle and fastener passing frequencies. It is assumed here that such coincidence is undesirable on high-speed rail systems. It is difficult to predict the combined effect of parametric excitation such as that generated by axle and fastener passing with that of excitation of track resonances by wheel-rail roughness. Where attempted, the approach is usually to estimate the effects separately and assume that they can be combined by linear addition.

Bogie-passing, axle-passing, fastener-passing frequencies always occur in this order and are generally well spaced in the measurements considered here. The loaded track resonance generally occurs between the axle passing and fastener-passing frequencies. This suggests that if we make the further assumption that the axle passing and fastener passing frequencies are given for a particular train and track design and target train speed, then upper and lower bounds can be set on the desirable track stiffness if a coincidence with either of the other two frequencies is to be avoided.

If a desirable minimum factor of $\sqrt{2}$ is applied to separate any two characteristic frequencies, then Table 3 below can be constructed showing the upper and lower bounds on track resonant frequencies and corresponding fastening stiffness. Figure 4 presents the same information graphically.

Table 3. Bounds on values to avoid coincidence

Train Type	Speed km/h	Frequency Hz		Stiffness kN/mm	
		Lower	Upper	Lower	Upper
Thalys	250	32.7	81.8	13	146
	300	39.3	98.2	21	261
	350	45.8	114.6	31	456
CRH2	250	39.3	81.8	18	86
	300	47.1	98.2	30	153
	350	55.0	114.6	43	245

For ballasted track, the stiffness indicated is the series dynamic stiffness of the ballast and fastening, for slab track, the dynamic stiffness of the fastening. The table suggests ranges of track stiffness that will avoid coincidence of characteristic frequencies. Interestingly, although the upper bound limits on loaded track resonant frequency are the same for the two train types because the fastener spacing has been assumed to be the same, the upper bound limits on track stiffness differ because of the different unsprung masses of the two train types. Clearly, there are other factors that need to be taken into account in selection of appropriate track stiffness for high-speed operation, such as safe operation of the track, durability of components, control of ground vibration and noise and so on. Track stiffness values exist within the preferred range that could accommodate different trains travelling at different speeds on the same track. The actual stiffness of the Belgian track lies closer to the upper limit at the lowest speed considered, and becomes closer to the centre of the range as train speed increases. For the Chinese track, the actual stiffness lies closer to the lower limit at the lower speeds and exceeds this at the higher speeds.

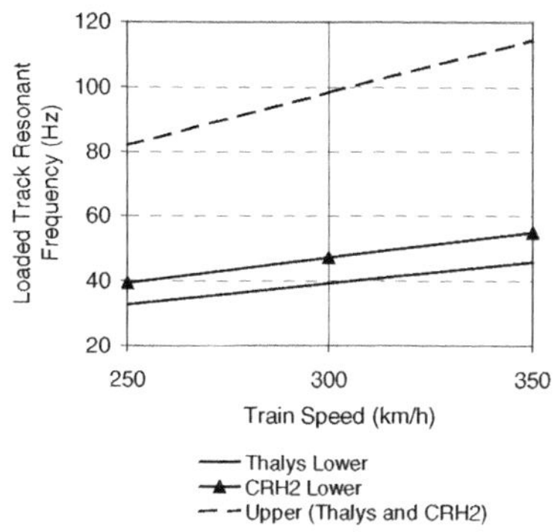

(a) Suggested ranges of loaded track resonance

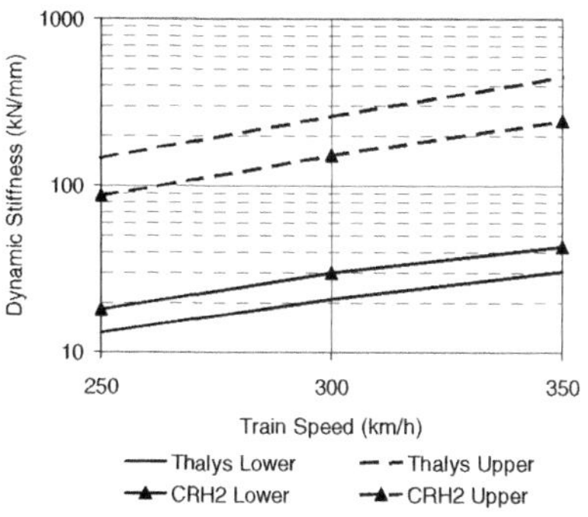

(b) Suggested ranges of track stiffness

Fig. 4. Suggested ranges of loaded track resonance and track stiffness

The potential for clashes between the loaded track resonance and the axle-passing frequency may be greater on track systems with significant added mass such as FST, or ballasted track with ballast mats, where there may also be the possibility of coincidence between the loaded track resonance and the bogie-passing frequency.

5 Conclusions

Characteristic frequencies that occur on all high-speed rail systems have been investigated, with a view to determining whether there is likely to be a coincidence of any of these that might lead to adverse behaviour of the vehicle–track system. No attempt has been made to predict what these effects might be if they do occur. Of the characteristic frequencies investigated, it has been found that for high-speed systems the loaded track resonance typically lies between the axle-passing frequency and the fastener-passing frequency. This suggests a suitable range for track stiffness that will avoid coincidence with either of these two parametrically executed frequencies. On this basis, recommended ranges of dynamic track stiffness have been suggested for both Thalys type trains and CRH2 type trains as a function of train speed.

References

[1] Cox, S.J., Wang, A., Adedipe, A.: Survey of metro excitation frequencies and coincidence of different modes. In: 9th International Workshop on Railway Noise, Munich, Germany (2007)

[2] Ju, S., Lin, H., Huang, J.: Dominant frequencies of train-induced vibrations. Journal of Sound and Vibration 319, 247–259 (2009)

[3] Degrande, G., Schillemans, L.: Free field vibrations during the passage of Thalys high-speed train at variable speed. Journal of Sound and Vibration 247(1), 131–144 (2001)

[4] Lombaert, G., Degrande, G., Bekaert, J.: The influence of the train speed on vibrations due to high speed trains. In: 9th International Workshop on Railway Noise, Munich (2007)

[5] Wang, A., Cox, S.J.: Noise Characteristics of high speed track with railpads of different stiffness. In: EuroNoise 1998, Munich, Germany (1998)

[6] Cox, S.J., Wang, A.: Effects of rail fastening on railway track noise. In: InterNoise 2000, France (2000)

[7] Wang, A., Cox, S.J.: Effect of railpad stiffness on rail roughness growth and wayside noise levels on high speed track. In: The 6th World Congress on Railway Research, Edinburgh, UK (2003)

[8] Letourneaux, F., Cordier, J.F., Poisson, F., Douarche, N.: High speed railway noise: Assessment of mitigation measures. In: 9th International Workshop on Railway Noise, Munich (2007)

[9] Koh, H.I., Kwon, H.B., You, W.H., Park, J.H.: A study on source mechanism in the interior noise problem of high speed trains. In: 9th International Workshop on Railway Noise, Munich (2007)

[10] Evaluation of human exposure to vibration in buildings (1 Hz to 80 Hz). BS 6472

[11] Cox, S.J., Wang, A.: Effect of track stiffness on vibration levels in railway tunnels. Journal of Sound and Vibration 267, 565–573 (2003)

[12] Bent, M.: Hexie Hao Overview. Railvolution 9(6), 22–25 (2009)

[13] Xi, Y.: Inspection report of deflection and vibration measurements in Dabieshan tunnel on He-Wu PDL. Report No. VS2009-101-1, China Ship Scientific Research Center (August 2009)

[14] Grassie, S.L., Gregory, R.W., Harrison, D., Johnson, K.L.: The dynamic response of railway track to high frequency vertical excitation. Journal of Mechanical Engineering Science 24, 77–90 (1982)

[15] Tunna, J.M.: Wheel/Rail forces due to wheel irregularities. In: Proceedings 9th International Wheelset Congress, Montreal (1988)

[16] Tests on SNCB High Speed Line in Belgium July 2002: Measurement of wave decay rate, track receptance, rail roughness of unloaded track and track responses under traffic. Pandrol Report No. 41019-8 (2002)

Railway Noise Reduction Technology Using a Damping Material

Günther Koller[1], M.T. Kalivoda[2], Martin Jaksch[2], Martin Muncke[3], Takashi Oguchi[4], and Yoshifumi Matsuda[4]

[1] koocoo technology & consulting gmbH
St.-Veit Gasse 28/1/5 1130 Wien, Austria
Tel.:+43 676 960 69 54
e-mail: guenther.koller@koocoo.eu
[2] psiA-Consult GmbH
Lastenstr. 38/1 1230 Wien, Austria
Tel.:+ 43 1 865 6755
e-mail: jaksch@psia.at
[3] OBB-Infrastruktur AG
Zieglergasse 6 1070 Wien, Austria
Tel.:+ 43 1 93000 31928
e-mail: martin.muncke@oebb.at
[4] Sekisui Chemical Co., Ltd.
2-2 Kamichoshi-cho Kamitoba Minami-ku Kyoto 601-8105, Japan
Tel.:+ 81 75 662 8524
e-mail: matsuda021@sekisui.jp

Summary

Two effective railway noise reduction technologies using a high-performance damping material have been developed. One is a rail noise reduction system for reducing rolling noise using a combination of damping material and sound isolation and absorption technologies. The other one is an application of a high-performance damping material on the surface of web within a railway bridge structure as a means for reducing structure noise. An actual track-based field test for demonstrating these technologies was performed using a track of Austrian Federal Railways. As a result, the rail noise reduction system was found effective for reducing noise by 2 to 4 dB in a normal section (ballast track) and for a bridge section the combined use of the rail noise reduction system and damping material reduced noise by 2 to 4 dB of which the rail noise reduction system provided 0.5 to 3.5 dB, and the damping material on the bridge web contributed 0 to 1.5 dB.

1 Introduction

Recently, for reason of energy saving, railway transportation has been considered a more effective mode of transportation than others though noise produced by a

T. Maeda et al. (Eds.): Noise and Vibration Mitigation for Rail Trans. Sys., NNFM 118, pp. 159–166.
springerlink.com

railway service has been a matter of high concern. Railway noise includes rolling noise, drive-system noise and structure noise [1]. Rolling noise is a typical type of noise that can occur within a normal section and is a consequence of the surface roughness of rails and wheels [2]. Structure noise is a radiation of sound from structures, e.g. steel-made bridge main girders directly below the sleepers that transmit railway vibration produced when a train is running [3]. There is a variety of products available for reducing these types of noise. They include a damping material Calmmoon Sheet developed by Sekisui Chemical and a rail noise reduction system Calmmoon Rail based on the damping material available for reducing rolling noise. This report shows a result of field test using an actual track of Austrian Federal Railways to demonstrate the performance of the rail noise reduction system and the damping material. The test used a normal section (ballast track) for testing the rail noise reduction system and bridge section for testing a combined configuration of the system and additionally installed damping material.

2 Overview of the Noise Reduction Technologies

2.1 Damping Material

The damping material consisting of two layers —– resin layer (t = 1.0 mm) effective for absorbing vibration and metallic layer (t = 0.3 mm) —– is a constrained layer damping material having a total thickness of 1.3 mm. The adhesive surface of the resin layer, after the separate paper is removed, can be directly attached to the surface of the material without special technology or tools, which leads to higher efficiency in installation work. In this experiment, the damping materials were installed on the surface of a web of a bridge structure to measure how the material could reduce the sound arising from the web.

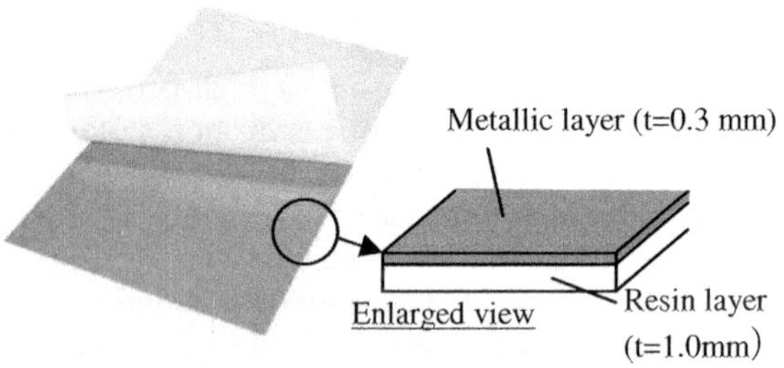

Fig. 1. Damping Material

Fig. 2. Damping Material Installed

2.2 Rail Noise Reduction System

A rail noise reduction system comprises the damping material and consists of a pair of two covers and two fixtures. To install the system, simply place the cover at each side of the rail and then secure it with the fixtures, thus requiring no special tool. The cover consists of metallic plate, sound absorption material, and the damping material. Each component has its own role for reducing noise, as shown below. The sound arising from the rail can be insulated by the metallic plate. However the metallic plate transmits rail vibration and resonates to radiate sound. The damping material reduces this vibration and vibration-caused sound. The sound absorption material effectively absorbs sound occurring between the metallic plate and rail.

An advantage of the system is that a pair of two covers and two fixtures measures as light as 2.3 kg in weight, which permits easier installation.

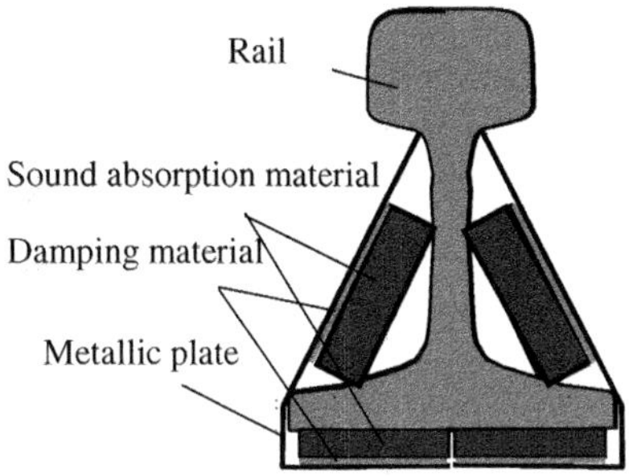

Fig. 3. Rail Noise Reduction System

Fig. 4. Rail Noise Reduction System Installed

Another feature of the rail noise reduction system is, contrary to an existing mass-rubber based rolling noise reduction system mounted on the side of the rail to reduce noise by damping rail vibration [5], that it does not directly affect the rail vibration. Figure 5 shows decay rates measured on an actual track section of Austrian Federal Railways. Decay rate is a measure of reduction of rail vibration in dB per m of rail. This figure shows that the rail noise reduction system has little effect on the decay rate. This is because this noise reduction system uses sound insulation and absorption technologies that are not affected by such vibrations.

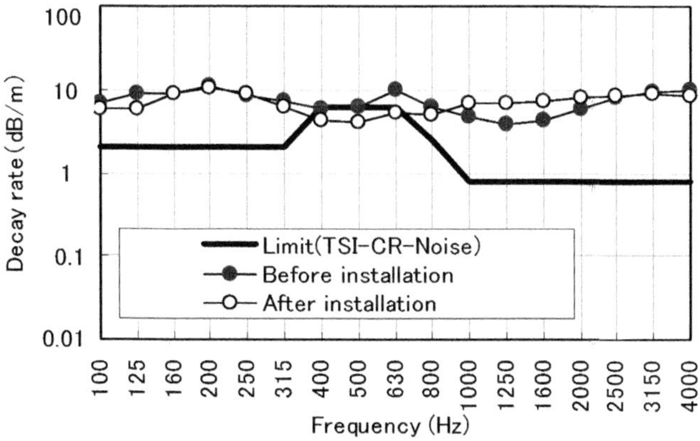

Fig. 5. Vertical Track Decay Rate before and after installation

3 Demonstration of the Effectiveness of the Rail Noise Reduction System for Normal Track Section

3.1 Overview of Experiment

The experiment was conducted on a normal track section at Deutsch Wagram, north of Vienna, of Austrian Federal Railways. Table 1 shows the track conditions of the section. The rail noise-reduction system was installed in a length of about 40 m in the section. The noise-measuring method per ISO 3095 was employed with a microphone placed 7.5 m horizontally from the track center and 1.2 m above the rail level to measure sound pressure level before and after the system was installed. For this experiment, four different types of trains — freight train, ÖBB S-Train (Class 4020), ÖBB Regional-Doubledeck-Train (Class 8033) and ÖBB S-Train (Class 4024) — were used.

Table 1. Track Conditions of the Normal Section Used for the Test

Track configuration	Rail	Sleeper	Train speed
Straight	S49	Concrete	40 to 120 km/h

3.2 Result

Figures 6, 7, 8 and 9 show the equivalent continuous A-weighted sound pressure level (LA, pb) as a function of train speed V in logarithmic presentation. Train speed V is standardized by 80km/h. The solid lines in these figures show approximated curves for measurements before installation of the rail noise-reduction system, while the dotted lines show those after installation of the system. The difference of these lines represents the system's contribution to noise reduction. These figures show that the noise level produced becomes greater as the train runs faster. The freight train and Class 4020 train produced more noise than Class 8033 and Class 4024 trains. Rolling noise depends on the surface roughness of the rails and wheels, which means that trains having wheels whose surface roughness is larger produce more noise when running on the same rail. This can be a reason why the relatively older freight train and Class 4020 train having larger surface roughness of their wheels than the relatively new trains Class 8033 and Class 4024 produce more noise.

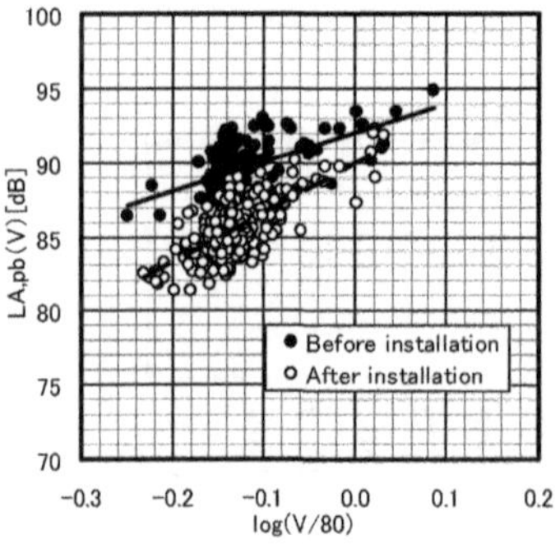

Fig. 6. Freight Train Speed vs. Noise Level

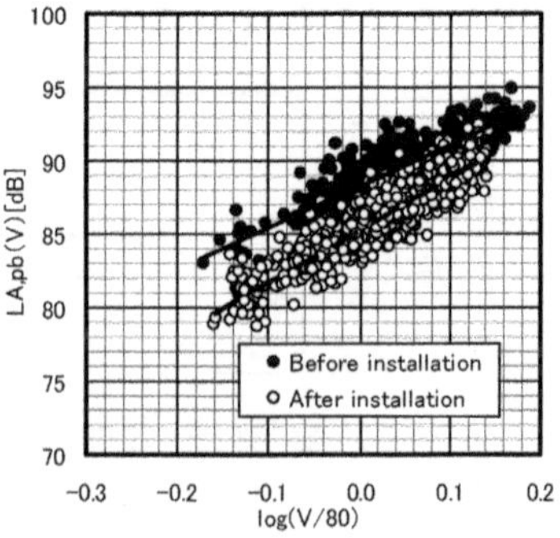

Fig. 7. Class 4020Train Speed vs. Noise Level

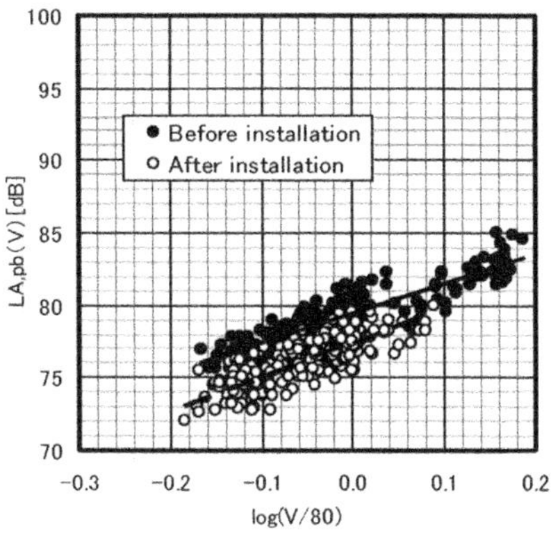

Fig. 8. Class 8033Train Speed vs. Noise Level

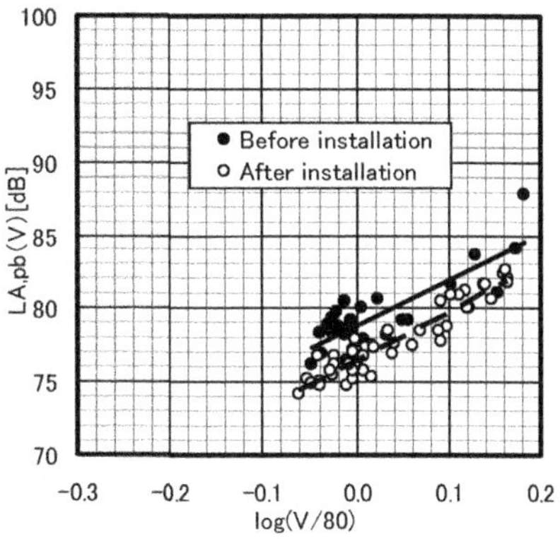

Fig. 9. Class 4024Train Speed vs. Noise Level

Figure 10 shows, for each train type, based on the approximate curves shown in Figs. 6 through 9 the difference in noise level between before and after installation of the rail noise-reduction system, and how the train speed affects the noise reduction of the system. Comparing train types, the freight train and Class 4020 train had larger noise reduction at speed of 60 to 70 km/h though the noise reduction effect reduced as the train speed increased. This result suggests that for the freight train and Class 4020 train, when they run at lower speeds, the dominant source of noise is the rail, though the wheel noise, another main source of noise, becomes relatively larger as the train speed increases. As for Class 8033 and 4024 trains, in contrast, the rail and wheel contribution for noise production does not much change throughout the entire range of train speed. And one possible reason for this result is that the Class 4024 train uses smaller wheels that produce smaller radiation sound.

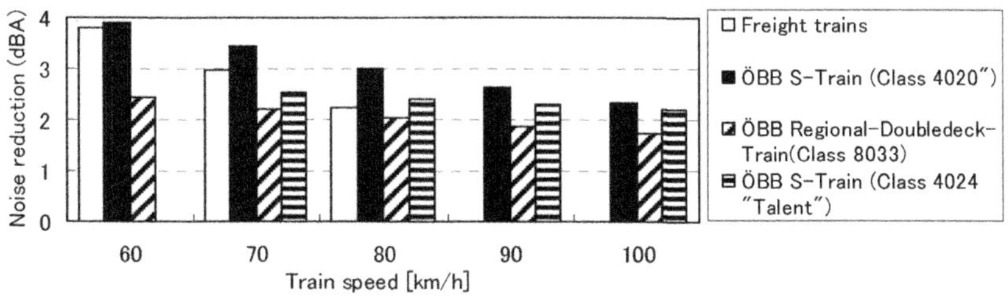

Fig. 10. Train Speed vs. Contribution of the Rail Noise Reduction System

Figures 11, 12, 13 and 14 show frequency analyses for noise measurements, respectively, for each train type under test. These figures show data for train speed of 80km/h. These figures show that for the freight train, the noise peak is at around 1 kHz while for other train types it is between 630 and 800Hz; and the contribution of the rail noise reduction system is largest at 3-7 dB between 630Hz and 1.25kHz.

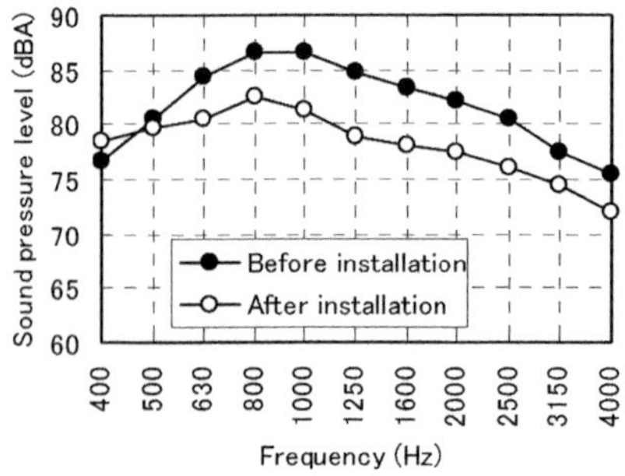

Fig. 11. Frequency Analysis for Freight

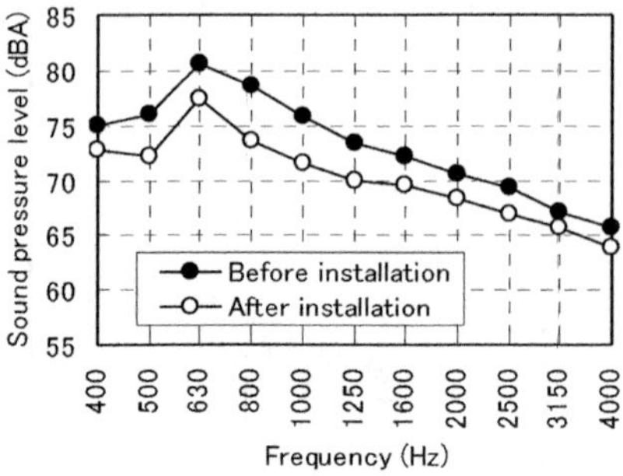

Fig. 12. Frequency Analysis for Class 4020

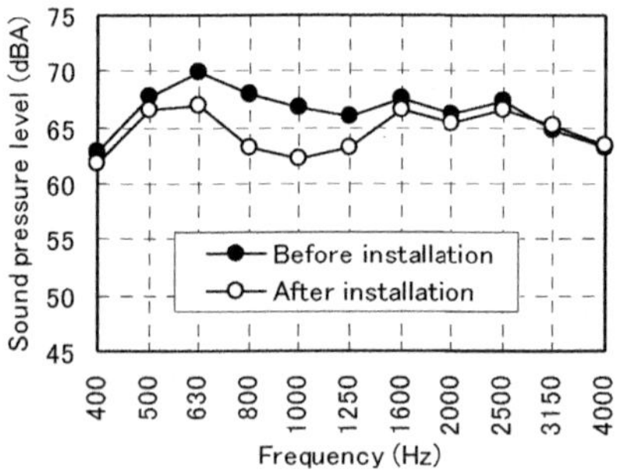

Fig. 13. Frequency Analysis for Class 8033

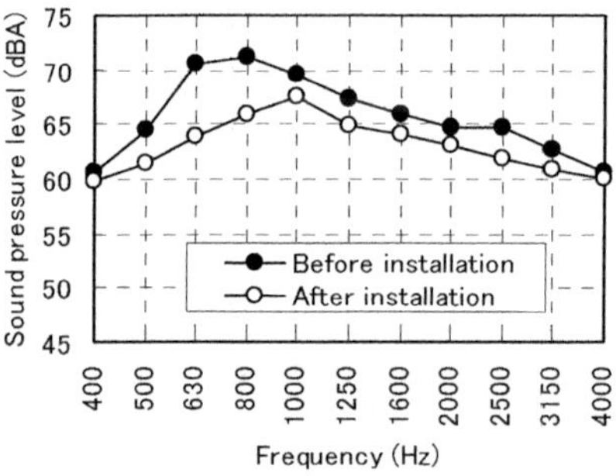

Fig. 14. Frequency Analysis for Class 4024

4 Demonstration of the Effectiveness of the Rail Noise Reduction System/Damping Material for Bridge Section

4.1 Overview of Experiment

The experiment was conducted on part of the line to Hainburg about 40 km east of Vienna, consisting of bridge sections alternating with normal track sections. Table 2 shows the track conditions for this experiment. The rail noise reduction system was installed at each of the three bridge sections to cover a total installation length of 60 m. The damping materials were installed on each side of about 15-m-long part of the bridge webs as "additional measures" to the rail noise reduction system. As with measurement at normal sections, the noise-measuring method per ISO 3095 was employed. The experiment used trains of ÖBB S-Train (Class 4020) and ÖBB S-Train (Class 4024).

Table 2. Track Conditions of the Bridge Section Used for the Test

Track configuration	Rail	Sleeper	Train speed
Straight	S49	Synthetic sleeper	30 to 65 km/h

4.2 Result

Figures 16 and 17 show the equivalent continuous A-weighted sound pressure level (LA, pb) as a function of the train speed in logarithmic presentation. The solid lines in these figures show approximated curves for measurements before installation of the rail noise-reduction system. The dotted lines represent those after installation of the system. The chain lines represent those after installation of both the rail noise reduction system and damping material. These figures show that the Class 4020 train produced more noise than Class 4024 trains. The reason for the difference in noise production level again is the abovementioned difference in wheel roughness between the two vehicle types. Class 4020 train sets are cast iron block braked with high surface roughness of wheels while modern class 4024 train sets are disc braked with thick small wheels.

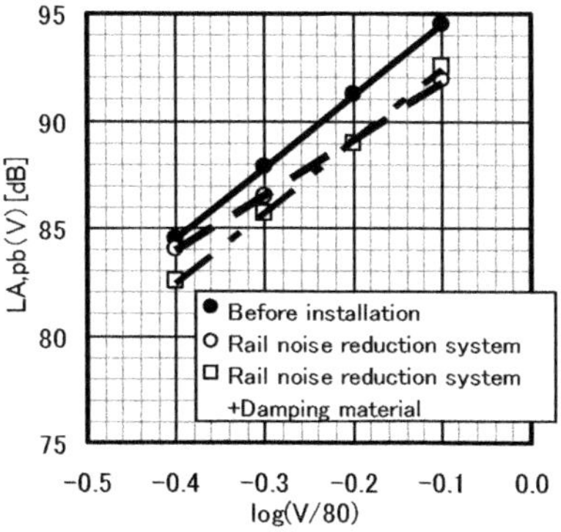

Fig. 15. Class 4020 Train Speed vs. Noise Level

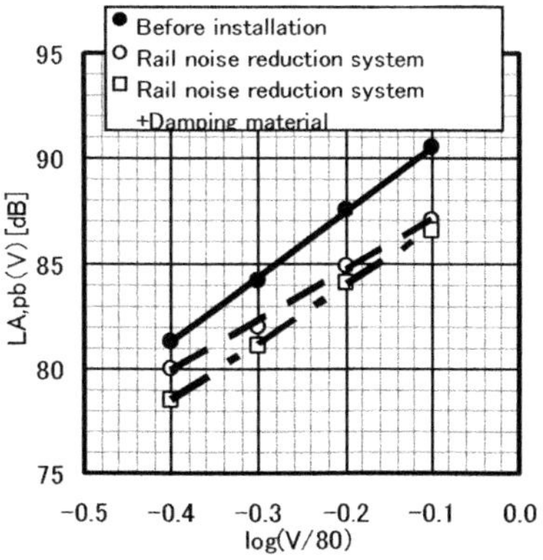

Fig. 16. Class 4024 Train Speed vs. Noise Level

Figure 17 and 18 show, for each train type, how the train speed affects the noise reduction of the rail noise reduction system and damping material. These figures show that the rail noise reduction system and damping material contribute to noise reduction by 0.5 to 3.5 dB and 0 to 1.5 dB, respectively; therefore, the combined use of these two measures leads to noise-reduction by 2 to 4 dB in total. These figures also show that the rail noise reduction system contributes to larger noise reduction when the train runs at higher speed; however, the damping material has less effect on noise reduction at higher train speed. This is probably because,the bridge structure generates more noise than the rail at low train speeds; however, the rail noise becomes greater as the train speed increases. Another finding is that the Class 4024 train that has small wheels that produce smaller sound radiation is given a larger noise reduction contribution of the rail noise reduction system, and the damping material's contribution is not affected by the particulars or operating characteristics of the trains.

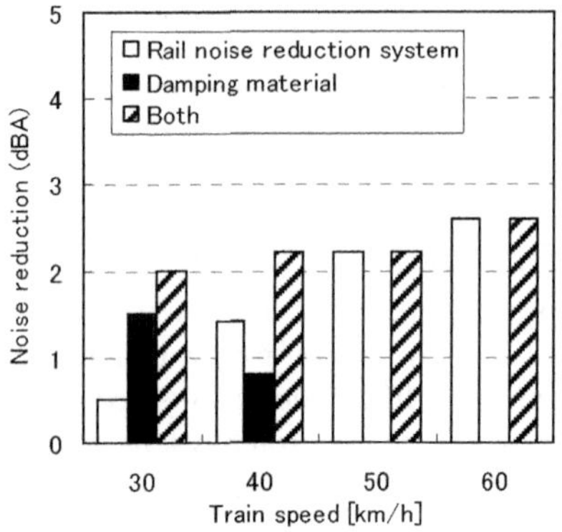

Fig. 17. Noise Reduction for Class 4020 Train

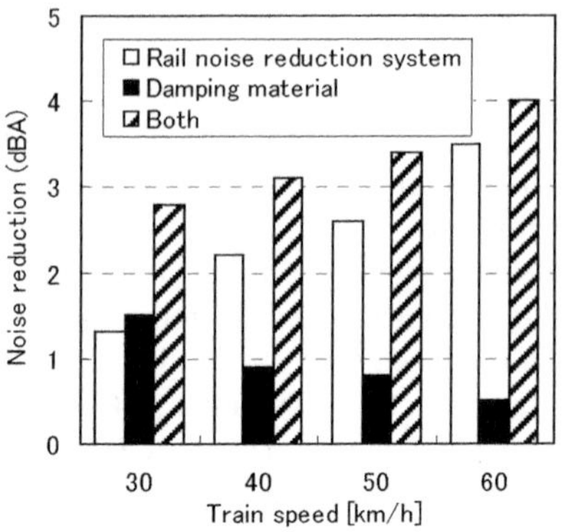

Fig. 18. Noise Reduction for Class 4024 Train

5 Conclusion

The high-performance damping material intended for railway noise reduction was tested in cooperation with Austrian Federal Railways using its actual tracks. The rail noise-reduction system based on this damping material has proved effective for reducing noise by 2 to 4 dB when it was applied for a normal section (ballast track). For a bridge section, combined use of the rail noise reduction system and damping material can reduce noise by 2 to 4 dB as a sum of 0.5 to 3.5 dB by the rail noise-reduction system and 0 to 1.5 dB by the damping material. To further reduce noise, train maintenance for keeping the wheel surface roughness at lower level and use of smaller wheels should be taken into consideration in addition to direct rail noise-reduction measures such as those mentioned here.

References

[1] Railway Technical Research Institute: Measuring and Evaluation Technologies for Railway Noise Annoyance (October 2003)
[2] Kitagawa, T.: An investigation on the Influence of Wheel and Track Parameters upon Rolling Noise. RTRI Report 22(5) (May 2008)
[3] Zenda, Y., Tanaka, S., Nagakura, K., Obara, T., Satou, K., Minami, H.: Effect of Rail/Wheel Roughness and Wheel Shape on Rolling Noise. RTRI Report 22(5) (May 2008)
[4] Asmussen, B., Stiebel, D., Kitson, P., Farrington, D., Benton, D.: Reducing the Noise Emission by Increasing the Damping of the Rail: Results of a Field Test. In: Noise and Vibration Mitigation for Rail Transportation Systems. NNFM, vol. 99, pp. 229–235 (2008)

Detection of Brake Type and Tread Surface Quality of Passing Trains Based on Rail-Sleeper-Force Measurements

Frank Mueller-Boruttau[1], Norbert Breitsamter[1], Holger Molzberger[1], and Alfons Buchmann[2]

[1] imb-dynamik,
D-82266 Inning, Breitbrunner Str. 5, Germany
Tel: +49 8143 992800, Fax: +49 8143 8767
e-mail: fmb@imb-dynamik.de
[2] Karlsruhe Institute of Technology, Institute for Road and Railway Systems
76131 Karlsruhe, Otto-Ammann-Platz 9, Germany
Tel: +49 721 608-8994, Fax: +49 721 608-8998
e-mail: alfons.buchmann@kit.edu

Summary

Tread surface quality and *tread defects* of the wheels of passing trains are the predominant influences on the noise radiated from wheels and rails. It is well known that the wheels of vehicles with cast iron tread brakes are much noisier than wheels with other brake types.

In many cases the train operator does *not* know which kind of brakes the vehicles of a train are equipped with. However, when analyzing trackside noise measurements it would be very helpful to have information on noise level relevant data like wheel tread state and brake type. But it is usually not possible to gather such data of arbitrary passing trains.

Recent imb-dynamik measurements with the rail-sleeper-force measurement system MONI from INNOtec Systems GmbH, Germany, have shown that this information can be gathered even for passing trains at arbitrary speed.

When evaluating simultaneous measurements of noise and rail-sleeper-force a very good correlation between tread state, rail-sleeper-force and radiated noise could be found. In addition the brake type – cast iron and K-block – could be identified by evaluating the roughness of the wheel tread. German railways DB confirmed the brake type identifications performed during the measurements so far.

Now the relationship between tread state and radiated noise can be determined even *for individual wheels*. It is possible to detect precisely how much noise is caused by a particular wheel tread quality or a particular tread defect.

Rail-sleeper-force measurements can be used to predict wheel-rail noise levels and to design countermeasures.

T. Maeda et al. (Eds.): Noise and Vibration Mitigation for Rail Trans. Sys., NNFM 118, pp. 167–173.
springerlink.com

1 Introduction

It is well known that vehicles equipped with cast iron tread brake blocks are much noisier than those with K blocks. But if one is performing standard train noise measurements one is *not* able to tell the brake block type from these measurements alone. A wheel with K blocks and tread defects can be as noisy as a cast iron wheel with a normal tread state. One is also not able to tell exactly which vehicle, and even less which wheel, has caused which amount of noise.

Often it is expected that a train operator knows – and thus can tell the experimenter – the kind of brakes the wheels of a train are equipped with. This is however not the case. One of the reasons is that even vehicles of the same type sometimes come with varying brake block types.

It would be very taxing to derive that information only from noise measurements. In fact, this would require a complex and expensive measurement technology like, for example, microphone array measurements and a highly sophisticated analysis procedure.

For many reasons it is highly desirable to know the brake block types *and* the tread state of each wheel *and* their respective noise emissions. It should of course, also be possible to obtain that knowledge without stopping the train, collecting information from the train operator, or the like.

Apart from noise measurements there are other standard measurement techniques at hand such as vibration measurements, force measurements, or one could perform optical analyses of train video or images.

If only vibration measurements are at hand it is difficult or even impossible to get quantitative tread data for single wheels since adjacent wheels interact via vibration transmission along the rail and the subgrade. Image analysis procedures are quite expensive and the authors are not aware of any existing techniques to identify brake types that way; tread state analysis that way is fully out of question.

It may come as some surprise that *trackside dynamic force measurements on passing trains* are able to solve the task at hand. It is well understood that these force measurements must comprise high frequencies up to many kHz since the acoustical phenomena encompasses these frequency ranges. So the force measurement system must be able to deal with highly dynamical forces.

2 Rail-Sleeper-Force Measurements

Ideally one would like to know the wheel-rail-force. Various measurement systems yielding this force exist; however, they are quite expensive, not mobile and versatile enough, and / or their frequency range is not large enough.

During recent years, the measurement system MONI has been developed by INNOtec Systems GmbH, Germany. Experiments have shown that its frequency range and other properties satisfy the requirements described above. In particular it allows the detection of the wheel tread state with extreme precision up to high frequencies; and, since the wheel tread roughness is in turn closely related to the brake type, it is possible to find out the brake type. (Actually additional information is derived from the measurements, such as the vehicle type; that information is also used in further analyses.) The first results gained during the start up of a German research project of wheel-rail noise including MONI-measurements are presented here. It is however to be expected that the data with this system of prove very valuable for in-depth analysis of the noise emission.

The measurement system uses sleeper pads equipped with force gauges. The modified sleeper pads are installed in the track like standard sleeper pads. Thus the system is mobile and the present superstructure is not at all tampered with; instead it is virtually identical with the state of the track before installation of the gauges. This requirement is essential for the validity of the gathered data.

Fig. 1. MONI measurement system: a) sensors, b) set for one sleeper, c) sensor position

3 Brake Type and Defect Detection

For many vehicles – especially cargo cars – brakes with two types of brake blocks are used: a) *cast iron*, and b) *K-block* (synthetic material). Cast iron blocks roughen and corrugate the tread while K-blocks, sort of polish the tread instead. With the measurement technique described above it is possible to detect the tread state via measuring rail-sleeper-forces.

Fig. 2, Fig. 3 and Fig. 4 show examples of original and unprocessed measurement time series of rail-sleeper-forces for cast iron brakes, K-block brakes and polygonalized wheels, respectively, as well as photos displaying treads of wheels with these features. (Photos are examples, since the trains – as is usually the case for such investigations or general measurements – could not be stopped. However, the investigations and checks of the results gained so far have proven that the assessment was correct in all cases.)

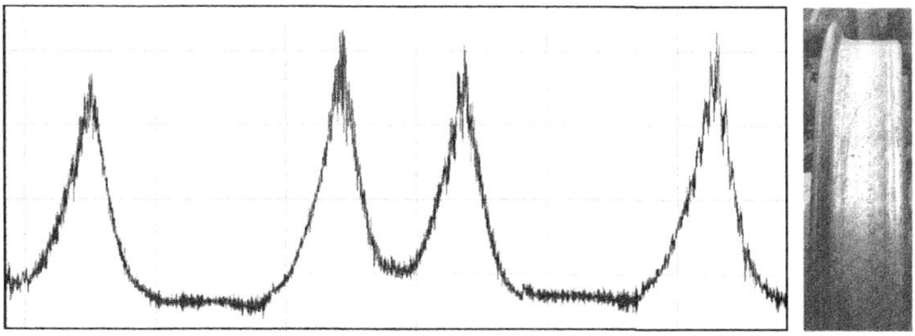

Fig. 2. Cast iron brakes (noisy wheels): a) rail-sleeper-force, b) example of wheel tread (color figure online)

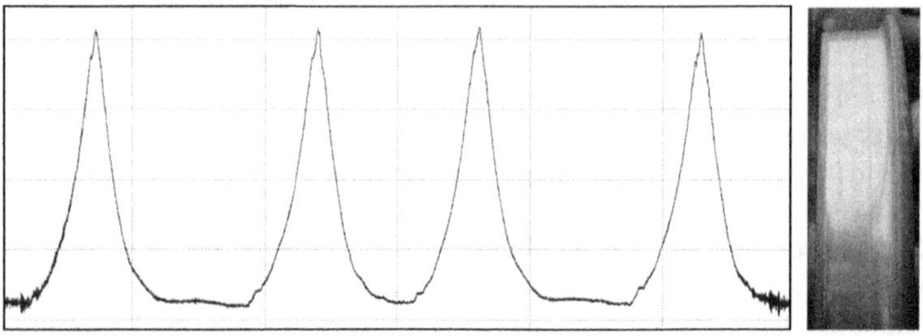

Fig. 3. K-block brakes (silent wheels): a) rail-sleeper-force, b) example of wheel tread (color figure online)

Thus, the *detection of the brake type* is possible, as well as an assessment of the *tread roughness and defects severity*. German railways DB explicitly confirmed the brake type assessments.

It will also finally be possible to automatically detect tread and wheel defects based on the measured time series. Fig. 4 shows as an example of the defect type *polygonalization.*

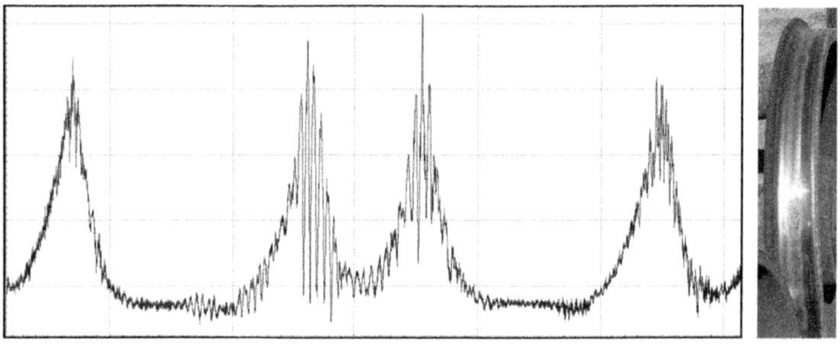

Fig. 4. Wheel with polygonalization (2nd and 3rd wheel): a) rail-sleeper-force, b) example of wheel tread (color figure online)

In a similar way each tread defect has a characteristic "signature" that can be used to identify the defect type.

4 Relation between Noise and Rail-Sleeper-Force Spectra

For a cargo train consisting of cargo vehicles of identical type with *cast iron* as well as *K-block brakes,* its noise levels and rail-sleeper-forces have been measured during one passage at the distances of 2.5 m, 6 m and 12 m.

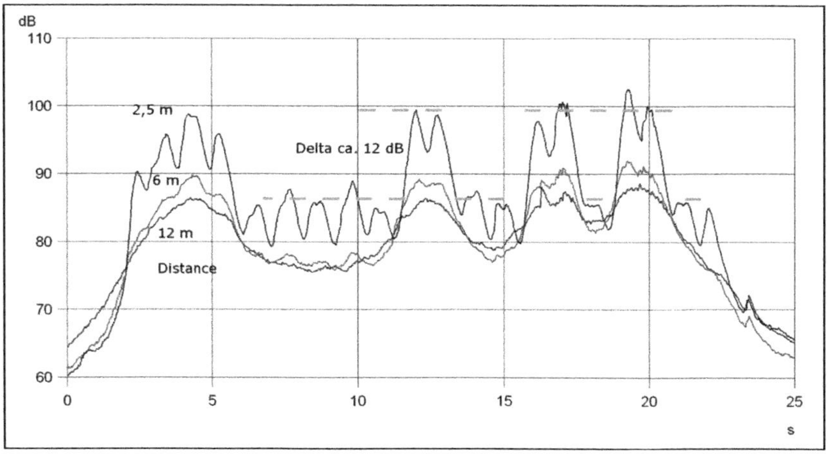

Fig. 5. Noise measurements at various distances 2.5 m, 6 m, 12 m (color figure online)

Fig. 5 shows that the difference between noisy and silent vehicles (Delta) is about 12 dB. But even noise measurements at a distance of 2.5 m do not reveal any information on the causes of peak levels or special signal characteristics at, e.g. 17 s or 20 s. Analysis of force-measurements however, do reveal the causes quite clearly. As of now (August 2010) such analyses must be performed manually; this will be updated in the near future.

To show the approximate interchangeability of noise and force measurements 1/3rd octave spectra of noise and rail-sleeper-force have been calculated for appropriately selected time intervals for both cast iron vehicles and K-block vehicles; they are displayed in Fig. 6. Force *and* noise vary similarly; thus – after the end of the research project – one spectrum can be used to predict the other one. (It must be taken into account that the data presented here is based on relatively few measurement time series. However, the measurements gathered show that the technique as such is working.)

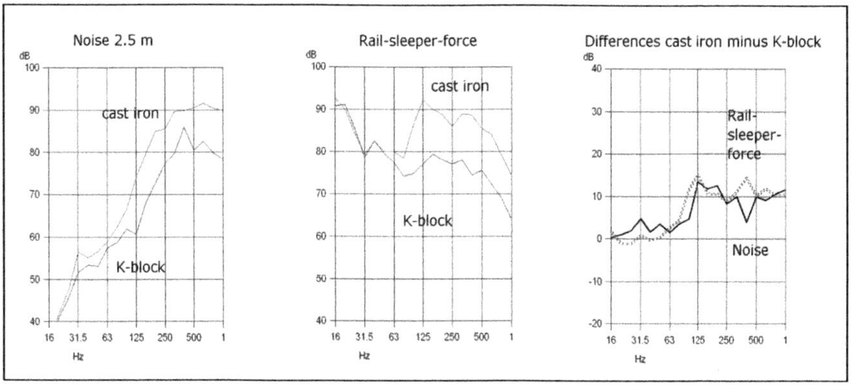

Fig. 6. Spectra of noise, rail-sleeper-force and their differences (color figure online)

The spectra in Fig. 6 show that for frequencies below 80 Hz there is only a small difference between cast iron levels and K-block levels. This is a hint that the brake type does not influence the *out-of-roundnesses*. (It will be investigated whether this is the case in general. As of now, no such data with a broad statistical database is thought to exist.) However, for higher frequencies a level of difference can be found that is quite similar for noise *and* force. Thus one can expect that it will be possible in the near future to predict the noise radiated from the wheels based on the rail-sleeper-force measurements. It is assumed that the 400Hz deviation was caused by sleeper eigen vibration.

The correlation between noise and rail-sleeper-force can also be seen in spectral time charts in more detail (Fig. 7). Of course there are large differences between noise and force in the *low frequency range* due to the fact that there is no noise radiation in the low frequencies.

The general idea is to formulate a transformation function to derive diagram 7a from diagram 7b. The results obtained until now (August 2010) are very promising; one can expect that the goals aimed at will be achieved.

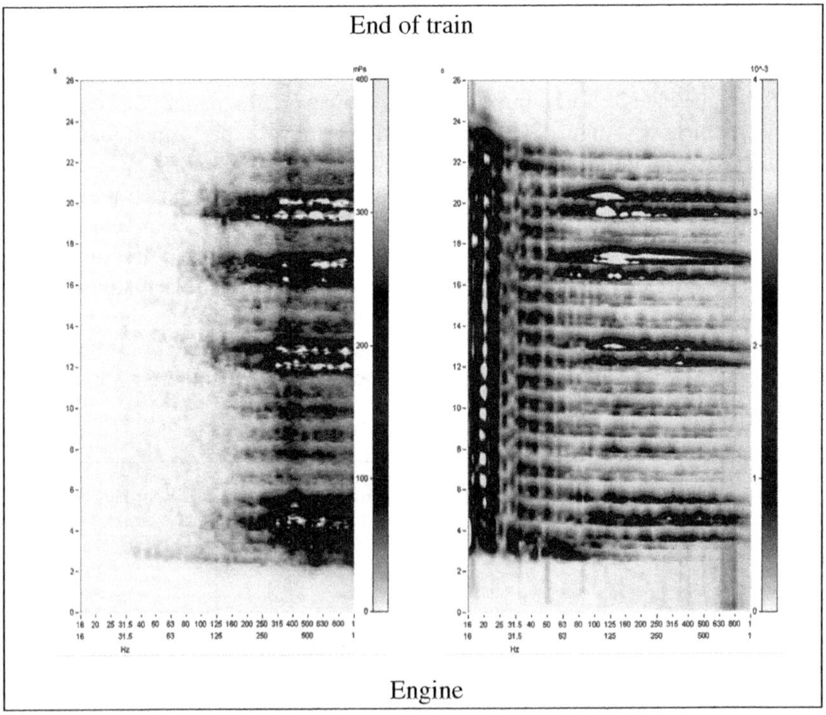

Fig. 7. Spectral time chart of a) noise and b) rail-sleeper-force

5 How to Deal with Rail Roughness

The analysis procedure described above works best when the rail surface at the measurement site has a good quality. It seems quite pointless to perform train-and-wheel influence oriented noise measurement at a site where the noise is predominantly influenced by rail roughness. However even a good quality rail surface influences the noise levels somewhat. The force measurement and software procedures are able to deal with that problem.

Describing the procedures in detail would vastly exceed the scope of this article, however, some basic information shall be given. The basic idea is that the rail surface is always the same for each passing wheel of a train, whereas the treads are varying statistically. Averaging procedures, averaging over, one and / or a number of trains can profit from this knowledge.

As an example a close look on the peak levels of Fig. 3 reveals that all peaks have a characteristic indention on the top left side as well as another one on the foot of the left slope. These repeating features are caused by small rail surface defects. Since they are *not* random but systematic their influence can be calculated and accounted for.

Also the rail-sleeper-forces of extraordinary smooth wheels – new or recently reprofiled ones – are predominantly influenced by the rail roughness. The rail-sleeper-force of an ideally round, quasi-polished wheel allows deriving the exact rail surface. Based on these facts the rail surfaces' influence can be corrected even for less than optimal measurement sites. It is also possible to detect the rail surface degradation during long-term measurement campaigns.

6 Outlook

It is the central aim of the research project to identify the relations between the various influences on the noise of a train such as tread state, vehicle type, speed, etc. A set of formulae shall be defined enabling one to calculate the noise levels and even the spectra based on the knowledge of these influences without actually measuring the noise.

Since the research project is a current one, and a large measurement campaign is being performed at the time of writing this article, more insight will be available at a future date.

Estimation of Vibration Power Flow to and Sound Radiation from a Railway Concrete Viaduct Due to Vehicle/Track Interaction

Z.G. Li and T.X. Wu

State key Laboratory of Mechanical System and Vibration,
Shanghai Jiao Tong University,
800 Dong Chuan Road, Shanghai 200240, China
Tel.: +86 21 34206332-819; Fax: +86 21 34206332-819
e-mail: txwu@sjtu.edu.cn

Summary

The structure-born noise from a railway concrete viaduct due to vehicle passage is predicted via analysis of vehicle-track interaction, vibration power flow into and sound radiation from the structure. The dominant component of the structure-born noise from a concrete viaduct is below 100 Hz, which is associated with the vehicle's unsprung mass oscillating on the track stiffness, and irrelevant to train speed. Use of soft rail pad/fastener can effectively reduce the structure-born noise via reducing the dynamic force transmission to the viaduct, and thus, the vibration power flow. A reduction of 15 dB (A) in sound power radiated from a concrete viaduct can be achieved by replacing 60 MN/m pad with 6 MN/m rail fastener.

1 Introduction

Concrete viaduct structure is widely used for the elevated lines in Shanghai Metro Light Rail System. Although its construction is easier and costs less than the underground tunnel, the sound radiated from the viaduct structure during train passage causes environmental noise problem. The predominant component of radiation from a concrete viaduct is usually below 100 Hz, and can propagate over long distance. As the size of a concrete viaduct is large, the sound radiation from the structure is difficult to control except by reducing its vibration energy level, e.g. using soft rail pad/fastener to decrease the dynamic force transmission from the wheel/rail interface to the structure.

In this work, the vehicle-track dynamic interaction due to a relative displacement excitation is simulated using a slab track and a simplified vehicle model. The force in the rail pad is calculated and used as the excitation to the viaduct structure. The mobility matrix of the viaduct structure is determined using the finite element

T. Maeda et al. (Eds.): Noise and Vibration Mitigation for Rail Trans. Sys., NNFM 118, pp. 175–183.
springerlink.com

software ANSYS. The vibration power flow injected into the structure caused by roughness excitation is calculated, provided the viaduct is composed of concrete plates, and the distribution of vibration energy among the plates is determined according to their mobility or thickness. The radiation ratio is calculated for each plate of the viaduct using the method described by Xie, et al.[1]. The sound power radiated is predicted from the viaduct structure. Using the model and the methodology developed, the effects of soft rail pad/fastener are estimated on reduction of the structure-born noise from a concrete viaduct.

2 Vehicle-Track Interaction and Excitation to Viaduct

Considering a slab track laid on a viaduct, the rail is modelled as an Euler beam on an elastic foundation. In the audible frequency range, as the vibration displacement is much smaller than the deformation of medium-soft (≤200MN/m) rail pad, the concrete viaduct can be simplified to a rigid foundation in the track model. To calculate the wheel-rail interaction force, an eighth of vehicle is employed to interact with the track, see Fig. 1.

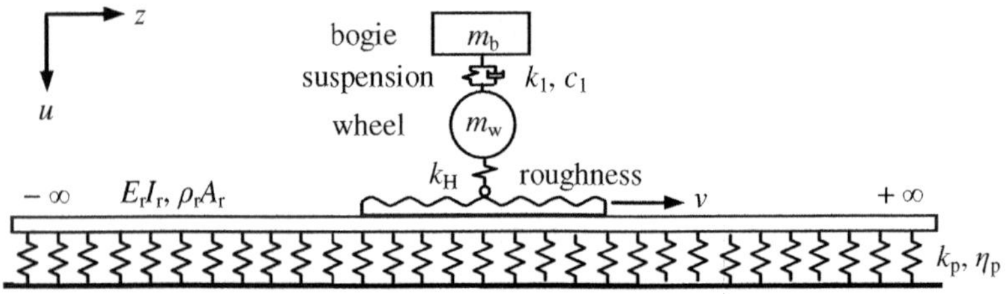

Fig. 1. Track model on viaduct

The wheel-track interaction force due to a relative displacement excitation is calculated using the following formula

$$F_c = -\frac{R}{\alpha_v + \alpha_c + \alpha_t}, \tag{2.1}$$

where R is the combined roughness spectrum on the wheel and rail tread, α_v is the receptance of vehicle, α_c is the contact receptance, which is inverse of the linearised contact stiffness, $1/k_H$ and α_t is the track receptance.

The force in the rail pad is calculated and regarded as the excitation to the viaduct structure, and can be determined according to the pad stiffness and rail displacement due to the roughness excitation.

Table 1. Parameters used in the model

Bogie	m_b	700 kg	Rail pad	K_p	60 MN/m
Wheel	m_w	900 kg		η_p	0.25
Vehicle suspension	k_1	525 kN/m	Concrete viaduct material property	E	36 kN/mm^2
	c_1	12 kN·s/m		μ	0.2
Contact stiffness	k_H	1.31 MN/mm		η	0.1
Rail	E_rI_r	6.6 MN·m^2		ρ	2500 kg/m^3
	P_rA_r	60 kg/m	Viaduct support	k_b	3.4 MN/mm
Span length	d	0.625 m		η_b	0.25

The parameters used for the vehicle, track and viaduct model are listed in Table 1. The combined wheel-and-rail roughness spectrum is assumed in Fig. 2 compared to the ISO limit for rail.

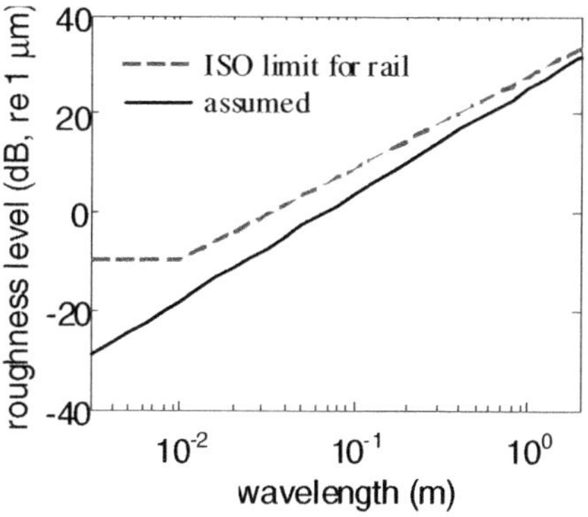

Fig. 2. Roughness spectrum (color figure online)

The calculated wheel-rail interaction force is shown in Fig. 3 due to the assumed roughness and at train speed 80 km/h. Also shown in Fig. 3 is the force transmitted to the viaduct, which is the root of the sum of the squared force in all rail pads for the vehicle at mid-span of the viaduct. The peak in wheel-rail force appears at about 60 Hz, at which the unsprung mass of the vehicle oscillates on the track support stiffness. The force transmitted to the viaduct is also shown as a peak there. The wheel-rail interaction force is smallest at about 200 Hz, at which the rail bounces on the pad stiffness.

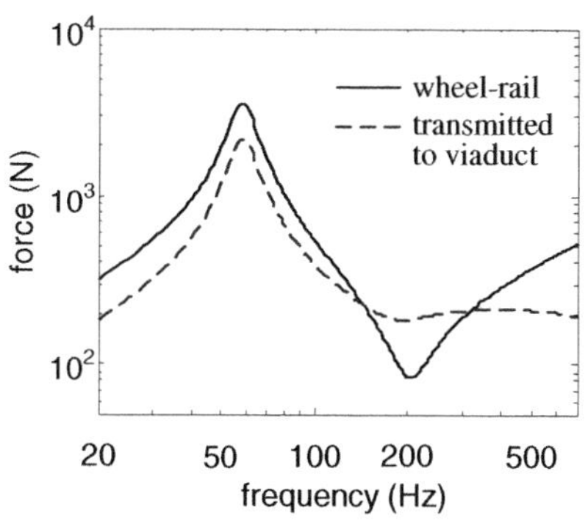

Fig. 3. Excitation force, wheel at mid-viaduct (color figure online)

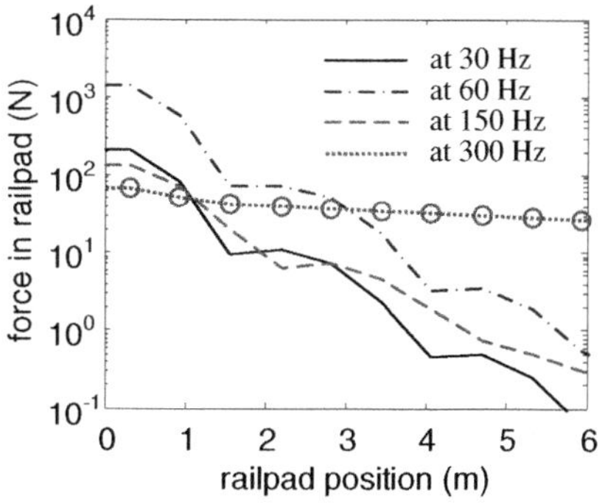

Fig. 4. Decay of force in pad, ○ pad position (color figure online)

Fig. 4 shows the variation of the force in rail pad along the track. The force transmitted to the viaduct can be seen to decay rapidly at low frequencies at an averaged rate 10 dB/m, whereas it decays slowly at high frequencies, for example, 1 dB/m at 300 Hz.

3 Power Flow to Viaduct Structure

The viaduct considered is of box-shaped structure. Its cross-section is shown in Fig. 5. There are two tracks on the top and the vehicle is supposed to run only on one track. At each end of a viaduct a vertical plate is placed to close the box and to transmit load from the deck to the support of the viaduct. In the middle of a viaduct a vertical plate is also placed to transmit load from the deck to the bottom.

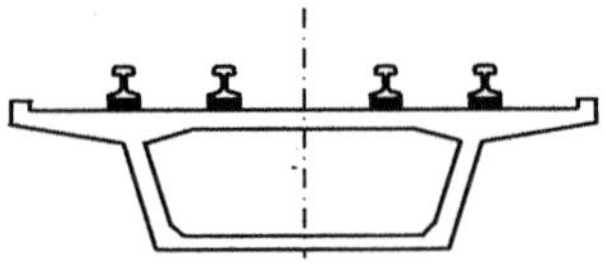

Fig. 5. Cross-section of viaduct

The viaduct structure is analysed using the FE software ANSYS. It is divided into deck, flange, web and bottom plates and they are modelled by shell elements. The minimal meshing size for the plates is 0.2 m to meet the requirement of six elements per wavelength at 700 Hz, the highest frequency of analysis.

The geometric size of the model is shown in Fig. 6. Each viaduct is 30-m long and its ends rest on the rubber-steel bearings for absorption of impact.

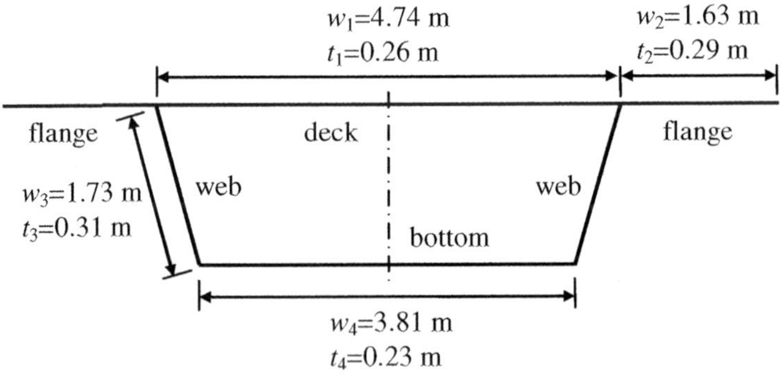

Fig. 6. Cross-sectional size of viaduct FE model

The natural frequency and vibration mode are calculated for the viaduct structure. As the structure is large, its vibration modes are plentiful even in low frequency region. Fig. 7 shows the modal density of the viaduct structure in one third octave band. It can be seen to be low only in a few bands at low frequency, and in most bands it is between 0.6 and 1.2. In 63 Hz band, in which the resonance occurs for the unsprung mass on the track stiffness, the modal density of the viaduct exceeds one.

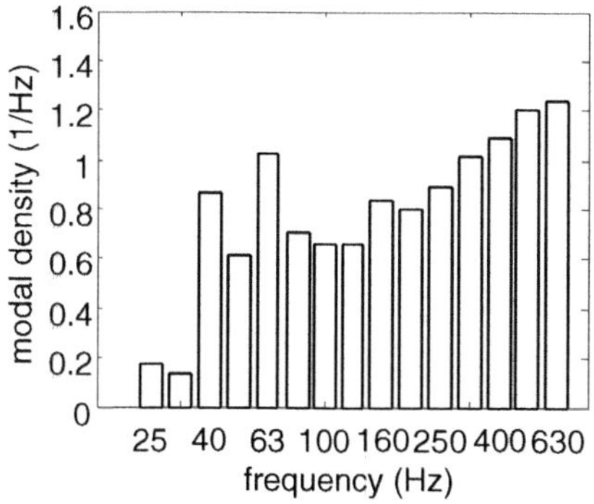

Fig. 7. Modal density of viaduct

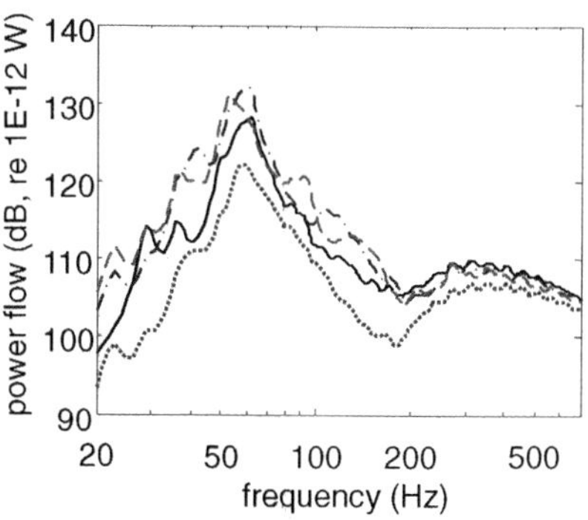

Fig. 8. Power flow to viaduct, wheel —at mid span, – –3/8 span, –·–1/4 span, ····end (color figure online)

The vibration power flow injected into the viaduct structure is predicted due to the vehicle-track interaction (two wheels and two rails) caused by the roughness spectrum assumed. It is calculated according to the excitation force in the viaduct and the vibration velocity at the forcing point, given by

$$P_{\text{in}} = \frac{1}{2}\text{Re}\left(\sum_{m=1}^{2N_{\text{p}}} F_{\text{p}m} V_m^*\right), \tag{3.1}$$

where F_{pm} is the force in the m-th rail pad, V_m is the velocity of structure vibration at the pad position, and * represents the complex conjugate. N_{p} is the number of the pad under one rail along the viaduct. The power flow P_{in} is also given by

$$P_{\text{in}} = \frac{1}{2}\text{Re}\left(\mathbf{F}_{\text{p}}^{\text{H}} \mathbf{Y}^{\text{H}} \mathbf{F}_{\text{p}}\right), \tag{3.2}$$

where $\mathbf{F}_{\text{p}}$ is the force vector and $\mathbf{Y}$ is the mobility matrix of the viaduct structure. Superscript H represents the complex conjugate transpose operation.

The power flow into the viaduct caused by the roughness spectrum in Fig. 2 is shown in Fig. 8 at train speed 80 km/h. The power flow can be seen to reach a peak at about 60 Hz, at which resonance of the unsprung mass on the track stiffness occurs, and the excitation force to the viaduct reaches maximum. Away from this frequency the power flow comes down. It is similar for the vehicle at different places except at the end of the viaduct, where the power flow into the viaduct is lowest because the excitation force caused by vehicle-track interaction is equally distributed to the two viaducts adjacent.

The power flow into the viaduct, averaged for the vehicle at different places of a viaduct, is shown in Fig. 9. It can be seen to increase with train speed and its total level increases at a rate of 7 dB per doubling the speed. However, the frequency, at which the power flow peaks, does not change with train speed.

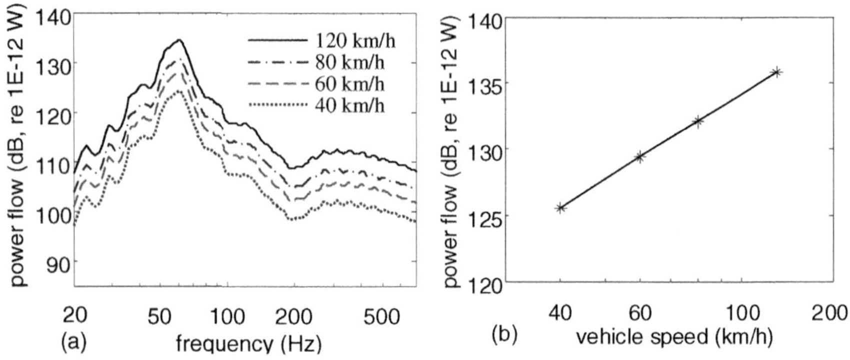

Fig. 9. (a) Power flow into viaduct at different train speed, (b) total power flow level (color figure online)

Fig. 10 displays the predicted results of the averaged power flow into the viaduct when different rail pads/fasteners are used at train speed 80 km/h. It can be observed that the power flow into the structure can effectively be reduced by use of soft rail pads/fasteners. This is because the soft pads/fasteners work as isolators to prevent transmission of vibration power from the rail to the viaduct. In addition, the frequency, at which resonance of the unsprung mass on the track stiffness occurs and the power flow peaks, decreases with rail pad stiffness and so does the peak value of the power flow. For the track using 60 MN/m rail fastener (WJ-2 in China), for example, the power flow can be reduced by about 10 dB when it is replaced with 6 MN/m rail fastener (Pandrol's Vanguard).

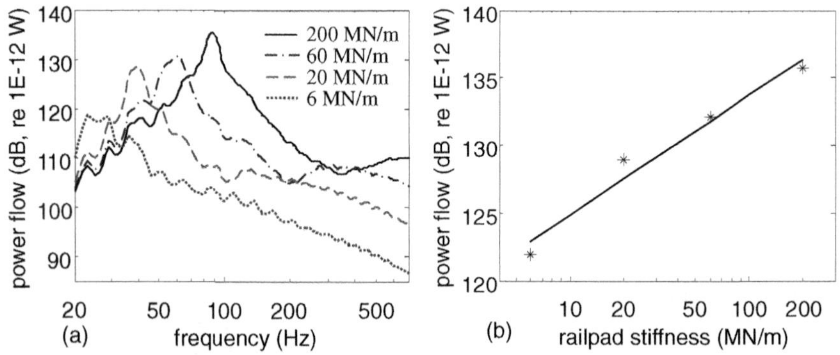

Fig. 10. (a) Averaged power flow into viaduct for different pad stiffness, (b) total power flow (color figure online)

4 Vibration Energy Distribution and Sound Radiation

Once the vibration power is injected into the viaduct, it transmits in the structure and radiates sound. Since the radiated power is much smaller than the dissipated power, the power injected into the structure, P_{in}, is approximately equal to the dissipated power by all plates of the viaduct structure [2]

$$P_{in} \approx P_{dis} = \omega\eta \sum_i m_i \left\langle v_i^2 \right\rangle, \tag{4.1}$$

where m_i and $\left\langle v_i^2 \right\rangle$ are the mass and spatially averaged velocity squared for plate i, respectively. For two well coupled sub-systems, if the modal overlap factor $M=\eta\omega n > 1$, where n is the modal density, the following relation holds [3]:

$$\frac{\left\langle v_1^2 \right\rangle}{\left\langle v_2^2 \right\rangle} = \frac{\mathrm{Re}(Y_1)}{\mathrm{Re}(Y_2)} = \frac{t_2^2}{t_1^2}, \tag{4.2}$$

where Y_1 and Y_2 are the mobilities, t_1 and t_2 are the thicknesses of the two coupled plates.

Applying equation (4.2) to (4.1) for the viaduct structure, the spatially averaged velocity squared for plate j of the viaduct can be determined by,

$$\left\langle v_j^2 \right\rangle = \frac{P_{in}}{\omega\eta\rho t_j^2 \sum_i S_i / t_i}, \tag{4.3}$$

where S_i and t_i are the area and thickness of plate i, respectively. The averaged velocity can be seen to be inversely proportional to the thickness of the plate.

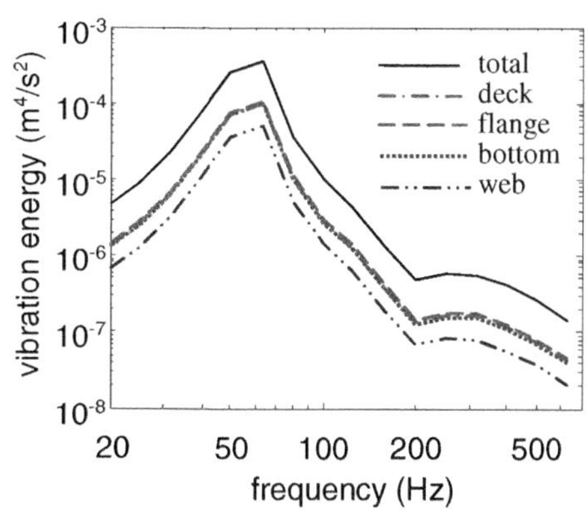

Fig. 11. Vibration energy of viaduct (color figure online)

The spatially averaged velocity squared of the sub-systems of the viaduct can be estimated using equation (4.3). Vibration energy in terms of product $S_j \left\langle v_j^2 \right\rangle$ is calculated for each plate of the viaduct due to the assumed roughness excitation at train speed 80 km/h, and the results are shown in Fig. 11. It can be seen that the vibration energy of the deck, flange and bottom is similar and higher, but it is lower for the web, although their area is different and the flange area is doubled as sound radiation comes out from both surfaces. The vibration energy of the viaduct structure reaches peak at about 60 Hz, the resonance frequency of the unsprung mass bouncing on the track stiffness.

Sound power radiated from each plate of the viaduct structure is predicted by,

$$P_{rad} = \rho_0 c_0 \sum_i \sigma_i S_i \left\langle v_i^2 \right\rangle, \tag{4.4}$$

where ρ_0 is the density of the air, c_0 is the sound speed in the air and σ_i is the radiation ratio of plate i of the viaduct.

The sound radiation ratio for each plate of the viaduct is calculated using the method described by Xie, et al [1]. It is an averaged radiation ratio for the excitation experienced at all positions of a plate. Fig. 12 shows the calculated results. The flange's radiation ratio is low at low frequencies because both surfaces of the flange radiate sound and thus acoustic short circuit is present.

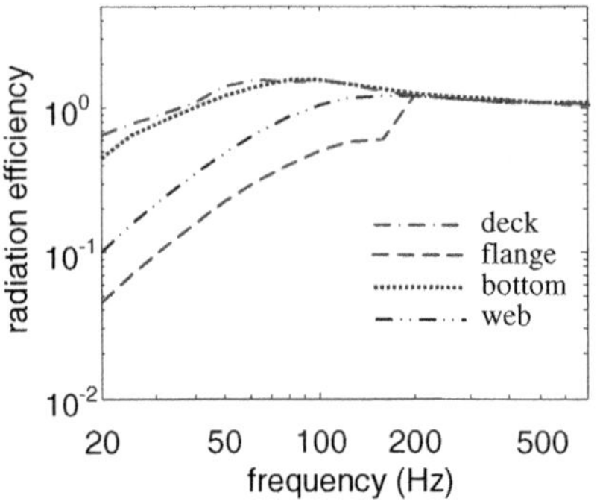

Fig. 12. Sound radiation ratio of viaduct parts (color figure online)

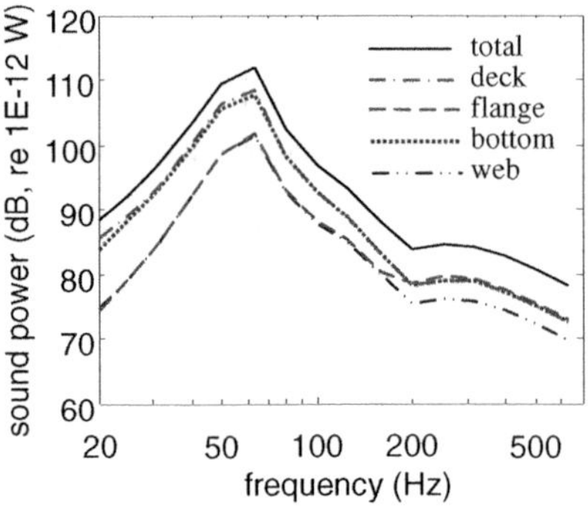

Fig. 13. Sound power radiated from viaduct (color figure online)

Fig. 13 shows the sound power level radiated by the structure at train speed 80 km/h. The main component is in frequency region 40-80 Hz. The component above 200 Hz comes down by 25-30 dB, compared to the peak. The major source of noise is the deck and bottom, whereas the flange and web are secondary.

Fig. 14 shows the sound power radiated at different train speeds. The sound power can be seen to increase by 7-8 dB when the train speed is doubled, but the frequency component distribution does not change with train speed.

The effects on structure-born noise reduction are shown in Fig. 15 by using soft rail pad/fastener. The sound power radiated is effectively reduced, for example, when the stiffness of the rail fastener decreases from 60 MN/m to 6 MN/m or from 200 MN/m to 20 MN/m, and also the sound power level radiated can be reduced by about 15 dB (A). This is because the vibration power flow into the viaduct is reduced and the resonance frequency of the unsprung mass on the track stiffness is shifted to the lower frequency region, refer to Fig. 10.

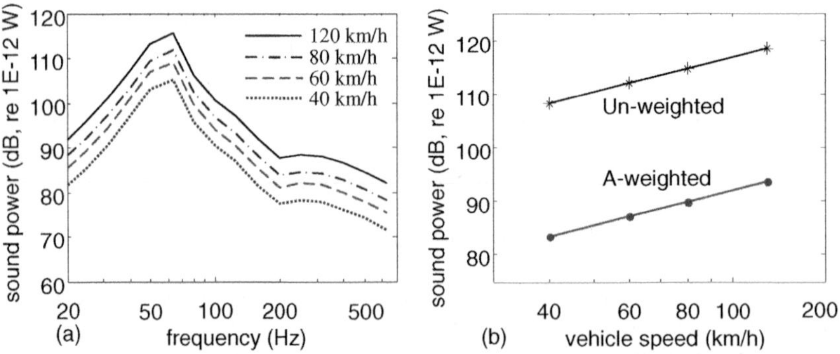

Fig. 14. (a) Sound power radiated at different train speed, (b) A-weighted sound power level (color figure online)

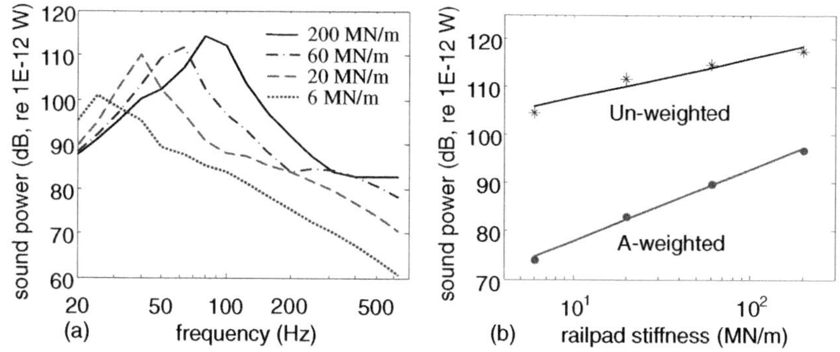

Fig. 15. (a) Sound power level using different pad stiffness, (b) A-weighted sound power level (color figure online)

5 Conclusions

The structure-born noise from a railway concrete viaduct is associated with vehicle-track interaction, dynamic force transmission and vibration power flow into the structure. The predominant frequency component of the noise is below 100 Hz, which is related to the unsprung mass oscillating on the track stiffness and irrelevant to the train speed. Soft rail pad/fastener can effectively reduce the structure-born noise by reducing the force transmission to the viaduct and thus the power flow. A reduction of 15 dB (A) in sound power radiated from a concrete viaduct can be achieved by replacing 60 MN/m pad with 6 MN/m rail fastener.

References

[1] Xie, G., et al.: The radiation efficiency of baffled plates and strips. Journal of Sound and Vibration 280, 181–209 (2005)

[2] Harrison, M.F., et al.: The calculation of noise from railway viaducts and bridges. Journal of Rail and Rapid Transit. 214(3), 125–134 (2000)

[3] Cremer, L., Heckl, M.: Structure Borne Sound, 2nd edn. Springer, Berlin (1988)

An Investigation on Vibratory and Acoustical Characteristics of Concrete Bridge for Shinkansen

Y. Kozuma and K. Nagakura

Railway Technical Research Institute,
2-8-38 Hikari-cho Kokubunji-shi, Tokyo 185-8540, Japan
Tel: +81 42 573 7353, Fax: +81 42 573 7418
e-mail: kozuma@rtri.or.jp

Summary

The railway bridge noise is generated by its vibration due to the wheel-rail interaction. In the vicinity of the bridge, the bridge noise has often offered greater contribution to the total wayside noise than other sources. Therefore, reduction of the bridge noise is an important issue for the wayside noise problem. In this article, vibratory and acoustical characteristics of the concrete rigid-frame bridge for Shinkansen have been investigated by field tests and numerical analysis to promote a better understanding of the bridge noise. Additionally, analysis of the girder concrete bridge has also been conducted.

1 Introduction

In the railway system, the structure borne noise consists of sound radiation from various elements of track and vehicle, such as rail, wheel and bridge. Fig. 1(a) shows the mechanism of the bridge noise generation. The bridge noise is generated by a vertical contact force between wheel and rail. The contact force is mainly induced by two factors: a relative displacement between wheel and rail due to the roughnesses on their surfaces, and a parametric excitation [1)] due to the wheel load. The contact force is transmitted from the rail into the bridge structure, and generates the bridge noise.

Fig. 1(b) shows an example of the spectra of the measured noise of Shinkansen at 300km/h around the concrete bridge. It is obvious that the concrete bridge noise is predominant in the frequency range of 63-400Hz. In order to reduce the noise components in this frequency range, a deeper appreciation of vibratory behaviour of the concrete bridge would be required, and a better understanding of the concrete bridge noise should be developed.

T. Maeda et al. (Eds.): Noise and Vibration Mitigation for Rail Trans. Sys., NNFM 118, pp. 185–192.
springerlink.com

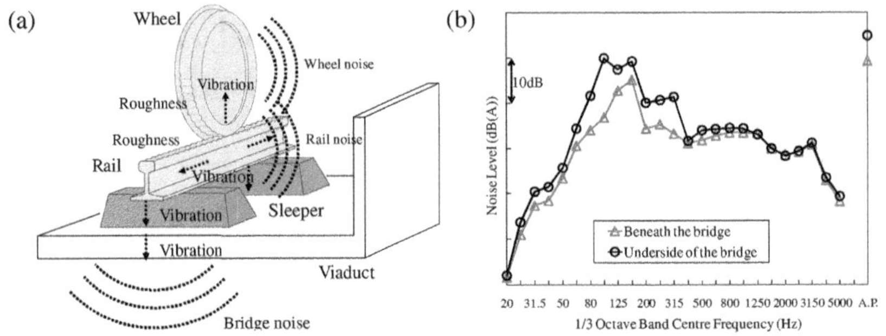

Fig. 1. Outline of the bridge noise. (a) Mechanism of bridge noise generation. (b) Example of the spectra of measured noise of Shinkansen at 300km/h near the bridge. (color figure online)

2 Vibratory Characteristics of the Concrete Rigid-Frame Bridge

2.1 Impact Test

The measurement was carried out by hitting 320 points on the concrete rigid-frame bridge shown as black markers in Fig. 2(b) with an instrumented impact hammer. By taking into account that the bridge has a symmetric profile, the excitation points were confined within a quarter of a single span. The vibratory behaviour of the bridge has been measured with an accelerometer installed on the underside surface of the centre slab.

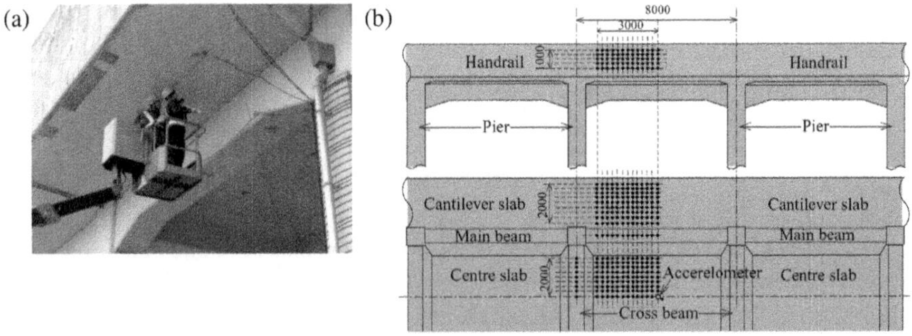

Fig. 2. Outline of the impact test on the concrete rigid-frame bridge noise. (a) Photograph of impact test. (b) Diagram of impact testing. (color figure online)

2.2 Vibratory Characteristics of the Bridge

Fig. 3(a) shows the point accelerance of the midsection of the centre slab. There are some peaks related to the resonance behaviour of the bridge.

In a previous study on the concrete rigid-frame bridge by Kitagawa [2)], the result showed similar trend in the frequency range below 200Hz. However, sharp peaks are also confirmable in high frequency range above 300Hz, which differs from the former study.

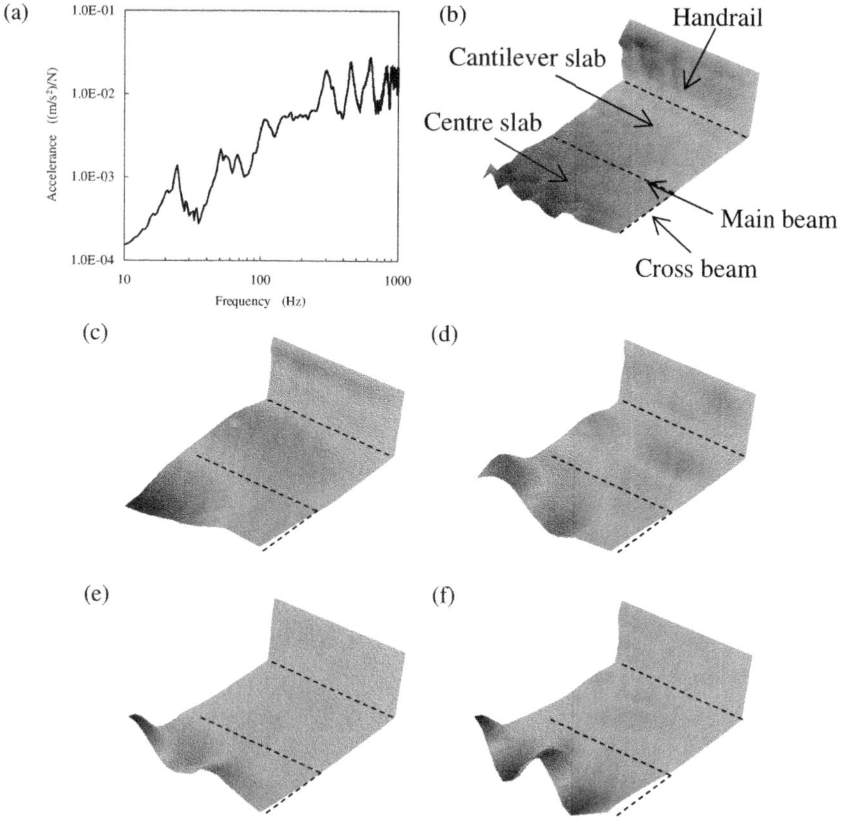

Fig. 3. Results of impact test. (a) Vertical driving point accelerance of the rigid-frame bridge. Modal shape of the rigid-frame bridge at (b) 24Hz, (c) 52Hz, (d) 194Hz, (e) 301Hz and (f) 451Hz.

The vibratory mode corresponding to each resonance frequency is confirmed by referring the measured frequency response functions. Fig. 3(b)-(f) show the modal shapes of the bridge.

The resonance peak at 24Hz in Fig. 3(a) seems to be related with the vibratory mode that the entire bridge vibrates in the same phase (Fig. 3(b)). At 52Hz (Fig. 3(c)) and 108Hz, the anti-node of the mode corresponds to the midsection of the centre slab. The centre slab and the cantilever slab vibrate in opposite phase at 52Hz.

The vibratory modes at 301Hz (Fig. 3(e)), 451Hz (Fig. 3(f)) and 625Hz seem to be associated with the local bending waves, which propagate in the centre slab.

Aforementioned vibratory modes are identified only by referring the peaks in the frequency response function at the midsection of the centre slab. Consequently, vibratory modes with nodes at the midsection of the centre slab have been neglected in this study. Therefore, it should be noted that not all of the vibratory modes are identified in this study.

3 Acoustical Characteristics of the Concrete Rigid-Frame Bridge

To investigate the acoustical characteristics of the bridge, both measurement and numerical analysis were carried out. The numerical analysis is based on the boundary element (BE) method. The boundary conditions were set by both measured results of a running test and the impact test described in the section 2.

3.1 Running Test

The running test was carried out to obtain actual displacements of the bridge and to evaluate bridge noise level distribution. In the measurement, 8 accelerometers were installed on the bridge (Fig. 4(a)), 13 microphones were set around the bridge (Fig. 4(b)). In order to avoid the contributions of other noise sources to the total noise, the results at low speed (70km/h) were used.

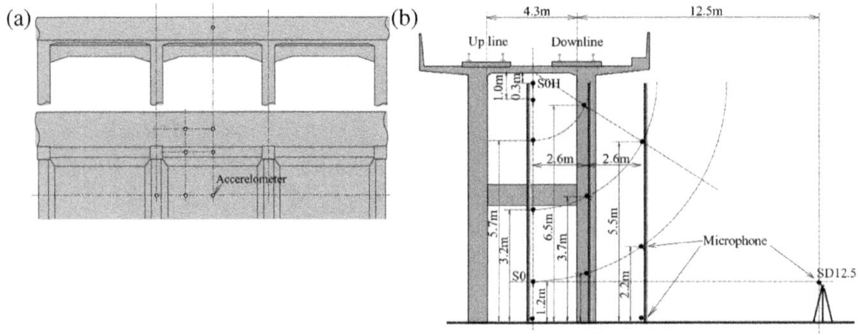

Fig. 4. Diagrams of (a) vibration measurement for the train pass-by and (b) sound radiation measurement.

3.2 Numerical Analysis

The bridge noise level distributions were estimated by using the BE analysis. Fig. 5 shows the BE analysis model. In the analysis, the bridge was modelled as a quarter of a single span by introducing symmetric boundary (full reflection) conditions on the cut planes. The symmetric boundary condition is also adopted at ground level to feature the ground effect.

The displacements at each node i in the direction j can be expressed as a linear combination of the first M modes of vibrations at frequency ω [3), 4)]:

$$u_{ij}(\omega) = \sum_{k=1}^{M} \xi_k(\omega)\Phi_{ijk}, \qquad (3.1),$$

where $u_{ij}(\omega)$ is the displacement at node i in direction j, $\xi_k(\omega)$ is the modal participation factor of mode k and Φ_{ijk} is the displacement at node i in direction j for eigenvector k.

If n displacement components and M modal shapes are known by measurements or calculations, the above relation becomes a set of n frequency domain equations with M unknown modal participation factors. Here, n was set as 8 corresponding to 8 accelerometers in Fig. 4(a) and M was set as 57. By solving this set of equations using singular value decomposition method [4], the modal participation factors were obtained. Once the modal participation factors are known, the displacements at all other points can be calculated by substituting $\xi_k(\omega)$ and Φ_{ijk} into equation (3.1), and the normal velocity distribution of the entire bridge model was obtained by:

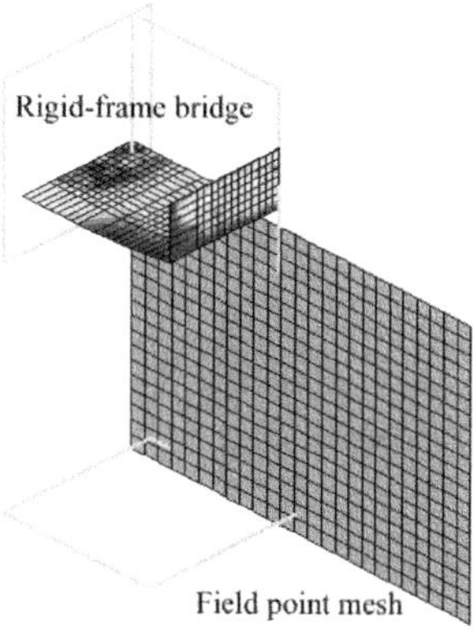

Fig. 5. BE analysis model.

$$v_{n,i}(\omega) = \sum_{k=1}^{M} i\omega\xi_k(\omega)\Phi_{n,ik}\,. \quad (3.2).$$

3.3 Comparison of the Noise Level Distribution between Measurement and Numerical Analysis

Fig. 6(a)-(c) show the calculated and measured results of noise level at the representative points. The results show similar trends and good agreement at the underside of the bridge (Fig. 6(a)) and beneath the bridge (Fig. 6(b)). At a point which is 12.5m away from the centre of the track, the measured result exceeds the calculated result in most frequency bands (Fig. 6(c)). This is due to the fact that the other noise sources, such as rolling noise, are included in the measured result.

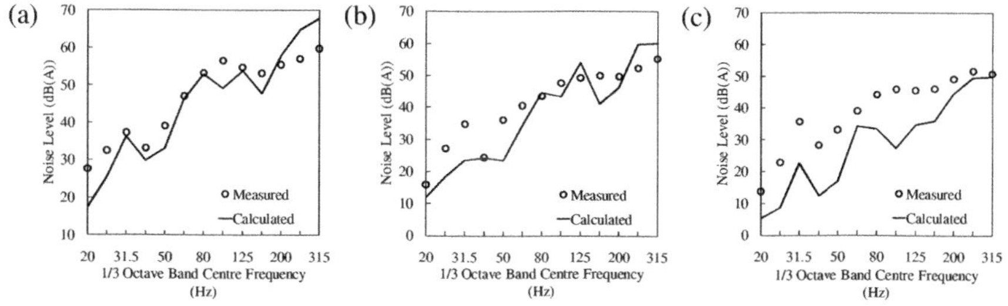

Fig. 6. Calculated and measured noise level of the bridge (a) at underside of the bridge (S0H), (b) beneath the bridge (S0) and (c) 12.5m away from the centre of the track (SD12.5). (The measurement points are detailed in Fig. 4)

Fig. 7(a)-(t) show the measured and calculated noise level distributions near the bridge. In both measured and calculated results, the bridge noise has strong directivity in vertically downward direction from the centre slab. Focusing on the cantilever slab and the handrail, the results of numerical analysis show strong radiation in the frequency range below 160Hz band, while their sound radiation is weaker and more indistinct than the centre slab in the measured result.

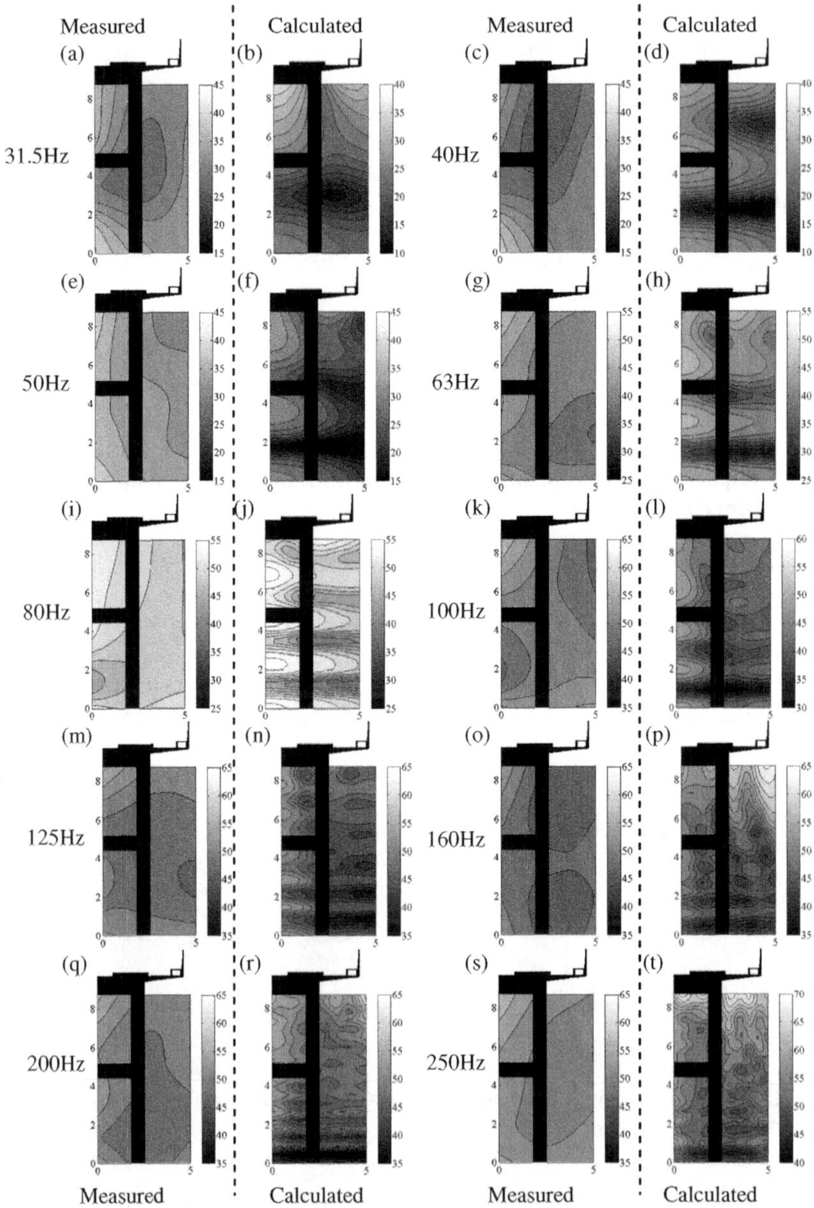

Fig. 7. Calculated and measured noise level distributions near the bridge. (a)(b) 31.5Hz, (c)(d) 40Hz, (e)(f) 50Hz, (g)(h) 63Hz, (i)(j) 80Hz, (k)(l) 100Hz, (m)(n) 125Hz, (o)(p) 160Hz, (q)(r) 200Hz, (s)(t) 250Hz.

Similar trend is confirmed in both measured and calculated results that noise level near the ground level tends to be high in the frequency range below 63Hz band. It suggests that some effect occurred due to acoustic resonance between bridge and ground.

4 Analysis on the Concrete Girder Bridge

4.1 Outline of Numerical Analysis

Analysis on the concrete girder bridge (Fig. 8) was also attempted to verify the aforementioned numerical analysis on the bridge. In order to set the boundary conditions for the girder bridge, modal shapes of the bridge derived by the finite element (FE) modal analysis below 500Hz are used instead of impact test. The actual displacements are obtained from the running test at the speed of 160km/h.

The girder bridge is supported at both ends, one is a fixed-end and the other is the movable end (movable in the directions longitudinal and vertical to rail). These constraint conditions were taken into account in the FE modal analysis model.

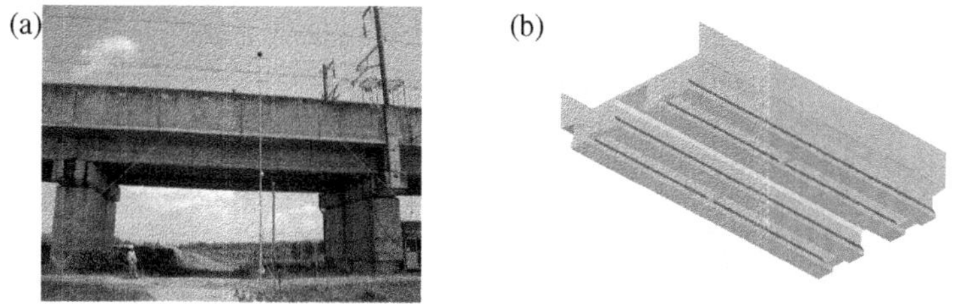

Fig. 8. (a) The concrete girder bridge and (b) diagram of numerical analysis model. (color figure online)

4.2 Comparison of the Noise Level between Measurement and Numerical Analysis

Fig. 9(a), (b) show the calculated and measured results of noise level at the representative points. The results of the girder bridge also show good agreement at the underside of the bridge (Fig. 9(a)) and beneath the bridge (Fig. 9(b)). These results indicate that the BE analysis seems to be applicable to any type of concrete bridges if their modal shape (whether measured or calculated) and the boundary conditions are set appropriately.

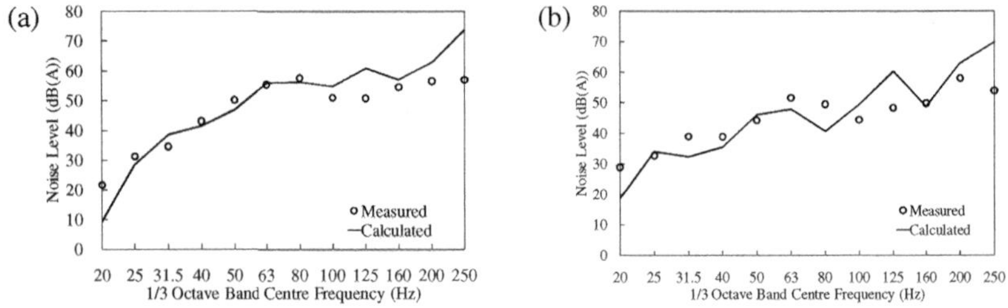

Fig. 9. Calculated and measured noise level of the girder bridge (a) at underside of the bridge and (b) beneath the bridge.

5 Conclusions

The vibratory characteristics of concrete rigid-frame bridge for Shinkansen have been revealed by detailed impact test and modal analysis.

The acoustical characteristics of the bridge have been investigated by both measurement and numerical analysis. For numerical analysis, the boundary condition was produced by coupling the modal shape and the actual displacement on the bridge, which are obtained in the impact test and running test. The results showed similar trends between measurement and numerical analysis.

The numerical analysis for the girder concrete bridge has also been attempted by adopting the calculated modal shapes as a part of boundary condition. The results showed good agreement with the measured noise.

Therefore, the bridge noise of any type of concrete bridges can be predicted rather accurately in the frequency range below 250Hz by the BE analysis if their modal shape (whether measured or calculated) and the boundary conditions are set appropriately.

Acknowledgements

The authors wish to express their gratitude to the staff of Technical Research and Development Dept., West Japan Railway Co. for their great support and collaboration to carry out various test campaigns.

References

[1] Thompson, D.J.: Railway noise and vibration. Elsevier, Amsterdam (2009)
[2] Kitagawa, T., Nagakura, K.: An investigation of vibratory properties of concrete railway bridges (in Japanese). In: Proceedings of 19th JSME Environmental Engineering Symposium, pp. 86–88 (2009)
[3] Nagamatsu, A.: Introduction to modal analysis. CORONA Publishing Co. Ltd. (2000) (in Japanese)
[4] LMS SYSNOISE, Revision 5.6 User's Manual, LMS (2003)

Local Vibration Properties of RC Viaduct under Passing Trains

K. Matsuoka[1], K. Kaito[2], T. Watanabe[3], and M. Sogabe[3]

[1] Department of Civil Engineering, Osaka University
2-1 Yamada-oka Suita Osaka, Japan
Tel: +81 6 6879 7598, Fax: +81 6 6879 7601
e-mail: k-matsuoka@civil.eng.osaka-u.ac.jp
[2] Frontier Research Center, Osaka University
[3] Railway Dynamics Division, Railway Technical Research Institute

Summary

In order to sustain the development of high-speed trains, it is necessary for railway bridges to understand the dynamic behavior of bridge members precisely, by identifying not only the whole bridge vibration properties but also local ones, and low-order to high-order mode vibration properties. Actually, there are common concerns over resonance and noises caused by the vibration of bridge members due to high-speed trains. However, there have been few vibration measurement experiments for carrying out identification of high-order member vibration properties. In this case, the authors conducted the experiment with passing trains, measuring vibration concurrently at several points on the intermediate and projecting slabs of a reinforced concrete (RC) viaduct for high-speed trains. Through the vibration measurement experiment, the authors verified the possibility of detection of high-order vibration modes of members and identified the dominant vibration mode when a high-speed train passed. This article also discusses the relation between vibration properties and the dominant frequency induced by the multi-axle load of trains. By accumulating the data of actual measurements of local vibration, it is possible to contribute to the refinement of finite element analysis.

1 Introduction

Recently, the speed of trains that run on arterial railroads across Japan has significantly increased, and the technology for higher speed is being developed.[1] The major problems with the structures caused by high-speed trains include resonance from the viewpoint of safety, and ground vibration, and noise from the viewpoint of usability.

Research relating vibration to noise has advanced the ability to diagnose noise sources. As the technology for identifying the sound source has been developed, it

T. Maeda et al. (Eds.): Noise and Vibration Mitigation for Rail Trans. Sys., NNFM 118, pp. 193–200.
springerlink.com

is now possible to identify the members that cause noise when a train passes. As for structure vibration, most of the previous studies are focused on resonance, because it is essential to secure safety. Resonance control has been discussed based on the measurements of actual bridges and numerical calculation, and most studies are focused on the first-order mode of the entire bridge. From now on, it will be necessary to tackle the usability problems also in the structure field, and so it is necessary to know the vibration properties of each member when a train passes. By knowing the characteristics of member vibration, it will become possible to conduct local structural condition monitoring, apply data to partial dynamic numerical calculation, and study member resonance when a train passes at a higher speed. These variables are difficult to understand with only the data on the entire system. However, a bridge is made up of several kinds of members, and all members interact with one another. At a lower order, the vibration mode of the entire system is excited. To study the vibration mode of a member it is necessary to discuss the high-order vibration components. However, there has been little research into high-order member vibrations.

In this case, the authors attempt to identify local vibration characteristics of members from the acceleration response to the loading of a passing train, and precisely understand the actual behavior of members in response to a passing train while focusing on a reinforced concrete (RC) bridge, which is the most common type in Japan. Chapter 2 describes the basic position of this study. Chapter 3 outlines the experiment on a RC viaduct under a passing train, and identifies, and discusses vibration characteristics.

2 Background for This Study

2.1 Local Vibration Problems

The structural vibration of a railway bridge is closely related to the "speed effect of multi-axle loading," and dates back to the research in the 1970s by Fryba, et al. [2] Theoretical studies, model experiments, and bridge measurements have been conducted continuously all over the world[3, 4]. To control dynamic responses, resonance has been taken into account in the designing process since 1983. Nomogram methods were proposed while considering high-speed running in 2004.[5] As an example of this research, Sogabe, et al, identified the resonance of the first-order mode of an entire simple girder bridge after numerical analyses and actual bridge experiments, and proposed the above dynamic designing method.[6] As for the vibration problems in the environment, Yoshioka and Hara et al. analyzed the relation between ground vibration and the vibration of a RC viaduct based on accumulated vibration measurements[7]. Recently, Nagakura summarized the characteristics of railroad noise and mitigation against it[8]. The level of structural noise is lower than that of wheel-rail noise. but its frequency spectrum is much lower. Therefore, it radiates in the vicinity of railways as low-frequency noise. In addition, it was pointed out that the power of structural noise is related to the cube of train speed so that the future high-speed trains will cause significantly more noise.

As a result, it is meaningful to analyze the structural vibration of a railway bridge from the structural viewpoint, to improve the environment. In addition, it is considered important to undestand high-order member vibrations due to a passing train. This study is aimed at providing data on structural element vibrations which are critical for coping with vibration problems due to a passing train. The authors measured vibrations due to a passing train, and studied the characteristics of member vibrations, focusing on general types of railway bridges. In addition, the major vibration modes are defined by comparing them with the predominant frequency when a train passes. The studied bridge type is a RC viaduct. This bridge was built after the construction of Sanin Shinkansen and a slab track was installed. It is assumed that this bridge has different vibration properties from those of previously-built railway bridges.

2.2 Identification Method and Prerequisites

This study is focused on the vibration properties of a viaduct due to a passing train. As a method for identifying vibration characteristics, the cross-spectral method, in which a measurement point is used as a reference point, was adopted, and the natural frequency and vibration modes were identified. With this method, the damping ratio cannot be identified, but by using the cross-spectral method, the effects of noise components can be removed, and so it is possible to identify vibration modes precisely. However, it is necessary to assume two conditions: (1) the damping ratio of the bridge is small, and modes are not close to one another, and (2) the input is a uniform spectrum which does not have pure tone frequency components. This study is conducted under the assumption that the above conditions are satisfied. The details of the identification method are described by Matsuoka, et al [9]. When vibration characteristics are identified using the acceleration response due to a passing train, the results do not represent the vibration characteristics of a bridge only, but it actually represents the complex system of a bridge and a train,, and the vibration state is like forced vibration. However, since the objective of this study is to understand the actual behavior of bridge members when a train passes, the above problems are disregarded. At each measurement point, a piezoelectric accelerometer was used, and only low-frequency components were measured, using a 2-kHz anti-aliasing filter.

3 Vibration Properties of a RC Viaduct

3.1 Description of the Subject Bridge and Vibration Measurement

This chapter is focused on a RC viaduct, whose aspect ratio per span is low and in which member vibrations would occur at a lower order. Fig. 1 shows the schematic diagram of the bridge. This bridge is a three-span RC bridge constructed in 1979, and is made up of two 7.88-m spans at both ends and a 8.1-m span at the center. The central span is longer than the other spans. This study is focused on the central span. This bridge is a symmetrical double track, except the shorter overhanging slab width at the downbound side.

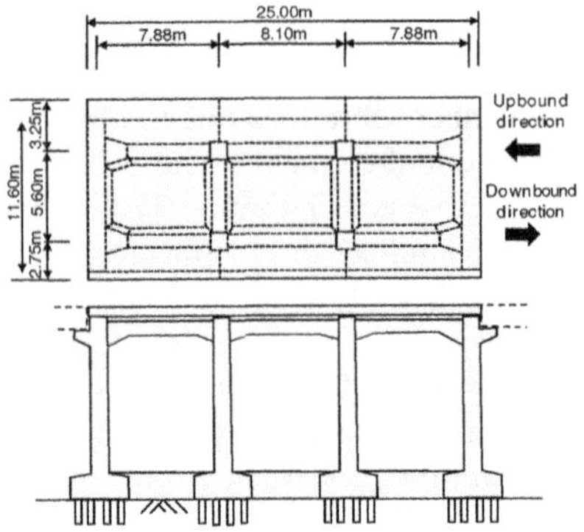

Fig. 1. Schematic diagram of the target bridge

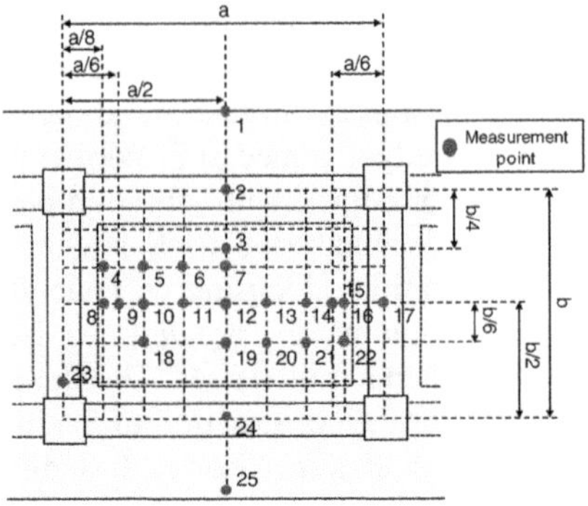

Fig. 2. Layout of accelerometers (color figure online)

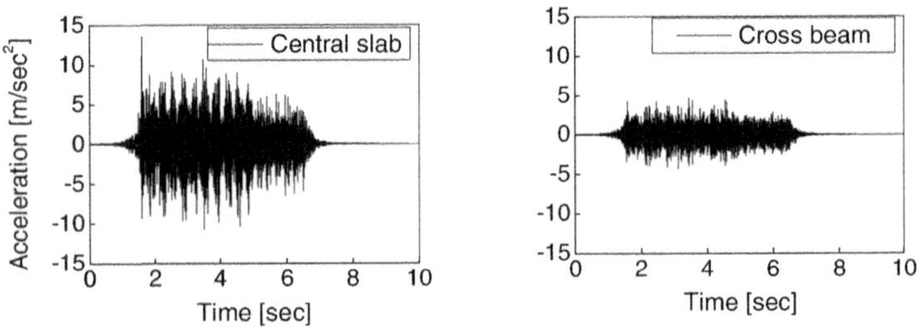

(a) Point 12

(b) Point 17

Fig. 3. Acceleration response due to a passing downbound train

The intermediate slab members were studied mainly. About 20 measurement points are set at the subject member. Measurement points are also set at other members to determine the differences in vibration amplitude and frequency characteristics. Fig. 2 shows the layout of the measurements points on the central slab. The layout of measurement points varied among target members, and the acceleration response of each member to a passing train was measured. The acceleration response of measurement point 12 (central slab) and 17 (cross beam) when train passed (down-train at a speed of 267km/h) is shown in Fig. 3. As shown in Fig. 3-(a), the acceleration amplitude due to a passing train reached 14 m/sec^2 in the central slab. The figure shows the change in the response from when the train entered the bridge to when it exited.

Table 1. Results of identification by the experiment with a passing train (central slab)

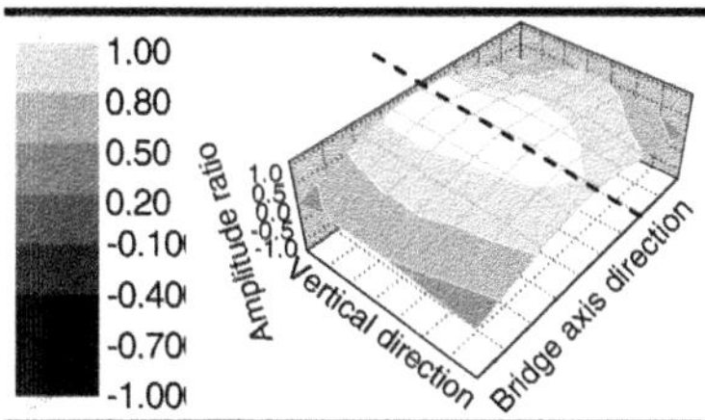	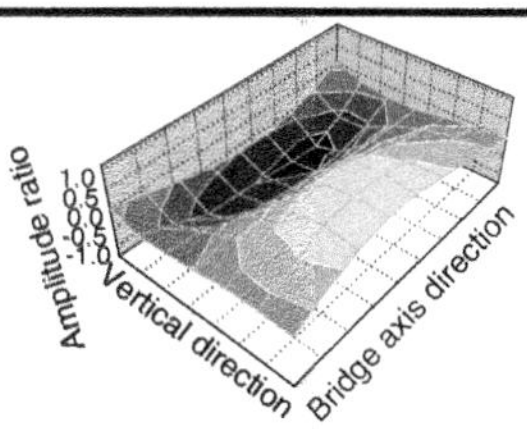
Contour map	Contour map
First-order deflection mode of the entire bridg	First-order anti-symmetric mode of the entire bridge
Natural frequency : 22.6Hz	Natural frequency : 30.3Hz
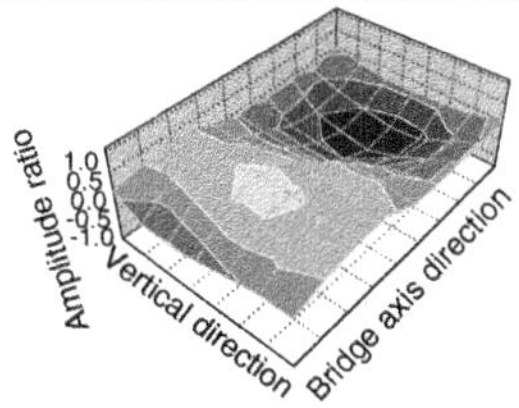	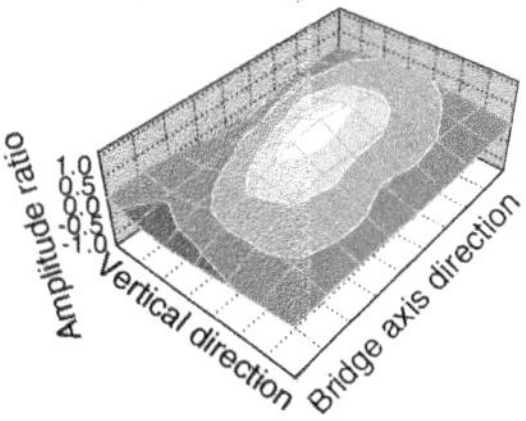
Contour map	Contour map
Second-order deflection mode of the entire bridge	First-order mode of the central slab
Natural frequency : 43.0Hz	Natural frequency : 52.7Hz
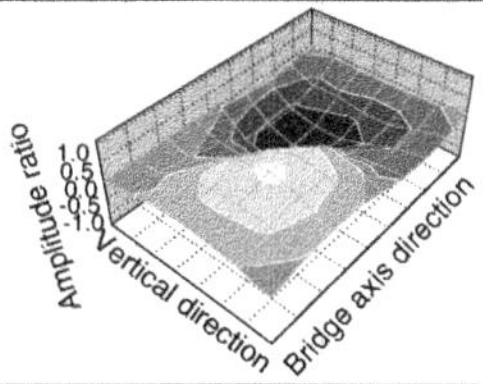	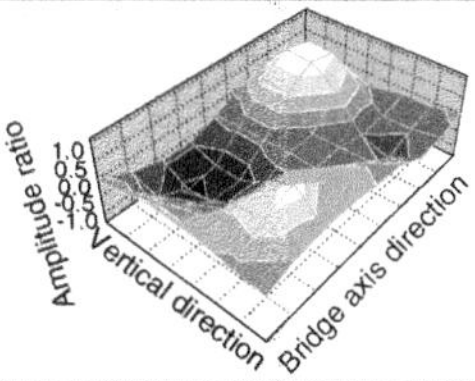
Contour map	Contour map
Second-order diagonal mode of the central slab	Second-order modes of the central slab in the bridge axis and vertical directions
Natural frequency : 60.6Hz	Natural frequency : 72.1Hz
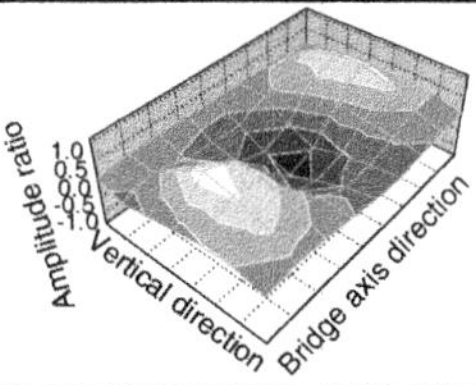	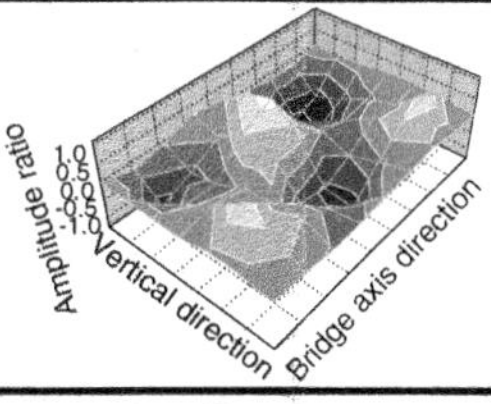
Contour map	Contour map
Third-order mode of the central slab	Third-order mode in the bridge axis and second-order in the vertical of central slab
Natural frequency : 93.7Hz	Natural frequency : 161.1Hz

3.2 Vibration Characteristics of the Central Slab

The vibration characteristics of the central slab are identified. The layout of measurement points is as shown in Fig. 2. Here, the mode of the flat plate whose 4 sides are simply supported was defined as the theoretical mode.

3.2.1 Identification of Vibration Characteristics

The natural frequencies and vibration modes of the central slab were identified, using the response to a passing train obtained at each measurement point. The results are summarized in Table 1. Table 1 includes the diagram representing the vibration modes of the central slab. For this diagram, the central slab was divided with 8 meshes in the bridge axis direction and six meshes in the vertical direction, and the value of each contact point was gridded with respect to measurement points. In addition, it was assumed that the mode displacement of the upper part of each column is zero. Table 1 indicates that the three low-order modes from 22.6Hz to 43.0Hz which can be said to be the entire modes with the amplitude of the main girder being large. As the mode of the central slab whose amplitude is outstanding, the first-order bowl-shaped mode was identified at 52Hz. The following modes can be considered to be the vibration modes of the members of the central slab, because the amplitudes of the main girder and the overhanging slab are smaller than the mode amplitude of the central slab. At 60.6Hz, the second-order mode whose maximums appear on the diagonal line of the central slab was identified. At 63.6Hz, the second-order mode in the vertical direction of the central slab was identified. Around 60Hz, the second-order mode in the bridge axis direction was also identified. Therefore, there is a possibility that the second-order modes of the central slab interact with each other. As for higher-order modes, there were the second-order modes in the bridge axis and vertical directions at 72.1Hz, the third-order mode of the central slab at 93.7Hz, the fourth-order mode at 120.8Hz, the third-order mode in the bridge axis direction and the second-order mode in the vertical direction at 161.1Hz. From a total of nine experiments, it was found that the variance of the mode was about 5% regardless of train direction and speed.

3.2.2 Outstanding Modes Due to a Train Passes

The speed of the trains passing on the bridge can be classified into two ranges: around 260 km/h, and around 140 km/h. The configuration of a power spectrum varied with train speed, but not with train direction. Fig. 4 shows the representative power spectrum due to a passing train for each train speed. The values in the figure are the averages of values obtained at measurement points 5, 7, 10, 12, 14, 19, and 21, to illustrate the outstanding components of the central slab. When train speed is around 140 km/h, the outstanding components can be found around 60Hz only. However, when train speed exceeds 260km/h, the power spectral value around 60Hz is over three times greater. In addition, outstanding components of mainly high-order modes can be found around 90Hz, 120Hz, and 150Hz. It is considered that the vibration around 60Hz is due to the second-order mode of the central slab regardless of the train speed. As mentioned above, the second modes of the central slab in the diagonal, bridge axis, and vertical directions were all identified around 60Hz, and it is inferred that the outstanding components are produced when these transit and interact, but these outstanding components could not be separated in this study. It is believed that the peak around 90Hz is caused by the third-order mode of the central slab and the peak around 120Hz is caused by the fourth-order mode. On the other hand, there were no peaks in the low-order components of less than 50Hz, which is caused by the entire mode. An exception was a small peak around 20Hz where the first-order deflection mode was excited when train speed was 266 km/h.

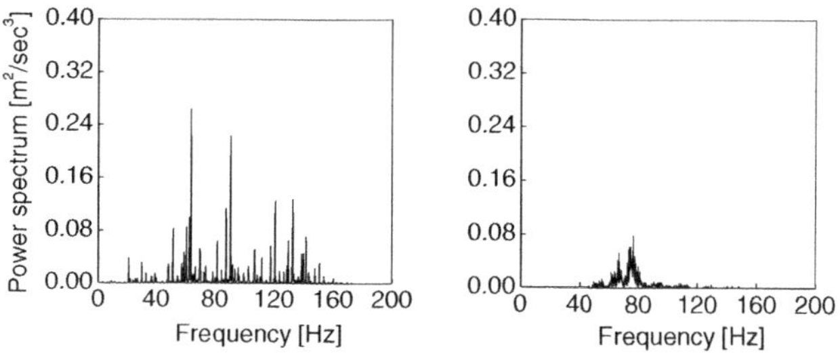

(a) Train Speed 266km/h (b) Train Speed 146km/h

Fig. 4. Power spectrum due to passing train

4 Conclusion

In this study, the authors summarized the local vibrations of a bridge due to a passing train from the viewpoints of safety and usability, and conducted vibration measurement and identification analysis for a central slab of the RC viaduct. As a result, the authors could observe the modes in which the vibration of central slab is dominant The conclusion is that the vibration of the RC bridge due to a passing train is mainly caused by the vibration of the central slab. It was also found that the second-order mode contributes to peak components.

Acknowledgment

Part of this study was carried out at Frontier Research Base for Global Young Researchers, Graduate School of Engineering, Osaka University based on the project of the Ministry of Education, Culture, Sports, Science, and Technology titled "Promotion of Suitable Research Environment for Young Scientists' Independent Research."

References

[1] Goto, H.: The pulling test run of a linear vehicle is begun in the Yamanashi linear experiment line. Journal of the Japan Society of Civil Engineers 82(2), 44 (1997) (in Japanese)

[2] Fryba, L.: Vibration of solids and structures under moving load, Noordhoff International (1972)

[3] Xia, H., Zhang, N., Guo, W.W.: Analysis of resonance mechanism and conditions of train-bridge system. Journal of Sound and Vibration 294, 810–822 (2006)
[4] Ju, S.H., Lin, H.T.: Resonance characteristics of high-speed trains passing simple supported bridges. Journal of Sound and Vibration 267, 1127–1141 (2003)
[5] Railway Technical Research Institute, Design standard and explanation such as railway structures – Concrete Structures, Maruzen (2004) (in Japanese)
[6] Sogabe, M.: Dynamic response and design of continuous concrete railway bridge in resonance area. Journal of Structure Mechanics and Earthquake Engineering 724(62), 83–102 (2003) (in Japanese)
[7] Hara, T., et al.: Development of a New Method to Reduce SHINKANSEN-Induced Wayside Vibrations Applicable to RigidI Frame Bridges. Journal of Structure Mechanics and Earthquake Engineering 766(68), 325–338 (2004) (in Japanese)
[8] Nagakura, K.: Present situation and future prospects of railway noise problem. The Journal of the Acoustical Society of Japan 64(10), 624–628 (2008) (in Japanese)
[9] Matsuoka, K., Kaito, K., Sugisaki, K., Watanabe, T., Sogabe, M.: Idetification of Vibration Properties of Open Deck Viaducts under Passing Train Loads. Journal of Applied Mechanics, JSCE 12, 983–994 (2009) (in Japanese)

Prediction and Measurement of Vibration on a Modern Concrete Railway Bridge Fitted with Resilient Rail Fasteners

D. Herron[1], C.J.C. Jones[2], D.J. Thompson[2], and D. Rhodes[1]

[1] Pandrol Limited, 63 Station Road, Addlestone, Surrey, UK
Tel: +44 1932 834500, Fax: +44 1932 850858
e-mail: d.herron@pandrol.com
[2] ISVR, University of Southampton, Highfield, Southampton, UK

Summary

The passage of a train over a railway bridge has been recognised as an important source of noise in urban environments. To help find means to reduce it, an accurate model for this noise is required. Prediction of the bridge vibration is the main challenge in determining the structure-borne noise. The computational cost of such an analysis using conventional Finite Element (FE) method is prohibitive, due to the enormous number of modes of vibration in the frequency range of interest. Statistical Energy Analysis (SEA) has therefore been used to model bridges in the past. However, this can be difficult to apply to bridges that have complex cross-sectional geometry.

Here, a Waveguide Finite Element (WFE) method is used to predict the vibration of a concrete box-section railway bridge. The computational cost of this analysis is acceptable and the bridge geometry can still be accounted fully. The approach taken to include the track structure found on this bridge in the WFE model has been developed from a separate study of this track-form.

1 Introduction

1.1 Bridge Vibration and Noise

When a train is travelling on a bridge, the vibration generated by the combined wheel-rail roughness is transmitted from the rail to the bridge, via the track support structure. This vibrational energy propagates through the bridge, causing it to radiate noise. This noise very often constitutes a significant addition to the wheel-rail rolling noise. Together with the prevalence of bridges in urban areas, the bridge noise is an important component of the noise impact of the railway. Consequently, there is a need to predict noise from proposed bridges and to study

T. Maeda et al. (Eds.): Noise and Vibration Mitigation for Rail Trans. Sys., NNFM 118, pp. 201–209.
springerlink.com

the principles of low-noise bridge design. It is the prediction of the bridge vibration that presents the main challenge in determining the structure-borne noise level and this is the subject of this article.

1.2 Models for Bridge Vibration

Due to the enormous number of modes of vibration of a railway bridge over the frequency range of interest for bridge noise (up to about 1.5kHz), the computational cost of a conventional finite element (FE) analysis is prohibitive. Statistical Energy Analysis (SEA) is more suitable for dealing with the response of large structures at high frequencies. The SEA model normally requires an estimate for the vibration power transmitted from the rail to the bridge, as an input parameter, to predict the bridge response. This power input estimate is usually made from an analytical model that is based on idealised structural elements, such as a single beam, two coupled beams or a plate. This approach is most readily applied to all-steel bridges, with simple cross-sectional geometry. It has also been applied to concrete box-section bridges [1] and concrete-steel composite bridges [2], but it is more difficult to account for the greater complexity of these structures using an analytical model for the power input to the bridge.

An advanced FE method, referred to here as Waveguide Finite Element (WFE) analysis, is used in this work to predict the vibration of a concrete box-section bridge on the Docklands Light Railway (DLR) in London. WFE differs from conventional FE only in the two-dimensional cross-section which is modelled using finite elements. The structure is assumed to be of infinite length, with a uniform cross-section. Many railway bridge structures satisfy these conditions reasonably well. Analytical wave functions are used to model the motion of the structure along its length, which brings a large reduction in the computational cost compared to conventional FE. Relative to the analytical models used to estimate the power input to the bridge in the SEA-based models, the use of finite elements to represent the cross-section of the bridge in a WFE model means that the analysis can be based on a much more complete description of its geometry.

1.3 Resilient Baseplate Track

The track structure on the bridge studied here is of the common two-stage resilient baseplate type, comprising a rail pad, cast iron baseplate ‘lid’, a soft baseplate pad and a lower baseplate that is fixed rigidly to the bridge deck. This promotes effective vibration isolation between the rail and bridge, and therefore low bridge vibration and noise. The stiffness of these baseplates is expected to show significant frequency-dependence. Here, a combined measurement and modelling study are used to determine its stiffness as a function of frequency.

2 Dynamic Behaviour of a Two-Stage Resilient Baseplate

2.1 Two-Stage Resilient Baseplates

The track structure on the concrete bridge under study here consists of Pandrol VIPA-SP two-stage resilient baseplates, shown below in Fig. 1.

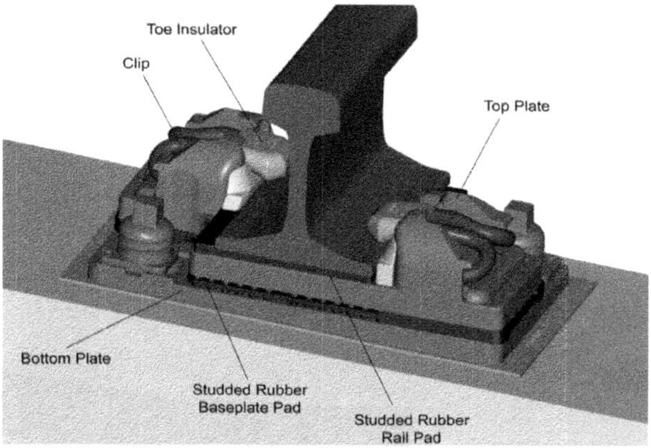

Fig. 1. Pictorial view of a Pandrol VIPA-SP resilient baseplate. (color figure online)

The bottom plate is fixed to the bridge deck with grout. The top plate; however, is resiliently mounted between the railpad and baseplate pad and is therefore free to vibrate. Various modes of vibration could be envisaged and its overall stiffness may therefore be frequency-dependent. The dynamic transfer stiffness, the force transmitted to a blocked termination per unit displacement on the input side, controls the transmission of vibration between the rail and bridge.

It was anticipated that reliable measurements for the dynamic transfer stiffness of the assembly would be difficult to achieve. Measurements were therefore made to characterise the vibration response of the key components in the assembly and used to develop a means to include the dynamic behaviour of the baseplates in the WFE analysis of the bridge and track.

2.2 Component Testing and Model Development

Three components of the baseplate are expected to control its overall transfer stiffness: the railpad, baseplate pad and top plate. The transfer stiffness and damping loss factor of the pads were measured using the method described in [3], in which the pad is positioned between two masses. The mass above the pad is excited by an impulse and accelerometers on both masses provide the information required to calculate the force transmitted across the pad, the displacement between them and therefore the transfer stiffness of the pad. For the top plate, point and transfer response measurements were made with the plate supported on resilient material, showing resonance frequencies of 685Hz and 1500Hz.

A conventional FE model was used to study the vibration of the Pandrol VIPA-SP baseplate and section of rail. First, a free vibration response analysis was run for only the top plate, modelled using plate elements, with no constraints applied. The predicted natural frequencies compared well with those found from the measurements made for the top plate on a resilient support. The railpad, baseplate pad and a mass to represent a section of rail were then added to the FE model. An array of nine linear spring elements was used to represent each pad.

A free vibration response analysis for the FE model of the baseplate assembly showed three modes that are expected to control its response to vertical excitation at the rail, shown in Fig. 2 below.

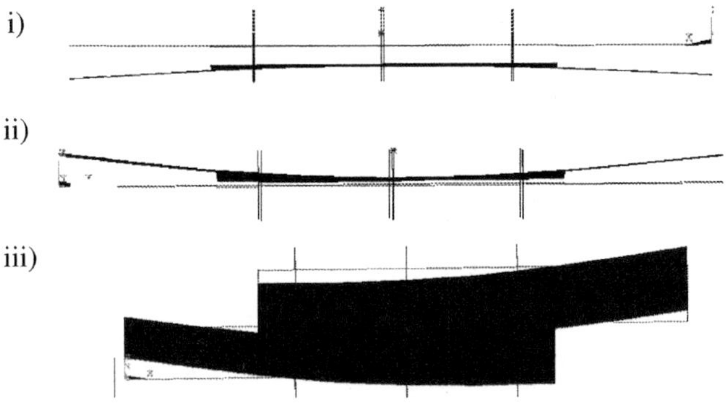

Fig. 2. Mode shape plots for three important modes of the baseplate assembly, with natural frequencies of: i) 350Hz, ii) 650Hz, iii) 1.2kHz. (color figure online)

The first mode of the assembly consists of the top plate and section of rail moving as a mass on the baseplate pad. In the second mode, the top plate moves vertically with larger amplitude at the outer parts of the plate, referred to here as the vertical-flapping mode. The third mode is the bending mode of the top plate. These modes indicate that the beam-bending motion across the width of the top plate (normal to rails) should be included in the WFE representation of the track.

3 WFE Analysis of a Concrete Box-Section Bridge

3.1 Theoretical Basis of WFE Method

A brief account of the basis of the WFE method is presented here, see [4, 5] for more details. The structure is defined in three-dimensional space, using x, y and z coordinates. The cross-section lies in the y-z plane, wave propagation is in the x direction in which the structure is taken to be of infinite extent. The structure undergoes steady state harmonic motion at circular frequency ω, such that the deformation is elastic. The displacement at any node in the cross-section, in response to a single wave of wavenumber κ, can be described by,

$$\mathbf{W}_i(x, y, z, t) = \overline{\mathbf{W}}_i(y, z)e^{-i\kappa x}e^{i\omega t}, \tag{3.1}$$

where $\overline{\mathbf{W}}$ indicates displacement of the cross-section. The virtual work principle can be employed to develop the following for the motion of the structure,

$$\left(\sum_{j=0}^{4} \mathbf{K}_j \left(\frac{\delta^j}{\delta x^j}\right) - \omega^2 \mathbf{M}\right) \overline{\mathbf{W}}(x) = \overline{\mathbf{F}}(x), \tag{3.2}$$

where $\mathbf{K}_j$ are a set of stiffness matrices, j takes the values 0, 1, 2 and 4 for the plate elements used in this work, $\mathbf{M}$ is the mass matrix and $\mathbf{F}$ is the force vector for the cross-section. The mass and stiffness matrices were found for a given cross-section geometry and a set of material properties using a computer program called WAveNumber Domain Software (WANDS), described by [5].

Equation (3.2) may be Fourier transformed to the wavenumber domain, referring to Equation (3.1) for the partial derivative terms. The excitation is assumed to be a concentrated load (a delta function in the spatial domain),

$$\left(\sum_{j=0}^{4} \mathbf{K}_j (-i\kappa)^j - \omega^2 \mathbf{M}\right) \tilde{\mathbf{W}}(\kappa) = \tilde{\mathbf{F}}(\kappa), \tag{3.3}$$

For free vibration, this represents a polynomial eigenvalue problem in κ. The eigenvalues obtained at a given frequency describe the dispersion characteristics of each wave and the eigenvectors describe the mode shape associated with each wave. For forced vibration, re-arranging Equation (3.3) for the cross-section displacements in the wavenumber domain gives,

$$\tilde{\mathbf{W}}(\kappa) = \left(\sum_{j=0}^{4} \mathbf{K}_j (-i\kappa)^j - \omega^2 \mathbf{M}\right)^{-1} \tilde{\mathbf{F}}(\kappa), \tag{3.4}$$

An inverse Fourier transform may be used to return the cross-section displacement vector to the spatial domain, with the integration performed using a method based on the method of residues [4]. This is substituted into Equation (3.1) to find the displacement of the structure at a given position along its length.

3.2 Application to a Concrete Box-Section Bridge on DLR

The WFE model is for only half of the bridge cross-section. Symmetric and anti-symmetric boundary conditions have been applied to the nodes on the bridge centre-line and used to recover the solution for the full bridge structure. A single plate element represents the track structure as a continuous resilient layer between the rail and the bridge deck (constant stiffness with frequency) in this first model. The nodes of this WFE model are shown in Fig. 3 below.

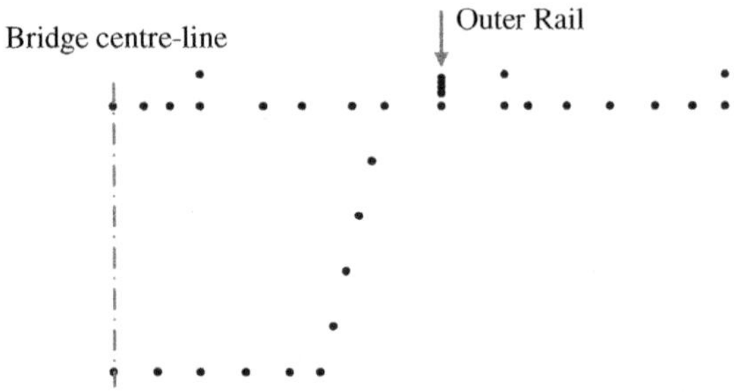

Fig. 3. Node positions in the WFE model for half of the concrete box-section bridge, showing the outer rail and constant stiffness track support. (color figure online)

In the second stage of the WFE modelling work, the outcome from Section 2 for the dynamic behaviour of the resilient baseplate track was implemented. The WFE representation of this track structure is shown in Fig. 4.

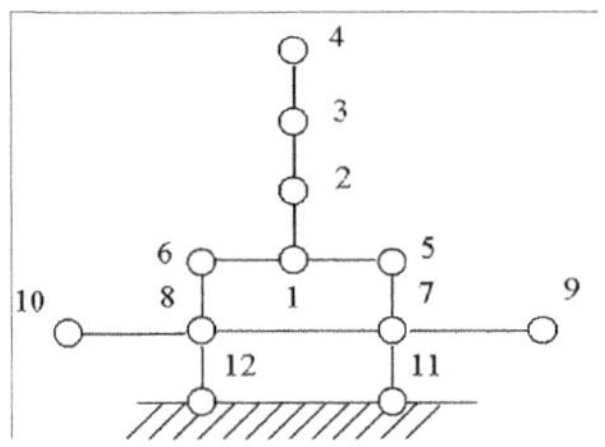

Fig. 4. WFE representation of two-stage resilient baseplate track.

Nodes 1 to 6 define the rail and 7 to 10 the top plate. The rail pad is represented by two elements, nodes 5 to 7 and 6 to 8. The baseplate pad by another two elements, nodes 7 to 11 and 8 to 12. Orthotropic material properties were used to minimise the effects of wave motion along the length of the track support, which does not occur physically because this support is not continuous.

3.3 Comparison of WFE Results to Measurement Data

The WFE model was first used to predict the transfer accelerance on the bridge deck as a function of distance, in one-third octave frequency bands. An appropriate damping loss factor for use in the WFE model was determined by comparison with measurements for transfer accelerance. Good agreement was found when the damping loss factor in the WFE model was set to 0.1.

A comparison between measured and predicted point mobility on the bridge deck, close to the outer rail seat, is shown in Fig. 5 below.

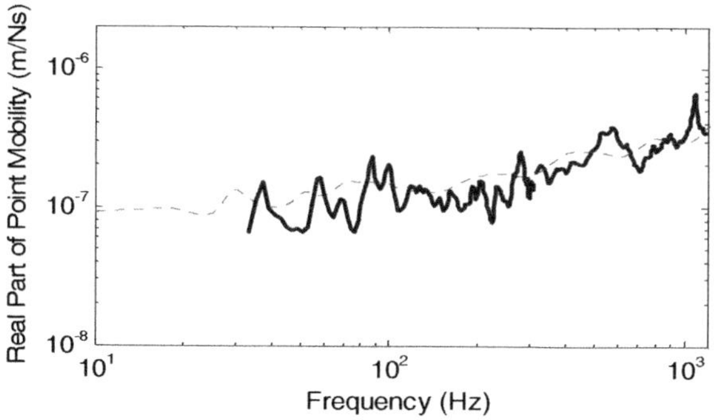

Fig. 5. Real part of point mobility on the bridge deck, close to the outer rail seat, shown versus frequency: ———, measurement; - - - - - - - , WFE.(color figure online)

The WFE result represents a frequency-average to the measurement, as expected for an infinite length model of a finite length structure [6].

The WFE representation of the two-stage resilient baseplate track was assessed by comparing the decay rates in the rail obtained from the WFE model with those from the measurements. The decay rates were found from the transfer accelerance using the procedure proposed in [7].

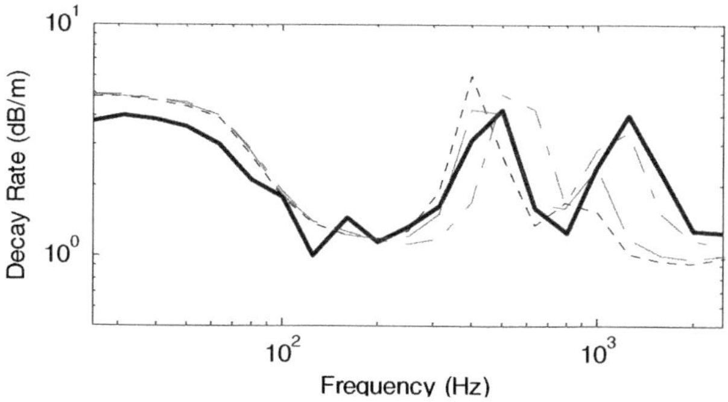

Fig. 6. Decay rate in the unloaded rail: ———, measured; - - - - - - -, WFE with railpad stiffness of 80MN/m; — — — — —, WFE with railpad stiffness of 100MN/m; —·—·—··, WFE with railpad stiffness of 200MN/m. (color figure online)

The high rates of decay in the 500Hz and 1.25kHz bands are expected to correspond to the vertical flapping mode and the bending of the top plate mode, respectively. The predicted decay rate is strongly dependent on the railpad stiffness, which increases significantly with frequency. This cannot be included in the WFE model directly, so the model was run separately for rail pad stiffness values appropriate to the lower, mid and upper parts of the frequency range. When this is considered, Fig. 6 shows that the predicted decay rates are in satisfactory agreement with the measurements.

In order to predict the bridge vibration during a train pass-by, an estimate is required for the wheel-rail roughness excitation force. This was made using a simple model for the vehicle, rail, contact spring between them and an assumed wheel-rail roughness spectrum. No roughness data specific to this location was available. The DLR rolling stock has resilient wheels, which introduces further difficulty to the modelling of the system excitation.

The average bridge deck velocity during a train pass-by is shown in Fig. 7 together with the predicted spatially-averaged velocity.

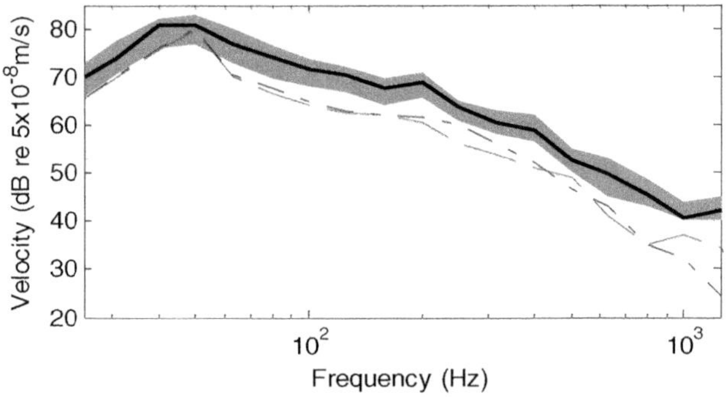

Fig. 7. Spatially-averaged velocity at outer rail position on bridge deck shown versus frequency: ———, measured average; ▇, measured range; —·—·—, WFE result for constant track stiffness; – – – –·, WFE result with proposed track model. (color figure online)

The predicted spectra have broadly similar shape to the measurement, but the deck velocity is under-predicted by typically 6dB, or a factor of 2. The results obtained using the two different track models in WFE are similar for the frequency bands up to 200Hz, but differ thereafter. The proposed track model gives only a result that is closer to the measurement than the constant stiffness track model in the 1kHz and 1.25kHz frequency bands.

4 Conclusions

The comparison made between the WFE results and measurements for point mobility on the bridge deck and the decay rate in the rail indicates that WFE is a good, 'mid-frequency' means to model the vibration of this class of structure. It has clear advantages over SEA or conventional FE. The lesser agreement shown for vibration under-traffic is likely to be due, at least in-part, to the estimates made for the wheel-rail roughness excitation force. The wheel-rail roughness is not known for this site and the resilient wheels on the rolling stock complicate the prediction of the excitation force from the assumed roughness spectrum. The proposed model for the two-stage resilient baseplate track predicts the decay rates in the rail well, but had little effect on the vibration under-traffic. It may therefore be of greater benefit in the prediction of rolling noise.

References

[1] Hardy, A.E.J.: Noise from Railway Bridges. In: Proc. Instn. Mech. Engrs, vol. 213, Part F (1999)

[2] Bewes, O.G., Thompson, D.J., Jones, C.J.C., Wang, A.: Calculation of Noise from Railway Bridges and Viaducts: Experimental Validation of a Rapid Calculation Model. Journal of Sound and Vibration 293, 933–943 (2006)

[3] Thompson, D.J., van Vliet, W.J., Verheij, J.W.: Developments of the Indirect Method for Measuring the High Frequency Dynamic Stiffness of Resilient Elements. Journal of Sound and Vibration 213(1), 169–188 (1998)

[4] Karassalo, I.: Exact Finite Elements for Wave Propagation in Range-Independent Fluid-Solid Media. Journal of Sound and Vibration 172(5), 671–688 (1994)

[5] Nilsson, C.-M., Jones, C.J.C.: Theory Manual for WANDS 2.1 Wavenumber Domain Software for Solids and Fluids. ISVR Technical Memorandum, No: 976 (2007)

[6] Skurdrzyk, E.: The Mean-Value Method of Predicting the Dynamic Response of Complex Vibrators. Journal of the Acoustical Society of America 67, 1105–1135 (1980)

[7] Jones, C.J.C., Thompson, D.J., Diehl, R.J.: The Use of Decay rates to Analyse the Performance of Railway Track in Rolling Noise Generation. Journal of Sound and Vibration 293(1), 475–484 (2006)

Development of Silent Steel Railway Bridge Equipped with Floating Ladder Track and Floating Reinforced Concrete Deck

T. Watanabe, M. Sogabe, K. Asanuma, and H. Wakui

Railway Technical Research Institute,
2-8-38, Hikari-cho, Kokubunji-shi, Tokyo, 185-8540, Japan
Tel: +81 42 573 7290, Fax: +81 42 573 7320
e-mail: tsutomuw@rtri.or.jp

Summary

As a countermeasure against structure-borne noise of steel railway bridge, we developed a new type silent steel railway bridge equipped with floating ladder track and floating reinforced concrete deck. As a result of the train running test, it was apparent that the new steel railway bridge with double floating system has reduced vibration velocity level of 10.5dB(A) at the main girder webs as compared with a steel railway bridge which has directly fastened tracks.

1 Introduction

A number of steel railway bridges have been constructed in Japan. Thin steel members used for the bridges are easy to vibrate and generate structure-borne noise. Therefore, according to these circumstances, the construction of steel railway bridges tends to be avoided in cities.

As a countermeasure against this structure-borne noise, various concepts such as installation of sound-walls and reinforced concrete decks that block the noise propagation paths from the web and flange to the surroundings, or attaching resilient materials to the web and flange, have been proposed previously. In addition to these treatments, we also suggested a method to intercept the transmission of vibrations due to the train running between the rails and each of the members.

Under such background, to improve the insulation performance of the vibration, we developed a new type of the steel railway bridge equipped with a floating ladder track [1] and a floating reinforced concrete (RC) deck.

Figure 1 presents the outline of a new type of the steel railway bridge, which we developed. We assume that this bridge is constructed in urban areas. In Japan, the type of the bridge constructed in the urban area tends to be a through bridge.

T. Maeda et al. (Eds.): Noise and Vibration Mitigation for Rail Trans. Sys., NNFM 118, pp. 211–219.
springerlink.com

Therefore, this bridge is a type of the through bridge. This bridge is equipped with a floating ladder track and a floating RC deck. The floating ladder track has a high bending rigidity of the ladder sleepers in the longitudinal direction which disperses the wheel load, and the low stiffness support of the track by resilient materials, reduces the transmission of vibrations. The floating RC deck supported by the resilient materials is a heavy mass spring system, and aims for a reduction of the transmission vibration, mainly in frequency ranges below 100Hz.

We quantitatively determined the structure-borne noise reduction effect of the double-floating steel railway bridge by performing impact experiments and train-running experiments. Further, applying a sound prediction technique [2, 3], we estimated the maximum noise level of a time-weighted characteristic ($L_{pA,\,Smax}$) and the equivalent noise level ($L_{Aeq,\,T}$).

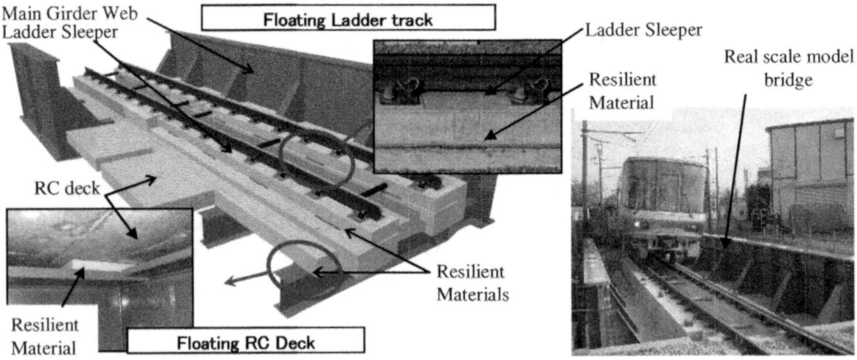

Fig. 1. Outline of silent steel railway bridge which we developed (color figure online)

2 Experiment of Structure-Borne Noise Reduction

2.1 Experiment Method

With the present level of technology, confirming the structure-borne noise reduction effects by empirically measuring the actual structures is the most accurate method. However, it is generally difficult to fix or change the operating conditions of the trains, and the conditions of the track supports and structures on an operating line. In addition, there are significant restrictions in carrying out quantitative evaluations from the point of view of train travel schedules and economics. Hence, a full-scale model girder was used in this study to grasp the basic properties of the structure-borne noise reduction effects.

Figure 1 shows the real scale model bridge and the installed resilient material. The girder length of the double-floating steel railway bridge is designed to be 35m and is intended for an operation line. However, the girder length of the model bridge was shortened to 12m, although the shape of cross-section remained the same as the full scale. The double floating structure was composed of floating ladder track and the floating RC deck. The RC deck was elastically supported by the resilient materials at the cross-over points of stringers and cross beams.

Tables 1 and 2 show the experiment cases and the material properties. Table 3 shows the natural frequency of each vibration system. In order to check the structure-borne noise reduction effect for the steel railway bridge, the impulse-vibration experiment and train-running experiment were conducted on differing supporting conditions for the ladder sleeper and RC deck. The train used in the experiment was composed of two standard traffic line vehicles. We installed the model's girder at a sharp curve at a section with 180m of curve radius, which limited the maximum speed of the train to 45km/h. In CASE 1 and CASE 3, we employed a rigid support for the ladder sleeper. However, the results from the ladder sleeper settlement measurements during running of the train showed about 1.0mm of settlement. The model uses a quasi-directly fastened track which is not of a completely rigid support.

Table 1. Experiment case

CASE	Ladder sleeper	RC deck
1	Rigid	Rigid
2	Floating ladder track	Rigid
3	Rigid	Floating RC deck
4	Floating ladder track	Floating RC deck(50MN/m)
5	Floating ladder track	Floating RC deck(13MN/m)

Table 2. Material properties

	Stiffness and Weight
Rail pad (vertical stiffness)	60 MN/m
Resilient material (vertical stiffness)	25MN/m (Floating ladder track) 50MN/m,13MN/m (Floating slab)
Floating ladder track	155 kN
RC deck	314 kN
Steel	290 kN

Table 3. Natural frequency of each vibration system

Vibration system	Natural frequency
Floating ladder track (rigid mode)	40Hz
Ladder + Floating RC deck (rigid mode)	10Hz(50MN/m) 6Hz(13MN/m)
Ladder + RC deck + Bridge (1st vibration mode)	17Hz

2.2 Evaluation of Noise Reduction

In the impulse-vibration experiment, we performed the Fourier transform on the vibration acceleration measured by the accelerometer attached to the girder members (the RC deck, the upper flange, the web and lower flange of the main girder) when we hit the rail head with a hammer. We applied the frequency response function (abbreviated as "accelerance" below) to assess the structure-borne noise reduction effect of the double-floating steel railway bridge. The accelerance is obtained from normalizing the vibration acceleration spectrum by the vibrating force spectrum. In the train-running experiment, vibration acceleration level L_a was calculated from the time-series data of the vibration acceleration obtained by the accelerometer attached to the girder members. We converted the vibration acceleration level into the vibration velocity level L_v based on equation (1).

$$L_v = L_a - 20\log(2\pi f) + 46, \qquad (1)$$

where f is the 1/3 octave-band frequency.

In general, the webs and flanges of main girder, stringer and cross beam are believed to be the sources of structure-borne noise. An additional 20 accelerometers were set in an array arrangement (a dense lattice-like arrangement). The point, which was deemed to be the center of the mode, was set into vibration with an impulse hammer, and then we analyzed vibration mode of each member. Specifically, vibration mode candidates were chosen from the cross spectrum between the reference point and each measurement point, and we created a mode diagram based on the Fourier amplitude and phase of each candidate.

2.3 Impulse Hammer Experiment Result

Figure 2 shows the measurement results of accelerance at the impulse experiment. There is a peak of main girder web accelerance around 50Hz. Compared with the CASE 1 with the rigid support structure, the CASE 3, 4 and 5 indicated the reduction effect.

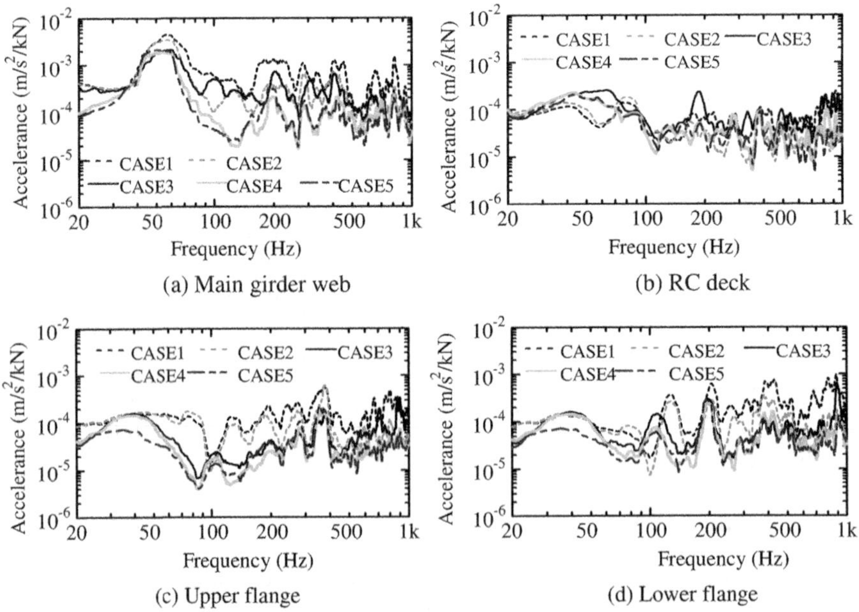

Fig. 2. Measurement results of accelerance at the impulse experiment (color figure online)

Figure 3 shows the result of vibration mode analysis at the main girder web. The accelerometers set in an array arrangement indicate that the vibration mode around 50Hz is the first mode of the main girder web.

In comparison with the CASE 1, the RC deck accelerance is decreased with 150Hz or higher in CASE 2 and with approximately 190Hz in CASE 4. No effect was seen in CASE 3, and the accelerance increases in the range of 30 to 80Hz and around 200Hz. The accelerance of the lower flange on the main girder decreases with 400Hz or higher in CASE 2 and with 50 to 600Hz in CASE 3 alongside to CASE 1. The CASE 4 includes the reduction effect of CASE 2 and CASE 3.

The accelerance of the flange under the main girder decreases with 200Hz or higher in CASE 2 and with 100Hz or higher in CASE 3 and CASE 4.

When the significance of the accelerance was examined by each member, the main girder web turned out to be the largest compared with other members. The accelerance reaches a peak around 50Hz, which is equivalent to the 1st natural frequency of the main girder web. The effect of noise reduction in the frequency band in CASE 4 became apparent.

There is no great difference between the CASE 4 and CASE 5.

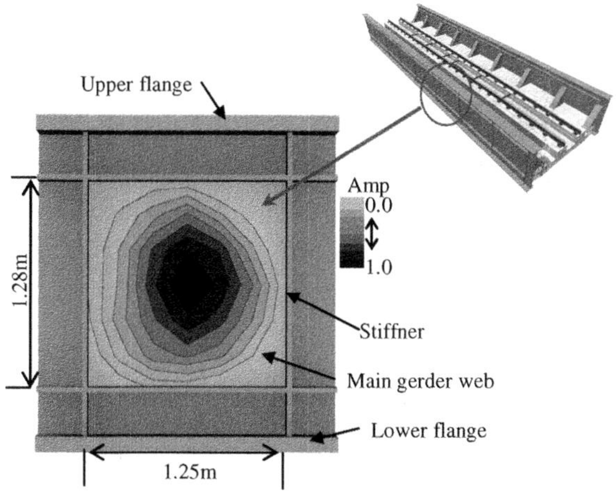

Fig. 3. Vibration mode at main girder web (50Hz) (color figure online)

2.4 Train Running Experiment Result

2.4.1 Vibration Acceleration Level

The comparison result indicates that the main girder web has the largest vibration acceleration level similar to the result of the impulse-vibration experiment and that the vibration of the main girder web was dominant. The following primarily describes the results from the main girder web.

Figure 4 shows the unweighted 1/3 octave-band vibration acceleration level of the train-running experiment (train speed 40km/h).

Looking at the vibration acceleration level of the main girder web, the noise reduction effect was evident with 80Hz or higher in CASES 2, 3, and 4 unlike CASE 1 with the rigid support structure. In the CASES 3 and 4 with the floating RC deck, the peak at 60Hz in CASE1 shifted to the lower frequency. In addition, the peak vibration acceleration level became smaller. The CASE 2 and CASE 4 have almost equal vibration acceleration level above 150Hz and CASE 2 and CASE 3 have equal vibration acceleration range below 100Hz. Therefore, the reduction effect of the floating RC deck for the main girder web is obtainable in the frequency band of 150Hz or lower. The reduction of the floating ladder track can be obtained in the frequency bands above 100Hz. Compared with the accelerance shown in the Fig.2, the tendencies of the frequency of vibration acceleration level peaks and the magnitude correlation of values in each case were found to be quite similar.

There is no great difference between the CASES4 and 5.

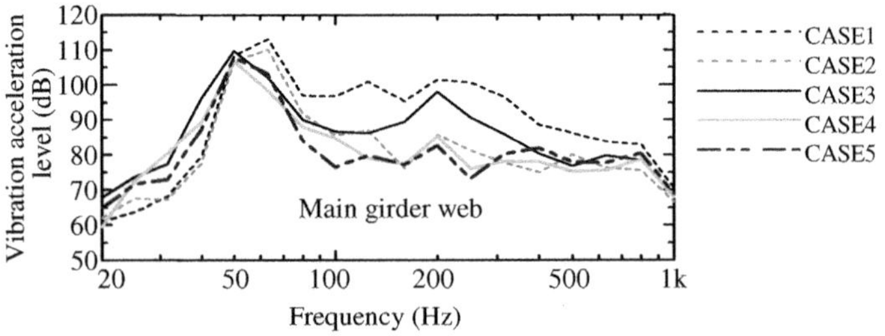

Fig. 4. Measurement result of vibration acceleration level at the train running experiment (color figure online)

2.4.2 Vibration Velocity Level

Figure 5 shows the measurement results of A-weighted vibration velocity level at the train-running experiment. We calculated the vibration velocity level from the vibration acceleration level obtained during train passage. The vibration velocity is a parameter directly associated with the sound radiation power W as shown by the equation (2), and considered as the proper index in the evaluation of structure-borne noise.

$$W = \sigma \rho c \bar{v}^2 S\,, \tag{2}$$

where σ is an acoustic radiation efficiency, ρ is a density of the air, c is sound velocity, $\bar{v}$ is A-weighted root-mean-square value of vibration velocity, S is an acoustic radiation area. When compared with CASE 1, with a rigid support structure and a train speed of 40km/h, the vibration velocity level of the main girder web decreased by 7.5dB(A) in CASE 2, by 6.1dB(A) in CASE 3, by 10.1dB(A) in CASE 4, and 9.7dB(A) in CASE5.

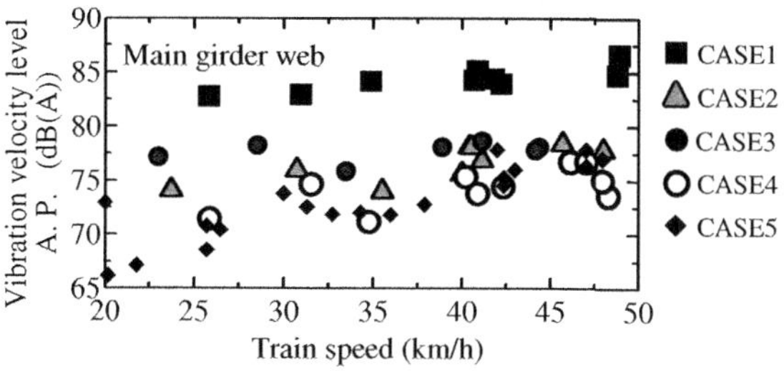

Fig. 5. Measurement result of vibration velocity level at the train running experiment (color figure online)

In this way, it is visible that reducing the vibration velocity of the main girder web, which makes the greatest contribution to the structure-borne noise of a steel bridge, can achieve significant noise reduction effects.

3 Estimation of Structure-Borne Noise Reduction

3.1 Estimation Method

The Ministry of Environment established the "Guidelines for Measures against Noise for New Construction or Large-scale Modification of Conventional Railways" in 1995. These guidelines have justifed an examination of equivalent sound levels to evaluate the influence of the new construction on or large-scale modification of the conventional railway to the environment along the railway line [4]. Focusing on the vibration velocity of the main girder web, which largely contributes the structure-borne noise of the steel railway bridge, in reference to the experiment results of the previous chapter and reference [5], we estimated the peak noise level and the equivalent sound level using the sound source analysis method [3] and evaluated each case.

3.2 Estimation Result

Figure 6 shows estimate results of $L_{pA,\,Smax}$ and $L_{Aeq,\,T}$ under the above-mentioned condition. The average value of the peak noise level decreased by 7.5dB(A) in CASE 2, 5.4dB(A) in CASE 3, and 10.5dB(A) in CASE 4 when compared with CASE 1, with the rigid support structure, as shown in the Fig.6. The average value of the equivalent sound level decreased by 7.6dB(A) in CASE 2, 5.4dB(A) in CASE 3, and 10.4dB(A) in CASE 4.

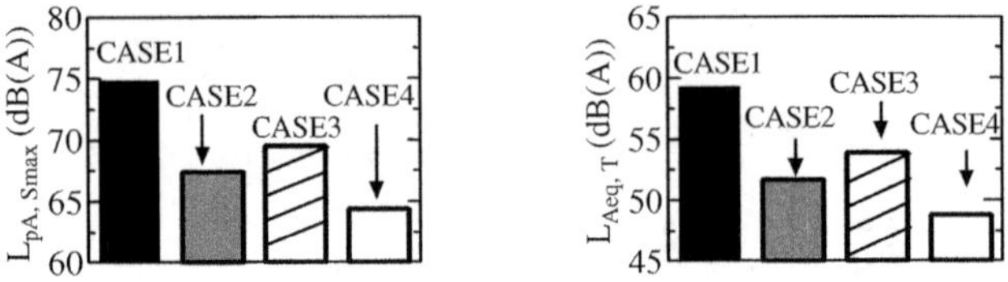

(a) $L_{pA,\,Smax}$ (b) $L_{Aeq,\,T}$

Fig. 6. Estimation results of LpA, Smax and LAeq, T (at the position 12.5 m away from the center of the bridge) (color figure online)

4 Conclusion

The following summarize the findings obtained in the study.

(1) The results of the impulse-vibration experiment with the full-scale model girder indicated that the double-floating structure could reduce the peak value of the accelerance, which is the natural frequency of the main girder web, when compared with a rigid support structure. The floating ladder track and floating RC deck obtained a noise reduction effect above 100Hz and in the range of 50Hz to 100Hz, respectively. The double-floating structure encompasses the effects of both of these.

(2) In the impulse-vibration experiment with the full-scale model, the main girder web was found to be the member which generated the largest accelerance. The dominant frequency of the main girder web is approximately 50Hz, which is found to be the first-order natural frequency in the mode analysis with accelerometers set in an array arrangement.

(3) The train-running experiment with the full-scale model girder quantitatively assessed the noise reduction effects from the standpoint of the vibration velocity level of the main girder web, which is closely relevant to the structure-borne noise. When compared with the rigid support structure, the vibration velocity level decreased by 7.5dB(A) with the floating ladder track, 6.1dB(A) with the floating RC deck, and 10.1dB(A) with the double-floating structure when the train ran at 40km/h.

(4) Based on the noise estimation method and the train-running experiment results, we estimated the peak noise level and equivalent sound level at a point 12.5m from the track center. The peak noise level and the equivalent sound level of the double-floating structure were lower than those of the rigid support structure by 10.5dB(A) and 10.4dB(A), respectively.

The authors completed this study with support from the Ministry of Land, Infrastructure and Transport.

References

[1] Okuda, H., et al.: Dynamic load, resistance and environmental performance of floating ladder track. Quarterly report of RTRI 45(3), 149–155 (2004)

[2] Hansaka, M., et al.: Analysis of Contribution of structure-borne noise radiated from each member and examination of materials to reduce noise. RTRI Report 21(2), 21–26 (2007)

[3] Kitagawa, T., et al.: Noise prediction method for conventional railways. RTRI Report 12(12), 41–46 (1998)

[4] Environmental Agency Committee for Guideline on Noise for Conventional line: Guidelines for measures against Noise for New Construction and Large-scale Improvement of Conventional line (1995)

[5] Koshiba, A., Sakamoto, K., Naitou, S., Shimura, Y.: Research of Low-noise Steel Bridges Using Laminated Damping Steel. RTRI Report 7(2), 27–34 (1993)

A Time-Domain Model for Coupled Vertical and Tangential Wheel/Rail Interaction - A Contribution to the Modelling of Curve Squeal

A. Pieringer and W. Kropp

CHARMEC/Applied Acoustics, Chalmers University of Technology
Sven Hultins gata 8a, 41296 Göteborg, Sweden
Tel: +46 31 7722209, Fax: +46 31 7722212
e-mail: astrid.pieringer@chalmers.se

Summary

Lateral forces due to frictional instability are seen as the main reason for the occurrence of curve squeal. Predicting squeal requires thus to describe the high-frequency wheel/rail interaction during curving including the coupling between vertical and lateral directions. In this article, a time-domain approach is presented which includes both vertical and lateral forces and takes into account the non-linear processes in the contact zone. Track and wheel are described as linear systems using pre-calculated impulse response functions. The non-linear, non-steady state contact model is based on an influence function method for the elastic half-space, includes a velocity-dependent friction coefficient and accounts for surface roughness. First results from the interaction model demonstrate the functioning of the approach.

1 Introduction

Curve squeal is a highly disturbing tonal sound generated by a railway vehicle negotiating a sharp curve. This type of noise is commonly attributed to self-excited vibrations of the railway wheel, which are either induced by stick-slip behaviour due to lateral creepage of the wheel tyre on the top of the rail or by contact on the wheel flange [1].

Although many curve squeal models have been proposed in the literature, e.g. the models [2-8], curve squeal remains difficult to predict. On one hand, this can be attributed to the lack of knowledge about important model parameters, such as, e.g. realistic friction coefficients. On the other hand, high-frequency wheel/rail interaction during curving is a complex phenomena, which poses a challenge in modelling. As curve squeal is intrinsically transient and non-linear, models aiming to predict squeal amplitudes have to be formulated in the time-domain. Due to the

T. Maeda et al. (Eds.): Noise and Vibration Mitigation for Rail Trans. Sys., NNFM 118, pp. 221–229.
springerlink.com

required computational effort of time-domain solutions, it is usually necessary to simplify wheel, rail and contact dynamics and, by consequence, the models might not include all the important features of the phenomena.

The aim of the study presented in this paper is to contribute to the modelling and understanding of curve squeal by proposing a detailed time-domain model for dynamic wheel/rail interaction that considers the coupling between vertical and tangential directions. The computational effort is reduced by representing vehicle and track by the impulse response functions that are calculated in advance. This technique, which has been proved efficient, for instance, in the area of tyre/road noise [9] and in vertical wheel/rail interaction [10], makes it possible to include a non-linear, non-steady state contact model that is solved at each time step in the interaction model.

2 Wheel/Rail Interaction Model

The wheel/rail interaction model is primarily intended for quasi-static curving of the leading inner wheel in a railway bogie. The model relies on the wheel/rail contact position and the angle of attack of the wheelset (i.e. the lateral creepage) as input parameters. These parameters can be pre-calculated with a vehicle dynamics programme.

Fig. 1 shows the reference frame for the wheel/rail interaction model, which is moving with the nominal contact point along the rail.

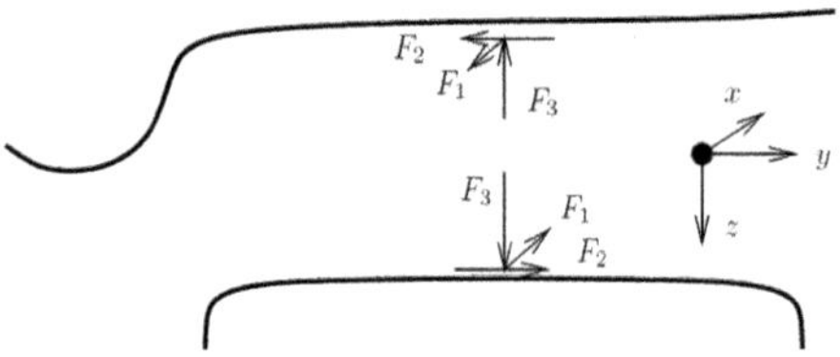

Fig. 1. Reference frame.

2.1 Wheel and Track Model

The vehicle is represented by a single flexible C20 wheel disregarding the influence of the axle. The wheel is modelled by axi-symmetric finite elements and represented by its modal basis. The receptances of the wheel at the wheel/rail contact point on the wheel tread are calculated by modal superposition.

The track consisting of one continuously supported BV50 rail is modelled with wave-guide finite elements using the software package WANDS [11]. This model takes advantage of the two-dimensional geometry of the rail, but nonetheless, considers the three-dimensional nature of the vibration by assuming a wave-type solution along the rail. Cross-sectional deformations of the rail, which are important for high-frequency applications and lateral dynamics are taken into account. Fig. 2 shows the vertical and lateral point receptances and the vertical/lateral cross receptances of the wheel and the track at the nominal contact point. On the wheel, the nominal contact point is assumed at the centre of the wheel tread. On the rail head, the nominal contact point is assumed at a distance of 1.2 cm from the centre. This offset introduces a coupling between vertical and lateral dynamics of the rail.

The impulse response functions (or Green's functions) of the wheel, g_{ij}^{W}, are obtained by inverse Fourier transform from the wheel receptances, G_{ij}^{W}. The subscripts i and j denote the excitation and response direction, respectively. The lateral and vertical displacements of the wheel at the contact point, $\xi_2^{\mathrm{W}}(t)$ and $\xi_3^{\mathrm{W}}(t)$, are then calculated by convoluting the contact forces with the Green's functions

$$\xi_j^{\mathrm{W}}(t) = -\int_0^t \sum_{i=2}^{3} F_i(\tau)\, g_{ij}^{\mathrm{W}}(t-\tau)\,\mathrm{d}\tau, \qquad j = 2,3. \tag{2.1}$$

The longitudinal dynamics of the wheel is not taken into account and the influence of wheel rotation is neglected.

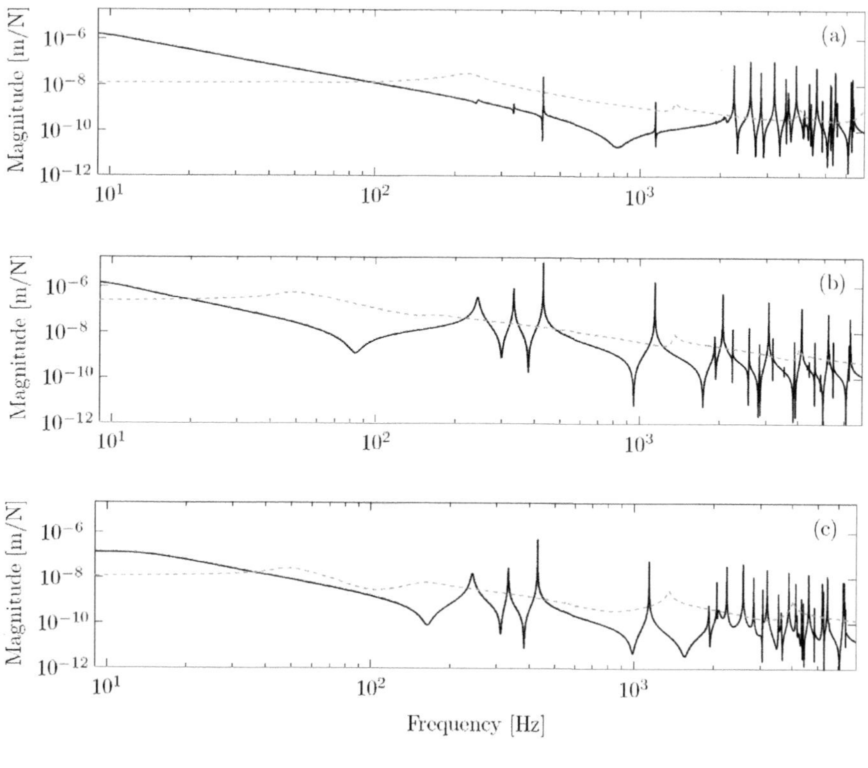

Fig. 2. Magnitudes of the wheel (solid line) and track (dashed line) receptances at contact: (a) vertical point receptance, (b) lateral point receptance, (c) vertical/lateral cross receptance.

The track is represented by a special type of Green's functions denoted by moving Green's functions, $g_{ij,v}^{\mathrm{R},x_0}(t)$, which include the motion of the nominal contact point along the rail [12]. The function $g_{ij,v}^{\mathrm{R},x_0}(t)$ describes, for excitation of the rail (index R) in i-direction at the position x_0 at time $t_0 = 0$, the displacement response of the rail in j-direction at a point, moving with train speed v away from the excitation, at the nominal contact point between wheel and rail. The discrete version of the moving Green's function $g_{ij,v}^{\mathrm{R},x_0}(t)$ is constructed from (ordinary) Green's functions $g_{ij}^{\mathrm{R},x_0,x_0+\alpha}(t)$, where the superscripts specify the excitation point x_0 and the response point $x_0+\alpha$ on the rail. The Green's functions $g_{ij}^{\mathrm{R},x_0,x_0+\alpha}(t)$ are obtained from the corresponding track transfer receptances by inverse Fourier transform. The lateral and vertical displacements of the rail at the contact point, $\xi_2^{\mathrm{R}}(t)$ and $\xi_3^{\mathrm{R}}(t)$, are calculated by convoluting the contact forces with the moving Green's functions.

$$\xi_j^{\mathrm{R}}(t) = \int_0^t \sum_{i=2}^{3} F_i(\tau)\, g_{ij,v}^{\mathrm{R},v\tau}(t-\tau)\,\mathrm{d}\tau, \qquad j=2,3. \tag{2.2}$$

The longitudinal dynamics of the track are not taken into account. In the case of continuously supported track used in this article, the moving Green's functions are independent from the excitation position x_0 on the rail.

2.2 Contact Model

The contact model is an implementation of Kalker's model CONTACT [13], which is a three-dimensional, non-steady state rolling contact model based on the assumption that wheel and rail can be locally approximated by elastic half-spaces. In addition to the parameters included in CONTACT, the contact model used in this article considers the combined roughness of wheel and rail on several parallel lines in the rolling direction, the contribution of the structural dynamics of wheel and rail to the creepage and a velocity-dependent friction coefficient.

The potential contact area is divided into N rectangular elements with side lengths Δx and Δy in x- and y-direction, respectively. Assuming that the wheel and rail are made of the same material, quasi-identity holds and, consequently, normal and tangential contact problem can be solved separately [13].

2.2.1 Normal Contact

The normal contact problem consists in determining which elements of the potential contact area are in contact, and in calculating the local vertical displacement u_{I3} and the contact pressure p_{I3} in every element I. The local vertical displacement, which is the displacement difference between rail and wheel

$$u_{I3} = u_{I3}^{\mathrm{R}} - u_{I3}^{\mathrm{W}}, \qquad I = 1,\ldots,N, \tag{2.3}$$

is related to the contact pressure according to,

$$u_{I3} = \sum_{J=1}^{N} A_{I3J3}\, p_{J3}, \qquad I = 1,\ldots,N, \tag{2.4}$$

where A_{I3J3} are influence coefficients for the elastic half-space. The total vertical contact force, F_3, is obtained by summing the contributions from the different elements

$$F_3 = \sum_{I=1}^{N} p_{I3} \Delta x \Delta y \ . \tag{2.5}$$

Introducing the variable d_I which describes the distance between the deformed bodies in each element, the contact conditions are formulated as

$$d_I \geq 0, \quad p_{I3} \geq 0, \quad d_I \, p_{I3} = 0 \quad . \tag{2.6}$$

If contact occurs in a surface element, the distance is zero and the contact pressure is positive. If contact does not occur, the distance is positive and the pressure is zero. Adhesion and penetration are excluded by (2.6). The distance d_I is obtained as,

$$d_I = -\delta + u_{I3} + z_I^{\mathrm{R}} - z_I^{\mathrm{W}} + r_I^{\mathrm{R}} - r_I^{\mathrm{W}} \ , \tag{2.7}$$

where z_I^{R} and z_I^{W} are the profiles of rail and wheel, r_I^{R} and r_I^{W} are the roughness of rail and wheel and δ is the approach of distant points

$$\delta = \xi_3^{\mathrm{W}} + \xi_3^{\mathrm{S}}(P) - \xi_3^{\mathrm{R}} \ . \tag{2.8}$$

The variable $\xi_3^{\mathrm{S}}(P)$ is the position of the primary suspension of the wheel corresponding to the nominal preload, P, which represents the vehicle components above the primary suspension.

The normal contact problem is solved with an active set algorithm [13].

2.2.2 Tangential Contact

In frictional rolling contact, the contact area is divided into a stick and a slip area. The tangential contact problem consists of determining which elements are in stick and in slip, and in calculating the local tangential displacements $u_{I\tau}$ and tangential stresses $p_{I\tau}$ at the surface. The relation between local tangential displacements and tangential stresses is given by,

$$u_{I\tau} = \sum_{\alpha=1}^{2} \sum_{J=1}^{N} A_{I\tau J\alpha} \, p_{J\alpha}, \qquad \tau = 1,2 \ , \tag{2.9}$$

where $A_{I\tau J\alpha}$ are influence coefficients for the elastic half-space. The tangential forces, F_τ, are obtained by summing the contributions from the different elements.

$$F_\tau = \sum_{I=1}^{N} p_{I\tau} \Delta x \Delta y \ , \qquad \tau = 1,2 \ . \tag{2.10}$$

A contact element belongs to the stick area, if the local shift, $S_{I\tau}$, vanishes,

$$S_{I\tau} = 0 \ , \qquad \tau = 1,2 \ . \tag{2.11}$$

Otherwise the contact element belongs to the slip area. The local shift is defined as,

$$S_{I\tau} = u_{I\tau} + W_{I\tau}^{*} - u'_{I\tau} \ , \qquad \tau = 1,2 \ . \tag{2.12}$$

The variable $u'_{I\tau}$ represents the local displacement at the previous time step. In Kalker's formulation, $W_{I\tau}$ is the rigid shift calculated as,

$$\begin{aligned} W_{I1} &= (\xi - y\phi)\Delta x \\ W_{I2} &= (\eta + x\phi)\Delta x \end{aligned}, \tag{2.13}$$

where ξ, η and ϕ are the longitudinal, lateral and spin creepage. In this article, the contribution of the structural dynamics of wheel and track is added to the rigid shift.

$$\begin{aligned} W_{I1}^* &= W_{I1} \\ W_{I2}^* &= W_{I2} + \left(\xi_2^{R} - \xi_2^{W}\right) - \left(\xi_2'^{R} - \xi_2'^{W}\right), \end{aligned} \tag{2.14}$$

where $\xi_2'^{R}$ and $\xi_2'^{W}$ are the lateral displacements of rail and wheel at the previous time step. In the slip area, the following relations hold,

$$\frac{p_{I\tau}}{\sqrt{p_{I1}^2 + p_{I2}^2}} = -\frac{S_{I\tau}}{\sqrt{S_{I1}^2 + S_{I2}^2}}, \qquad \tau = 1,2 \tag{2.15}$$

$$p_{I1}^2 + p_{I2}^2 = \left(\mu_I\, p_{I3}\right)^2, \tag{2.16}$$

where μ_I is the friction coefficient. Equation (2.15) assures that the slip occurs in the direction opposite to the tangential stress. Equation (2.16) states that the tangential stress in the slip zone is equal to the traction bound $\mu_I p_{I3}$. The friction coefficient is either assumed constant $\mu_I = \mu_S$ or dependent on the local slip velocity $s_\tau \approx S_\tau / \Delta t$ according to [14],

$$\mu_I = \mu_S \left[\frac{50}{100 + \left(s_{I1}^2 + s_{I2}^2\right)} + \frac{0.1}{0.2 + \sqrt{s_{I1}^2 + s_{I2}^2}} \right]. \tag{2.17}$$

The tangential contact problem in the case of a constant friction coefficient is solved with an active set algorithm [13] combined with the Newton-Raphson method. In case of a velocity-dependent friction law, an additional iteration loop is needed at each time step. The iterative procedure proposed by Croft [14] has been implemented for this purpose.

3 Simulation Results

In this section, first results from the interaction model are presented to demonstrate the functioning of the approach. Three simulations have been carried out, which have the following characteristics:

I. constant friction coefficient $\mu_l = \mu_S$; smooth wheel and rail ($r_l^W = r_l^R = 0$),
II. friction law (2.17); smooth wheel and rail ($r_l^W = r_l^R = 0$),
III. friction law (2.17); rough wheel ($r_l^W \neq 0$) and smooth rail ($r_l^R = 0$).

The wheel roughness used in simulation III has been measured on a wheel with sinter block brakes in 25 parallel lines with a spacing of 2 mm across the width of the running surface [15]. The roughness level is low over the whole frequency range of interest. Common parameters used in all the three simulations are a static preload of $P = 65$ kN, a friction coefficient of $\mu_S = 0.3$, a train speed of $v = 100$ km/h, a spatial resolution of $\Delta x = \Delta y = 1$ mm and a lateral creepage of $\eta = 5 \cdot 10^{-3}$. Wheel and rail profiles are assumed cylindrical with wheel radius $R_W = 0.39$ m and rail head radius $R_R = 0.3$ m. The longitudinal creepage, ξ, and the spin creepage, ϕ, are set to zero. The train speed used is considered as high in the context of curve squeal, and will be reduced in future application of the simulation model.

The results of simulations I-III are presented in Fig. 3 in terms of time series of the total lateral contact force F_2. The first 72 ms correspond to the period of preload application and stepwise increase of the lateral creepage to the prescribed value. While the contact force goes towards a steady-state solution in simulation I, stick-slip oscillations are observed in the two cases with velocity-dependent friction law. In simulation II (Fig. 3a) a first stick-slip oscillation emerges at 0.29 s, which then gradually gives way to a second, more high-frequent stick-slip oscillation. The main frequency component of this oscillation is identified as 6320 Hz (see Fig.4), which corresponds to the (8,0)[1] axial mode of the wheel. In the case of the rough wheel, the high-frequency stick-slip oscillation builds up faster (Fig. 3b). As an effect of the roughness excitation, the spectrum of the contact force shows generally higher levels as comparison with simulation II, but the main peaks occur still at the same frequencies and have the same amplitudes (Fig. 4).

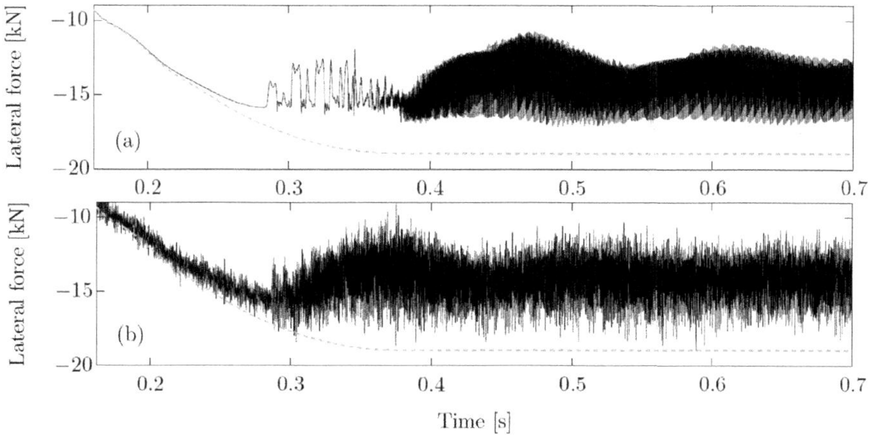

Fig. 3. Time series of the lateral contact force of simulation I (*dashed line*) in comparison to (a) simulation II (*solid line*); (b) simulation III (*solid line*).

[1] The (n,m) mode has n nodal diameters and m nodal circles.

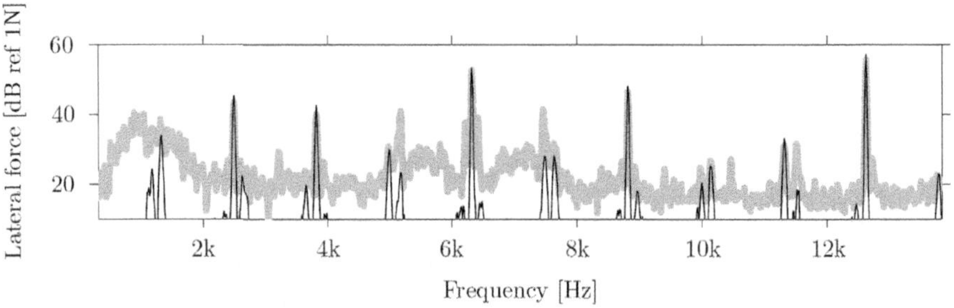

Fig. 4. Auto-spectrum of the lateral contact force of simulation II (black line) and III (grey line).

4 Conclusions

A time-domain model has been presented, which simulates the coupled vertical and tangential wheel/rail interaction at high frequencies. Simulation results showed the build-up of stick-slip in the case of a velocity-dependent friction law and illustrated the applicability of the model in the context of curve squeal.

Acknowledgements

The authors are grateful for the support by Prof. David Thompson (ISVR, University of Southampton) who supplied us with the wheel model used in this article. Furthermore, we would like to thank Dr. Briony Croft (former PhD student at ISVR) for her help with the implementation of the wheel model. Finally, the access to the Wave Guide Finite Element toolbox WANDS (developed by the Dynamics Group at ISVR), which has made the modelling of the track an easy and efficient task, is greatly acknowledged.

References

[1] Thompson, D.: Railway Noise and Vibration: Mechanisms, Modelling and Means of Control. Elsevier, Oxford (2009)

[2] Rudd, M.J.: Wheel/rail noise - Part II: Wheel squeal. J. Sound Vib. 46(3), 381–394 (1976)

[3] Fingberg, U.: Ein Modell für das Kurvenquietschen von Schienenfahrzeugen, Fortschrittberichte VDI, Reihe 11, Nr. 140, VDI-Verlag, Düsseldorf (1990)

[4] Périard, F.: Wheel-Rail Noise Generation: Curve Squealing by Trams, PhD Thesis, Technische Universiteit Delft (1998)

[5] Heckl, M.A., Abrahams, I.D.: Curve squeal of train wheels, Part I: Mathematical model for its generation. J. Sound Vib. 229(3), 669–693 (2000)

[6] de Beer, F.G., Janssens, M.H.A.: Squeal noise of rail-bound vehicles influenced by lateral contact position. J. Sound Vib. 267, 497–507 (2003)

[7] Chiello, O., Ayasse, J.-B., Vincent, N., Koch, J.-R.: Curve squeal of urban rolling stock - Part 3: Theoretical model. J. Sound Vib. 293, 710–727 (2006)

[8] Huang, Z.Y., Thompson, D.J., Jones, C.J.C.: Squeal Prediction for a Bogied Vehicle in a Curve. In: Schulte-Werning, B., et al. (eds.) Noise and Vibration Mitigation. NNFM, vol. 99, pp. 313–319 (2008)
[9] Wullens, F., Kropp, W.: A three dimensional contact model for tyre/road interaction in rolling conditions. Acta Acust. United Ac. 90(4), 702–711 (2004)
[10] Pieringer, A., Kropp, W., Nielsen, J.C.O.: A Time Domain Model for Wheel/Rail Interaction Aiming to Include Non-Linear Contact Stiffness and Tangential Friction. In: Schulte-Werning, B., et al. (eds.) Noise and Vibration Mitigation. NNFM, vol. 99, pp. 285–291 (2008)
[11] Nilsson, C.-M., Jones, C.J.C., Thompson, D.J., Ryue, J.: A waveguide finite element and boundary element approach to calculating the sound radiated by railway and tram rails. J. Sound Vib. 321, 813–836 (2009)
[12] Nordborg, A.: Wheel/rail noise generation due to nonlinear effects and parametric excitation. J. Acoust. Soc. of Am. 111(4), 1772–1781 (2002)
[13] Kalker, J.J.: Three-Dimensional Elastic Bodies in Rolling Contact. Kluwer Academics Publishers, Dordrecht (1990)
[14] Croft, B.E.: The Development of Rail-head Acoustic Roughness, PhD Thesis, ISVR, University of Southampton, UK (2009)
[15] Thompson, D.J., Remington, P.J.: The effects of transverse profile on the excitation of wheel/rail noise. J. Sound Vib. 231(3), 537–548 (2000)

A Comparison of Predicted and Measured Ground Vibrations due to High Speed, Passenger, and Freight Trains

G. Lombaert, G. Degrande, P. Galvín, E. Bongini, and F. Poisson

[1] K.U. Leuven, Department of Civil Engineering, Kasteelpark Arenberg 40, 3001 Leuven, Belgium
e-mail: {geert.lombaert,geert.degrande}@bwk.kuleuven.be
[2] Escuela Técnica Superior de Ingenieros, Universidad de Sevilla, Camino de los Descubrimientos s/n, 41092 Sevilla, Spain
e-mail: pedrogalvin@us.es
[3] SNCF, Direction de l'Innovation et de la Recherche, rue de Londres 45, 75379 Paris Cedex 8, France
e-mail: {estelle.bongini,franck.poisson}@sncf.fr

Summary

In March 2008, the SNCF has launched a blind prediction test for for railway-induced ground vibration. The aim was to compare predictions from different numerical models to measured ground vibrations at a site along the LGV Atlantique and a site along the Paris-Bordeaux line. Apart from vibration measurements, SNCF has also performed tests and collected data for the determination of the input parameters of the prediction models. Based on these input parameters, the free-field vibrations have been predicted by means of a numerical model that has been developed at K.U.Leuven. This model takes into account the dynamic interaction between the train, the track and the soil. The track geometry is assumed to be invariant with respect to the longitudinal direction, allowing for an efficient numerical solution of the dynamic track–soil interaction problem in the frequency-wavenumber domain. Although a good qualitative agreement has been obtained between the predicted and measured vibration velocities, the ratio between the predicted and measured running RMS values of the vibration velocity is sometimes as high as 3 or 5.

1 Introduction

Vibrations and re-radiated noise in buildings due to railway traffic are a matter of growing environmental concern. Vibration due to railway traffic is generated by dynamic vehicle-track interaction induced by wheel and rail unevenness. The dynamic axle loads are transferred by the track to the soil and generate elastic waves that excite nearby buildings through their foundations. In the frequency range between 1 Hz and 80 Hz, the vibration is experienced as mechanical vibrations of the body, while at higher frequencies between 16 Hz and 250 Hz, the vibration is perceived as low-frequency noise radiated by building parts.

T. Maeda et al. (Eds.): Noise and Vibration Mitigation for Rail Trans. Sys., NNFM 118, pp. 231–238.
springerlink.com

Several numerical models have been developed for the prediction of free-field vibration due to railway traffic. As an alternative to three-dimensional (3D) models that have a very high computational cost, several two-and-a-half dimensional (2.5D) models have been proposed. The basic assumption underlying the 2.5D methodology is that the geometry of the coupled track-soil system is invariant in the longitudinal direction of the track. This allows for a Fourier transform with respect to the coordinate along the track and leads to an efficient solution in the frequency-wavenumber domain. Aubry et al. [1] have applied a 2.5D procedure to study the response of an infinitely long beam, coupled to an elastic halfspace, due to a moving load. The methodology has been applied by Sheng et al. [2] to an infinite layered beam model for the track, coupled to a layered halfspace. This model has also been elaborated to account for dynamic train-track interaction [3]. A similar methodology has been followed by Auersch [4]. Lombaert et al. [5, 6] have validated a 2.5D model for the prediction of railway-induced vibrations by means of extensive field tests that have been performed within the frame of homologation tests of the high-speed line between Brussels and Köln. Other validations based on data from homologation tests on the line between Brussels and Paris [7] have been presented by Paolucci and Spinelli [8] and Galvín and Domínguez [9].

In March 2008, the SNCF has launched a blind prediction test for railway-induced ground vibration. In the test, a site along the LGV Atlantique and a site along the Paris-Bordeaux line were considered. At K.U.Leuven, the ground vibration at both sites has been predicted with a numerical model that accounts for quasi-static and dynamic excitation [5, 6]. In these computations, a simplified analytical track model was considered and the presence of the embankment was disregarded. Recently, the computations have been repeated with a more elaborate 2.5D coupled FE-BE model [10] that includes the embankment. The present paper discusses the results that have been obtained for the site along the Paris-Bordeaux line. In the next section, data provided by SNCF [11] are used to define the input of the numerical model. Next, the predicted free-field vibration and experimental free-field vibration are compared for three train passages: a TGVA high-speed train at a speed of 200 km/h, an AGC TER passenger train at a speed of 150 km/h, and a freight train at a speed of 40 km/h.

2 The Characteristics of the Train, the Track and the Soil

2.1 The Track Model

The railway track on the Paris-Bordeaux line is a classical ballasted track, with UIC60 rails that are supported by rail pads and fixed with clips on twin block concrete sleepers [11]. The track is supported by a ballast layer with a thickness h_b of about 0.30 m on top of an embankment with a width $w_{e1} = 5$ m at the top, a width $w_{e2} = 10$ m at the soil's surface, and a height $h_e = 1.5$ m.

The track–soil system has been modelled by a coupled 2.5D FE-BE model. In Fig. 1, the cross section of the model is shown with the mesh of the 2.5D solid elements that are used to model the ballast and the embankment. In the following,

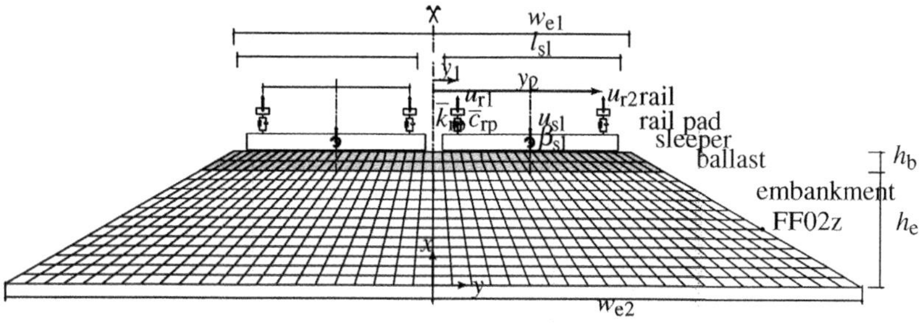

Fig. 1. Cross section of model 1 of the ballasted track on the embankment

the input parameters of the model are discussed. The reader is referred to Lombaert et al. [6, 5] and François et al. [10] for the theoretical background of the model.

The rails are represented by Euler-Bernoulli beams with a bending stiffness $E_rI_r = 6.45 \times 10^6\,\text{N/m}^2$ and a mass per unit length $\rho_rA_r = 60.34\,\text{kg/m}$ for each rail. The positions of the rail are determined by $y_1 = 1.145\,\text{m}$ and $y_2 = 2.580\,\text{m}$, with $y_2 - y_1$ equal to the track gauge r_d. The internal energy dissipation in the rail is modelled by a loss factor $\eta_r = 0.05$. The rail pads are modelled as continuous spring-damper connections. The rail pad stiffness k_{rp} of a single rail pad and the sleeper spacing $d = 0.6\,\text{m}$ have been used to calculate an equivalent stiffness $\bar{k}_{rp} = k_{rp}/d = 130 \times 10^6\,\text{N/m}^2$. A loss factor $\eta_{rp} = 0.23$ is used for the rail pad. The concrete sleepers are assumed to be rigid in the plane of the track cross section and modelled as a uniformly distributed mass $\overline{m}_{sl} = m_{sl}/d$ of 417 kg/m. The sleeper's rotational inertia $\rho_{sl}\bar{I}_{sl} = \rho_{sl}I_{sl}/d$ is equal to $298\,\text{kgm}^2/\text{m}$. The ballast bed is modelled as an elastic continuum. with a Young's modulus $E_b = 111 \times 10^6\,\text{N/m}^2$, Poisson's ratio $\nu_b = 0.36$, density $\rho_b = 1550\,\text{kg/m}^3$, and loss factor $\eta_b = 1.00$. The embankment is modelled as an elastic continuum with a Young's modulus $E_e = 269 \times 10^6\,\text{N/m}^2$, Poisson's ratio $\nu_e = 0.257$, and a density $\rho_e = 1350\,\text{kg/m}^3$.

In order to identify the dynamic soil characteristics at the measurement site, SNCF has commissioned a Spectral Analysis of Surface Waves (SASW) test. The results show that the soil can be represented by a single layer with a thickness of 2.5 m and a shear wave velocity of 281 m/s on a halfspace with a shear wave velocity of 493 m/s [11]. Furthermore, a value of $1350\,\text{kg/m}^3$ is given for the density of the top layer, while the density of the underlying halfspace is equal to $1050\,\text{kg/m}^3$. The Poisson's ratio has been estimated as 0.257 for the top layer and 0.179 for the halfspace. The material damping ratio β in deviatoric and volumetric deformation has a value of 0.05 for both layers. Although the values for the density and Poisson's ratio seem too low, they have been maintained in the calculations to ensure that the results can be compared with those of other prediction models in the benchmark test.

3 Passage of a TGVA at a Speed of 200 km/h

The TGVA is an articulated train with two traction cars and ten passenger cars. At frequencies of more than a few Hertz, the vehicle's primary and secondary suspensions isolate the body and the bogie from the wheelset [12]. A simplified vehicle model that only considers the motion of the bogie and the wheelsets is therefore used to predict the dynamic axle loads. The model contains 4 degrees of freedom for each bogie: the vertical displacement and the rotation about the center of gravity, and the vertical displacement of both wheelsets. When the displacements of the wheelsets are fixed, a vertical translational mode with a natural frequency of 8.8 Hz and a rotational mode with a natural frequency of 13.7 Hz are found for the bogie.

The dynamic axle loads are computed for a simulated track unevenness generated from a single-sided power spectral density $\tilde{S}_{\text{rzz}}(n_y)\,[\text{m}^3]$ of the following type:

$$\tilde{S}_{\text{rzz}}(n_y) = \tilde{S}_{\text{rzz}}(n_{y0}) \left(\frac{n_y}{n_{y0}} \right)^{-w} \tag{1}$$

where $\tilde{S}_{\text{rzz}}(n_{y0})$ is the reference value of the PSD at $n_{y0} = 1/(2\pi)\,\text{m}^{-1}$ and w is the exponent that determines the decrease with increasing cyclic wavenumber n_y. The coefficients $S(n_{y0}) = 9.59 \times 10^{-7}\,\text{m}^3/\text{rad}$ and $w = 4.40$ have been determined by a fit with experimental values from a measurement trolley and a measurement coach.

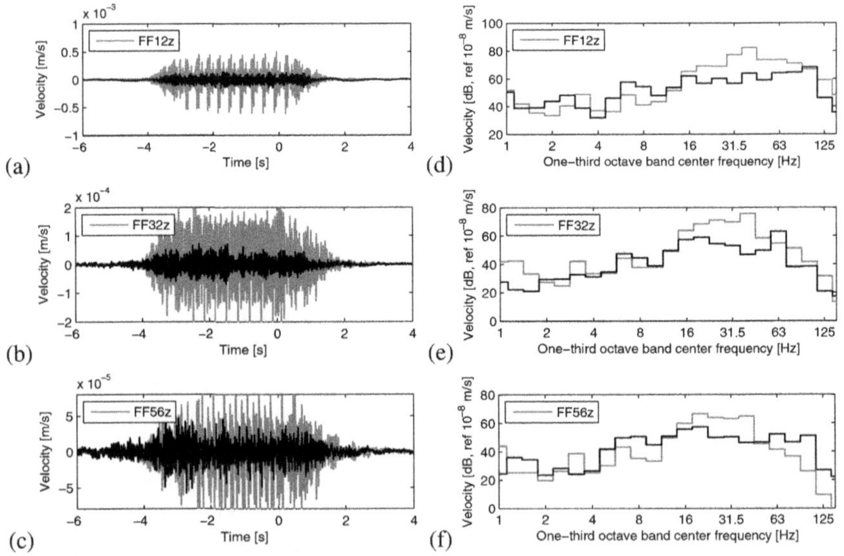

Fig. 2. (a)-(c) Time history and (d)-(f) one-third octave band spectra of the predicted (grey line) and experimental (black line) free-field velocity at 12 m, 32 m, and 56 m from the outer rail during the passage of the TGVA at a speed of 200 km/h.

In Fig. 2, the the time history and one-third octave band spectra of the free-field velocity during the passage of the TGVA at a speed of 200 km/h are shown. The one-third octave band spectra of the free-field velocity are computed according to the German standard DIN 45672-2 [13] on a reference period T_2. The time history of the free-field velocity at 12 m from the outer rail is shown in Fig. 2a. In the predicted time history, the passage of individual cars can clearly be distinguished, whereas the measured time history shows a nearly stationary vibration level during the passage of the train. The measured vibration level is largely overestimated by the prediction. Comparing the predicted and measured one-third octave band spectra of the free-field velocity shows that the overestimation is the most pronounced in the one-third octave bands between 16 Hz and 63 Hz. In the low-frequency range a good agreement is found. The overestimation of the free-field velocity in the higher one-third octave bands is probably due to an overestimation of the dynamic axle loads.

The time history of the free-field velocity at 32 m (Fig. 2b) and at 56 m (Fig. 2c) shows that the overestimation decreases with increasing distance from the track. This is mainly due to the fact that the free-field velocity is increasingly underestimated in the one-third octave bands above 63 Hz (Fig. 2e and 2f). The overestimation in the one-third octave bands between 16 Hz and 63 Hz is still present. In the predictions, the attenuation with distance of the high-frequency components is stronger than in the measurements. This is probably due to an overestimation of the material damping ratio in the soil.

4 Passage of a AGC TER at a Speed of 150 km/h

The AGC TER train is an articulated train with two traction cars and one central car. In the following, only the motion of the bogie and the wheelsets is taken into account. The vertical mode of the bogie has a natural frequency of 4.0 Hz, while the rotational mode has a natural frequency of 7.6 Hz. In Fig. 3, the time history and one-third octave band spectra of the free-field velocity during the passage of the AGC TER at a speed of 150 km/h are shown. Comparing the measured and predicted free-field velocities shows that the same trends are observed as in the case of the TGVA (Fig. 2). At 12 m from the outer rail, the vibration level is overestimated (figure 3a), primarily in the one-third octave bands between 16 Hz and 63 Hz (Fig. 3d). With increasing distance from the track, the agreement between the measured and predicted vibration levels in the free-field (Fig. 3b and 3c) improves. At 56 m from the track, the overestimation in the one-third octave bands around 31 Hz is almost completely compensated by an underestimation in the one-third octave bands around 8 Hz and at 63 Hz.

5 Passage of a Freight Train at 40 km/h

The freight train is composed of one traction car BB 22200, 22 Hopper wagons, and four high-sided wagons. The traction cars and freight wagons have two bogies. The

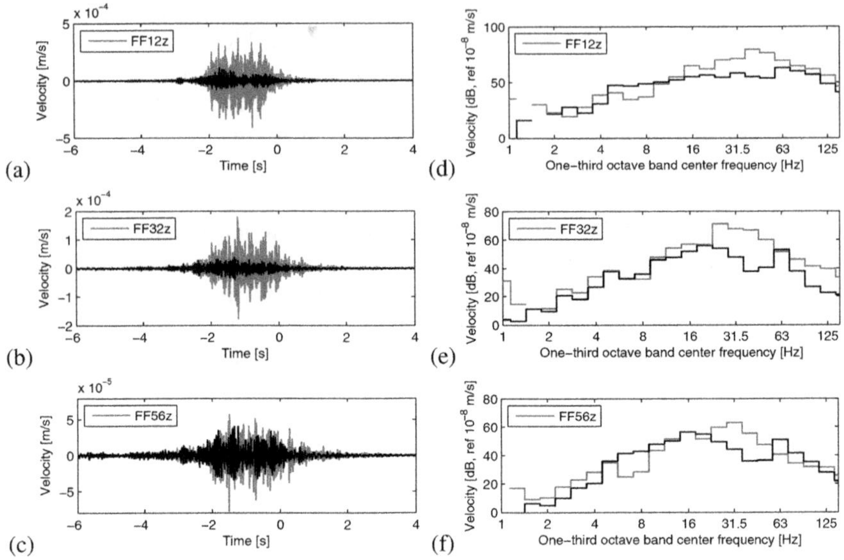

Fig. 3. (a)-(c) Time history and (d)-(f) one-third octave band spectra of the predicted (grey line) and experimental (black line) free-field velocity at 12 m, 32 m, and 56 m from the outer rail during the passage of the AGCTER at a speed of 150 km/h.

traction car has primary and secondary suspensions that isolate the carriage from the track vibrations. The freight wagons only have a primary suspension system. The bogies of these cars are connected to the body by a relatively stiff pivot. A simple multi-body model of the entire car has therefore been used to compute the dynamic axle loads. Since the speed of the freight train is only 40 km/h and very low compared with the Rayleigh wave velocity in the soil, the quasi-static contribution to the free-field velocity is very small and has been disregarded in the computations. In figure 4, the time history and one-third octave band spectra of the free-field velocity during the passage of the freight train are shown. In contrast to the previously discussed results for the passages of the TGVA (Fig. 2) and AGC TER (Fig. 3), the vibration velocity at 12 m from the track (Fig. 4a) is now underestimated. The one-third octave bands around 31.5 Hz where the free-field response was previously overestimated (Fig. 2d and 3d) are the only bands where the response is not underestimated. With increasing distance from the track, the underestimation in the one-third octave bands below 31.5 Hz and above 31.5 Hz increases considerably. The attenuation with distance of the predicted and measured vibration velocities is therefore similar as in the case of the TGVA and the AGC TER.

When the results for the three trains are compared, it is observed that the measured vibration levels are less sensitive to the train speed than the predicted vibration levels. In the present case where only track unevenness characterized by the PSD function in equation (1) is accounted for, the influence of the train speed on the dynamic axle loads is mainly determined by the value of the exponent w in equation (1) [6]. A lower value of w might therefore lead to more consistent results for different train passages. The lack of consistency might also be due to the fact that other excitation mechanisms less dependent on the train speed are not accounted for. The relatively strong variation in time of the measured vibration level for the freight train suggests that wheel unevenness plays an important role in this case.

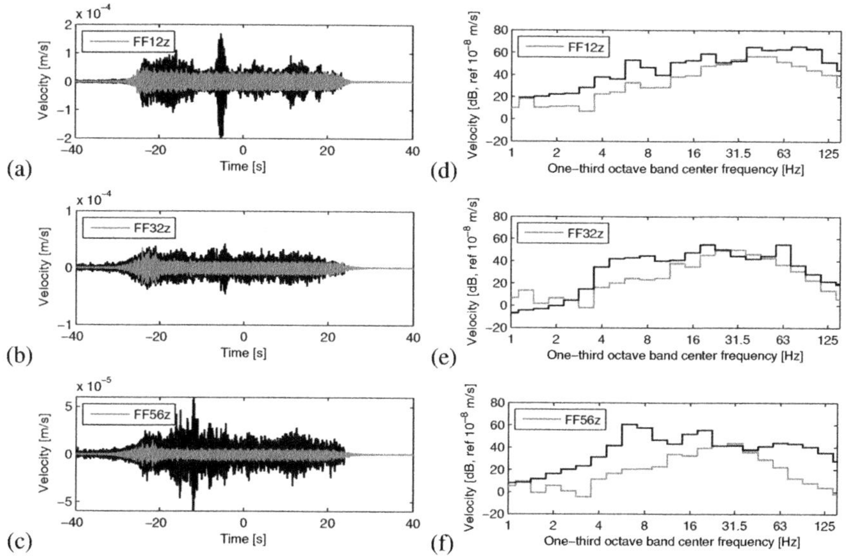

Fig. 4. (a)-(c) Time history and (d)-(f) one-third octave band spectra of the predicted (grey line) and experimental (black line) free-field velocity at 12 m, 32 m, and 56 m from the outer rail during the passage of a freight train at a speed of 40 km/h.

6 Conclusion

The present paper discusses predictions of ground vibrations that have been made within the frame of a blind prediction test of the SNCF. A comparison of the predicted and measured vibration velocity shows that a good qualitative agreement has been obtained, although the ratio between the predicted and measured vibration level amounts to 3 or 5. The main reason for the lack of a closer quantitative agreement seems to be an overestimation of the dynamic axle loads. Moreover, when results for different train types are compared, it is observed that the measured vibration velocity in the free-field is less sensitive to the train speed than the predicted vibration velocity. This could be due to inaccuracies in the measured track unevenness or the fact that other excitation mechanisms are not accounted for.

References

[1] Aubry, D., Clouteau, D., Bonnet, G.: Modelling of wave propagation due to fixed or mobile dynamic sources. In: Chouw, N., Schmid, G. (eds.) Workshop Wave 1994, Wave propagation and Reduction of Vibrations, Ruhr Universität Bochum, Germany, pp. 109–121 (December 1994)

[2] Sheng, X., Jones, C.J.C., Petyt, M.: Ground vibration generated by a harmonic load acting on a railway track. Journal of Sound and Vibration 225(1), 3–28 (1999)

[3] Sheng, X., Jones, C.J.C., Thompson, D.J.: A comparison of a theoretical model for quasi-statically and dynamically induced environmental vibration from trains with measurements. Journal of Sound and Vibration 267(3), 621–635 (2003)

[4] Auersch, L.: The excitation of ground vibration by rail traffic: theory of vehicle-track-soil interaction and measurements on high-speed lines. Journal of Sound and Vibration 284(1-2), 103–132 (2005)

[5] Lombaert, G., Degrande, G., Kogut, J., François, S.: The experimental validation of a numerical model for the prediction of railway induced vibrations. Journal of Sound and Vibration 297(3-5), 512–535 (2006)

[6] Lombaert, G., Degrande, G.: Ground-borne vibration due to static and dynamic axle loads of InterCity and high speed trains. Journal of Sound and Vibration 319(3-5), 1036–1066 (2009)

[7] Degrande, G., Schillemans, L.: Free field vibrations during the passage of a Thalys HST at variable speed. Journal of Sound and Vibration 247(1), 131–144 (2001)

[8] Paolucci, R., Spinelli, D.: Ground motion induced by train passage. ASCE Journal of Engineering Mechanics 132(2), 201–210 (2006)

[9] Galvín, P., Domínguez, J.: High-speed train-induced ground motion and interaction with structures. Journal of Sound and Vibration 307, 755–777 (2007)

[10] François, S., Schevenels, M., Lombaert, G., Galvín, P., Degrande, G.: A 2.5D coupled FE-BE methodology for the dynamic interaction between longitudinally invariant structures and a layered halfspace. Computer Methods in Applied Mechanics and Engineering 199(23-24), 1536–1548 (2010)

[11] Bongini, E., Poisson, F.: Ground vibrations simulation cases parameters. Technical report, SNCF, France (2009)

[12] Knothe, K., Grassie, S.L.: Modelling of railway track and vehicle/track interaction at high frequencies. Vehicle Systems Dynamics 22, 209–262 (1993)

[13] Deutsches Institut für Normung. DIN 45672 Teil 2: Schwingungsmessungen in der Umgebung von Schienenverkehrswegen: Auswerteverfahren (1995)

Verification of an Empirical Prediction Method for Railway Induced Vibration

H. Verbraken*, H. Eysermans, E. Dechief, S. François,
G. Lombaert, and G. Degrande

Department of Civil Engineering, K.U. Leuven
Kasteelpark Arenberg 40, 3001 Heverlee, Belgium
e-mail: Hans.Verbraken@bwk.kuleuven.be

Summary

The Detailed Vibration Assessment is an empirical procedure developed by the U.S. Federal Railroad Administration (FRA) for the prediction of railway induced vibration and re-radiated noise. The vibration velocity level in the free-field is predicted with a force density, characterizing the source, and a line transfer mobility, characterizing the transfer of vibration due to a line load. The line transfer mobility is determined with in situ measurements of transfer functions. The force density is obtained by subtracting the line transfer mobility from the vibration velocity level due to a train passage. It is assumed that the resulting force density can be used to predict the vibration velocity level at other sites with similar train and track characteristics. In this paper, the influence of the soil characteristics on the force density and the resulting vibration velocity level predicted with the FRA procedure is investigated. Numerical simulations are used to compute the vibration velocity level and the line transfer mobility at three sites with different soil characteristics. From these results, the force density due to a train passage is determined for each site. Finally, the three force densities are used to investigate the influence of the soil characteristics on the predicted vibration velocity level due to a train passage.

1 Introduction

The introduction of new transportation infrastructure in an urban environment can generate vibration and re-radiated noise in buildings that cause discomfort to people. An accurate prediction of the vibration is required so that effective mitigation measures can be taken.

Several numerical models have been developed for the prediction of vibration in the free field due to rail traffic at grade [1, 2] and in tunnels [3, 4]. The validation of these models shows that a good knowledge of all relevant input parameters such as the dynamic soil characteristics is required to accurately predict ground vibration due to running trains. In practice, empirical prediction methods such as the prediction procedure proposed by Nelson and Saurenman [5] are often used. This

* Corresponding author.

T. Maeda et al. (Eds.): Noise and Vibration Mitigation for Rail Trans. Sys., NNFM 118, pp. 239–247.
springerlink.com

method is described as the Detailed Vibration Assessment procedure in the guidance manuals of the Federal Railroad Administration (FRA) [6] and the Federal Transit Administration (FTA) [7] of the U.S. Department of Transportation. The FRA procedure predicts the vibration velocity level L_v $[\mathrm{dB\,ref\,10^{-8}\,m/s}]$ in one-third octave bands as

$$L_v = L_F + TM_L \tag{1}$$

where L_F is the force density $[\mathrm{dB\,ref\,1N/\sqrt{m}}]$ and TM_L is the line transfer mobility $[\mathrm{dB\,ref\,10^{-8}\,\frac{m/s}{N/\sqrt{m}}}]$.

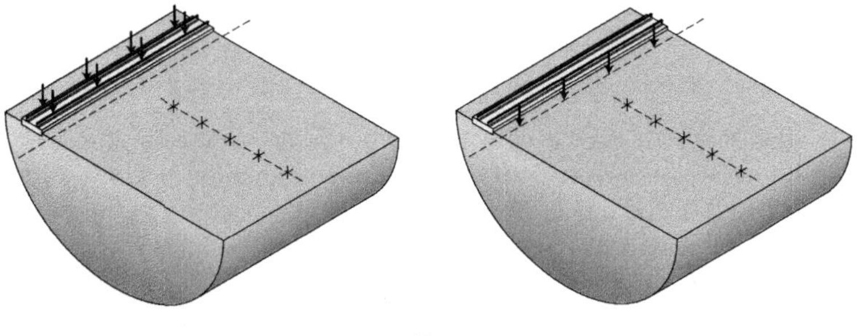

(a) (b)

Fig. 1. Setup for vibration propagation tests at grade with impacts (a) on the rail and (b) adjacent to the track

The line transfer mobility TM_L is a measure for the vibration energy that is transmitted through the soil relative to the power per unit length radiated by the source. It is determined experimentally with the setup shown in Fig. 1a. The point transfer mobilities TMP_k $[\mathrm{dB\,ref\,10^{-8}\,\frac{m/s}{N}}]$ due to single load impacts on the rails are integrated along the rail alignment to obtain a line transfer mobility TM_L.

The force density L_F is a measure for the power per unit length radiated by the source. It is obtained indirectly from equation (1) by subtracting the line transfer mobility TM_L from the vibration velocity level L_v measured during a train passage (site 1), resulting in a normalized force density of the train passage. The vibration velocity level due to a similar train passage can then be predicted at another site with similar soil characteristics (site 2) by combining the force density measured on site 1 with the line transfer mobility measured on site 2. In situations where a track is present on site 2, e.g. when a prediction of the vibration level for new rolling stock is required, the line transfer mobility on both sites can be determined with impacts on the rails (Fig. 1a). In case of new railway lines, however, no track is present on site 2 and the line transfer mobility on this site is determined with impacts directly on the soil. The line transfer mobility on site 1 should then also be determined with impacts on the soil, adjacent to the track (Fig. 1b). In the last approach, an equivalent force density is obtained that represents the energy that has to be introduced into the soil to obtain the same vibration level as for the train passage.

The force density is influenced by several parameters such as the train, track and soil characteristics and can therefore only be used to predict the vibration velocity level on sites with similar conditions. In the present paper, the influence of the soil characteristics on the force density and the resulting prediction of the vibration velocity level is investigated. First, the force density due to a train passage is determined on three sites with different soil characteristics. For this purpose, the vibration velocity level and the line transfer mobility are computed on each site by means of a numerical model. Second, the force densities are used to predict the vibration velocity level at another site. The influence of the soil on the predicted vibration velocity levels is investigated. Two cases are considered. In the first case, the force densities are determined with transfer mobilities derived from impacts on the rails (Fig. 1a). In the second case the force densities are determined with transfer mobilities derived from impacts at the soil's surface adjacent to the track (Fig. 1b).

2 Vibration Velocity Level in the Free Field due to a Train Passage

The vibration velocity level in the free field due to a train passage is computed by means of a numerical model. The free-field vibration is caused by dynamic axle loads, generated by a train running over an uneven track. In this case, a ballasted slab track and UIC 60 rails with an unevenness according to FRA class 6 is considered. The velocity is computed at three sites with different soil characteristics, representing a soft, medium and stiff soil (Table 1).

Table 1. Dynamic soil characteristics

Soil type	C_s [m/s]	C_p [m/s]	ν [-]	ρ [-]	β [-]
Soft	100	200	0.33	1800	0.025
Medium	150	300	0.33	1800	0.025
Stiff	300	600	0.33	1800	0.025

Numerical results in this paper have been obtained by means of a 2.5D coupled finite element - boundary element (FE-BE) prediction method [8]. A finite element model for the track is coupled to a boundary element method for the supporting soil. The longitudinal invariance of the geometry is exploited by using a Fourier transform resulting in an efficient solution in the frequency-wavenumber domain.

The vibration velocity level due to the passage of a Thalys HST at a speed of 80 m/s is computed by calculating the one-third octave band RMS values of the stationary part of the time history of the vibration velocity due to the train passage. The Thalys train has a length of 200.18 m and has 26 axles. Fig. 2 compares the vibration velocity level at 8 m and 32 m from the track at the three different sites. A peak around 60 Hz is found, which is due to the resonance of the wheels on the track.

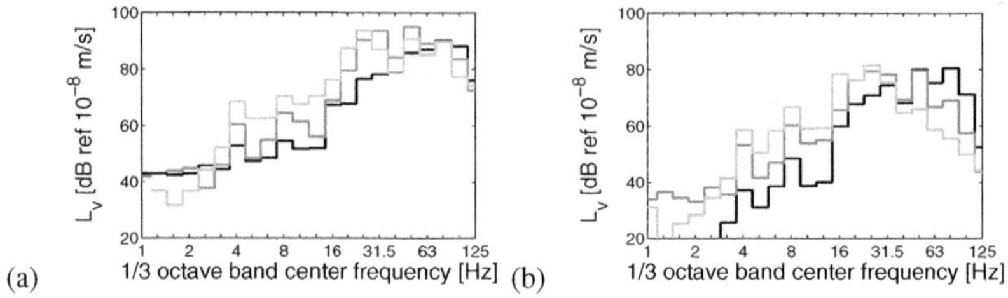

Fig. 2. Vibration velocity level at (a) 8 m and (b) 32 m from the track on soft (light grey line), medium (dark grey line) and stiff (black line) soil

Close to the track, the highest response is obtained for the soft soil. The response is increasingly attenuated not only at higher frequencies but also with increasing distance from the track. As this attenuation is stronger for soft soils, the high-frequency response at larger distances is the highest in case of the stiff soil.

Below 8 Hz the results should be interpreted with care, as the stationary part of the response only contains a few cycles due to the short duration of the train passage.

3 Line Transfer Mobility

In this section, the line transfer mobility is determined numerically by simulating the experimental measurements of the transfer function (Fig. 1).

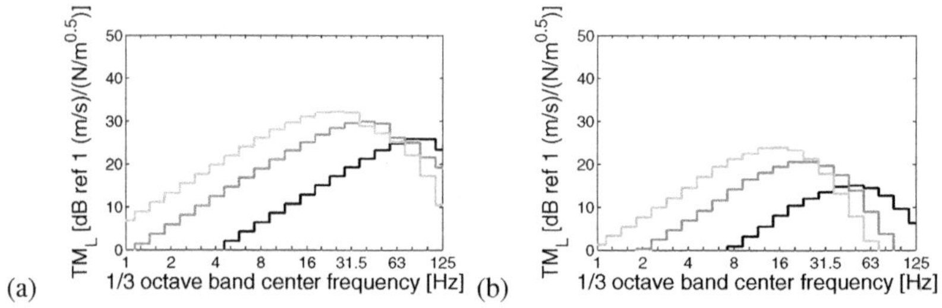

Fig. 3. Line transfer mobility from the rails to a point in the free field at (a) 8 m and (b) 32 m from the track on soft (light grey line), medium (dark grey line) and stiff (black line) soil

First, the case is considered where the line transfer mobility is determined with impacts on the rails (Fig. 1a). Fig. 3 compares the transfer mobilities at 8 m and 32 m from the track for the three sites. The stiffness of the soil has a large influence on the transfer mobility. At low frequencies and close to the track, the highest response is obtained for the soft soil. With increasing frequency, the vibration levels decrease due to two mechanisms. First, vibration is attenuated due to energy dissipation in the soil characterized by material damping. The attenuation depends on the number of wavelengths between the source and the receiver and is therefore more important for higher frequencies and soft soils. Second, high-frequency vibration is partially

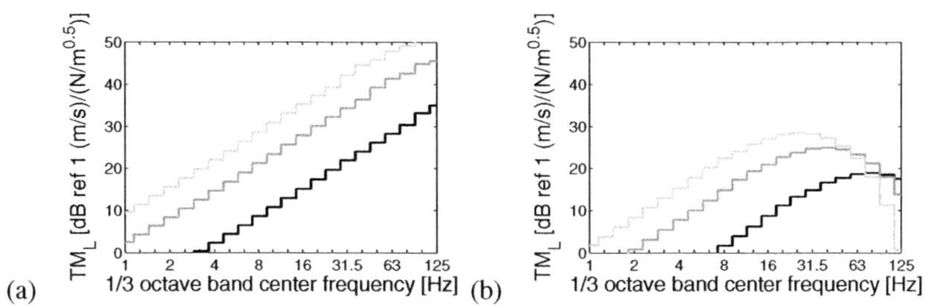

Fig. 4. Line transfer mobility from the soil adjacent to the track to a point in the free field at (a) 8 m and (b) 32 m from the track on soft (light grey line), medium (dark grey line) and stiff (black line) soil

filtered by the track system and is transmitted into the soil at a reduced level. The track filtering effect is determined by the ratio of the wavelength and the width of the track and is therefore more important for soft soils as well.

Next, the case is considered where the line transfer mobility is determined with impacts on the soil adjacent to the track (Fig. 1b). Fig. 4 shows the line transfer mobility at 8 m and 32 m from the track for the three sites. The most important difference with the results based on impacts on the rails (Fig. 3) is that the track filtering effect is not present, so that the attenuation in the high-frequency range is less prominent.

4 Force Density of a Train Passage

The force density is now determined by subtracting the line transfer mobility at 8 m and 32 m (Fig. 3 and 4) from the vibration velocity level at 8 m and 32 m (Fig. 2), respectively.

First, in Fig. 5 the force densities are obtained by applying equation (1) with a line transfer mobility determined with impacts on the rails. The force densities are different for the three sites, as the axle loads are affected by dynamic train-track-soil interaction. The high values of the force densities below 8 Hz are not fully reliable

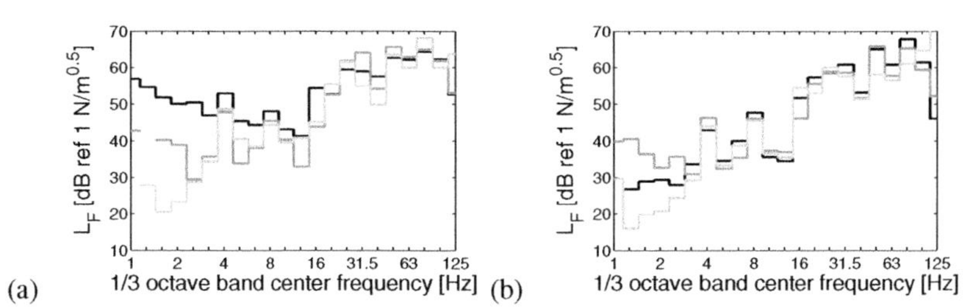

Fig. 5. Force density derived from impacts on the rails and vibration velocity level at (a) 8 m and (b) 32 m from the track on soft (light grey line), medium (dark grey line) and stiff (black line) soil.

due to the inaccurate results for the vibration velocity level in this frequency range. In a prediction of the vibration velocity level, however, these high values of the force density are added to low values of the line transfer mobility in the low-frequency range and are therefore unimportant. The force densities all obtain a maximal value near the resonance frequency of the wheels on the track. Above 8 Hz, they differ by, at most, 10 dB.

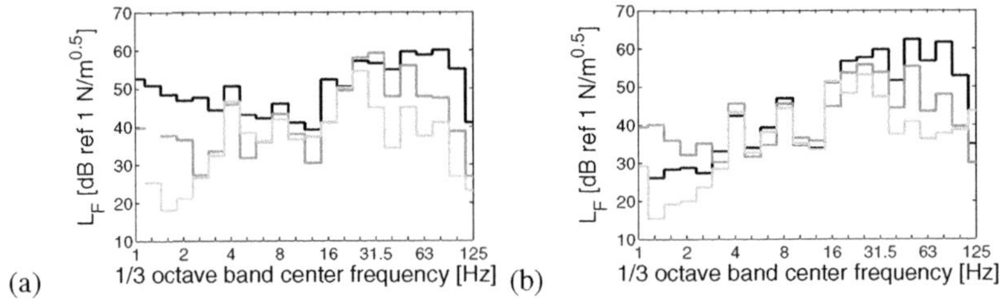

Fig. 6. Force density derived form impacts adjacent to the track and vibration velocity level at (a) 8 m and (b) 32 m from the track on soft (light grey line), medium (dark grey line) and stiff (black line) soil.

Second, in Fig. 6, the force density is obtained by means of a line transfer mobility determined with impacts on the soil adjacent to the track. In this case, an equivalent force density is obtained that represents the force that has to be applied adjacent to the track to obtain the same vibration velocity in the free field as for the train passage. Above 8 Hz, large differences are found between the force densities, as the track filtering effect is not present in the line transfer mobility and is therefore compensated by the force density. Since the track filtering highly depends on the soil characteristics, force densities of different soils differ up to 30 dB.

5 Prediction of the Vibration Velocity Level

A prediction according to the FRA procedure is now made for the vibration velocity level due to the passage of the Thalys HST on a site with medium soil. This prediction is made by combining the line transfer mobility on the medium soil with the force densities of the train passage on three different soils. In this way, the influence of the soil characteristics on the predicted vibration velocity is investigated. The force densities determined at 8 m from the track are chosen. The predictions are compared with a 12 dB interval ($\pm$6 dB) around the exact solution, that has been previously shown in Fig. 2.

First, the force densities determined with impacts on the track are used. Fig. 7 compares the predicted vibration velocity levels for the three force densities with the exact solution. At 8 m from the track (Fig. 7a), the prediction obtained from the force density of the medium soil coincides with the exact solution. Above 8 Hz, the predictions obtained from the force densities of the soft and the stiff soil generally differ less than 6 dB from the exact solution and are therefore relatively good.

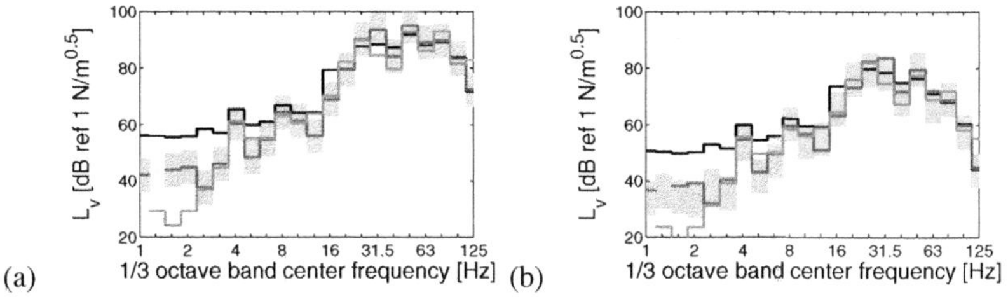

Fig. 7. Prediction of the vibration velocity level at a point in the free field at (a) 8 m and (b) 32 m from the track, using a force density determined with impacts on the rails on a soft (light grey line), medium (dark grey line) and stiff (black line) soil compared with a 12 dB interval around the exact solution (grey region).

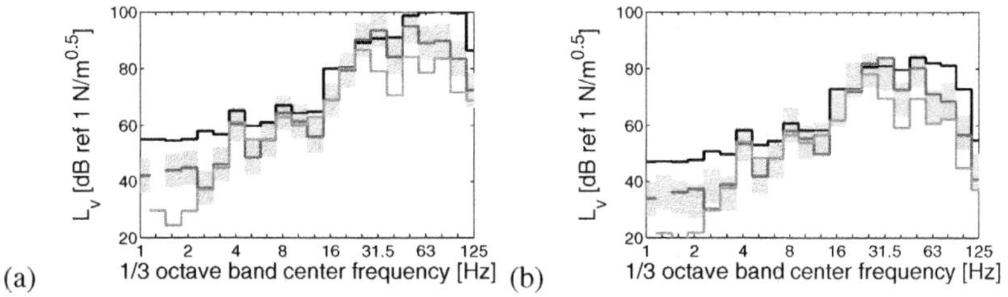

Fig. 8. Prediction of the vibration velocity level at a point in the free field at (a) 8 m and (b) 32 m from the track, using a force density determined with impacts adjacent to the track on a soft (light grey line), medium (dark grey line) and stiff (black line) soil compared with a 12 dB interval around the exact solution (grey region).

Second, the results computed by means of the equivalent force densities determined with impacts adjacent to the track are compared with the exact solution (Fig. 8). In this case, a poor prediction of the vibration velocity level is obtained with the force density from the soft and the stiff soil, both at low and high frequencies. At frequencies above 8 Hz, differences up to 15 dB are observed. This is due to the fact that the equivalent force densities strongly depend on the soil characteristics, as described in section 4. As a consequence, a prediction of the vibration velocity level cannot be made with a force density obtained on a site with significantly different soil parameters.

When no appropriate experimental force density is available, e.g. in case of new track or new train types or in case of significantly different soil characteristics, the force density can alternatively be obtained by numerical simulations. For this purpose, analytical expressions for the force density L_F and the line transfer mobility TM_L have been derived [9]. A hybrid numerical–experimental prediction is then obtained, where a computed force density is combined with experimental line transfer mobilities.

6 Conclusions

The FRA procedure is an empirical procedure that predicts railway-induced vibrations using a force density, characterizing the source, and a line transfer mobility, characterizing the transfer of vibrations. In this paper, the FRA procedure is verified by means of numerical simulations using a 2.5D coupled finite element – boundary element method. The force density of a train passage is computed for three sites with different soil characteristics. By predicting the vibration velocity level with each force density, the influence of the dynamic soil characteristics on the predicted vibration velocity level is investigated. Two cases are considered in the analysis. In the first case, the line transfer mobility is determined with impacts on the rails, while in the second case it is determined with impacts adjacent to the track.

In the first case where impacts are given on the rail, the predictions of the vibration velocity level in the free field are relatively good, with differences generally below 6 dB, even with force densities determined at sites with significantly different soil characteristics. In the second case where impacts are given on the soil adjacent to the track, the influence of the soil characteristics on the force density is very large with differences up to 15 dB. The prediction of the vibration velocity level is only accurate when a force density determined on a similar soil is used.

As the second approach is the most convenient, and even the only possibility in the case of new infrastructure, it is concluded that accurate predictions according to the FRA procedure can only be made when an appropriate force density is available, determined at a site with similar conditions. Alternatively, the force density can be computed numerically and used in a hybrid numerical–experimental approach.

Acknowledgements

The first author is a Research Assistant and the fourth author a Postdoctoral Fellow of the Research Foundation - Flanders (FWO). Their financial support is gratefully acknowledged.

References

[1] Lombaert, G., et al.: The experimental validation of a numerical model for the prediction of railway induced vibrations. Journal of Sound and Vibration 297(3-5), 512–535 (2006)
[2] Sheng, X., et al.: Ground vibration generated by a load moving along a railway track. Journal of Sound and Vibration 228(1), 129–156 (1999)
[3] Andersen, L., Jones, C.J.C.: Coupled boundary and finite element analysis of vibration from railway tunnels – a comparison of two- and three-dimensional models. Journal of Sound and Vibration 293(3-5), 611–625 (2006)
[4] Degrande, G., et al.: A numerical model for ground-borne vibrations from underground railway traffic based on a periodic finite element – boundary element formulation. Journal of Sound and Vibration 293(3-5), 645–666 (2006)

[5] Nelson, J.T., Saurenman, H.J.: A prediction procedure for rail transportation ground-borne noise and vibration. Transportation Research Record 1143, 26–35 (1987)
[6] Hanson, C.E., et al.: High-speed ground transportation noise and vibration impact assessment. HMMH Report 293630-4, U.S. Department of Transportation, Federal Railroad Administration, Office of Railroad Development (2005)
[7] Hanson, C.E., et al.: Transit noise and vibration impact assessment. Report FTA-VA-90-1003-06, U.S. Department of Transportation, Federal Transit Administration, Office of Planning and Environment (2006)
[8] François, S., et al.: A 2.5D coupled FE-BE methodology for the dynamic interaction between longitudinally invariant structures and a layered halfspace. Computer Methods in Applied Mechanics and Engineering 199(23-24), 1536–1548 (2010)
[9] Verbraken, H., et al.: Verification and extension of an empirical prediction method for subway induced vibration with a coupled FE-BE model. In: Proceedings of the 16th International Congress on Sound and Vibration, Kraków, Poland (2009)

Track-Based Control Measures for Ground Vibration – The Influence of Quasi-Static Loads and Dynamic Excitation

N. Triepaischajonsak[1], D.J. Thompson[1], C.J.C. Jones[1], and J. Ryue[2]

[1] Institute of Sound and Vibration Research,
University of Southampton, Southampton SO17 1BJ, UK
e-mail: djt@isvr.soton.ac.uk
[2] School of Naval Architecture and Ocean Engineering,
University of Ulsan, Ulsan, 680-749, Korea

Summary

Ground vibration from surface trains is induced by two main excitation mechanisms, due to the moving quasi-static loads and dynamic loads due to track roughness. The relative importance of these two mechanisms has been investigated using a semi-analytical model of the coupled vehicle, track and ground system. The track is represented as an infinite, layered beam resting on one or more elastic soil layers, overlying a three-dimensional half-space of ground material. The train is modelled as a multi-body system with both primary and secondary suspensions and its motion is included. The quasi-static and dynamic loads have been investigated over a wide range of conditions for locations close to the track and further away. Results are shown in terms of insertion gains for both dynamic and quasi-static components for a wide variety of conditions. It is confirmed that an assessment of the effect of a change in support stiffness should not be based on measurements too close to the track – in the case considered the distance should be at least 10 m to give representative results.

1 Introduction

When a train runs on the track, vibration is induced at the wheel-rail interface which is transmitted through the ground into neighbouring properties. Here it may be experienced as feelable vibration or as audible rumbling noise [1]. Although situations may occur where high-speed trains exceed the speed of surface waves in the ground, more often ground-borne vibration and noise occur in situations where the train speed is considerably less than this wave speed. Indeed, vibration problems on surface railways are often associated with heavy freight trains while ground-borne noise is commonly associated with metro systems operating in tunnels; neither of these operate at a particularly high speed.

T. Maeda et al. (Eds.): Noise and Vibration Mitigation for Rail Trans. Sys., NNFM 118, pp. 249–257.
springerlink.com

Several previous studies [2, 3] have considered the relative importance of the moving quasi-static loads and dynamic loads due to track roughness in the generation of vibration. The quasi-static loads produce large displacements under the track but, for train speeds lower than the wave speeds in the ground, these form a near field which does not propagate into the far field. Nevertheless for locations close to the track and low frequencies the quasi-static loads may still be important. An important question is whether measurements made relatively close to the track can be used to determine the 'source strength' for a particular vehicle/track combination or to interpret the effects on vibration in the far field of changes to the track or the vehicles.

The purpose of the present paper is to study the relative importance of these two excitation mechanisms, moving quasi-static loads and dynamic loads due to track roughness, over a wide range of conditions for locations close to the track and further away. Results are focused on the effects of track-based mitigation measures such as resilient rail fasteners, sleeper soffit pads and ballast mats.

2 Model

Calculations have been performed using the semi-analytical model for the ground vibrations induced from surface trains, Train-Ground Vibration (TGV), developed by Sheng *et al.* [4-6]. The track is represented as an infinite, layered beam resting on one or more elastic soil layers, overlying a three-dimensional half-space of ground material. The track is assumed to be in contact with the ground over a width corresponding to the base of the embankment and the dynamic load on the ground is assumed to be constant over this width. The embankment is represented by a stiffness layer and consistent mass is included for both the ballast and the embankment. The ground is modelled in the wavenumber domain in the directions parallel and perpendicular to the track. The response in the various layers is then transformed back to the spatial domain by an inverse Fourier transform.

The vertical dynamic behaviour of the train is modelled using a multi-body system with both primary and secondary suspensions. Train motion is included, allowing the prediction of vibration due to both dynamic and quasi-static loads. This model has been validated against measurements at a number of sites [6, 7].

The ground model consists of a single weathered layer overlying a stiffer half-space. To consider a range of ground stiffnesses four soil types are considered for the upper layer and assigned nominal values, as listed in Table 1.

Table 1. The parameters used for various types of the ground

Parameters of ground	Soil type				
	Soft clay	Clay/Sand	Sand	Chalk	Half-space
P-wave speed (m/s)	1700	1700	1700	2000	2100
S-wave speed (m/s)	120	220	350	1100	1200
Density (kg m^{-3})	2000	2000	2000	2000	2000
Layer depth	3.0 m	3.0 m	3.0 m	3.0 m	infinite

The parameters used to represent the vehicles are typical values applying to a five-coach electric multiple unit train, as listed in Table 2. Unless otherwise stated the train speed is 25 m/s. The parameters used for the track system are typical values for a ballasted track, as listed in Table 3.

Table 2. Properties used to represent the vehicles

	Mass (kg)		Stiffness (MN/m)	Damping (kNs/m)
Wheelset	1200	Contact stiffness	2420	-
Bogie	4700	Primary suspension	1.6	20
Vehicle body	30000	Secondary suspension	0.47	33.6

A number of track design variants are also considered as listed in Table 3. These include changing the rail pad stiffness, the effects of an embankment and various soil conditions. The effects of track-based vibration control measures such as examples of ballast mats and sleeper soffit pads are investigated in terms of insertion gain at different distances from the track.

Table 3. Parameters used to represent different track variants (values are for two rails)

Parameters	original	variants	
Rail bending stiffness, N/m^2	1.26×10^7	-	
Rail mass per unit length, kg/m	120	-	
Rail pad stiffness, N/m^2	3.5×10^8	3.5×10^7	
Sleeper mass per unit length, kg/m	490	-	
Ballast stiffness, N/m^2	3.15×10^8	-	
Ballast mat stiffness, N/m^2	-	40×10^6	
Sleeper soffit pad, N/m^2	-	247.5×10^6	22.5×10^6
Embankment height, m	-	4.0	0.75

3 Results for Reference Situation

Fig. 1 shows the predicted vibration for the reference situation on the soft clay soil at 0 m (directly under the track) and at 10 m. It can be seen that the response beneath the track is dominated by the quasi-static loads up to 12.5 Hz whereas at 10 m away the dynamic loads are dominant above about 3 Hz. The peaks in the response at low frequency are related to the axle spacing.

The dynamic excitation can be understood in terms of the mobilities of the vehicle and track [1]. These are plotted in Fig. 2 for the reference situation and also for the softer rail pad. The model for the track includes the ground beneath it. The vehicle is predominantly mass-controlled with two strongly damped resonances whereas the track is stiffness-controlled. Where the two curves cross, in this case at around 65 Hz, a resonance of the coupled vehicle-track system occurs and this can be seen to correspond to a peak in the vibration in Fig. 1.

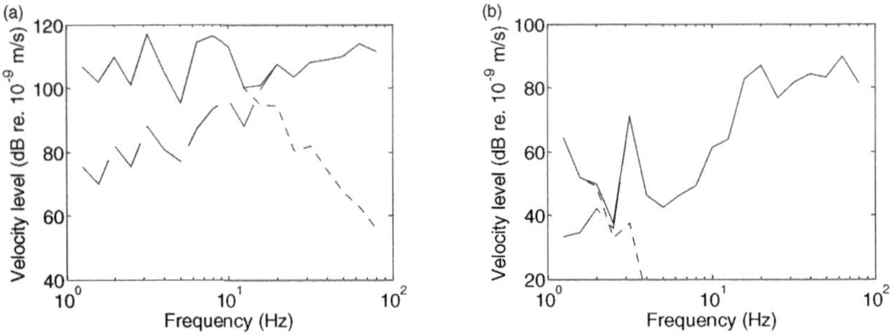

Fig. 1. Ground vibration for reference case on soft clay (a) at 0 m and (b) 10 m from the track. —, total; ······, quasi-static loads; – – –, dynamic loads.

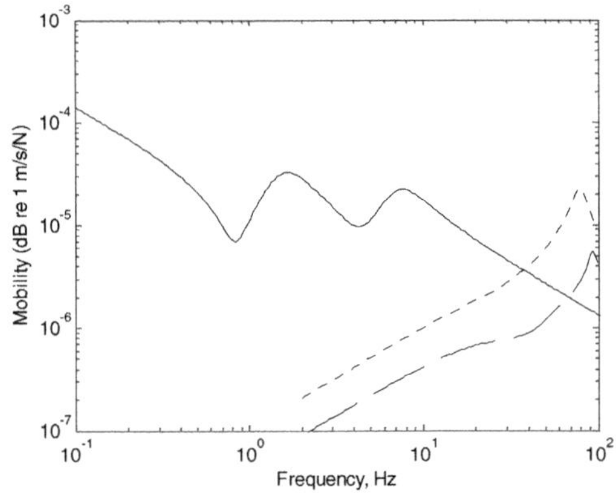

Fig. 2. Point mobilities of vehicle and track for reference situation. —, vehicle; – – –, track; ······, soft rail pad.

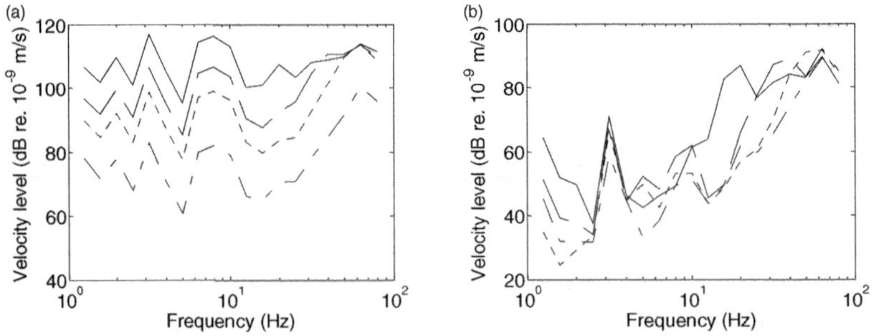

Fig. 3. Ground vibration for reference case on different soils (a) at 0 m and (b) 10 m from the track. —, soft clay; – – –, clay/sand; ······, sand; – · – ·, chalk.

Fig. 3 shows the overall responses at 0 m and 10 m for the different ground parameters. The responses at 0 m are significantly affected by the ground stiffness, whereas the response at 10 m is affected much less.

4 Results for Modified Track Parameters

Introducing a softer spring in the track support lowers the coupled resonance frequency and reduces the ground response at higher frequencies. Fig. 4 illustrates this by showing the difference between the results for soft and stiff rail pads.

Results are shown in terms of the difference in level at various distances for total, dynamic component and quasi-static component. Although the stiffness of the rail pad has been reduced by a factor of 10, the overall track stiffness is only reduced by a factor of 2.4 at low frequencies, see Fig. 2. Consequently, the resonance frequency is lowered to about 40 Hz. The dynamic component shows the classic trend of an increase in vibration at the new resonance frequency, followed by a maximum reduction at the old resonance before tending to a constant reduction at high frequency.

On the other hand, the quasi-static component is reduced at all frequencies above 3 Hz. The softer track support leads to a smoothening of the deflection under each bogie so that the higher frequency content of the deflection pattern is reduced. The insertion gain for the quasi-static load is virtually identical at all distances (it is shown only for 0 m).

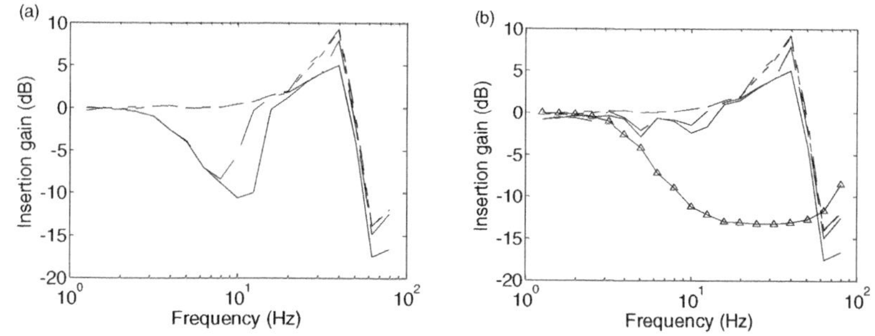

Fig. 4. Insertion gain due to changing the stiffness of rail pad: (a) total response: —, 0 m; – – –, 3 m; ······, 10 m; – · – ·, 25 m; (b) –Δ–, quasi-static component at 0 m; dynamic component: —, 0 m; – – –, 3 m; ······, 10 m; – · – ·, 25 m.

The insertion gain for the overall response follows that of the dynamic excitation for locations far from the track, but at 0 m it follows the quasi-static curve up to about 10 Hz, above which it tends towards the curve for dynamic excitation. This can be understood from the contributions shown in Fig. 1. It is clear from these results that an assessment of the insertion gain due to a change in rail pad stiffness should not be based on measurements too close to the track — in this case the distance should be at least 10 m to give representative results.

Fig. 5 shows the influence of the different ground properties on these insertion gains for a soft rail pad. The insertion gains for the dynamic and quasi-static components are virtually unaffected by the ground stiffness, although the resonance peaks shift up in frequency slightly with increasing ground stiffness. More significantly, the relative importance of the quasi-static component reduces with increasing ground stiffness, so that on the stiff ground the insertion gain follows that for the dynamic load more closely even at 3 m.

Fig. 6 shows the insertion gains for the soft ground site for a different train speed. As the speed increases the relative importance of the quasi-static excitation increases [1]. The insertion gain corresponding to the quasi-static component shifts towards higher frequencies as speed increases whereas that for the dynamic component is independent of train speed. Thus the overall results are slightly different from those in Fig. 4.

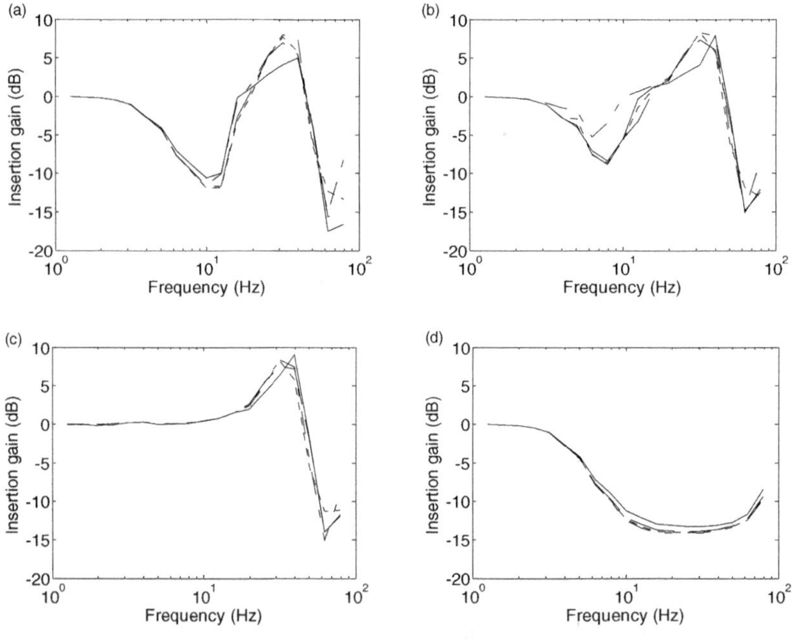

Fig. 5. Insertion gain of the ground vibration due to change in rail pad stiffness on different soil types: (a) all four sites total at 0 m; (b) all four sites total at 3 m; (c) all four sites total at 10 m; (d) all four sites quasi-static at 0 m. —, soft clay; – – –, clay/sand; ······, sand; – · – ·, chalk.

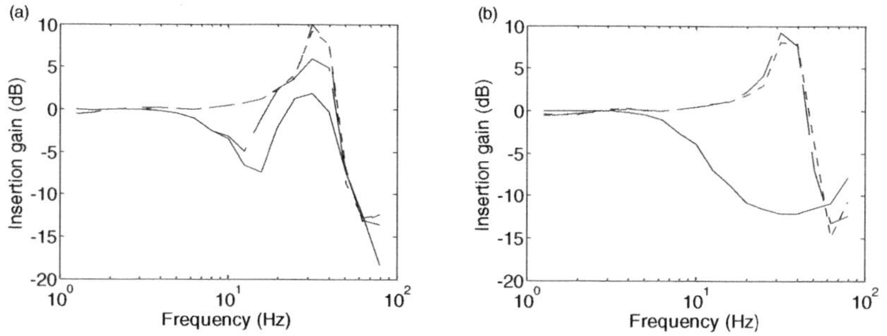

Fig. 6. Insertion gain of the ground vibration due to change in rail pad stiffness for soft clay soil at 50 m/s; (a) total response: —, 0 m; – – –, 3 m; ······, 10 m; – · – ·, 25 m; (b) —, quasi-static component at 0 m at 50 m/s; dynamic component at 10 m. – – –, 50 m/s; ······, 12.5 m/s.

Fig. 7 shows the insertion gains for a ballast mat and a sleeper soffit pad. These show similar trends to the results for the rail pad stiffness. The stiff soffit pad gives little benefit. The softer soffit pad gives very similar results to the ballast mat and is therefore not shown.

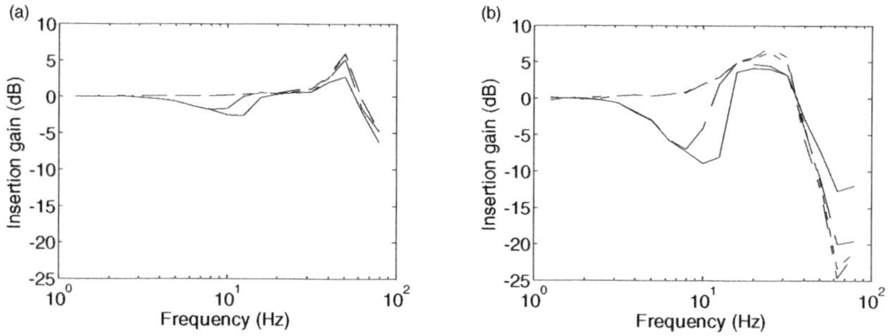

Fig. 7. Insertion gain of the ground vibration with the case of inserting; (a) stiff soffit pad and (b) the ballast mat for the total response: —, 0 m; – – –, 3 m; ······, 10 m; – · – ·, 25 m.

It is clear from these various results that softer rail pads, ballast mats and soffit pads all lead to significant increases in ground vibration at 10 m and beyond in the region 20-40 Hz due to the lowered vehicle/track resonance. Reductions in vibration only occur above about 40 Hz for the parameters considered in this paper. However, a measurement close to the track would suggest that the benefits extend to lower frequencies.

Fig. 8 shows the effects of introducing an embankment into the model. Two different heights of embankment are considered as listed in Table 3. The embankment is modelled only as a layer of damped springs with consistent mass. Inserting the embankment gives a similar trend in terms of the insertion gain as for the rail pad stiffness. However, for the larger embankment the width of the connection with the ground is also increased. This leads to a reduction in level at

0 m (as the load is spread over a wider area) and an increase at 10 m (as this position is now closer to the base of the embankment).

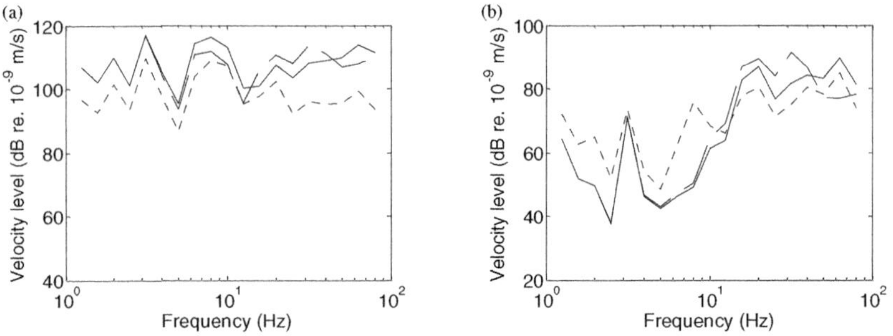

Fig. 8. Ground vibration for the case of: —, none; – – –, small; ······, large embankment on soft clay (a) at 0 m and (b) 10 m from the track.

5 Conclusions

The ground vibration generated by both quasi-static and dynamic excitation mechanisms has been investigated using a semi-analytical model. The quasi-static loads dominate the ground vibration just beneath the track up to about 12.5 Hz at 25 m/s. Above this frequency the response is clearly due to the dynamic load. At the distances of 10 m and further away from the track, the dynamic loads are dominant across almost the whole frequency range. The effects of changing track support stiffness are similar on the different ground types considered, at least at distances of at least 10 m. The train speeds affect the quasi-static load by shifting its frequency components toward higher frequencies while the insertion gain for dynamic loads is unaffected.

It is confirmed that an assessment of the insertion gain of a change in rail support stiffness should not be based on measurements too close to the track. A distance of at least 10 m is required to give representative results.

References

[1] Thompson, D.J.: Railway noise and vibration: mechanisms, modelling and means of control. Elsevier, Oxford (2008)

[2] Sheng, X., Jones, C.J.C., Thompson, D.J.: A comparison of a theoretical model for quasi-statically and dynamically induced environmental vibration from trains with measurements. Journal of Sound and Vibration 267, 621–635 (2003)

[3] Lombaert, G., Degrande, G.: Ground-borne vibration due to static and dynamic axle loads of InterCity and high-speed trains. Journal of Sound and Vibration 319, 1036–1066 (2009)

[4] Sheng, X., Jones, C.J.C., Thompson, D.J.: A theoretical model for ground vibration from trains generated by vertical track irregularities. Journal of Sound and Vibration 272, 937–965 (2004)

[5] Sheng, X., Jones, C.J.C., Thompson, D.J.: A theoretical study on the influence of the track on train-induced ground vibration. Journal of Sound and Vibration 272, 909–936 (2004)

[6] Sheng, X., Jones, C.J.C., Thompson, D.J.: A comparison of a theoretical model for quasi-statically and dynamically induced environmental vibration from trains with measurements. Journal of Sound and Vibration 267, 621–635 (2003)

[7] Triepaischajonsak, N., Thompson, D.J., Jones, C.J.C., Ryue, J., Priest, J.A.: Ground vibration from trains: experimental parameter characterisation and validation of a numerical model. Proceedings of Institution of Mechanical Engineers, Part F 225, 140–153 (2011)

Ground Vibration Induced by Railway Traffic – The Influence of Vehicle Parameters

A.A. Mirza[1], A. Frid[1], J.C.O. Nielsen[2], and C.J.C. Jones[3]

[1] Bombardier Transportation,
Östra Ringvägen 2, 721 23 Västerås, Sweden
Tel.: +46 10 852 8229; Fax: +46 10 852 7111
e-mail: adam.mirza@se.transport.bombardier.com
[2] CHARMEC/Department of Applied Mechanics,
Chalmers University of Technology, Sweden
[3] Institute of Sound and Vibration Research,
University of Southampton, U.K.

Summary

Measurements of ground vibration induced by a two-car EMU railway vehicle have been made at speeds up to 250 km/h. The results show an increasing ground vibration level with increasing speed and a clear excitation generated by each passing wheelset and the discrete sleeper support. Simulations with a multibody vehicle model on a railway track, including a ground model consisting of isotropic elastic layers, show good agreement with the measurement results.

A study is carried out using simulations to investigate the influence of vehicle parameters on ground vibration levels. Nine parameters including vehicle mass, suspension properties and the geometric properties of the vehicle are studied using a fractional factorial design (FFD) method of parametric study. The results show an increase in vibration level for a stiffer primary suspension and a heavier unsprung mass. In general, the interactions between different parameters are low.

1 Introduction

Railway-induced ground vibration can have a negative environmental impact on inhabitants living close to a railway line and, in extreme cases, cause damage to adjacent buildings. Ground vibration problems are mainly associated with freight traffic but may also occur for passenger traffic. In particular, for high-speed trains running on soft soils there is a risk of exceeding the critical speed of ground wave propagation. Observations of severe vibration of this kind were reported in [1]. Much of the previous work in the field has studied the influence of ground properties; this study investigates the influence of rolling stock properties.

T. Maeda et al. (Eds.): Noise and Vibration Mitigation for Rail Trans. Sys., NNFM 118, pp. 259–266.
springerlink.com

2 Theory

A moving railway vehicle excites vibration in two ways: (1) the ground experiences the vehicle as a set of static loads, one load for each wheel-rail contact, travelling at the speed of the vehicle. This excitation is referred to as quasi-static excitation and can be fully described by the nominal axle load, the spacing between the axles and the speed of the vehicle [2]. The quasi-static excitation generally dominates vibration at the track. However, for vehicles travelling at speeds below that of the vibration waves in the soil, this only excites near-field waves and so only contributes significantly to the total vibration within a few metres from the track. (2) Dynamic forces arise from track irregularities and wheel out-of-roundness. These excite propagating waves in the soil and therefore usually dominate the vibration transmitted to locations farther from the track. This component of vibration is determined by the amplitude and wavelength of the wheel/rail imperfections as well as the dynamic properties of the vehicle and the track [2]. The axle loads of trains travelling faster than the wavespeeds in the soil excite propagating waves directly, i.e. via the quasi-static mechanism and this can lead to higher levels of vibration at a distance from the track [3].

The propagation of ground vibration is strongly influenced by the properties of the ground. In particular, in the frequency range of interest in railway vibration the layered structure of the first few metres of soil affects a characteristic rise in the frequency response of ground around 10 to 20 Hz [3]. This effect is essential to the characterisation of railway vibration at a particular site.

3 Vehicle Parameter Study

The vehicle influence is investigated in a parameter study using the 'Train Ground Vibration (TGV) model' [3]. This software includes a multibody model of the vehicle covering nine different parameters according to Fig. 1a. The base vehicle model represents a modified two-car Bombardier Regina EMU ('Gröna Tåget'), which is used as a test bench in a Swedish research programme to investigate rolling stock requirements to enable a speed increase from 200 km/h to 250 km/h on existing tracks.

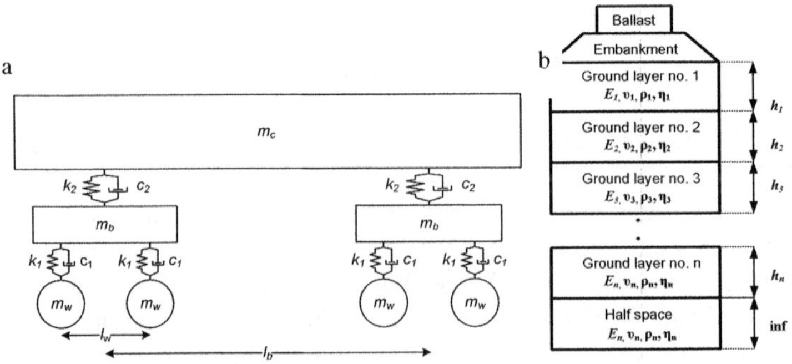

Fig. 1. a) Multibody vehicle model, b) layered ground model.

Table 1. Vehicle parameters

Parameter	Coded variable	High value	Low value
Carbody mass, m_c [kg]	X1	4.5×10^4	3.5×10^4
Bogie frame mass, m_b [kg]	X2	6.0×10^3	4.0×10^3
Wheelset mass, m_w [kg]	X3	2.0×10^3	1.6×10^3
Half distance between bogie centres, l_b [m]	X4	12.0	7.0
Half distance between axles, l_w [m]	X5	1.45	1.25
Primary suspension vertical dynamic stiffness per axle, k_1 [N/m]	X6	2.8×10^6	2.0×10^6
Primary suspension vertical viscous damping per axle, c_1 [Ns/m]	X7	4.0×10^4	2.0×10^4
Secondary suspension vertical dynamic stiffness per bogie, k_2 [N/m]	X8	8.0×10^5	4.0×10^5
Secondary suspension vertical damping per bogie, c_2 [Ns/m]	X9	3.0×10^4	1.0×10^4

The parameters were given a high and a low value realistic for a design of the current vehicle type, see Table 1. The central parameter value is the arithmetic average of the high and low levels in each case. With two levels for each parameter, a total of $2^9 = 512$ different parameter configurations are possible. Hence, to fully resolve the effects of single parameters ('main effects') in a formal parametric study and their interactions (cross terms), 512 simulations would be required.

The track model consists of an Euler beam (both rails) resting on distributed springs (rail pads). The sleepers are modelled as a continuous mass layer resting on ballast and embankment; two continuous vertical layers of springs with consistent mass. The ground is modelled as a number of soil layers with dynamic properties and thickness according to Fig 1b. Three ground models used in the simulations are specified in Table 2. The parameter study was performed on the 'Greby' ground model which has properties similar to the field measurement site. The two additional ground models named 'Steventon' and 'via Tedalda' were used for simulations with the vehicle to study effects of the ground properties.

To study the interaction between vehicle parameters within reasonable computing times, a fractional factorial design (FFD) [4] approach was chosen. This involves a strategic selection of the 512 possible parameter combinations. In limiting the number of simulations, the trade-off is a 'confounding' of the parameter effects. If a large fraction of the full factorial design is run, a less severe confounding is achieved. A small fraction results in poorer resolution of the main and low-order interactions of parameters. An FFD of resolution IV was chosen. This requires only 32 runs but leads to confounding between main effects and three-factor interactions, and between different two-factor interactions. The FFD was performed with respect to the vibration velocity level in 1/3 octave frequency bands averaged over 0 – 2 m from the track. The vehicle speed was set to 200 km/h in all simulations and the quasi-static and dynamic vibration responses were treated separately.

Table 2. Ground properties

Model (site)	Layer	Young's modulus, E [MPa]	Poisson's ratio, υ	Density, ρ [kg/m^3]	Loss factor, η	Thickness, h [m]
Greby	1	50	0.300	1800	0.2	2.0
	2	70	0.300	1800	0.2	2.0
	Half-space	500	0.300	2000	0.2	inf
Steventon	1	86	0.498	2000	0.2	0.7
	2	239	0.493	2000	0.2	2.0
	3	86	0.498	2000	0.2	3.0
	Half-space	941	0.471	2000	0.2	Inf
via Tedalda	1	467	0.450	1800	0.1	10
	Half-space	1879	0.450	1800	0.1	Inf

With each parameter at two levels, the result of the FFD can be represented by a linear model according to Eq. 1.

$$y = \beta_0 + \sum_{i=1}^{9} \beta_i x_i + \sum_i \sum_j \beta_{ij} x_i x_j, \quad i < j \qquad (1)$$

where β_i and β_{ij} are linear regression coefficients representing the main effects and the two-parameter interactions, respectively. Each parameter (variable) is coded and normalised such that its corresponding numerical value is $x_i = -1$ at low value and $x_i = 1$ at high value.

4 Field Measurements

To validate the model and to study the effect of vehicle speed, vibration measurements were conducted in Sweden in 2009 using the Gröna Tåget trainset at speeds up to 250 km/h. The measurements took place on the track from Skövde to Töreboda with BV50 (50 kg/m) rails on concrete sleepers. This is a straight part of the line comprising double track on an approximately 2 m high embankment surrounded by farm land. The site was instrumented with accelerometers on the rail head and at 3, 10 and 20 m from the track centre-line. Three parallel cross-sections of the track were instrumented, separated by 20 m.

The accelerometers were fixed to the rail head and to the ground using a concrete plate at 3 m from the track centre-line. Accelerometers were mounted in the ground at further distances from the track using metal spikes. Measurements were performed for the two-car EMU at speeds of 100, 150, 200 and 250 km/h, in both travelling directions.

Rail roughness was measured at the test site (along a 100 m section) within one month of the vibration measurements. The Swedish railway authority Trafikverket provided vertical track irregularity data for wavelengths up to 100 m for the same track section measured a few months earlier. These data were used as input for the simulations of the ground vibration response using the TGV model.

5 Results

In Fig. 2, the measured vibration velocity response of the rail head is shown for pass-by runs of the Regina at speeds of 100, 150, 200 and 250 km/h. In the figure, peaks related to the excitation by sleeper passing (40 – 100 Hz), wheelset passing (10 – 25 Hz) and passing of bogies in adjacent cars (3 – 8 Hz) are highlighted with trend lines. The overall level of the response is clearly increasing with increasing vehicle speed.

Figs 3a and b show the calculated ground vibration response at 10 m from the track for different vehicle speeds and using different ground models. Fig. 4a shows the same pattern as Fig. 3, with peaks shifting towards higher frequencies for increasing speeds. The response at frequencies below 10 Hz is strongly attenuated at 10 m compared with the response measured on the rail. The largest variations between ground models are observed between 3.16 and 10 Hz. The via Tedalda model results in a different response compared with the other models above 50 Hz. At high frequencies the vibration transmission is controlled by the uppermost ground layer, which is stiffer in the via Tedalda model.

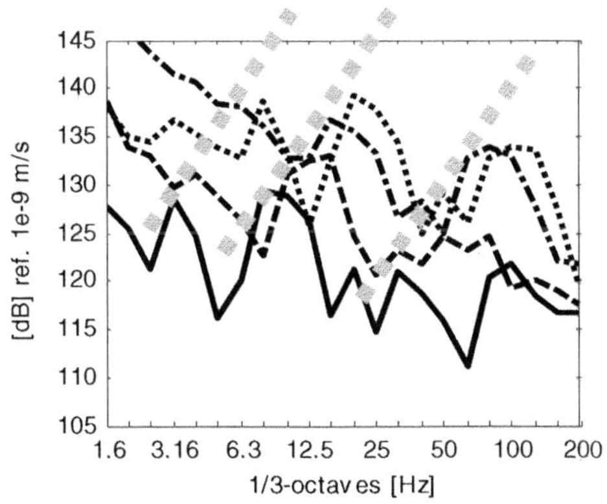

Fig. 2. Spectra of rail head vibration for train at different speeds. —— 100, – – – 150, – • – 200, • • • 250 km/h.

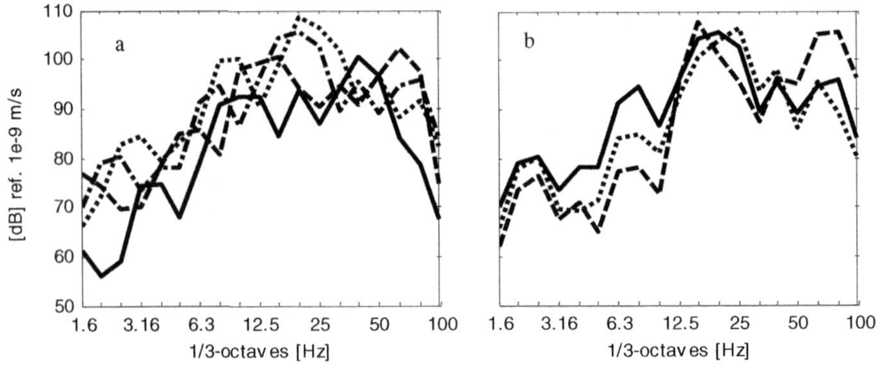

Fig. 3. Calculation results. a) Regina on Greby ground model, response at 10 m from centre of track. —— 100, — — — 150, — • — 200, • • • 250 km/h. b) Regina at 200 km/h on three different ground models, response at 10 m from the track centre-line . —— Greby, — — — via Tedalda, • • • Steventon.

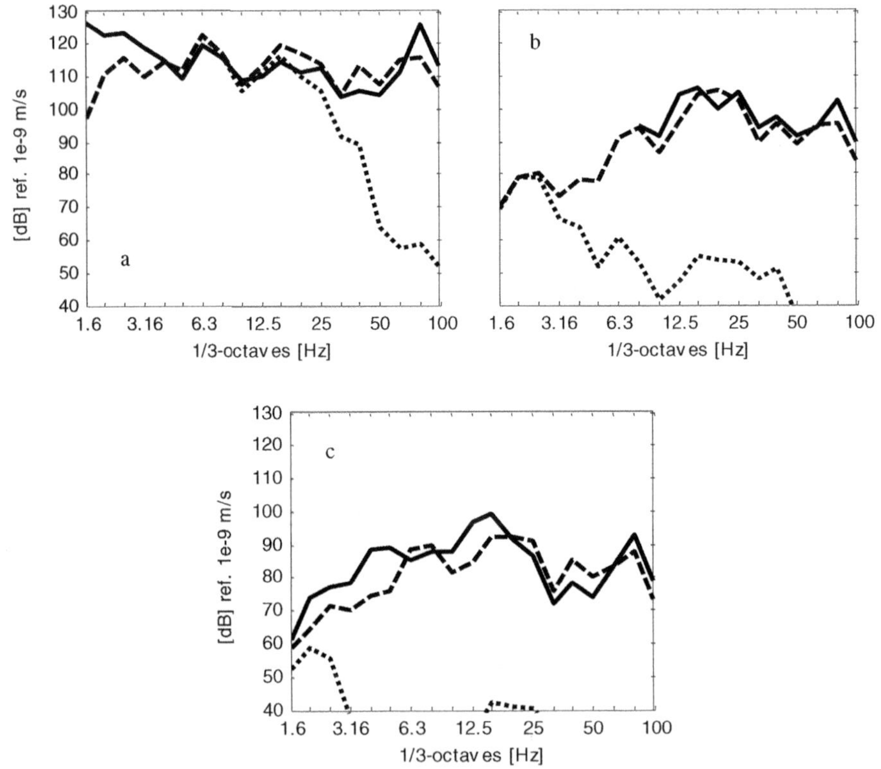

Fig. 4. Ground vibration velocity excited by the Regina vehicle running at 200 km/h at a) 3, b) 10 and c) 20 m from the track centre-line. —— measured, — — — calculated total response, • • • calculated quasi-static response.

Fig 4a-c shows comparisons between the measured and simulated ground responses at different distances from the track. In general, good agreement between the measured and calculated spectra is seen. The level of the calculated response has been adjusted by choice of ground loss factors of 0.2. This is in-line with the findings for surface vibration in soft saturated soils in [5].

Both measured and calculated results at 10 and 20 m from the track show a strong attenuation of the low-frequency response in-line with the character of quasi-statically excited vibration. The sensors mounted at 10 m suffered a poor signal-to-noise ratio below 8 Hz. These data are omitted. The quasi-static vibration response is influenced only by the static wheel load. Hence, only the vehicle mass, (apart from speed) governs this response.

Fig. 5 illustrates the result from the parameter study of the dynamic vibration response. The linear regression coefficients in Eq. 1, evaluated in different octave bands, are displayed for each main effect and interaction effect. The parameters and their interactions are represented by their coded variables (Table 1). Here X*X denotes an interaction between two parameters, while X+X denotes a confounding. A positive regression coefficient indicates that an increase in the vehicle parameter value leads to an increase in vibration level.

The regression coefficients with the largest magnitudes are related to the bogie distance (X4) and the primary suspension stiffness (X6). Changing the bogie distance causes a frequency shift of the response and hence the regression coefficient varies between octave bands. A stiffer primary suspension leads to an increased vibration level especially in the 8 Hz octave band, (X6 at 8 Hz). In general, the interaction effects are small.

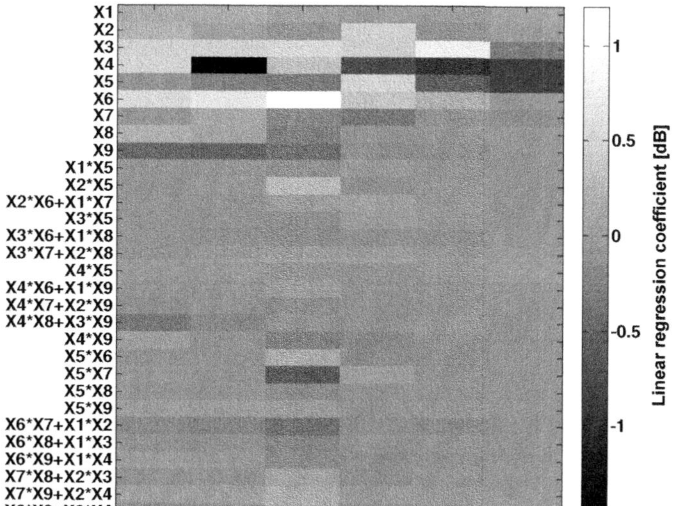

Fig. 5. Results from the parameter study. Linear regression coefficients for the main and interaction effects given for different octave frequency bands.

6 Concluding Remarks

Measurements of ground vibration from a vehicle clearly show increasing vibration levels with speed. Vehicle properties such as the wheelbase can be identified in the measured spectrum close to the track. At further distances from the track, single excitation mechanisms are difficult to distinguish and the spectrum shape is more determined by the properties of the ground.

Ground vibration response simulated with the TGV model shows good agreement with the spectral content of the measurements, and also a satisfying prediction of the attenuation of ground vibration at low frequencies. The attenuation at high frequencies, which is governed by the damping in the ground, has been more difficult to model. The damping measured during geotechnical investigations of the soil had to be significantly increased to achieve agreement with measurement results. This could indicate that geotechnical measurement techniques are inappropriate for this kind of application.

The results of the parameter study show the highest influence on ground vibration from the primary suspension stiffness and the unsprung mass. Changes in the geometric relations of the vehicle mainly shift the excitation into a different octave band.

Acknowledgements

Discussions with, and input from, Dr Martin Li and Mr Alexander Smekal of Trafikverket are gratefully acknowledged.

References

[1] Seminar on High Speed Lines on Soft Ground, Dynamic soil-track interaction and ground borne vibration, Gothenburg (March 16-17, 2000)

[2] Jones, C.J.C., Sheng, X., Thompson, D.J.: The roles of track roughness and axle-load time history in the generation of ground vibration from surface-running trains. In: Grundman, Schuëller (eds.) Structural Dynamics Eurodyn 2002, Lisse, The Netherlands, Balkema, pp. 467–472 (2002)

[3] Sheng, X., Jones, C.J.C., Thompson, D.J.: A comparison of a theoretical model for quasi-statically and dynamically induced environmental vibration from trains with measurements. Journal of Sound and Vibration 267(3), 621–635 (2003)

[4] Montgomery, D.C.: Design and analysis of experiments. Wiley, New York (1984)

[5] Triepaischajonsak, N., Thompson, D.J., Jones, C.J.C., Ryue, J., Priest, J.A.: Ground vibration from trains: experimental parameter characterisation and validation of a numerical model, to be published in Proceedings of Institution of Mechanical Engineers, Part F

Vibration Source Localization along Railway Tracks

Carl Wersäll, Anders Bodare, and K. Rainer Massarsch

Geo Risk & Vibration Scandinavia AB,
Solna Torg 13, SE-17145 Solna, Sweden
Tel.: +46 707169037
e-mail: carl.wersall@georisk.se

Summary

Ground-borne vibration from railway traffic is an increasing problem in urbanized areas and measures are often needed to minimize its effects on the environment. An important question when dealing with railway problems is to identify the source(s) of vibration emitted along the railway track. Once this information is available, it is often possible to mitigate the problem by improving stiffness of the railway track and/or to upgrade worn-out or damaged rail sections and turnouts.

This paper describes a method which makes it possible to determine the locations of track sections which are likely to emit strong ground vibration. A purpose-built track-bound vehicle which can be vibrated continuously at different frequencies can identify track sections having unfavorable dynamic foundation conditions. A theoretical concept is proposed to calculate the potential of energy emission from the vehicle moving along the track.

Further, an innovative method is presented which makes it possible to determine the location of vibration sources by measurement of ground vibrations from existing railway traffic. This information can be used to determine the location of track sections where remedial measures are needed. Results are presented, illustrating application of the concepts, which can also be applied to other types of vibration problems.

1 Introduction

Railway systems offer environmentally friendly and efficient transportation for freight and passenger traffic. However, in urban areas, railway traffic can have negative-environmental consequences such as disturbance from ground vibration. Commonly, negative effects are related to the living environment and human

T. Maeda et al. (Eds.): Noise and Vibration Mitigation for Rail Trans. Sys., NNFM 118, pp. 267–274.
springerlink.com

disturbance. Vibration can also affect structural elements of buildings or sensitive equipment (e.g. in hospitals, offices, factories or research institutions). The consequences of vibration have also increased due to more economic structural design which makes buildings more sensitive to vibration. Further, general public awareness of environmental effects is increasing, which is reflected by more restrictive vibration standards and specifications.

In some cases it is possible to consider potentially-negative effects at the design stage, for instance by incorporating vibration-reducing measures. However, prediction of ground vibration caused by railway traffic and its effects on buildings and installations is often a complex problem. Therefore, vibration problems are often neglected at the design stage and only noticed once buildings have been constructed. At this stage it is more difficult to mitigate vibration problems. In order to apply remedial measures along railway tracks the location of the source(s) of vibration needs to be identified.

This paper describes prediction of ground vibration caused by railway traffic along tracks with variable conditions. A theoretical model is presented which makes it possible to identify track sections which have high-potential vibration emission, using a purpose-built vibrating vehicle. In addition, vibration generated by railway traffic can be used to accurately determine locations of vibration sources along the track.

2 Location of Vibration Sources

Dynamic track stiffness and vibration energy emission can be investigated by a track-bound, instrumented vehicle with a large vibrator [1, 2]. The Swedish Transport Administration together with the Royal Institute of Technology (KTH) in Stockholm has a long history of research in train-induced vibration (e.g. [3, 4]) and has developed a vibrating train car, called Rolling Stiffness Measurement Vehicle (RSMV). Similar vehicles exist in several other countries.

The RSMV is normally used for determining the condition of existing tracks and thereby directing maintenance actions. The vehicle can be vibrated at different frequencies (2 to 50 Hz) while driving at low speed (usually less than 10 km/h, up to 50 km/h). By measuring the induced dynamic force and acceleration it is possible to calculate vertical displacement (double-integration of acceleration) and dynamic stiffness of the track support. A sketch of the different components of the RSMV is shown in Fig. 1.

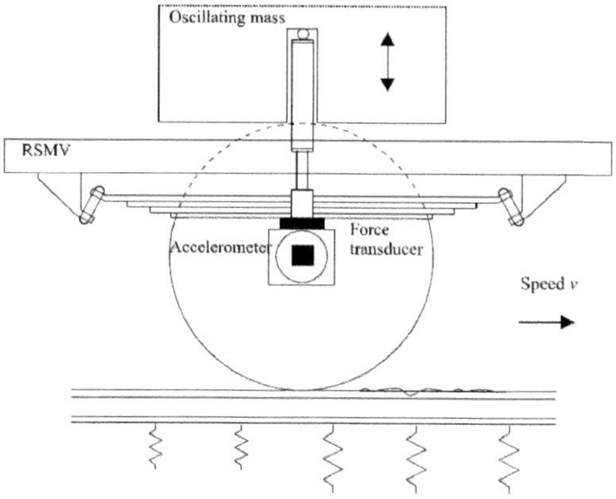

Fig. 1. Principal sketch of the RSMV measurement procedure, from [1]

From the calculated dynamic stiffness and phase angle, emitted energy from the vibrator to the track support and into the ground can be calculated by a new theoretical model, the calculation steps of which will be described in the following section. Thus, RSMV measurements make it possible to determine the location of potential sources of strong vibration. High potential of vibration energy emission does not necessarily mean that strong ground vibration will be generated by regular traffic at that location, as this also depends on other factors such as unevenness of the track, track changes, turnouts and the speed and dynamic characteristics of the train. If, however, such areas of high-potential energy emission coincide with unevenness along the railway track (e.g. a turnout) or a sudden change in track stiffness, strong vibration will be emitted and propagated to the surrounding soil.

Ground-vibration measurements from regular train traffic can also be used to locate the actual sources of vibration. By installing photocells at different track sections, it is possible to determine the location of the train in the time history record which can be linked to the time history of the vibration signal.

Based on information obtained by these two methods, it is possible to identify track sections where strong ground vibration is emitted and mitigation measures may be implemented, for example adjustment of tracks, modification of rail support, upgrading worn-out turnouts or ground improvement measures at locations showing strong vibration emission.

3 Energy Emission

The RSMV can be operated at specific frequencies, measuring acceleration and applied force. Normally, the purpose of such measurements is to determine the condition of the track by calculating displacement and dynamic stiffness. Dynamic track stiffness is normally presented along with phase angle between force and displacement, see Fig. 2, and is considered as a measure of the static condition of the ballast. These quantities are, however, also related to dynamic properties of the rail, ballast and subgrade and, based on the theoretical model below, they can be used to locate sources of high-potential vibration emission.

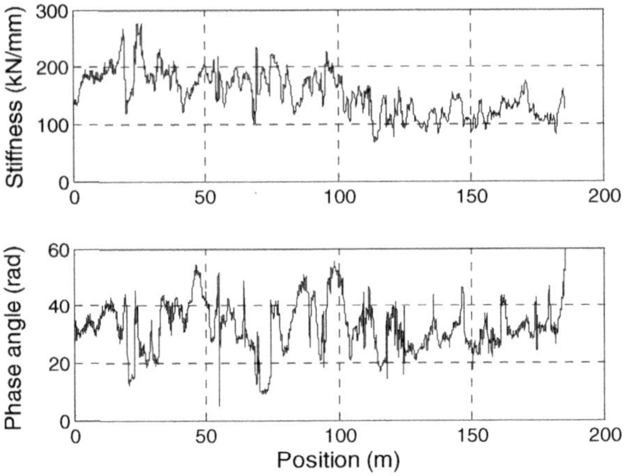

Fig. 2. Example of dynamic stiffness and phase angle as determined by the RSMV at 8 Hz

3.1 Dynamic Model

One of the most commonly used concepts for modeling train-induced vibration is a moving point load on a Bernoulli--Eulerbeam on a Kelvin material (BEK-model, Fig. 3, e.g. [5]). In case of the RSMV, the speed is very slow (normally less than 10 km/h) compared to the critical speed (greater than 200 km/h, e.g. [6, 7]). The model can therefore be analyzed with zero point load velocity.

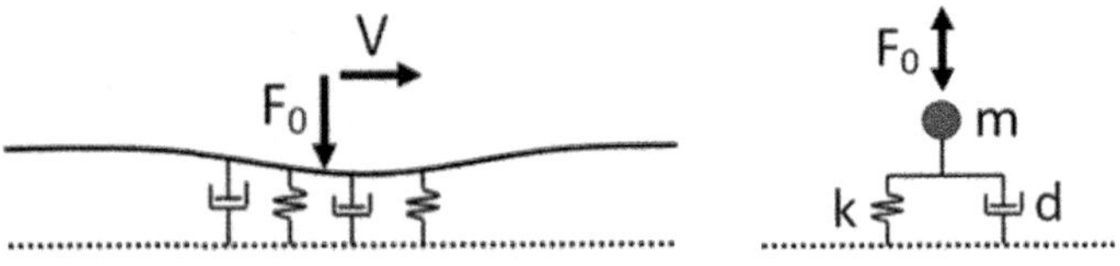

Fig. 3. BEK-model (left) and SDOF model (right)

A BEK-model with zero velocity yields results similar to that of the simpler mass-spring-damper SDOF model, shown in Fig. 3. Such a model, having a sinusoidal force with amplitude F_0, mass m, spring stiffness k and a damper with impedance d, is therefore used to analyze the RSMV results. The complex dynamic stiffness, $K(\omega)$, can by described by

$$\mathrm{Re}(K(\omega)) = k - m\omega^2, \mathrm{Im}(K(\omega)) = -d\omega . \tag{3.1}$$

Derivation of Eq. (3.1) can be found in various texts on acoustics or vibration analysis, e.g. [8].

3.2 Dissipated Energy

The RSMV measures particle acceleration and force amplitude. Acceleration is integrated twice to obtain particle displacement from which dynamic stiffness is calculated. Measurements are commonly presented in stiffness amplitude, K_0, and phase angle, ϕ. The following relationship is obtained from Eq. (3.1)

$$d \cdot \omega = K_0 \cdot \sin(\phi) \quad (3.2)$$

Of the three elements in the model (Fig. 3), only the damper dissipates energy. The dissipated energy per cycle is

$$E = \pi \cdot d \cdot \omega \cdot u_0^2, \quad (3.3)$$

and using Eq. (3.2),

$$E = \pi \cdot K_0 \cdot \sin(\phi) \cdot u_0^2. \quad (3.4)$$

The physical stiffness, $K_0(\omega)=F_0(\omega)/u_0(\omega)$, together with Eq. (3.4) yields

$$E = \pi \cdot \frac{F_0^2}{K_0} \cdot \sin(\phi). \quad (3.5)$$

The power, i.e. dissipated energy per unit time, then becomes

$$P = E \cdot f = \frac{\omega}{2} \cdot \frac{F_0^2}{K_0} \cdot \sin(\phi), \quad (3.6a)$$

or,

$$P = \frac{1}{2} \cdot F_0 \cdot v_0 \cdot \sin(\phi), \quad (3.6b)$$

where v_0 is the particle velocity (integrated from acceleration) measured on the RSMV. The emitted power can be calculated from Eq. (3.6a), using the information shown in Fig. 3. Assuming that the dynamic force and frequency are constant, the potentially-emitted energy increases with increasing phase angle and decreasing dynamic stiffness. Equation (3.6b) presents emitted power in terms of measured quantities and shows that only force, particle velocity and phase angle govern vibration emission.

3.3 Emitted Power

The energy from the RSMV vibrator that is emitted into the ground can be calculated at every location using Eq. (3.6). Since sub-grade response, and thus emitted power, varies with frequency, investigations are carried out at different frequencies. An example is shown in Fig. 4, where the RSMV was driven along the same track four times, operating at four different frequencies (5--8 Hz). Darker color indicates track sections of higher energy emission. The ballasted track in the example is underlain by 4--7 m of fill, followed by a thin peat layer (less than 1 m), 5--10 m of clay and 1--4 m of granular soils (glacial till) on bedrock.

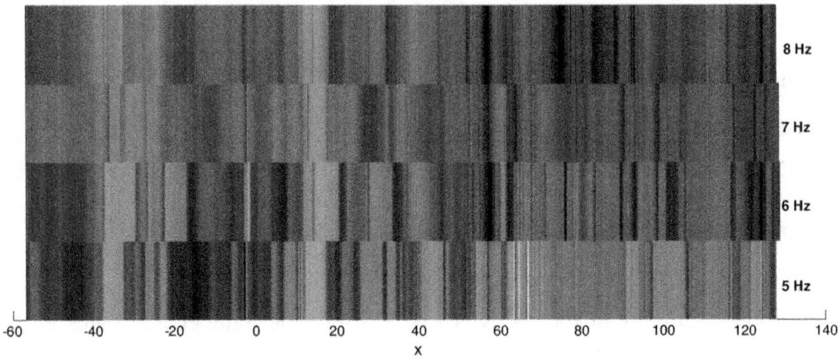

Fig. 4. Power emission from the RSMV for different frequencies. Darker color implies higher vibration emission

High-energy emission from the RSMV does not necessarily mean that large vibration is generated when regular trains pass that location. It is merely a quantification of *potential* vibration generation. If there is no unevenness in the track, an impulse is not generated. If, however, a sudden change in stiffness occurs or there is, for instance, a turnout or damaged rail,and the generated impulse will easily propagate into the subgrade and the surrounding soil.

4 Train-Induced Ground-Vibration Measurement

While the RSMV can determine potential vibration sources, ground vibration measurements from regular traffic can be used to locate vibration sources which are generated by trains. With the use of at least two photocells installed in different positions along the railway track, the train velocity and its position at any given time can be determined. By identifying measured vibration peaks and determining the train position when peaks occur, a relationship between train location and high vibration is obtained.

4.1 Determining Peaks

The train engine, which is usually heavier than the following cars, is a more distinct and stronger source of vibration. In addition, it is easier to determine the location of the engine than individual cars. Therefore, locating vibration sources is preferably carried out using recordings of vibration caused by the engine. These are easily identified since they are the first to arrive at a measuring point and generally have a larger amplitude. Once the engine has passed the measuring point, vibration is emitted from different train sections (cars), arriving simultaneously from different directions, making the analysis more uncertain.

Figure 5 shows a typical time history of vibration velocity (vertical component) measured approximately 12 m from the railway track. The train consists of one engine, followed by nine cars and a second engine at the back. One can clearly distinguish the engines and individual cars in the vibration record. In this example, only the peaks around $t = 20$ s were used for analysis.

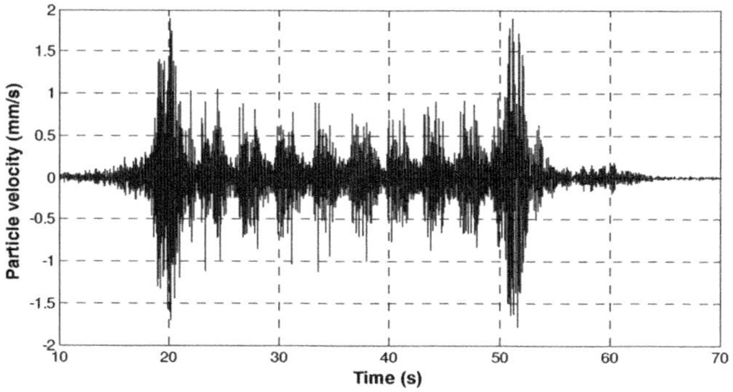

Fig. 5. Vertical vibration time history from a train recorded in the vicinity of the track (note that the train has two engines, one at the front and the other at the rear)

Figure 6 shows the location of the engine and the corresponding peak-particle velocity (PPV) when high peaks occur in the ground-vibration measurements. Obvious vibration sources are located at $x = -33$ m and $x = 45$ m.

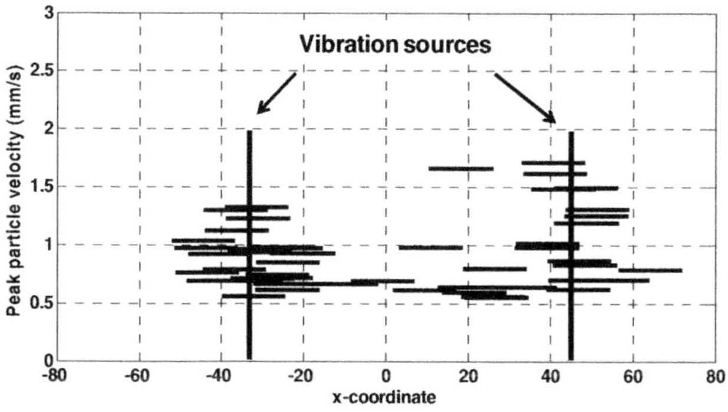

Fig. 6. Location of vibration sources determined by ground-vibration monitoring

5 Conclusions

Two methods have been presented to identify vibration sources along railway tracks. The first method uses a purpose-built track-bound measurement vehicle (RSMV), equipped with a vibrator. The vehicle moves with constant, low speed, vibrating with a prescribed frequency. The measured dynamic force and acceleration can be used to determine the dynamic track stiffness and phase angle between force and displacement. A theoretical model is presented, from which the power emitted into the subgrade is calculated by a simple relationship. The potential of vibration emission increases with increasing phase angle and decreasing dynamic stiffness. Measurements can be used to identify track sections with high-potential of vibration emission.

The source of vibration emission from operating trains can be determined by measuring ground vibration. The position of the train at any point in the vibration time records can be determined by installing photocells. High-vibration peaks in the records are determined by inspection of time histories, indicating track locations where high ground vibration is generated.

Acknowledgements

This paper presents results of research on railway-traffic-induced ground vibration at the Royal Institute of Technology (KTH) in Stockholm. The contributions by Dr. Kent Lindgren at Marcus Wallenberg Laboratory for Sound and Vibration Research are acknowledged. The RSMV unit was developed by Dr. Eric Berggren at the Swedish Transport Administration (Trafikverket).

References

[1] Berggren, E.: Railway Track Stiffness – Dynamic Measurement and Evaluation for Efficient Maintenance, Doctoral Thesis, Royal Institute of Technology, Stockholm (2009)

[2] With, C.: Train-Induced Vibrations on Embankments and in Buildings - Prediction and Validation of Some Models, Doctoral Thesis, Royal Institute of Technology, Stockholm (2008)

[3] Hall, L.: Simulations and Analyses of Train-Induced Ground Vibrations – A Comparative Study of Two- and Three-Dimensional Calculations with Actual Measurements, Doctoral Thesis, Royal Institute of Technology, Stockholm (2000)

[4] Bahrekazemi, M.: Train-Induced Ground Vibration and Its Prediction, Doctoral Thesis, Royal Institute of Technology, Stockholm (2004)

[5] Wenander, K.: Models of Train Induced Vibrations in Railway Embankment - Analytical Solutions and Practical Applications, Master of Science Thesis. Royal Institute of Technology, Stockholm (2004)

[6] Krylov, V.V.: Generation of Ground Vibration Boom by High-speed Trains. In: Krylov, V.V. (ed.) Noise and Vibration from High-speed Trains, pp. 251–283. Thomas Telford Publishing, London (2001)

[7] Bodare, A., Petek, K.: Sub and Super Seismic Train Induced Ground Vibrations – Theoretical Considerations and Test Results. In: Proceedings of Environmental Vibrations – Prediction, Monitoring, Mitigation and Evaluation 2005, Okayama, pp. 3–10 (2005)

[8] Kinsler, L.E., Frey, A.R., Coppens, A.B., Sanders, J.V.: Fundamentals of Acoustics. John Wiley & Sons, Inc., Chichester (1982)

Analysis and Design of New Floating Slab Track for Special Trackwork Using Finite Element Analysis (FEA)

J.E. Phillips and J.T. Nelson

Wilson, Ihrig & Associates, Inc.,
6001 Shellmound Street, Suite 400, Emeryville, CA 94608, U.S.A.
Tel.: +011 1 510 658 6719; Fax: +011 1 510 652 4441
e-mail: jphillips@wiai.com

Summary

This paper discusses Finite-Element Analysis (FEA) of a vibration-isolation system for double crossovers and turnouts in a future sub-way rail line extension. The previous design included ballasted track on continuous concrete slabs supported by natural rubber vibration isolators. The goal of the new design was to replace the ballast and tie track with rails fastened directly to discrete slabs similar to the vibration isolation system utilized on running portions of track within existing tunnels. FEA was used to evaluate the dynamic interaction between the vehicle, rails, resilient rail fasteners and the vibration isolation system. The results of the analyses guided the design in determining the size and thickness of the concrete slabs, the stiffness and design of the rubber vibration isolation pads and the rail fastener selection.

1 Introduction

A transit agency requested the authors' firm to develop a new vibration-isolation system for double crossovers and other special trackwork in a future sub-way rail line extension. Floating-Slab Track (FST) vibration-isolation systems were to be used as standard track on all new tunnel segments of the rail system to control ground-borne noise [1]. The rails at the crossovers in the existing rail system were not fastened directly to the FST, but were supported on ballasted track. The transit agency requested that ballasted tie track be eliminated from the crossovers in the future rail line extension. Switching to direct fixation of the rails at the crossovers would make them consistent with the rest of the rail lines in the system, maintain rail alignment and reduce maintenance costs.

The effectiveness of FST for reducing train vibration transmitted to the soil surrounding a tunnel is dependent upon a complex interaction between the dynamics of the bogie (truck) components, the rails, fasteners, concrete slab and the

T. Maeda et al. (Eds.): Noise and Vibration Mitigation for Rail Trans. Sys., NNFM 118, pp. 275–282.
springerlink.com

vibration-isolation pads. Therefore, a numerical model of the various components was developed to study this interaction and to guide the design of the vibration-isolation system to be implemented in the future rail extension.

2 Numerical Approach

The double crossover vibration-isolation performance was analyzed with a Finite-Element Analysis (FEA) method, using NEiNastran, provided by NEi Software [2]. The switch was analyzed separately of the crossing diamond and other switches, though adjoining sections of tangent track with double-ties were included. The center crossing diamond was analyzed separately of the switches as well.

The floating slab portions were modeled with linear-solid elements and the rails were modeled with linear-beam elements. The floating-slab support pads and side pads were modeled with discrete spring elements. Rigid links and springs were combined to represent direct fixation fasteners with vertical, lateral and longitudinal dynamic stiffness. Viscous damping elements were included at the ends of the modeled rails to simulate transmitting boundary conditions.

The isolation slab configurations considered are listed in Table 1.

Table 1. Numerical Models for Crossover Switch Vibration Isolation

Case	Slab Thickness – mm	Slab Size	Fastener Vertical Dynamic Stiffness – kN/mm -	Fastener Lateral Dynamic Stiffness - kN/mm -
1	300	Short	15.9	6.83
2	400	Short	15.9	6.83
3	500	Short	15.9	6.83
4	400	Short	26.3	10.0
5	500	Short	26.3	10.0
6	300	Short	123	14.7
7	400	Short	123	14.7
8	500	Short	123	14.7
9	400	Long	26.3	10.0

The slab size identified as “long” represents a cast-in-place slab divided up into sections of nominal lengths of 11.5 to 18 meters. The term “short” refers to the cast-in-place slabs divided into short sections of nominal length 5.5 to 9 meters. The portion supporting the closure rails was maintained with a nominal length of 18 m to avoid problems with uplift of the switch point due to loading of the heel block.

The performance of the floating-slab track was expected to depend to some extent on the dynamic characteristics of the resilient rail fasteners. The fastener vertical dynamic stiffnesses considered were for a "soft" fastener with a vertical dynamic stiffness of 15.9 kN/mm, "soft-to-medium" at 26.3 kN/mm and "stiff" at

123 kN/mm. The soft-to-medium fastener represents a typical modern-bonded direct-fixation fastener such as supplied by L.B. Foster or Lord Corporation for rail-transit application. The "stiff fastener" assumption was intended to represent the standard fastener currently in use on the transit system.

A graphical representation of one of the FEA models used in the analyses is provided in Fig. 1. Elements representing the rubber perimeter seals and vibration-isolation pads are omitted from the figure for clarity. For simplification, the intersection of the rails was modeled as a simple crossing of the beam elements representing the rails. Additionally, the closure rails, or points, were not included in the model.

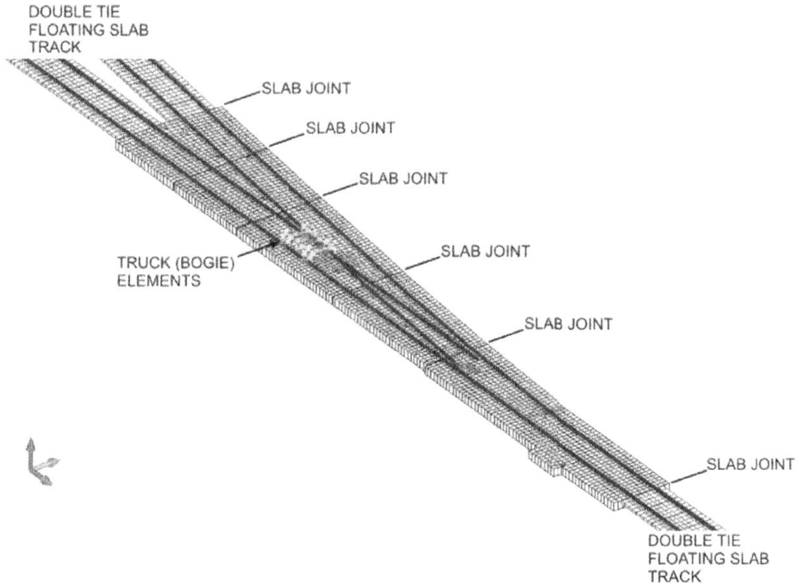

Fig. 1. Graphical representation of a Finite-Element Analysis model of a cross-over turnout (color figure online)

The vehicle was modeled as a system of springs, dashpots and masses with various rigid links designed to represent a rigid H-shaped bogie (truck) frame, as represented in Fig. 2. Axle bending was included in the model to represent the flexible axle supporting the gear unit, which is supported directly on the axle by bearings. The resonance frequency of the gear unit on the axle is within the range of ground-borne noise, and was believed to be a factor in the characteristics of ground-borne noise and vibration observed along the existing transit system.

A model of the vehicle bogie (truck) was positioned with one wheel at the "frog" gap. A relative displacement, or roughness, was induced by applying equal and opposite forces of very high magnitude across a spring with very high- artificial stiffness (orders of magnitude higher than the contact stiffness) such that the applied force divided by the artificial stiffness produced a specified relative displacement between the wheel and rail. The vertical complex (phase- dependent) forces transmitted to the invert by each floating slab support spring pad were summed to compute the net transmitted force.

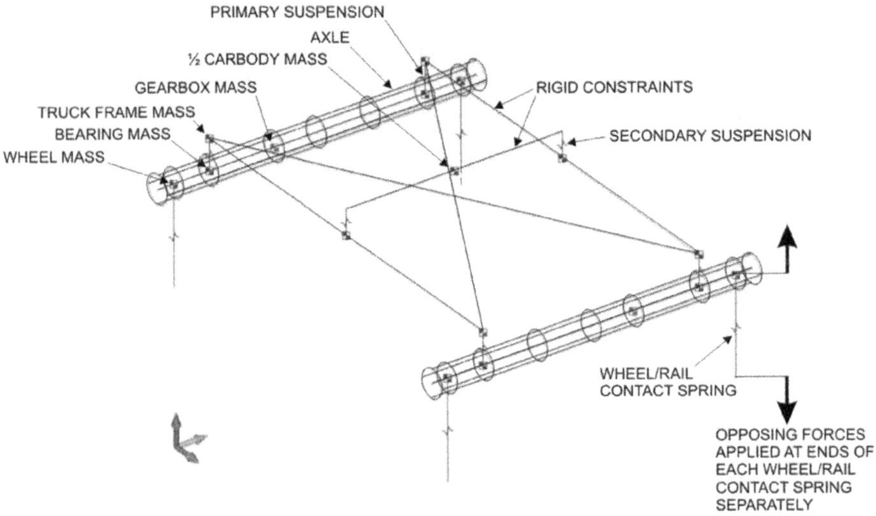

Fig. 2. Finite-element model of the bogie (truck) (color figure online)

3 Numerical Results

Figures 3 through 5 show the predicted effects of the floating slab with respect to a standard concrete invert with DF track in reducing ground vibration. These figures are plots of the insertion gain of the floating slab in one-third octave bands. Eachone-third octave band has a bandwidth of approximately 23% of its nominal center frequency.

The FST should have a natural frequency no greater than 16 Hz to provide sufficient vibration reduction at ground-borne noise frequencies. The above requirements, along with the slab weight and the transfer of stiffness from adjacent isolated slabs through the rails determined the stiffness of the isolation pads.

The predicted responses show amplification of vibration at the floating slab resonance frequency at about 16 Hz and a reduction of vibration at frequencies greater than 16 Hz. The amplification is controlled by the assumed loss factor of the isolation system. In practice, the predicted amplification has rarely been observed, probably due to greater damping than assumed in the models and the effects of moving loads. Another reason is that the floating slab pad stiffness is nonlinear, and static loading of floating slabs would cause the isolator pad stiffness to vary with position, which in turn would spread the resonance frequencies. Finally, the floating slab pads tend to smooth out invert roughness, with the result that the floating slabs tend to provide vibration reductions at frequencies below resonance, contrary to model predictions. Therefore, the resonance amplification shown is expected to be about 3 to 6 dB, and can also be nil.

A secondary peak in the response, typically in the range of 40 to 60 Hz, is due to the resonance of the bogie unsprung mass and rail on the resilient rail fastener supports. Without the floating slab, the vibration levels in this frequency range would contribute significantly to ground-borne noise and vibration.

In this study, the effects of the following parameters were investigated: isolation slab planar dimension, isolation slab thickness and rail fastener stiffness.

3.1 Effect of Slab Planar Dimension

Floating slabs of large lateral and longitudinal dimensions have been observed to behave poorly due to bending waves. The effects show up as increased in-tunnel and in-car rumble noise. The ground-borne noise reduction at the ground surface or in buildings is also compromised. Slab lengths and widths were purposely reduced to a minimum to avoid this problem. The shorter the slab, the higher the bending resonance frequency, and thus the more likely the slab would perform as a rigid mass at frequencies of ground-borne noise.

Figure. 3 illustrates the effect of the horizontal dimension parallel with the track of the cast-in-place floating slab on vibration reduction. The slab thickness was assumed to be 400 mm thick in both cases. The direct fixation fastener stiffness was assumed to be 26 kN/mm.

The results indicated in Fig. 4 indicate that there would be no significant reduction of vibration isolation performance with the longer isolation slab for the assumed thickness. This suggested that joint location would not affect slab performance significantly.

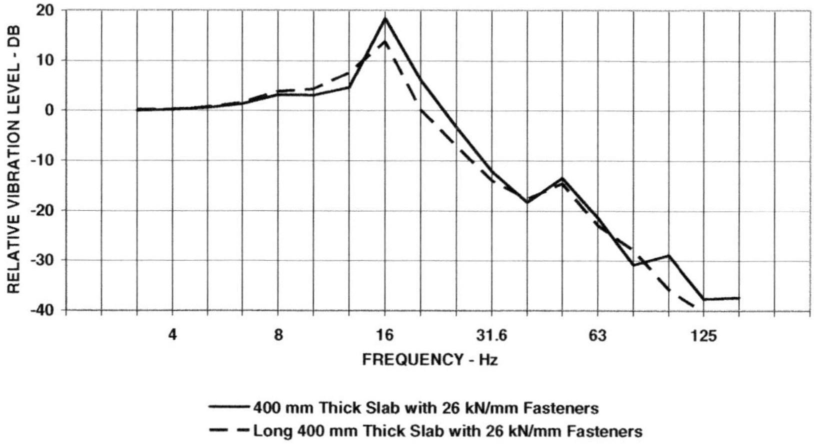

Fig. 3. Comparison indicating predicted effect of slab planar dimension (length) on the insertion gain of the floating-slab track.

3.2 Effect of Slab Thickness

Increasing the slab thickness increases the bending resonance frequency of the slab, thus broadening the range of frequencies over which the slab acts as a rigid mass. Therefore, increasing the slab thickness improves the vibration isolation performance of the slab.

The effects of slab thickness were predicted for three rail fastener stiffnesses. Figure. 4 illustrates the effect of slab thickness of 300 mm, 400 mm and 500 mm on vibration isolation with 123 kN/mm fasteners. The predicted response at 63Hz of the 300mm thick slab is slightly higher than that of either the 400 mm or 500mm thick slabs, and virtually identical at lower frequencies that is, the 300mm slab would be slightly less effective than the 400 or 500 mm thick slabs.

A comparison of 400 and 500 mm thick slabs with soft-to-medium 26 kN/mm fasteners indicated little difference between the 400 mm and 500 mm thick slabs. A comparison with "soft" fasteners of nominal dynamic stiffness 16kN/mm indicated a higher response at 63 Hz with the 500 mm thick slab relative to those of either the 300 mm or 400 mm thick slabs. This result was opposite to the results shown in Fig. 4 for the 123 kN/mm fasteners, contrary to expectation.

The mass per unit area of a 500 mm thick concrete slab would have roughly the same mass per unit area as the floating slabs with ballast-and-tie track. The results of the FEA analysis indicated that a 400 mm thick slab would work at least as well as the 500 mm thick slab.

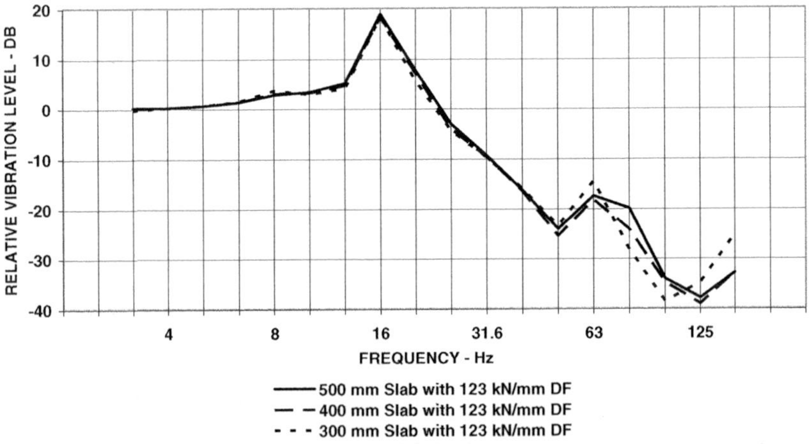

Fig. 4. Comparison indicating predicted effect of slab thickness on the insertion gain of the floating-slab track.

3.3 Effect of Fastener Stiffness

The responses with various rail fastener stiffnesses are compared in Fig. 5 for the 500 mm thick floating slab. These results indicate little difference in insertion loss at and below 25 Hz. The insertion loss is least for the "soft" fastener at 40 Hz, due to the unsprung mass and track resonance. This resonance frequency increases with increasing fastener stiffness, so that for the "stiff" fastener at 123 kN/mm stiffness, the resonance is about 63 Hz or higher. The 123 kN/mm fastener would appear to provide the best performance at lower frequencies. The "moderately soft" fastener of 26 kN/mm stiffness would appear to provide acceptable vibration reduction when used with the 500 mm thick slab. Similar characteristics in the predicted insertion gains were observed with the 400 mm thick slab.

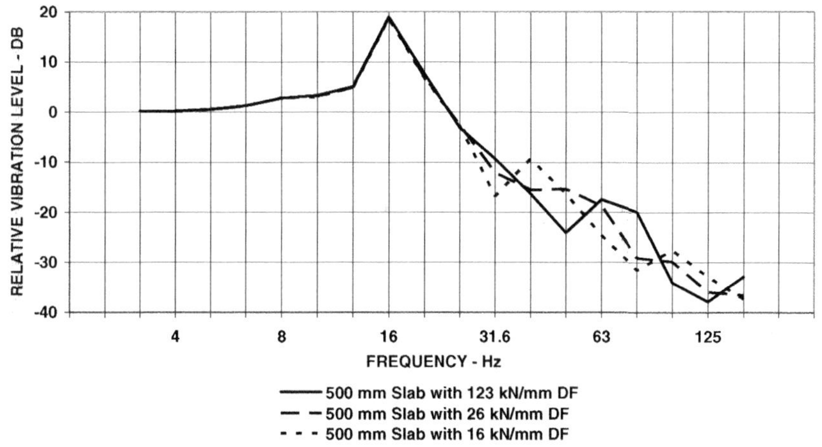

Fig. 5. Comparison indicating predicted effects of rail fastener stiffness on the insertion gain of the floating-slab track. (color figure online)

4 Conclusion

Based on the FEA results, a 400 mm thick slab with a moderately soft direct fixation fastener stiffness of 26 kN/mm appeared to provide a reasonable trade-off between vibration isolation and slab thickness. The nominal length of the slabs was 9 m, with the exception of the 18 m long slab under the closure rails to control rail uplift.

During the analyses, it was observed that the stiffness transferred between slabs through the rails had a significant effect on the overall stiffness of the isolation system. Adjustments were made in the support pad stiffness that would have ordinarily been missed by simpler models used for designing floating slab track.

References

[1] Saurenman, H.J., Nelson, J.T., Wilson, G.P.: Handbook of Urban Rail Noise and Vibration Control, U.S. Department of Transportation, Urban Mass Transportation Administration, UMTA-MA-06-0099-82-1, DOT-TSCUMTA 81 72 (1982)
[2] NEiNastran User's Manual, Version 10.0., NEi Software Inc., Westminster, CA, U.S.A (2010)

Evaluation of the Ground-Borne Vibration Reduction Properties of Tire Derived Aggregate Installed on the Denver RTD Light Rail System

D.A. Towers

Harris Miller Miller & Hanson Inc.
77 South Bedford Street, Burlington, MA 01803, USA
Tel.: +1 781 229 0707; Fax: +1 781 229 7939
e-mail: dtowers@hmmh.com

Summary

This paper presents the results of field tests performed to evaluate the ground-borne vibration reduction properties of tire derived aggregate (shredded tires, also known as TDA) as installed beneath ballast-and-tie track on the Denver Regional Transit District (RTD) Southeast Corridor (T-REX) Light Rail Transit (LRT) line in Denver, CO (USA). The vibration tests were carried out at four (4) locations where TDA has been installed beneath the tracks as well as at one control site without TDA underlayment. At each of the sites, ground-borne vibration measurements were made during multiple operations of a dedicated two-car train at three different speeds. In addition, impact tests were conducted to determine the local vibration propagation characteristics of the ground in an effort to normalize the train vibration data for the effects of local geology. A comparison of the test results at the TDA sites with the results at the control site provided a measure of the TDA vibration reduction.

The results of the study indicate that, on average, the TDA underlayment at RTD provides a ground-borne vibration reduction of about 3 dB in the one-third octave frequency bands centered at 25 Hz and 31.5 Hz and reductions of 8 to 14 dB at frequencies between 40 Hz and 160 Hz. These results are similar to those obtained from pre-operation vibration tests at the T-REX corridor and to those obtained from recent TDA vibration tests conducted at the Santa Clara Valley Transportation Authority (VTA) light rail system in San Jose, CA (USA). Compared to other types of vibration mitigation measures tested at the San Francisco, CA (USA) Muni light rail system, the vibration reduction performance of the TDA treatment at RTD was found to be equal or superior to that of a ballast mat and not as effective as a floating slab track installation.

T. Maeda et al. (Eds.): Noise and Vibration Mitigation for Rail Trans. Sys., NNFM 118, pp. 283–290.
springerlink.com

1 Introduction

This study was commissioned by RTD in response to concerns by the U.S. Federal Transit Administration (FTA) about the use of TDA underlayment as a vibration mitigation treatment for rail transit projects. The objective of the study was to document the vibration reduction properties of the TDA underlayment at locations along the T-REX corridor for comparison with the results of TDA tests conducted during project design, with TDA performance measured elsewhere and with the performance of similarly-appropriate vibration mitigation measures.

The original vibration analysis report prepared for the Southeast Corridor Project Final Environmental Impact Statement (FEIS) [1] recommended the installation of ballast mat beneath 1,280 meters of LRT track. Several years later, an updated analysis report prepared during final design [2] concluded that 2,330 meters of vibration mitigation treatment would actually be required. The final design study also included tests of TDA performed at a key location along the corridor using a vibratory roller as a vibration source.

Based on the results of the design analysis, ten segments of TDA underlayment were installed at seven locations during project construction. This treatment has been in place since the line opened in November 2006, encompassing three years of revenue-service operations at the time this study was conducted.

The sections of track with TDA underlayment are composed of a 305 mm thick layer of TDA (maximum 102 mm size tire shreds wrapped in a polypropylene fabric) beneath 406 mm of sub-ballast, with 203 mm of ballast above the sub-ballast to the base of the ties. Figure 1 presents a cross-section of the track construction details. The T-REX line uses continuous-welded rail (CWR) and concrete ties, and a typical alignment configuration includes two tracks located in between a sound wall on the southbound (community) side and a retaining wall on the northbound (highway) side.

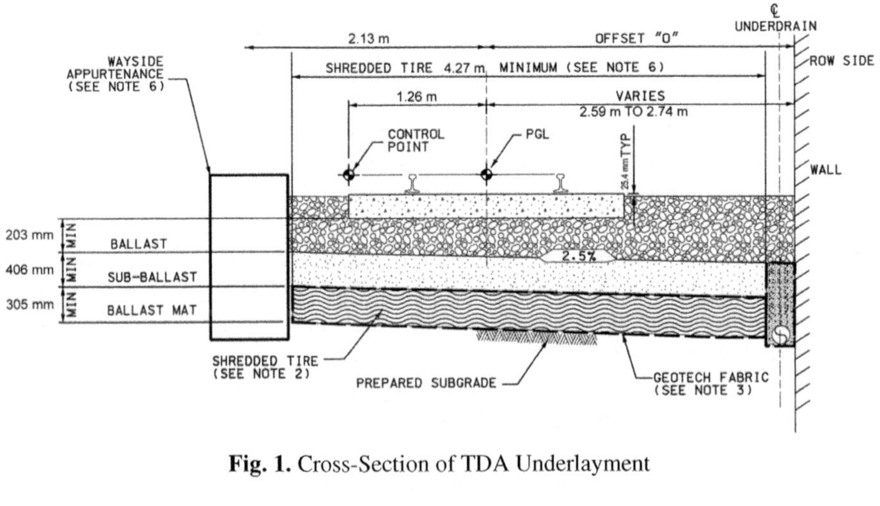

Fig. 1. Cross-Section of TDA Underlayment

2 Vibration Test Methodology

Vibration tests were carried out at multiple sites along the T-REX corridor during November 2009. For consistency, ground-borne vibration was measured at all sites during controlled operations of a dedicated test train including two of the newer Siemens Model SD160 high-floor, bi-directional, six-axle, articulated light rail vehicles in the RTD fleet. The vehicles were empty during the tests, each with a total weight of approximately 40,600 kg. In addition to the train tests, ground vibration propagation measurements were also conducted at each site. The test locations, procedures, instrumentation and analysis are described below.

2.1 Measurement Locations

Based on a survey of the corridor, it was determined that measurements would be feasible at five sites, including four sites where TDA was installed (designated by segment as TDA-1, TDA-5, TDA-8 and TDA-9) plus one control site with standard ballast-and-tie track construction. The measurement positions at a representative site (TDA-1) are shown in plan view in Fig. 2. All of the test locations were on the southbound side of the tracks and the adjacent sound wall.

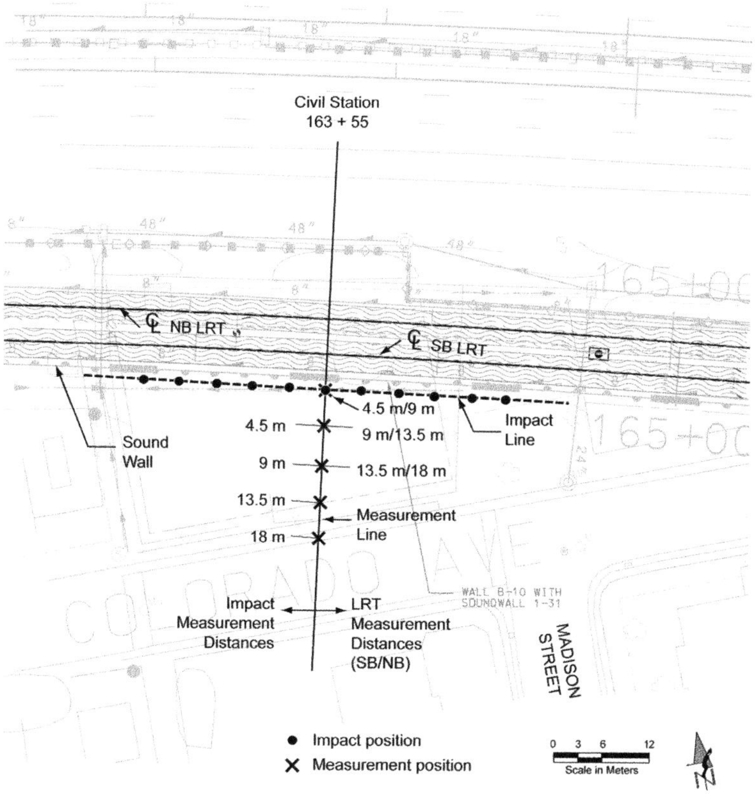

Fig. 2. Vibration Measurement Configuration at Site TDA-1

2.2 Test Procedures

Ground-borne vibration from LRT operations was measured at three positions at each site, using vertically-mounted accelerometers. As shown in Fig. 2, these were located at 4.5, 9 and 13.5 meters from the near-track (southbound) centerline and 9, 13.5 and 18 meters from the far-track (northbound) centerline. At each site, measurements were made during multiple operations of the dedicated two-car train at constant speeds of 40, 64 and 88 km/h through the test areas. The tests typically included two runs in each direction on both tracks at each speed, amounting to a total of 24 runs at each site.

The vibration propagation test procedure consisted of dropping a 27 kg weight from a height of about one meter to produce an impulsive force on the ground, and simultaneously measuring the force and the vibration response at specific locations. The relationship between the input force and the ground surface vibration, called the transfer mobility, characterized the vibration propagation at each test site. The force was measured using a load cell and the vibration responses from the impacts were measured using accelerometers. As shown in Fig. 2, the impacts were typically generated at 11 points spaced 4.5 meters apart along a line parallel to the tracks. The vibration response was measured at a series of positions along a line perpendicular to the tracks, at distances of 4.5, 9, 13.5 and 18 meters from the impact line.

2.3 Instrumentation and Data Analysis

The vibration signals from the accelerometers and load cell were conditioned and stored using digital recording equipment, and vibration acceleration levels were converted to one-third octave band velocity levels using digital signal processing software. For the LRT tests, an energy-average maximum vibration level was computed for each event. For the vibration propagation tests, the data were processed into point source transfer mobilities (PSTM), and these were then integrated into a line source transfer mobility (LSTM).

3 Measurement Results

3.1 LRT Vibration

A representative example of the LRT vibration measurement results is provided in Fig. 3, which compares the average ground vibration level spectra for all sites at a distance of 9 meters and a train speed of 88 km/h. The results indicate that while there is considerable variation at the low frequencies, the vibration levels at 20 Hz and above are similar at all sites, including the control site. Based on these results, one might conclude that the TDA is ineffective. However, the measured LRT vibration levels do not account for site-specific differences in ground-borne vibration propagation as discussed below in Sect. 3.2.

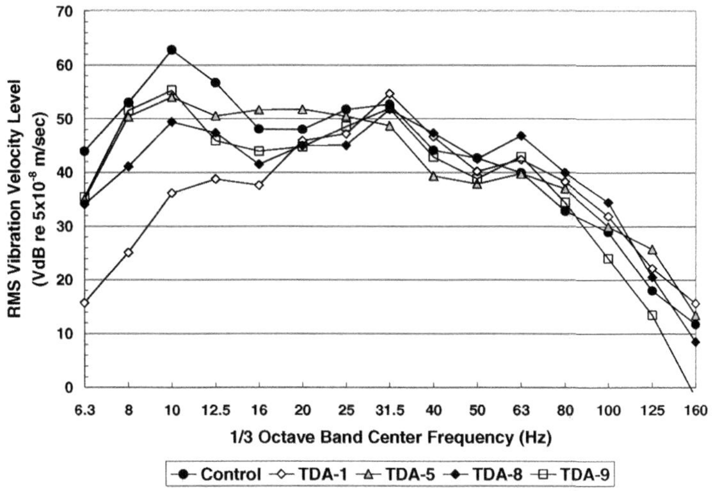

Fig. 3. Average Measured LRT Vibration Levels (9 m, 88 km/h)

3.2 Vibration Propagation

The results of the vibration propagation tests at each site were summarized in terms of LSTM for each of the four measurement distances. A representative example of these results is given in Fig. 4, which compares the LSTM at a distance of 9 meters for each of the sites. Most significant are the results for the control site, which has a lower response at frequencies above 20 Hz compared to the TDA sites. Thus, to evaluate TDA effectiveness, it is necessary to combine the propagation data with the LRT vibration data as described below in Sect. 4.

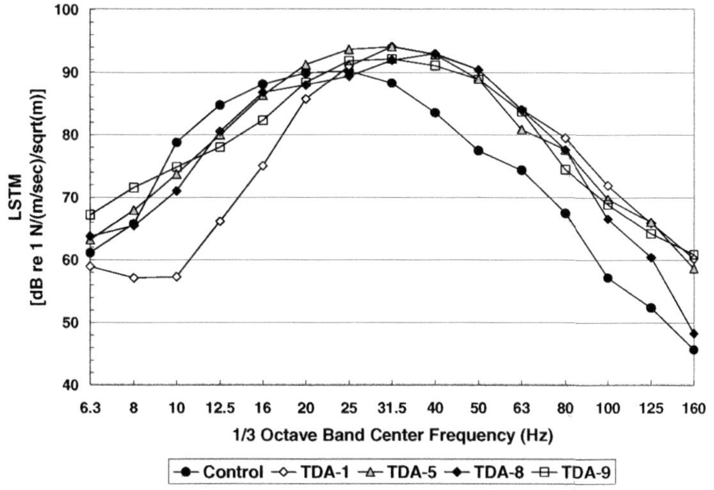

Fig. 4. Vibration Propagation Results (Line Source Responses) at 9 Meters

4 Evaluation

TDA effectiveness has been evaluated by comparing the Force Density Level (FDL) at each of the TDA sites with the FDL at the control site. The FDL describes the force that excites the ground; it includes the effects of both the vehicle and track structure at a given location but excludes the vibration propagation effects of the soil. The FDL is calculated on a one-third octave band frequency basis from the measured LRT vibration level (Lv) and the measured Line Source Transfer Mobility (LSTM) at each site using the following equation: FDL = Lv – LSTM.

As a representative example, the calculated FDL spectra at the measurement sites for a LRT speed of 88 km/h are provided in Fig. 5. In general, the FDL values represent averages of the results at all measurement distances. The results show that the FDL values for the control site typically exceed the FDL values for the TDA sites at frequencies above 20 Hz, which suggests that the TDA is effective in reducing ground-borne vibration in this frequency range.

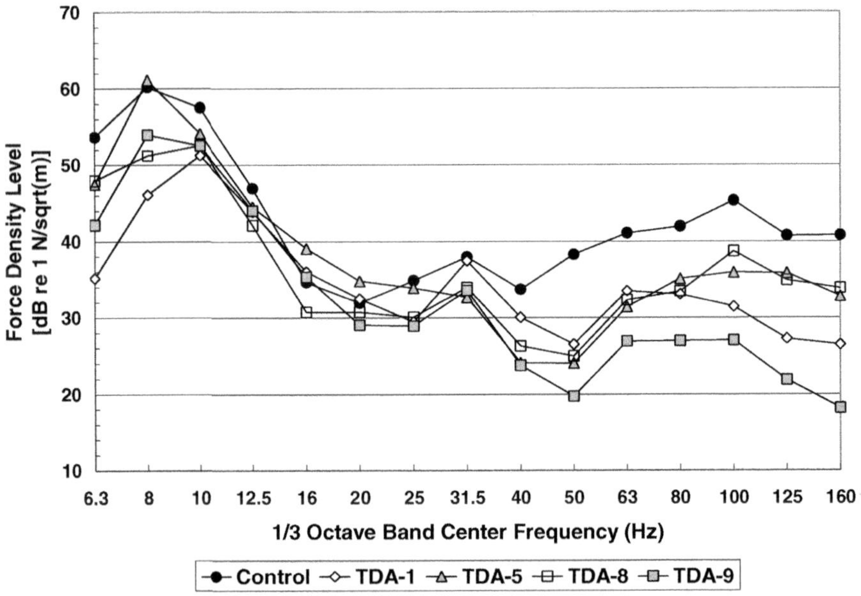

Fig. 5. LRT Force Density Levels at 88 km/h

The effectiveness of the TDA underlayment was determined by first comparing the FDL results at each TDA site with the FDL results at the control site, and then by averaging the results for each of the three LRT speeds. The resulting one-third octave band vibration reductions at each site are shown in Fig. 6, along with the overall average TDA vibration reduction. The results at the four TDA sites are reasonably consistent in the mid-frequency range from 10 Hz to 50 Hz.

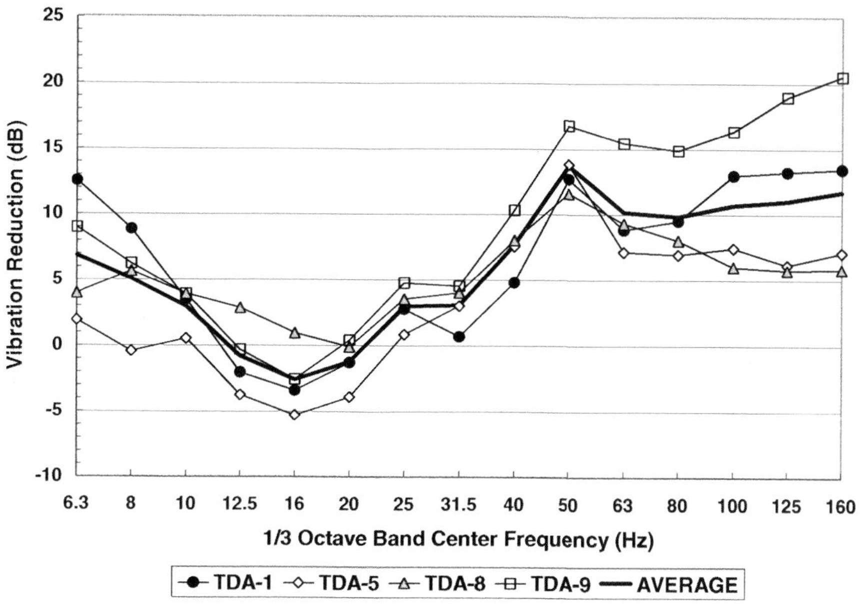

Fig. 6. Average TDA Vibration Reduction at the Denver RTD Southeast Corridor

On average, the TDA provides a vibration reduction of about 3 dB at 25 Hz and 31.5 Hz, and reductions of 8 to 14 dB between 40 Hz and 160 Hz. Although the results show overall average vibration reductions of 3 to 7 dB at frequencies between 6.3 Hz and 10 Hz, it is not expected that a consistent reduction of this very low-frequency ground-borne vibration will be achieved due to the potential variability at these frequencies as found in a recent study of TDA effectiveness [3].

Comparisons of the current findings with the results of previous TDA measurements are provided in Fig. 7. A comparison of the current results with the results of pre-operation tests using a vibratory roller [2] indicates roughly similar vibration reductions at frequencies of 25 Hz and above. A comparison of the current results with recent TDA vibration tests results at the Vasona Line of the Santa Clara Valley Transportation Authority (VTA) light rail system in San Jose, CA (USA) [3] indicates fairly similar vibration-reduction performance for both tests, with the VTA installations showing 3 to 5 dB greater reductions in the 12.5 Hz to 20 Hz frequency range and the RTD installations showing about a 6 dB greater reduction at 50 Hz.

Finally, Fig. 8 provides a comparison of the 2009 TDA vibration test results in Denver with vibration test results for ballast mat and floating slab track vibration mitigation installations at the San Francisco, CA (USA) Muni light rail system [4]. The comparison suggests that the vibration reduction effectiveness of the TDA treatment is equal to or greater than that of the ballast mat. In the frequency range from 31.5 Hz to 80 Hz, the TDA vibration reductions fall in between the reductions measured for the ballast mat and floating slab treatments.

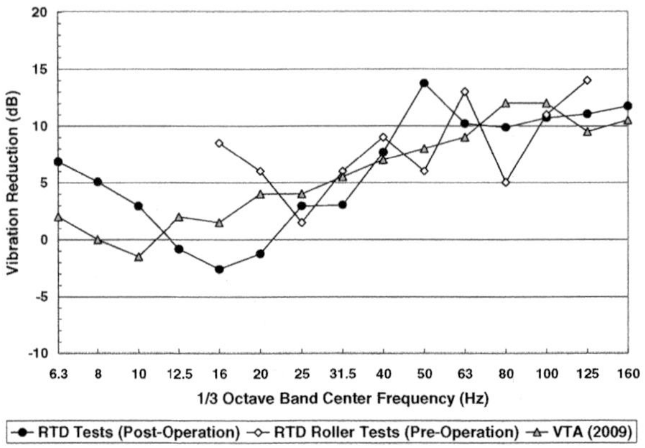

Fig. 7. Comparison of TDA Vibration Test Results

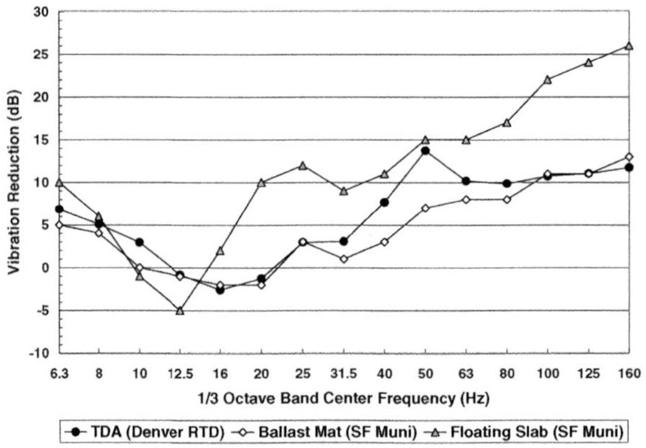

Fig. 8. Comparison of Measured Performance of Vibration Mitigation Treatments

References

[1] Harris Miller Miller & Hanson Inc., Vibration Impact Assessment - Southeast Corridor Light Rail Transit (LRT) Line - Denver, Colorado Metropolitan Area, HMMH Report No. 296650-01 (December 1999)

[2] Southeast Corridor Constructors, Final Southeast Corridor Multi-Modal Project-LRT Final Design Vibration Analysis and Mitigation Report, (January 20, 2004)

[3] Wilson, Ihrig & Associates, Inc., Evaluation of Tire Derived Aggregate as Installed Beneath Ballast and Tie Light Rail Track–Results of 2009 Field Tests, draft report to Santa Clara Valley Transportation Authority (May 2009)

[4] Wilson, Ihrig & Associates, Inc., Evaluation of Tire Derived Aggregate as Installed Beneath Ballast and Tie Light Rail Track – Results of 2006 Field Tests, final report to Dana N. Humphrey, Consulting Engineer (February 2007)

Specificities of Ground Vibration Induced by Freight Traffic, from the Track to a Building

E. Bongini[1], C. Guigou[2], C. Petit[3], F. Poisson[1], P. Ropars[2], and M. Villot[2]

[1] SNCF, Innovation and Research Department,
45 rue de Londres, 75739 Paris cedex 8, France
Tel.: +33153422356; Fax: +33153429784
e-mail: estelle.bongini@sncf.fr
[2] CSTB,
24 rue Joseph Fourier, 38400 St Martin d'Hères, France
[3] SATEBA,
31 place ronde, 92986 Paris la Défense cedex, France

Summary

CSTB, SATEBA and SNCF have started a project called VibsolFret dedicated to the specificity of ground vibrations induced by freight traffic. The first step of this project consists in determining freight specificities in terms of induced ground vibration. A measurement campaign is carried out, on an urban site, near a classic line. Commercial freight and passenger trains run on the track. Measurements are realised in free-field and in a building close to the track.

This measurement campaign allows us to determine some specificities of freight traffic in comparison with passenger train traffic. It particularly shows that global levels are more or less the same between freight rolling stock and passenger rolling stock (with realistic speeds in this case 80km/h for freight versus 120km/h for passenger trains). However, freight traffic induces longer time-signals with very high short-term levels. It is observed for free-field acceleration as well as for the acceleration measured within a building.

Measurements indicate that vibration inside a building is mainly driven by the building structure. To propose vibration reduction solutions by acting on the building itself, a numerical approach is used to simulate induced vibrations with several variations of building structure parameters.

At the same time, a numerical parametric study is carried out to evaluate the impact of mitigation measures installed on the track. This study will allow us to propose the most efficient combination of these parameters in terms of vibration reduction.

1 Introduction

With the intention of examining ground vibration induced by railway traffic, SNCF (French Railway Company), CSTB (French Scientific and Technical Centre

T. Maeda et al. (Eds.): Noise and Vibration Mitigation for Rail Trans. Sys., NNFM 118, pp. 291–298.
springerlink.com

for Building) and SATEBA (concrete sleepers manufacturer) have started in 2009 a project called VibsolFret.

This project aims:

- To quantify the specificities of freight traffic, in comparison to passenger traffic, in terms of ground vibration induced in the free field and in buildings near the track,

- To develop reduction solutions on the track, adapted to freight traffic and particularly to heavy axle loads.

The measurement campaign was carried out, on a dedicated site where both freight and passenger traffic occur. This paper will present, first, the measurement campaign results. Comparisons will be made between freight and passenger traffic in terms of classic indicators as well as the acceleration time-signal. The second part of this paper will be dedicated to a numerical approach and a parametric study that allows first, to consider different building configurations and secondly, to optimize track parameters within the scope of vibration reduction.

2 Measurement Campaign: Freight Specificities with Regard to Induced Ground Vibrations

The first step of this project consists in determining freight specificities in terms of the induced ground vibration. A measurement campaign was carried out, on an urban site, near a classic line. Freight and passenger rolling stocks pass-by on the track. Measurements were conducted in free-field and in a building close to the track.

A line of accelerometers was used to measure vibration propagation in the pseudo free-field, near the building. The track was also equipped with sensors (Fig.1). Inside the building, vertical and lateral accelerations and structure borne noise were also monitored (Fig.1).

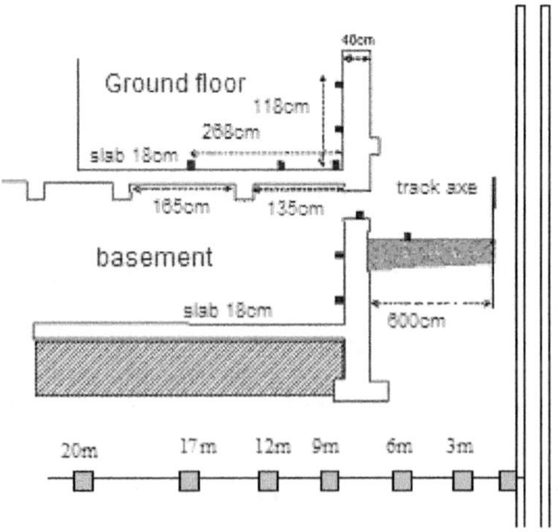

Fig. 1. Accelerometers in pseudo free-field and accelerometers distribution inside the building (color figure online)

Several post-processing techniques have been carried out on the measured accelerations, following the European standards such as ISO 2631 [1] or the German standard DIN 4150. The comparisons presented below are based on the indicator $(a_w)_{rms}$ suggested in the ISO 2631 standard. The post-processed signals in the building correspond to the sensor in the center of the ground floor slab. In fact, they were the highest-acceleration levels that had been measured.

The graphs presented in Figure 2 demonstrate that freight pass-bys cause higher $(a_w)_{rms}$ levels in the building than passenger train pass-bys: a linear mean value of 3.65 m/s^2 that corresponds to 71 dB is shown for the freight pass-by against a linear mean value of 2.45 m/s^2 i.e. 68 dB for passenger trains. However, the measurements were carried out near a passenger train station which implies that the freight pass-by speeds are globally higher than the passenger train speeds. Therefore, it can be seen that, for the freight pass-by at 28 km/h, numbered 6 on the left graph of Figure 4, the level is equivalent and even lower than the levels observed for passenger trains. The fastest passenger train (130km/h) corresponding to pass-by 7 on the right graph, presents a level of the same order of magnitude as the most contributing freight trains.

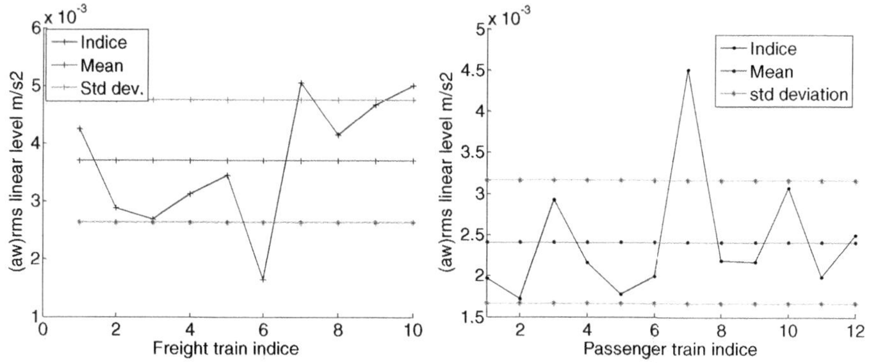

Fig. 2. $(a_w)_{rms}$ of the standard ISO 2631, from accelerations measured at the center of the ground floor slab. On the left, freight pass-bys, on the right, passenger train pass-bys. $(a_w)_{rms}$ level for each pass-by in black, mean value in blue and standard deviation in red. (color figure online)

A comparison of free-field measurement (not illustrated in this paper) shows the same trends: freight pass-bys cause higher $(a_w)_{rms}$ levels in the free-field than passenger trains pass-bys, whatever the distance between the sensor and the track, but it seems to be mostly driven by the pass-by speed.

Figure 3 shows that the spectral distribution is not discriminating between freight and passenger train-induced vibrations in the free-field. Both rolling stocks induce a peak around 50-60 Hz at distances beyond 6m from the track.

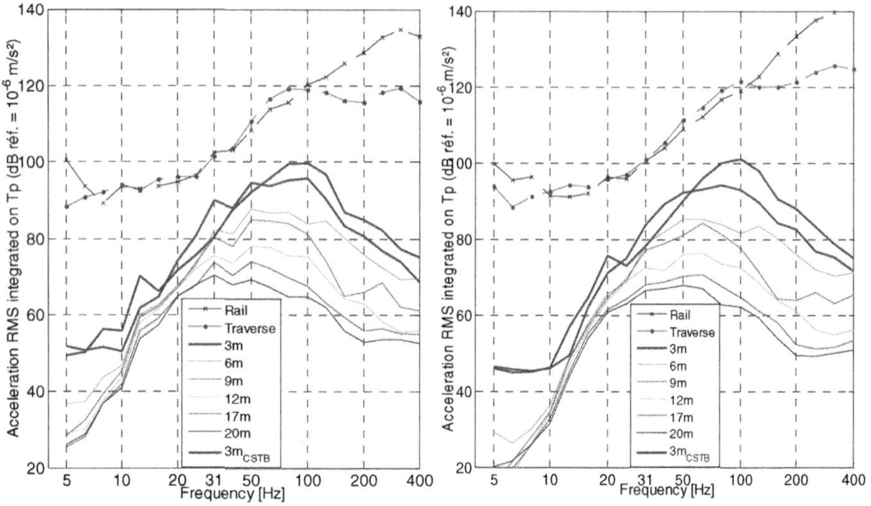

Fig. 3. Spectra of the RMS acceleration, integrated on Tp, for freight train at 30km/h on the left graph and a passenger train at 40km/h on the right graph. (color figure online)

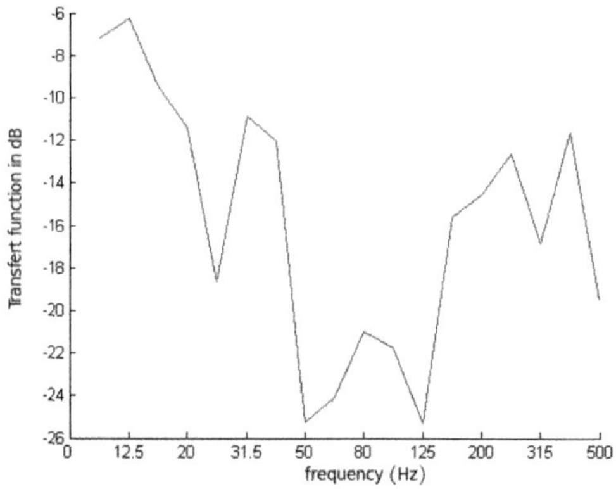

Fig. 4. Transfer function between the free field sensor at 6m and the center of the slab inside the building (at 6m from the track), for a given pass-by (color figure online)

Moreover, Figures 3, 4 and 5 demonstrate that the spectral distribution of the velocity inside the building is completely determined by the building modal behavior illustrated in Figure 4. In fact, whereas in the free-field the major contribution of ground vibration appears around 50 Hz beyond 6m (Figure 3), inside the building (Figure 5), the highest contribution appears at the one-third octave around 31.5 Hz, corresponding to one of the most energetic modes of the building, as shown on Figure 4.

Actually, the measurement campaign indicated that rolling stock pass-by speed is the most influent parameter on the acceleration magnitudes, in the free-field environment as well as in the building. Freight rolling stock and passenger trains with the same pass-by speed had the same order of magnitude of the $(a_w)_{rms}$ levels. Same spectral shapes were observed for both rolling stock types.

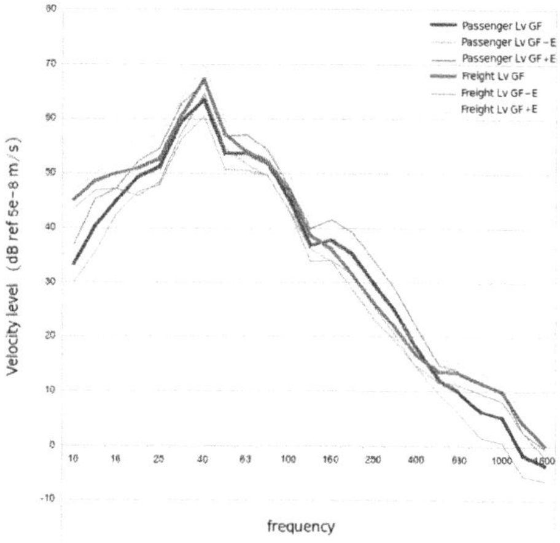

Fig. 5. Velocity one-third octave band spectra at the centered sensor on the ground floor slab, measured inside the building (color figure online)

However, as shown in Figure 6, freight traffic induces much longer time-signals with time-occurrences at a very high level, contrary to passenger trains (Fig 6) that induce a smoother vibration time-signal. It is observed for free-field accelerations as well as for building ones. A perception-based study should be carried out to define the effects of the signal's duration and its time-dependent variability on the trackside resident annoyance.

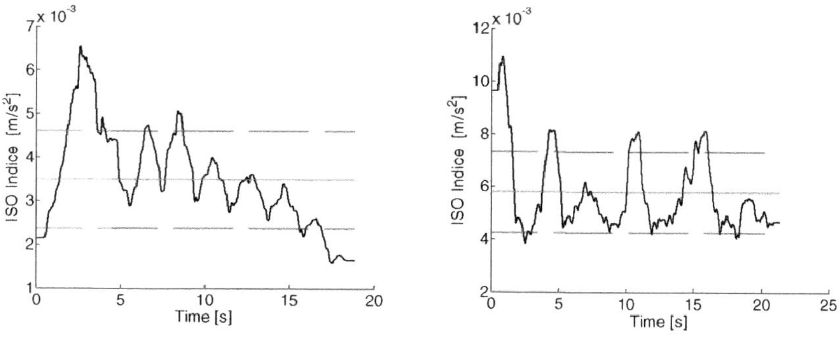

Fig. 6. $a_{w,rms}(t)$: slip average (integration on 1s) with the w_m weighting from ISO2631-2; sensor on the center of the ground floor slab. Left: passenger train (3), right: freight train (2) (color figure online)

3 Parametric Study

It has been demonstrated that the dynamic behavior of the building completely drives the spectral vibration distribution inside it. Therefore, the first part of the parametric study consists in varying the building characteristics. Calculations are carried out with a software named MEFISSTO [3] developed by the CSTB, based on a 2-D boundary element approach for the transmission of the vibrations in the ground and a 2-D finite-element model for the building response. Transfer functions between a free-field point near the building and the centered sensor on the ground floor slab are calculated. Variations of the ground floor slab thickness, thickness of the façade wall (including the buried basement wall) in front of the track (see Fig 1) and span of the building slabs are simulated.

The results demonstrate that:

- minor variations of the ground floor slab thickness (thickness between 15 cm and 25 cm, real thickness being equal to 18 cm) do not impact the vibration levels on the slab but shift the eigenfrequencies of the structure (to a lower frequency range when thickness is increased) therefore modifying the building ground floor modal behavior
- variations of the façade wall thickness (from 20 cm to 40 cm, real one being equal to 40 cm) impact the vibration inside the building: this change implies a modification of transmission path between ground and building; the thinner façade wall reacts much more to the ground vibration (because of the mass effect) and therefore transmits higher energy to the entire building
- an increase of the floor slabs span results in higher-vibration levels inside the building, in the lower-frequency range as the slab becomes more flexible.

This parametric study allows one to propose some directions for building improvement in case of annoying vibrations while keeping in mind that each structure has its intrinsic dynamic behavior, depending on its geometry and material properties.

A second parametric study is carried out using the CASC.voie software, developed by CSTB, to evaluate the impact of vibration-reduction solutions that could be installed on the track. Typically, several values of rail pad stiffness, under sleeper pad stiffness, under sole-plate pad stiffness, ballast height, rail mass, sleeper mass, sleeper spacing or soil stiffness have been tested. Cross variations between all the parameters have also been considered. The software used within this parametric study is based on a model of combined mass and stiffness/damping layers (Figure 7). The ground is represented with a line input impedance, calculated through a 2-D elastic half-space ground model based on the wave approach [2].

The excitation force is applied at the same location as the unsprung mass on the rail. The calculations are carried out with 2 different unsprung masses, representative of passenger train or freight rolling stock.

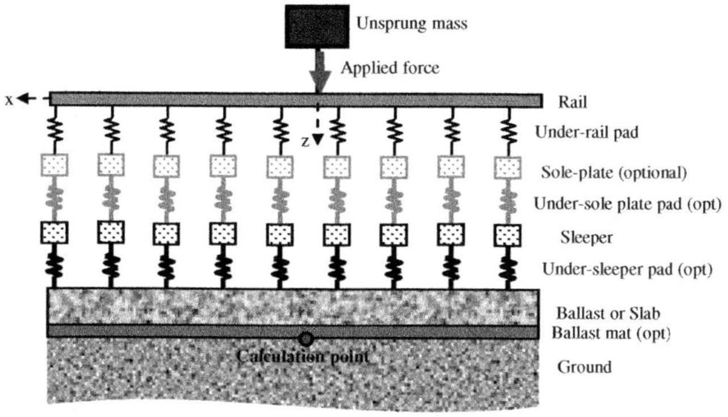

Fig. 7. Diagrammatic representation of the model (color figure online)

The software allows calculating the insertion loss factor between a nominal situation and the tested one, at the calculation point that corresponds to the interface between ground and track. Static deflexion is also calculated to ensure that the track deflexion corresponds to the standardized one.

Typical results are presented in the Table 1.

Table 1. Insertion loss factor for under-sleeper pad on various ground stiffness Reference: regular ground with a USP stiffness equal to 200MN/m

	Freight				Passenger train			
One-third oct band (Hz)	40	50	63	80	40	50	63	80
Regular ground / USP 50MN/m	-1.3	-0.5	3.3	9.2	-1.9	-3.3	-3.8	3.0
Regular ground / USP 100MN/m	-0.6	-0.8	-0.3	2.6	-0.6	-1.0	-1.9	-2.2
Regular ground / USP 400MN/m	0.3	0.5	0.8	0.1	0.3	0.5	0.9	1.6
soft ground / USP 50MN/m	-3.0	-4.2	-1.3	6.1	-3	-4.3	-6.4	-1.3
soft ground / USP 100MN/m	-1.0	-2.1	-3.4	-1.5	-1.5	-1.8	-2.6	-4.1
soft ground / USP 400MN/m	0.5	0.5	0.2	-1.6	-0.4	-0.1	0.4	1.2
stiff ground / USP 50MN/m	2.1	6.2	10.0	14.4	0.8	-1.4	1.2	8.8
stiff ground / USP 100MN/m	-0.3	2.1	5.6	8.7	3.2	0.7	-0.4	2.0
stiff ground / USP 400MN/m	0.7	0.4	2.3	4.3	4.8	3.0	2.0	2.6

A significant effect is considered with a positive insertion loss, greater than 3dB. This parametric study highlights that no solution has a significant effect on the entire frequency range between 40Hz and 100Hz, and the impact of each one is closely dependent to the type of ground and the rolling stock running on the track. The most effective solution is the under-ballast mat, for freight as for passenger trains, from 31.5 Hz and beyond. Except this solution, the best reductions are obtained with under sleeper pad for which the insertion loss is positive only for freight trains (see Table 1), for specific frequency ranges and preferentially for stiff ground.

The solution must therefore be chosen regarding to the ground characteristics and the most energetic frequency range of the excitation determined by the unsprung mass, the roughness spectrum and the pass-by speed.

In parallel to this study, calculations are, at that moment, carried out with the software TRAFFIC © developed by KUL (Katholieke Universiteit Leuven) based on a 2.5-D boundary-element model. This approach allows taking into account the effect of the propagation into the ground (several ground types are considered), that is, combined with the effect of the track parameters.

The last part of the project will consist in carrying out test bench measurements with some vibration mitigation measures that appear to be the most efficient within the numerical approach. Under sleeper pad with different stiffness will be tested, as well as sleeper shoes with various stiffness on a test bench taking into account the ballast, the sleepers, the rail and the fastening system. The experimental setup allows applying a harmonic load as well as a static load on two rail heads, at various frequencies and to quantify the accelerations of the whole sleeper components of the track.

4 Conclusion

The results of the measurement campaign show that, in the free field, acceleration global levels induced by freight or passenger trains are equivalent, with the same spectral distribution. In buildings, the equivalence between global levels remains and the spectral distribution is completely determined by the building modal behavior. For a given measurement site, the main parameter affecting global acceleration levels is the pass-by speed. Differences between freight and passenger traffic can be observed on time-signal because of high variability of freight-induced vibration.

The parametric study highlights the difficulty of finding a generic vibration mitigation measure: it has to be developed by taking into account: the building structure, the ground stiffness, the particular excitation driven by roughness, unsprung mass and pass-by speed.

Acknowledgements

This project is subsidised by the ADEME and by RFF for the SNCF part. We would like to thank in particular A. Guerrero from RFF.

References

[1] ISO 2631-1:1997 Mechanical vibration and shock – Evaluation of human exposure to whole-body vibration (1997)

[2] Guigou-Carter, C., Villot, M., Guillerme, B., Petit, C.: Analytical and experimental study of sleeper SAT S 312 in slab track Sateba system. Journal of Sound and Vibration 293(3-5), 878–887 (2006)

[3] Jean, P.: Boundary and finite elements for 2D soil-structure interaction problems. Acta Acustica 87, 56–66 (2001)

Ground Vibration on High-Speed Railway Tunnel

T. Watanabe[1], M. Sogabe[1], H. Yokoyama[1], T. Yonezawa[2], and S. Kiyota[2]

[1] Railway Technical Research Institute,
2-8-38, Hikari-cho, Kokubunji-shi, Tokyo, 185-8540, Japan
Tel.: +81 42 573 7290; Fax: +81 42 573 7320
e-mail: tsutomuw@rtri.or.jp
[2] Japan Railway Construction, Transport and Technology Agency

Summary

We developed 3-D dynamic interaction analysis method that was able to model the entire railway system including the vehicle, the track, the structure and the ground. We conducted ground vibration analyses in which ground properties and track structure were assumed to be analytical parameters. As a result, under the condition of this analysis, it became apparent that ground vibration decreased by 25 dB when Vs (shear wave velocity of the ground) was increased to 1.3 Vs. On the other hand, it increased by 15 dB when Vs was decreased to 0.7 Vs. Moreover, we made clear the vibration-reduction effect of vibration-reduction track.

1 Introduction

Ground vibration induced by running trains is the result of the phenomena that vibration caused by vehicles propagates to tracks, structures and the ground. Because a large number of parameters are involved in the propagation path, and because the analysis of the propagation of vibration requires a large-scale numerical model which can reproduce the propagation of high-frequency vibration even in the distance, an analytical method to predict an absolute value of ground vibration accurately cannot be easily established and has not been developed yet. Practically, ground vibration is predicted by empirical models based on measurements, or semi-empirical models based on numerical analysis aided by measurements [1]. These models are convenient and effective tools on a practical level. For example, a method to predict ground vibration at a target site using equivalent excitation forces obtained by the finite-element method has already achieved satisfactory results. However, empirical prediction is difficult in a case of railway vehicles, tracks or structures for which measurement data have not been accumulated.

T. Maeda et al. (Eds.): Noise and Vibration Mitigation for Rail Trans. Sys., NNFM 118, pp. 299–306.
springerlink.com

With this background, in order to complement the conventional methods described above, we accumulated information on the characteristics of railway vehicles, tracks, structures and the ground, built a three-dimensional model of the propagation path of the vibration, and attempted to establish a method that analytically predicts the ground vibration. Since this proposed method involves dozens of input parameters, the conventional prediction method and measurements must be used jointly with this new method. However, since this method enables one to model each component arbitrarily, various numerical experiments can be executed which are difficult to be reproduced in reality. The goals of our study are shown as follows:

(1) To establish a three-dimensional dynamic interaction analysis method that is able to arbitrarily take into consideration effects of vehicles, tracks, structures and the ground, aiming at examination of the ground vibration on a high-speed railway tunnel.
(2) To evaluate effects of each parameter on the ground vibration regarding the above-mentioned numerical analysis method.

2 Analysis Method

2.1 Outline of Analysis Method

In this study, the analysis target was a high-speed railway tunnel with overburden of 9.2 m depth and invert bottom of 19m depth. Vibration-reduction tracks were laid in the tunnel, and train speed was from 200 to 250 km/h. Table 1 shows characteristics of the ground.

As previously described, a large-scale analysis model which can deal with arbitrary structural forms is required for accurately reproducing ground vibration caused by train passage on the analysis target tunnel with numerical analysis. However, present computer capability has not reached to the level where a practical solution can be obtained with such a large-scale model, and some sort of engineering approach is needed.

Figure 1 shows an outline of the three-dimensional numerical analysis method developed in this study. At first, to predict the ground vibration, we obtained the vibration force from an analysis model of the vehicle-track-tunnel interaction system. The vibration force was loaded to the tunnel-ground interaction system to calculate the ground vibration. We used DIASTARSII for the vehicle-track-tunnel interaction system and SuperFLUSH/3D for the tunnel-ground interaction system.

A program called DIASTARSII [2], which analyzes dynamic interaction between vehicles and structure, was used in the analysis of the vehicle, track and a tunnel. The tunnel was modeled as supported by appropriate ground springs. Thus, the computed vibration force was considered to be equivalent to the one computed with consideration of interaction force in the entire system.

SuperFLUSH/3D, a program that analyzes dynamic interaction between the ground and structure was employed in the analysis of the tunnel-ground interaction system. As the vibration force used here is the value to which interaction with vehicles and tracks has been added already, it can be regarded that the numerical analysis executed takes into consideration all the simulated-interaction forces.

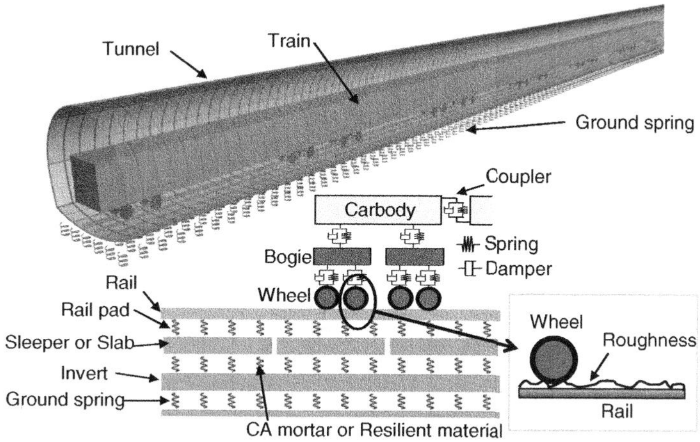

(a) Vehicle-track-tunnel interaction system

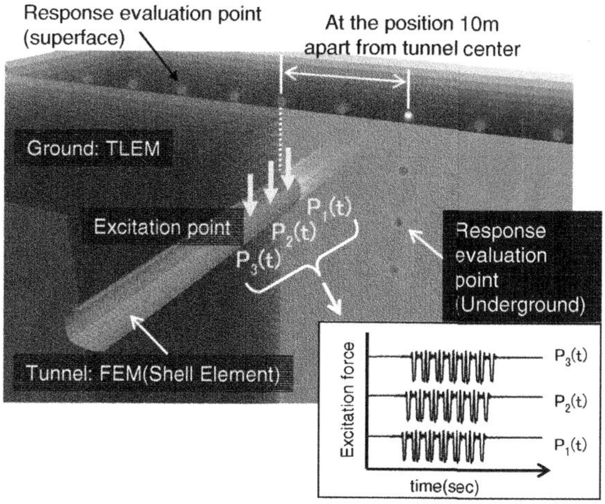

(b) Tunnel-ground interaction system

Fig. 1. Outline of the three-dimensional numerical analysis method (color figure online)

2.2 Analysis of Vehicle-Track-Tunnel Interaction System

Table 1 and 2 show characteristics of the ground and tunnel. The vehicle model was a one-car, 31 degrees of freedom Multi-Body Dynamics model, consisting of rigid mass points representing a carbody, truck frames and wheelsets mutually linked with springs and dampers. A model of the rail was constructed by providing node points at the one-fifth length of the rail-fastening system interval, modeled by beam elements. Sleepers and track slabs were modeled by beam elements. Rail-fastening systems and track support springs (CA mortar or resilient materials) were modeled by scalar-spring elements. A tunnel building frame was modeled by a lattice model of beam elements. A model of a tunnel support ground spring was constructed by using scalar-spring elements. As constitutive law between wheels and rails, a *Hertz* contact spring and *Kalker* linear creep theory were used for vertical and horizontal directions, respectively.

A method to solve simultaneous motion equations of vehicles and structure on the modal coordinate system was employed. However, as both of the equations had nonlinear terms, iterative calculation was conducted within Δt according to the *Newmark* average acceleration method.

The time history reaction waveform of the track support spring elements during train passage becomes vibration force of the next section.

Table 1. Characteristics of the ground

Layer	Soil property	Depth (m)	N-value	Shear wave velocity (m/s)
1	Sandy soil	0.0 to 2.2	6	145
2	Silt (ground water level shallow)	2.2 to 5.6	7	191
3	Silt (ground water level deep)	5.6 to 5.8	7	191
4	Sandy soil	5.8 to 14.7	7	153
5	Sandy soil	deeper than 14.7	36	264

Table 2. Characteristics of the tunnel

	Thickness (m)	Young's modulus (kN/mm^2)	Poisson's ratio	Unit weight (kN/m^3)	Damping ratio (%)
Tunnel lining	0.5	24.2	0.2	23	3.0
invert	0.45	24.2	0.2	23	3.0

2.3 Analysis of Tunnel-Ground Interaction System

In the tunnel-ground interaction system, we modeled the tunnel by shell elements, and the ground by thin-layer elements (TLE) [3]. The extension and depth of the model were 150 m both. The analysis frequency was below 100 Hz.

Figure 2 shows the analysis model of SuperFLUSH/3D. In this analysis, several excitation points were set in the tunnel invert section, and a transfer function from these points to ground response measurement points was computed. The aforementioned exciting force was converted to frequency domain by using Fourier transformation. Then, it was multiplied by the transfer function between the excitation points and ground response measurement points in order to obtain a response of evaluation points in the frequency domain. By applying inverse Fourier transformation to the product, the response of the response measurement points in the time domain was found.

Considering the structural periodicity of the track and tunnel, the excitation points of the tunnel invert section could be decreased in the unit of repeats in the same track structure at the model center, not at all the rail node points. By setting several analysis measurement points on the ground and adding the response from each measurement point while taking its time lag into consideration, the response during train-running can be simulated.

Figure 3 shows a relation between the excitation points and the measurement points. When structures are even, the response R_{A2} of the measurement point 2 to the excitation point A is equivalent to the response R_{B3} of the measurement point 3 to the vibration point B. In other words, R_{A2} is equal to R_{B3}, and similarly R_{C2} is equal to R_{B1}. The response $R_2(t)$ of the measurement point 2 when the train moves to the points A, B and C at the speed of V by the unit of Δt, can be expressed by the equation (1).

$$R_2(t) = R_{A2}(t) + R_{B2}(t) + R_{C2}(t) = R_{B3}(t - \Delta t) + R_{B2}(t) + R_{B1}(t + \Delta t) \tag{1}$$

Validity of this method has been verified by analysis of another site where actual measurement results were obtained [4].

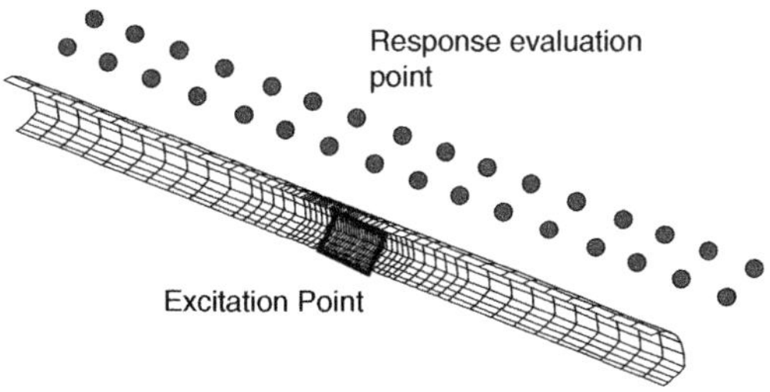

Fig. 2. Analysis model of SuperFLUSH/3D (color figure online)

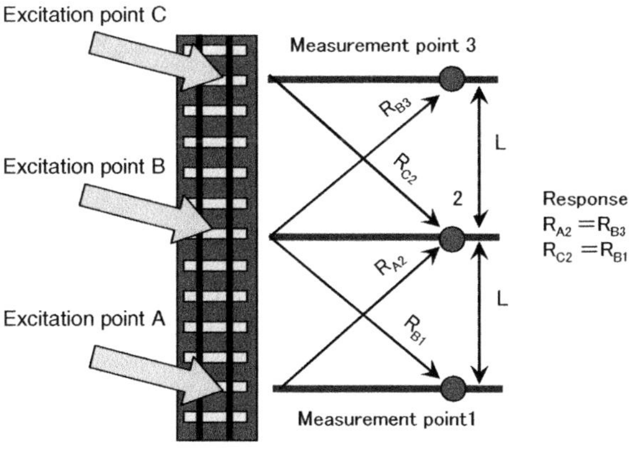

Fig. 3. Relationship of excitation points and measurement points (color figure online)

3 Analysis Result

3.1 Basic Case

Figure 4 shows the analysis results of vibration acceleration level (a_0=10^{-5}m/s^2) for each 1/3 octave band at each measurement point. The actual measurement results are also indicated for comparison. These results were obtained by using a piezoelectric accelerometer (PV-87 or PV-85). Average, maximum and minimum values of the measurements for several trains are shown considering the data dispersion of the measurement results.

The analysis results are generally in good agreement with the measurement results. Therefore it can be concluded that most of the actual phenomena could be reproduced by the three-dimensional dynamic interaction analysis method, which enabled modeling of the entire railway system of vehicles, tracks structures and the ground, established by us. Using this analysis results as a basic case, the following sections examine the effects of each parameter.

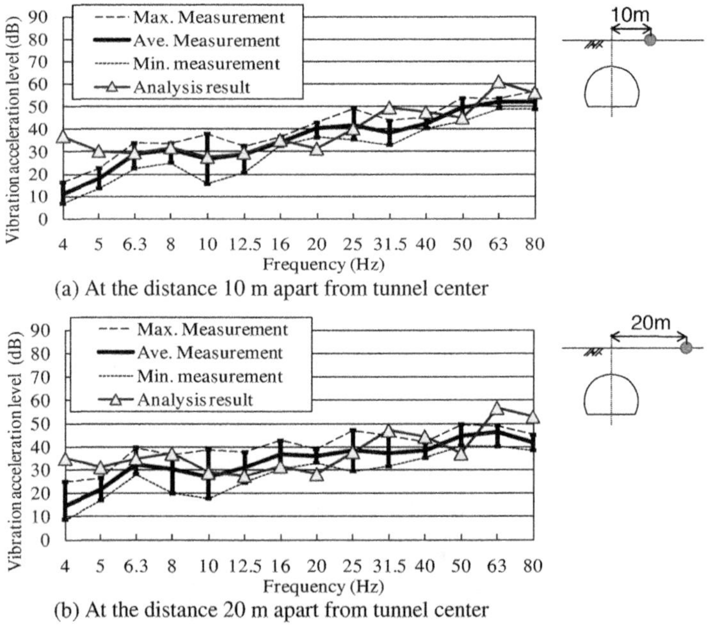

Fig. 4. Analysis results (color figure online)

3.2 Effect of Shear Wave Velocity of the Ground

Figure 5 shows the effects of ground shear wave velocity *Vs* on the ground vibration. Considering that *Vs* varies very much at a certain *N* value [5], *Vs* in the Table 1 was multiplied by 1.3 or 0.7. Comparison with the basic case revealed that multiplying *Vs* by 0.7 increased the ground vibration. This increased tendency was found at not just specific frequencies band, but in the entire frequency band.

Meanwhile, it was found that multiplying *Vs* by 1.3 decreased the ground vibration. This is most noticeable in the lower-frequency bands. It was discovered that multiplying *Vs* by 1.3 or 0.7 caused fluctuation of ground vibration from -15 dB to 25 dB under the conditions of this analysis.

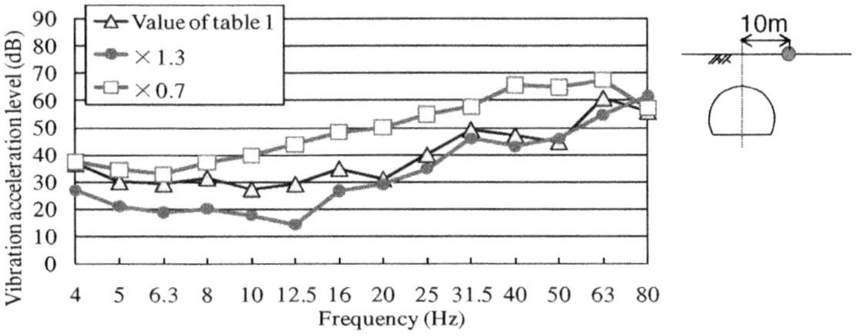

Fig. 5. Effect of shear wave velocity of the ground at the point 10m apart from the tunnel center (color figure online)

3.3 Effect of Damping Constant of the Ground

Figure 6 shows the effect of damping constant *h* of the ground on the ground vibration. The damping constant *h* was set to 5%. Compared with the basic case (*h*=2%), ground vibration varies for each frequency when the damping constant *h* was set to 5%. In some frequency bands, the response decreased by 15dB at the maximum.

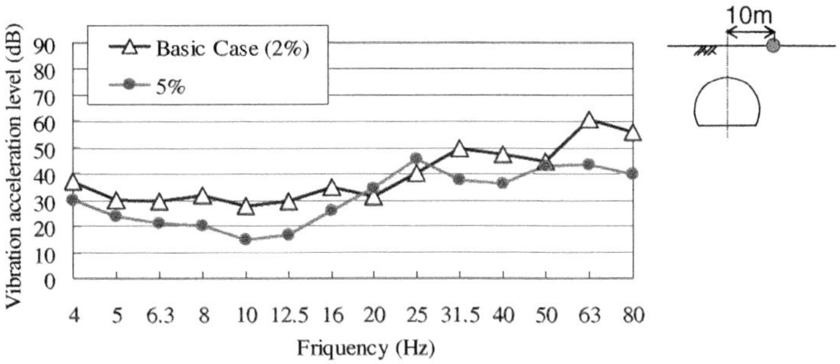

Fig. 6. Effect of damping constant of the ground at the point 10m apart from the tunnel center (color figure online)

3.4 Effect of Track Structure

Figure 7 shows the effect of track structure on the ground vibration. As described above, vibration-reduction tracks were laid in the analysis target section. As a comparison, a track structure in which the spring constant of the resilient material was hardened enough to be considered as the directly-fastened track, was installed. With the directly-fastened tracks, vibration increased by 5 to 25 dB in the frequency bands 10 Hz or higher, showing the effect of the vibration-reduction tracks in the target section. However, in the frequency bands lower than 10 Hz, the response was almost equal.

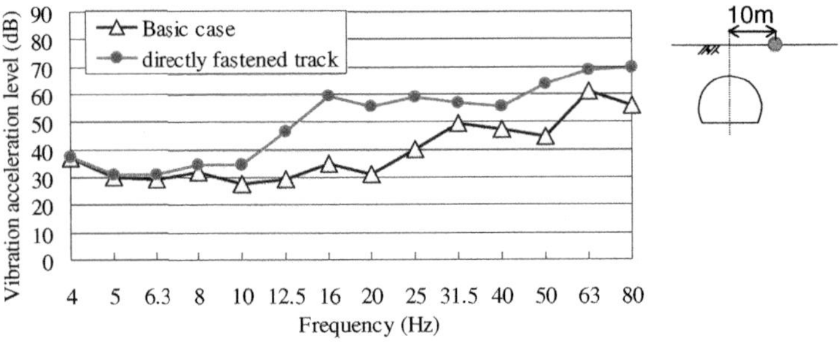

Fig. 7. Effect of track structure at the point 10m apart from the tunnel center (color figure online)

4 Conclusion

(1) A three-dimensional dynamic interaction analysis method which enabled modeling the entire system of vehicles, tracks, structures and the ground was established, and most of the actual phenomena could be reproduced.
(2) The results of the ground vibration analysis using the ground properties and track structures as parameters revealed that multiplying Vs by 1.3 or 0.7 caused fluctuation of ground vibration from -15 dB to 25 dB and a change of the damping constant from 2 to 5% reduced the response by 15dB at the maximum in a certain frequency band under the conditions of this analysis. In the comparison between the vibration-reduction track and the directly-fastened track, the vibration-reduction track turned out to reduce the ground vibration by 5 to 25 dB in the frequency band 10 Hz or higher.

We completed this study under commissioned research by Japan Railway Construction, Transport and Technology Agency.

References

[1] Yosioka, O., et al.: Prediction Analysis of Train Induced Ground Vibrations Using the Equivalent Excitation Force. RTRT Report 10(2), 41–46 (1996)
[2] Wakui, H., et al.: Dynamic Interaction Analysis for Railway Vehicles and Structures. Journal of Structural Mechanics and Earthquake Engineering 513(I-31), 129–138 (1995)
[3] Tajimi, H., et al.: Dynamic Analysis of Soil-Structure Interaction by the Thin Layered Element Method. Transactions of the Architectural Institute of Japan 243, 41–51 (1976)
[4] Takemiya, H.: Simulation of track-ground vibrations due to high-speed train: the case of X-2000 at Ledsgard. Journal of Sound and Vibration 261, 503–526 (2003)
[5] Imai, T.: Relation and the use example of N value and S wave velocity. The Foundation Engineering and Equipment, Monthly, 70–76 (1982)

The Effects of a Second Tunnel on the Propagation of Ground-Borne Vibration from an Underground Railway

K.A. Kuo[1], M.F.M. Hussein[2], and H.E.M. Hunt[1]

[1] University of Cambridge Engineering Department
Trumpington St., Cambridge, United Kingdom
Tel.: +44 1223 332730; Fax: +44 1223 332662
e-mail: kan26@cam.ac.uk
[2] University of Nottingham

Summary

Accurate predictions of ground-borne vibration levels in the vicinity of an underground railway are greatly sought in modern urban centers. Yet the complexity involved in simulating the underground environment means that it is necessary to make simplifying assumptions about this environment. One such commonly-made assumption is to model the railway as a single tunnel, despite many underground railway lines consisting of twin-bored tunnels.

A unique model for two tunnels embedded in a homogeneous, elastic full space is developed. The vibration response of this two-tunnel system is calculated using the superposition of two displacement fields: one resulting from the forces acting on the invert of a single tunnel, and the other resulting from the interaction between the tunnels. By partitioning of the stresses into symmetric and anti-symmetric mode number components using Fourier decomposition, these two displacement fields can by calculated with minimal computational requirements.

The significance of the interactions between twin-tunnels is quantified by calculating the insertion gains that result from the existence of a second tunnel. The insertion-gain results are shown to be localized and highly dependent on frequency, tunnel orientation and tunnel thickness. At some locations, the magnitude of these insertion gains is greater than 20dB. This demonstrates that a high degree of inaccuracy exists in any surface vibration-prediction model that includes only one of the two tunnels.

1 Introduction

The propagation of ground-borne vibration from underground railways can be irritating to those working and living in the vicinity of a railway. While these vibrations may not be significant enough to result in structural damage, their effect

T. Maeda et al. (Eds.): Noise and Vibration Mitigation for Rail Trans. Sys., NNFM 118, pp. 307–314.
springerlink.com

on the health of occupants and the day-to-day operation of vibration-sensitive premises has been recognized as an important environmental issue. There is a strong demand for accurate, user-friendly models that can be used by railway designers to predict underground-railway vibration levels based on train, track, soil and building parameters. Many of the models that exist today make simplifying assumptions about this environment, and one such assumption which is commonly made is to ignore the presence of a twin tunnel.

In most cases twin tunnels are oriented side by side, but occasionally the tunnels are piggy-back, with one on top of the other. Numerical models and scale models of twin tunnels exist in the literature [1-3] for the purposes of determining the static stresses and strains produced during and post- excavation. To date, the only evidence in the literature of a dynamic model which accounts for the vibration interaction between neighboring tunnels is a wave number FE-BE model [4]. This model compares the response of a large, single-bore, double-track tunnel with a pair of single, twin-track tunnels embedded in an elastic half space.

This paper describes the formulation of a novel solution for two parallel, infinite tunnels of circular cross-section embedded in a homogeneous, elastic soil full space. This twin-tunnel model is an extension of the single-tunnel model developed by Forrest [5] and furthered by Hussein [6]. Whilst various single-tunnel models exist, this model is unique in that it has the accuracy associated with being a 2.5D model yet at the same time does not require the extended computation times associated with coupled FE-BE approaches. This single-tunnel model has been proved to be an efficient and accurate method for calculation of vibration in a full space from underground railways [7]. Methods exist for adapting this model to obtain vibration predictions in a homogeneous half space or layered half space without significant loss of computational efficiency [8]. These same methods can be applied to the two-tunnel model described here.

2 Modeling of Twin Tunnels

The twin-tunnel model consists of two tunnels with circular cross-sections, of mean radius a_1 and a_2 and thickness h_1 and h_2 , respectively embedded in an elastic full space. The tunnels are separated by a distance c, with tunnel 2 located at an angle α from tunnel 1. Two right-handed cylindrical coordinate systems are defined: (r,θ,z) and (s,φ,z) defined clockwise on tunnel 1 and tunnel 2, respectively.

2.1 Dynamic Train Forces

Each of these tunnels is subjected to a set of known dynamic train forces acting directly on the tunnel invert. These known dynamic train forces result in stresses $\tilde{\boldsymbol{P}}_1$ and $\tilde{\boldsymbol{P}}_2$, where the subscript indicates that the stresses act on the inverts of tunnel 1 and tunnel 2, respectively, and the tilde indicates that the variable is defined in the wave number (ξ) domain. All variables, unless otherwise noted, are defined in the frequency domain. The known dynamic forces result in up to three stress components acting on the inner surface of the tunnel in the direction of the coordinate axes. Furthermore, using Fourier decomposition these stresses can be expressed as a combination of N symmetric and anti-symmetric cylindrical mode shapes, where superscripts 1 and 2 denote symmetry and antisymmetry, respectively, and subscript n is used to denote the nth mode shape contributions. For example, the known traction vectors acting on the inside of tunnel 1 are $\tilde{\boldsymbol{P}}_{1n}^{1}$ and $\tilde{\boldsymbol{P}}_{1n}^{2}$, and can be written as

$$\tilde{\boldsymbol{P}}_{1n}^{1} = \sum_{n=0}^{N} \begin{Bmatrix} P_{rn}^{1} \cos n\theta \\ P_{\theta n}^{1} \sin n\theta \\ P_{zn}^{1} \cos n\theta \end{Bmatrix} e^{i(\omega t+\xi z)}$$
$$\tilde{\boldsymbol{P}}_{1n}^{2} = \sum_{n=0}^{N} \begin{Bmatrix} P_{rn}^{2} \sin n\theta \\ P_{\theta n}^{2} \cos n\theta \\ P_{zn}^{2} \sin n\theta \end{Bmatrix} e^{i(\omega t+\xi z)}. \tag{2.1}$$

2.2 Superposition of Displacement Fields

The vibration response of the twin-tunnel system can be written as the superposition of two displacement fields. One displacement field is the result of stresses acting on the invert of a single tunnel (similar to a direct field), while the other displacement field is the result of interactions between the twin tunnels (similar to a scattered field). Hence the tractions resulting from the known dynamic forces acting on the invert of each tunnel can be written as the sum of two contributions: those tractions acting on the invert of a single tunnel $\tilde{\boldsymbol{F}}_1$ and $\tilde{\boldsymbol{F}}_2$, and those tractions that represent the motion induced by the neighboring tunnel $\tilde{\boldsymbol{G}}_1$ and $\tilde{\boldsymbol{G}}_2$. By apportioning the known tractions in this way, it is only necessary to consider a one-tunnel system at any time. This is expressed in Equation 2.2:

$$\begin{aligned} \tilde{\boldsymbol{P}}_{1n}^{1} &= \boldsymbol{F}_{1n}^{1} + \boldsymbol{G}_{1n}^{1} \\ \tilde{\boldsymbol{P}}_{1n}^{2} &= \boldsymbol{F}_{1n}^{2} + \boldsymbol{G}_{1n}^{2} \\ \tilde{\boldsymbol{P}}_{2n}^{1} &= \boldsymbol{F}_{2n}^{1} + \boldsymbol{G}_{2n}^{1} \\ \tilde{\boldsymbol{P}}_{2n}^{2} &= \boldsymbol{F}_{2n}^{2} + \boldsymbol{G}_{2n}^{2}. \end{aligned} \tag{2.2}$$

These equations must include every coordinate direction, every mode number n, every wave number ξ and every frequency. To solve these equations, the right-hand terms must be reduced to a function of the variable set $\tilde{\boldsymbol{F}}$. This is done by writing the tractions that represent the motion induced by the neighboring tunnel as a function of the tractions acting directly on the inside of the neighboring tunnel. In order to calculate the tractions that represent the motion induced by the neighboring tunnel, the equations for the one-tunnel model are used to determine the displacements around the virtual surface of the second tunnel. Thus to calculate $\tilde{\boldsymbol{G}}_2$, the traction vector $\tilde{\boldsymbol{F}}_1$ is applied to the invert of tunnel 1, the one-tunnel model is used to calculate the resulting displacements around the virtual surface of tunnel 2, and then the one-tunnel model is used to transform these displacements into the equivalent traction vector ($\tilde{\boldsymbol{G}}_2$) acting on the invert of tunnel 2.

2.3 Calculating Displacements around a Virtual Surface

In order to calculate the displacements at some point in the soil, it is first necessary to calculate the tractions at the tunnel-soil interface. Referring to the single-tunnel model in Forrest [5] and Hussein [6], the equation for a thin-walled cylinder can be used to relate the force acting on the tunnel invert and the tunnel-soil interface stresses. Once this relationship has been established, the equations of the soil continuum are used to calculate the displacements resulting at a point on the virtual surface of tunnel 2. These displacements are defined in the (r,θ,z) directions, but are now converted to the (s,φ,z) directions to obtain compatibility with the tunnel 2 coordinate system. The equations for converting between these coordinate systems are obtained by considering vector geometry.

At this stage the displacements in the (s,φ,z) directions have been calculated as a function of $\tilde{\boldsymbol{F}}_1$ at a single point on the virtual surface of tunnel 2. It is necessary to calculate these displacements at a series of M evenly-spaced points around the virtual surface. Once they have been determined, Fourier decomposition is used to partition the displacements into symmetric and anti-symmetric mode number contributions. These displacement contributions are finally converted to the equivalent tractions $\tilde{\boldsymbol{G}}_2$ acting on the inner surface of tunnel 2 using the equation for a thin-walled cylinder. Hence the end-result of this formulation is that the traction vectors $\tilde{\boldsymbol{G}}_{2n}^1$ and $\tilde{\boldsymbol{G}}_{2n}^2$ are expressed as linear functions of the traction vectors $\tilde{\boldsymbol{F}}_{1n}^1$ and $\tilde{\boldsymbol{F}}_{1n}^2$: $g_3(\tilde{\boldsymbol{F}}_{1n}^1,\tilde{\boldsymbol{F}}_{1n}^2)$ and $g_4(\tilde{\boldsymbol{F}}_{1n}^1,\tilde{\boldsymbol{F}}_{1n}^2)$, respectively. Using the same procedure, the traction vectors $\tilde{\boldsymbol{G}}_{1n}^1$ and $\tilde{\boldsymbol{G}}_{1n}^2$ are expressed as linear functions of the traction vectors $\tilde{\boldsymbol{F}}_{2n}^1$ and $\tilde{\boldsymbol{F}}_{2n}^2$: $g_1(\tilde{\boldsymbol{F}}_{2n}^1,\tilde{\boldsymbol{F}}_{2n}^2)$ and $g_2(\tilde{\boldsymbol{F}}_{2n}^1,\tilde{\boldsymbol{F}}_{2n}^2)$, respectively. Equation. 2.2 can now be written as a series of simultaneous equations:

$$
\begin{aligned}
\tilde{\boldsymbol{P}}_{1n}^1 &= \boldsymbol{F}_{1n}^1 + g_1(\tilde{\boldsymbol{F}}_{2n}^1,\tilde{\boldsymbol{F}}_{2n}^2)\\
\tilde{\boldsymbol{P}}_{1n}^2 &= \boldsymbol{F}_{1n}^2 + g_2(\tilde{\boldsymbol{F}}_{2n}^1,\tilde{\boldsymbol{F}}_{2n}^2)\\
\tilde{\boldsymbol{P}}_{2n}^1 &= \boldsymbol{F}_{2n}^1 + g_3(\tilde{\boldsymbol{F}}_{1n}^1,\tilde{\boldsymbol{F}}_{1n}^2)\\
\tilde{\boldsymbol{P}}_{2n}^2 &= \boldsymbol{F}_{2n}^2 + g_4(\tilde{\boldsymbol{F}}_{1n}^1,\tilde{\boldsymbol{F}}_{1n}^2).
\end{aligned}
\tag{2.3}
$$

The solution of these simultaneous equations represents the traction vectors $\tilde{\boldsymbol{F}}_1$ and $\tilde{\boldsymbol{F}}_2$ applied to each of the tunnel inverts in the single-tunnel models to replicate the displacement field produced by the known traction vectors $\tilde{\boldsymbol{P}}_1$ and $\tilde{\boldsymbol{P}}_2$ applied to the tunnel inverts in the twin-tunnel model. The displacement field of the twin-tunnel model is thus the superposition of the displacement field produced by the traction vector $\tilde{\boldsymbol{F}}_1$ acting on tunnel 1 in a single-tunnel model, and the displacement field produced by the traction vector $\tilde{\boldsymbol{F}}_2$ acting on tunnel 2 in a single-tunnel model.

3 Results and Discussion

The above formulation is implemented in Matlab, and the displacement fields are presented in terms of the vertical or horizontal displacements, always in terms of $dB_{ref}(1m)$. Unless otherwise specified, the tunnel and soil parameters used to calculate the results are shown in Table 1.

Table 1. Parameter values used for the twin-tunnel model

Tunnel Parameters	Soil Parameters
Young's Modulus E = 50 GPa	Lame's First Parameter λ = 360 MPa
Density ρ = 2500 kg m^{-3}	Lame's Second Parameter μ = 90 MPa
Poisson's ratio υ = 0.3	Density ρ = 2250 kg m^{-3}
Separation distance c = 10 m	Shear Modulus Damping Ratio η_G = 0.06
Separation angle $\alpha = \pi/2$ (side-by-side)	-
Radius $a_1 = a_2$ = 3 m	-
Thickness $h_1 = h_2$ = 0.25 m	-

It is necessary to determine the number of mode numbers (N) and points around the virtual tunnel surface (M) required to capture the dynamic behavior of the system. Convergence in both the near-field and the far-field is achieved by using the first 11 mode numbers, and 44 points around the virtual tunnel surface.

The main aim of this modeling is to quantify the inaccuracy that exists in vibration-prediction models that include only one of the two tunnels. This inaccuracy is expressed here in terms of insertion gain, which represents the ratio of the vibration response of a twin-tunnel model to that of a single-tunnel model. Insertion gain is illustrated in Fig. 1, and can be seen to be particularly useful at locating regions which are susceptible to inaccuracies resulting from a single-tunnel assumption. In the case shown in Fig. 1, in the region above the right-hand tunnel the twin-tunnel model predicts vibration levels in the order of 20dB higher than those predicted by the single-tunnel model. A significant shadow region to the left of the left-hand tunnel is also observed.

Figure 2 presents the vertical-displacement insertion gain for two different tunnel orientations as a function of frequency and position along a horizontal line located 15m above the centre of the loaded tunnel. In both cases a unit harmonic point load has been applied radially to the base of tunnel 1. For the side-by-side tunnels, the insertion gain is seen to be highly dependent on both frequency and measurement position. There is no apparent trend in the distribution, which is due to the scattering effect of the second tunnel. For the piggy-back tunnels, it can be seen that the maximum negative insertion gains occur in the region stretching directly above the tunnel invert.

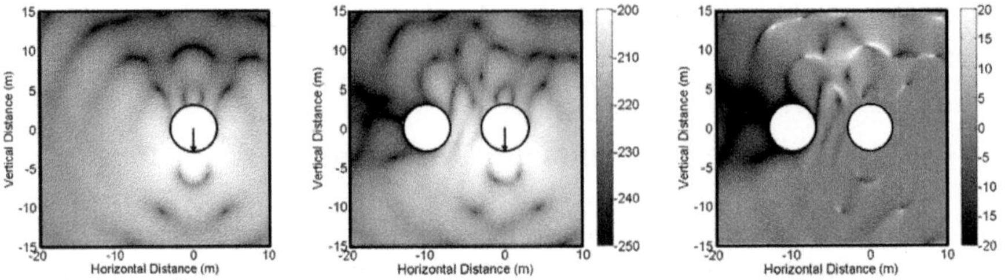

(a) (b) (c)

Fig. 1. The vertical displacement field (dB) produced at 60Hz by a unit vertical point force applied to a tunnel invert calculated using (a) a single-tunnel model, and (b) a twin-tunnel model. The difference between these two displacement fields is shown in (c), the insertion gain.

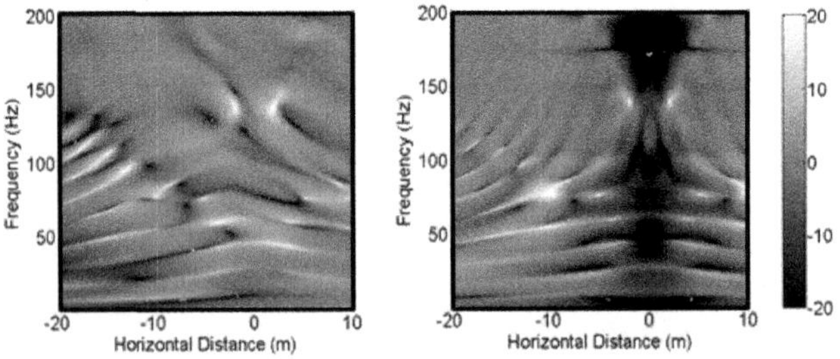

(a) (b)

Fig. 2. Insertion gains as a function of frequency and position (indicated by the dash-dot line), in terms of vertical displacements for: (a) side-by-side tunnels; and (b) piggy-back tunnels.

Figure 3 presents the vertical displacement insertion gains at 20Hz and 60Hz calculated as a function of α, where α is the angle of tunnel 2 measured from tunnel 1. The loading is again a unit harmonic point load applied radially to the base of tunnel 1. A number of vertical lines of large insertion gain can be seen in this figure. By examining the displacement plots of the single-tunnel model, the location of these lines can be seen to correspond with the troughs in the response calculated using the single-tunnel model. Thus the addition of a twin-tunnel has effectively smoothed the single-tunnel displacement field to create a more uniform response. Three additional lines are plotted on these graphs: the solid line indicates the location of the measuring point lying on the line joining the centers of the two tunnels, while the dashed lines on either side indicate the locations of the measuring points lying on the lines joining the extreme edges of the two tunnels. The overall trend in the region enclosed by these lines is large, negative insertion gains, indicating that the second tunnel acts as to shield this region from the vibrations propagated from tunnel 1. Apart from this trend, the insertion gains are again seen to be highly dependent on frequency and measurement position.

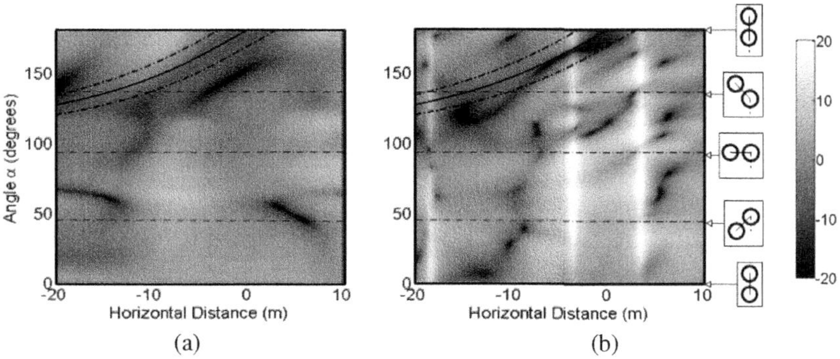

Fig. 3. Vertical insertion gains as a function of angle α and position at: (a) 20Hz; and (b) 60Hz.

The final figure presented here, Fig. 4, compares the vertical insertion gains for twin tunnels of varying invert thickness h. At low frequencies, such as at 20Hz shown in Fig. 4(a), there is little separation between the responses and hence the tunnel thickness has little influence on the general shape of the insertion-gain plot. At frequencies above approximately 40Hz, the tunnel thickness has a stronger influence on the general shape of the insertion gain plot, and there is no consistent correlation observed between the thickness of the tunnels and the level of insertion gain. For example, at 60Hz shown in Fig. 4(b), there is greater than 10dB variation at given measurement positions.

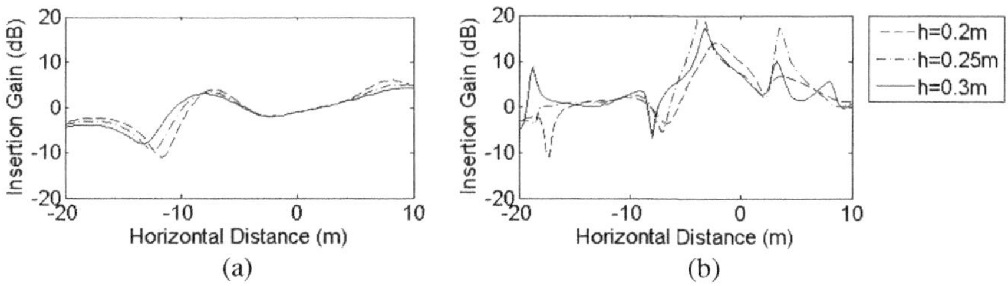

Fig. 4. Insertion gains as a function of position and tunnel thickness calculated at: (a) 20Hz; and (b) 60Hz.

4 Conclusions

This paper presents the formulation of a unique model for underground-railway vibration that includes the interaction between neighboring tunnels. Each of these tunnels is subject to a set of known dynamic train forces that can be written as the sum of those tractions acting on the invert of a single tunnel and those tractions that represent the motion induced by a neighboring tunnel. By apportioning the forces in this way, the vibration response of a two-tunnel system can be written as a linear combination of displacement fields produced by a single-tunnel system. Using

Fourier decomposition, stresses are partitioned into symmetric and anti-symmetric mode number contributions to minimize computation times.

Analysis of the vibration fields produced over a range of frequencies, tunnel orientations and tunnel geometries is conducted, and it is observed that the interaction between neighboring tunnels is highly significant, at times in the order of 20dB. This demonstrates that a high degree of inaccuracy exists in any surface vibration-prediction model that includes only one of the twin tunnels. The results of this modeling also indicate that the presence of other underground inclusions, such as buried services, can have a significant effect on the ground-vibration field. It is recommended that all future model predicting vibration levels from underground railways include the interaction between neighboring tunnels.

References

[1] Ng, C.W.W., et al.: Three-dimensional numerical investigations of New Austrian Tunneling Method (NATM) twin tunnel interactions. Canadian Geotechnical Journal 41, 523–539 (2004)

[2] Addenbrooke, T.I., Potts, D.M.: Twin tunnel construction – ground movements and lining behaviour. Geotechnical Aspects of Underground Construction in Soft Ground, 441–446 (1996)

[3] Chu, B.L., et al.: Mechanical behavior of a twin-tunnel in multi-layered formations. Tunnelling and Underground Space Technology 22, 351–362 (2007)

[4] Sheng, X., et al.: Modelling ground vibration from railways using wavenumber finite- and boundary-element methods. Proceedings of the Royal Society A – Mathematical Physical and Engineering Sciences 461, 2043–2070 (2005)

[5] Forrest, J.A., Hunt, H.E.M.: A three-dimensional tunnel model for calculation of train-induced ground vibration. Journal of Sound and Vibration 294, 678–705 (2006)

[6] Hussein, M.F.M., Hunt, H.E.M.: A numerical model for calculating vibration from a railway tunnel embedded in a full-space. Journal of Sound and Vibration 305, 401–431 (2007)

[7] Gupta, S., et al.: A comparison of two numerical models for the prediction of vibrations from underground railway traffic. Soil Dynamics and Earthquake Engineering 27, 608–624 (2007)

[8] Hussein, M.F.M., et al.: An efficient model for calculating vibration from a railway tunnel buried in a half-space. In: Proceedings of ICSV14, Cairns, Australia (2007)

A Numerical Investigation of Railway Ground Vibration Mitigation Using a Trench in a Layered Soil

A. Garcia-Bennett[1], C.J.C. Jones[2], and D.J. Thompson[2]

[1] Garcia-BBM,
2 Los Arcos, 28770 Colmenar Viejo, Madrid, Spain
Tel.: +34 918 450 441; Fax: +34 918 453 499
e-mail: ana.garcia@garcia-bbm.com
[2] Institute of Sound and Vibration Research,
University of Southampton, Highfield, Southampton, UK

Summary

Ground vibration from railway lines at grade often has its highest amplitudes at frequencies around or even below 10 Hz. A ground vibration 'barrier' (*i.e.* an open or in-filled trench) would be an attractive means of mitigation of vibration since it would not interfere with the engineering or operation of the track. However, before detailed research, a preliminary theoretical study was conducted to test the idea since simple considerations suggest that barriers might have to be unfeasibly deep to be effective. However, the layered ground structure is known to be important in determining the spectrum of transmitted vibration and recent practical investigations suggest that trenches of reasonable depth may be effective. As a prelude to more detailed investigations, an investigation has been carried out using a two-dimensional coupled finite element (FE) and boundary element (BE) model. The results indicate that, in the peak frequency range of vibration transmission, for a typical alluvial valley site, reductions of 4 dB at 10 Hz and 8 dB at 14 Hz might be achieved with a 5 m deep trench.

1 Introduction

The railways for a long time have been concerned about ground vibration from trains running on surface lines. This type of vibration has significant frequency components from about 4 Hz to 80 Hz coinciding with the frequency range of 'whole body' vibration perception. Vibration propagating from the track to line-side properties as surface waves often has its highest amplitude components around or even below 10 Hz, especially in the case of heavy axle-load freight trains. It has been shown that annoyance to line-side residents may be caused

T. Maeda et al. (Eds.): Noise and Vibration Mitigation for Rail Trans. Sys., NNFM 118, pp. 315–322.
springerlink.com

when the vibration level is just perceptible and surveys find that up to 35% of those living within 100 m of railway lines can indeed feel vibration [1, 2].

One possible measure to mitigate vibration is to construct a vibration-reflecting barrier in the ground next to the track. This idea is attractive to railways as the interference with track operations and maintenance would be minimal. Such a barrier must present a change of characteristic impedance in the propagation medium, the soil, to cause reflection of waves. This might be achieved in practice either with an open trench or with an in-filled trench of material either stiffer or softer than the soil. Although there are practical difficulties of soil stability, an open trench presents the best reflective properties and the controlling geometric parameter is merely its depth. Simple open trenches have therefore been chosen for this initial study, aimed at testing whether barrier measures might be effective.

1.1 Background

From field experiments in the 1960s, and considering the ground as a homogeneous half-space, Richart, Hall and Woods [3] stated an empirical 'rule of thumb' that a trench could reduce surface vibration propagation significantly at frequencies for which the ratio of the depth of the trench to the Rayleigh wave length of propagating vibration (d/λ_R) was greater than about 0.6.

For railway vibration at around 10 Hz, as stated above, and for normal soft soils with shear velocities of the order of, say, 250 ms^{-1}, efficient barriers would have to be around 15 m deep. This is hardly feasible in engineering terms and therefore certainly not from an economic point of view. However, more recent research on railway vibration based on measurements and computer-modelling techniques have shown that railway-vibration spectra owe much of their character at the offending frequencies to a cut-on in vibration propagation due to the layered structure of the soil in the first few metres of depth [4].

At very low frequency, wavelengths of vibration are long in comparison to the upper soil layer depths. The wave shape therefore penetrates deep into the soil and its propagation characteristics are controlled by the stiffer substratum material. As the wavelengths become shorter at higher frequencies, a propagating mode of vibration appears within the upper layers. For this mode of wave propagation, the wave speed and vibration response amplitudes are controlled by the material properties of the softer material. Thus a rise in the response-amplitude spectrum takes place at some frequency. Although the frequency of this cut-on depends on the properties of the site, it often occurs between about 10 and 20 Hz and controls the peak in the spectrum of vibration propagated to the line-side. Although this effect is important for railway vibration, few previous studies of ground barriers take it into account.

There is some hope from recent practical investigations by railways [5] (and [6] for a tramway) that barriers of reasonable depth could be effective. However, it is the nature of *ad hoc* practical tests that they are difficult to control and their conclusions should be backed by theoretical explanation for their evidence to be taken as reliable.

1.2 Previous Studies

Only a few studies of ground-vibration barriers have been conducted to examine Richart, Hall and Woods' guidance using modern numerical-modelling techniques. Beskos *et al.* [7] used the two-dimensional boundary element method assuming homogeneous soil conditions to study the effect of rectangular trenches. Their parametric study confirmed that a depth according to $d/\lambda_R > 0.6$ led to an amplitude reduction to 25%, *i.e.* a 12 dB reduction. They found open trenches to be more effective than in-filled trenches and that, in the latter case, the width of the trench was also important. However for an open trench, the depth was the only important geometric factor. Ahmad and Al-Hussaini [8] carried out a similar study covering the frequency range from 5 to 200 Hz, deriving more detailed design rules for in-filled trenches. They also extended the study to a ground consisting of a layer over a half-space parameterized by the shear wave velocity ratio, v_{s1}/v_{s2}. The studies were limited to trenches placed close to the receiver rather than the source. Yang and Hung [9] used a two-dimensional model based on FE coupled with infinite FE to study trenches close to the source, more suited to the considerations of railway companies. They found that open trenches were better at isolating vertical than lateral vibration and also that mitigation could be compromised when the trench was very close to the source. Their explanation was that body waves carrying a significant fraction of the energy close to the source could pass below the trench.

2 The Approach Taken in the Current Work

In the light of the past work, and the need to contain the scope of the current work, it was decided only to study open trenches defined by the geometrical parameters of distance between the source and the depth. Two layered ground types were studied based on soil properties and layer depths validated by comparison of measured transfer frequency response functions and calculations. Here, for reasons of space, only results for a site comprising a 2 m layer of alluvial soil on deep river gravels are presented. This situation provides a good test of the effects of high-shear velocity contrast in the layer and substratum to compare with half-space simplification. The properties of the soils are presented in Table 1.

A coupled two-dimensional FE-BE modelling approach [10] was chosen for the study. This is an appropriate choice from the various modelling methods available to the authors since low-computing times and the ability to incorporate arbitrary geometry are required. For the stated aims of the work, train and track modelling and three-dimensional propagation are not so important and would add complexity to the study. Since a two-dimensional model is used, no investigation is made of the necessary extent of a trench along the railway.

Figure 1 shows the basic geometry of the model implemented. The ground is modelled with a 2 m deep layer to a distance of 50 m to the left and 110 m to the right of the centre-line of a track. This model size ensures negligible effects of the boundaries and allows a suitable extent of distance over which the vibration response is examined. Only responses on the right-hand side of the track were used. The model error due to element size, edge effects *etc*. has been verified to be negligible by comparison with an exact model for a layered half-space ground based on the theory in [11].

The study examined various depths of trench to 10 m. This is around the limit of practicable engineering. Even then, it would require retaining structure and/or sloping sides, but studies are agreed that width, geometry and structures are of secondary importance to depth for open trenches. For the present aim of examining feasibility, therefore, a constant width of 1 m is used and no retaining structure modelled.

Different possible positions of the trench have been considered as shown in the figure. It was found that the 5.75 m position produced better results than the closer positions. The further trench positions (at 10 and 40 m) showed similar benefits for distances beyond them to the 5.75 m position. This means that a 5.75 m position would treat the further tracks of multiple-track lines. However, positions further from the near track would not be accommodated within railway land. For brevity therefore, only results from the trench position, 5.75 m from the near track centre-line, are presented here.

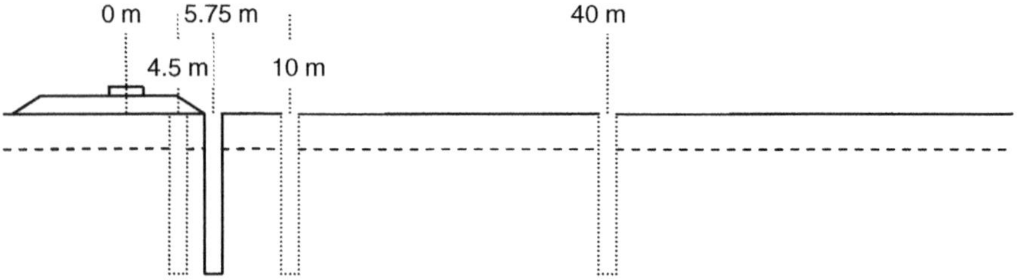

Fig. 1. The geometry of the two-dimensional FE-BE model indicating alternative trench locations.

Table 1. Ground layer properties.

Layer	Depth (m)	P-wave speed m/s	S-wave Speed m/s	Density (kg/m^3)	Loss Factor	Young's Modulus (MNm^{-2})	Poisson's ratio
1	2.0	343	78	1500	0.1	26.88	0.473
2	-	1653	245	2000	0.1	357.5	0.489

3 Results

The model produces the complex amplitude of the response in the vertical and lateral directions for a unit amplitude load at the track. In order to simplify the presentation of vibration response and the insertion gain (amplitude reduction factor), the response has been reduced to a 'pseudo-response', the root mean square of the vertical and lateral vibration velocity amplitudes [10]. A summary presentation of frequency and distance variation of this and the insertion gain is shown in Fig. 2 for a 10 m deep trench in the layered alluvial site.

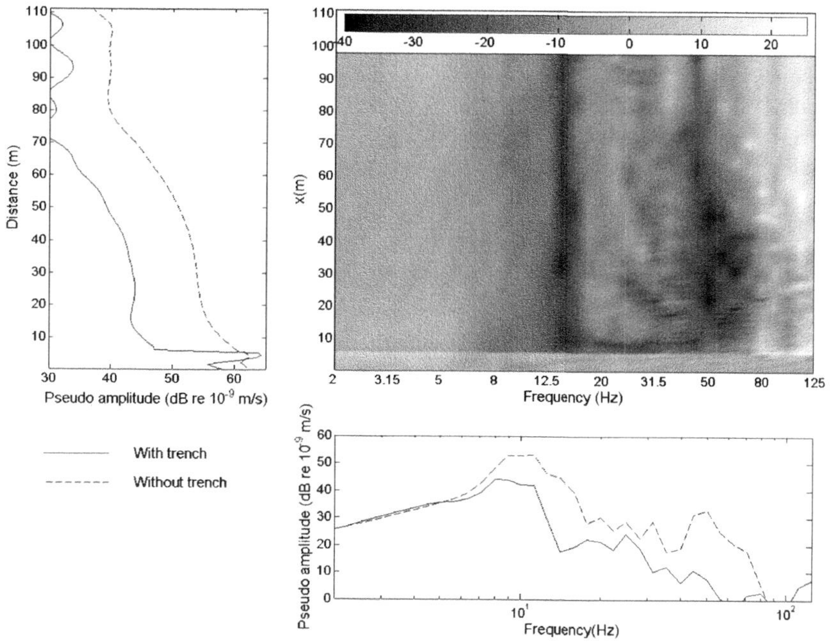

Fig. 2. The summary of results comparing the vibration response with and without a 10 m deep trench at 5.75 m from the track centre-line. Top left: variation with distance at 10 Hz; top right: insertion gain; bottom right: variation with frequency at 40 m

The contour plot of Fig. 2 presents the insertion gain due to the trench with the graph below showing the comparative frequency response, and to the side, the variation with distance of the pseudo-response.

The figure shows increase in vibration due to the cut-on of propagation in the upper layer between about 5 and 10 Hz; leading to a broad peak in response between about 9 and 12 Hz. This range is likely to be reflected as the most important components in a spectrum of train vibration. The 10 m trench causes a worthwhile reduction of around 10 dB at these frequencies. For higher frequencies up to 80 Hz, the vibration is also reduced. Above 80 Hz the response is more than 40 dB down on the peak levels and apparent amplifications may be due to numerical error and, in any case, do not matter at this very low level. The mitigation of vibration by the trench is maintained for all distances up to 110 m from the trench although this aspect should be investigated more carefully with a three-dimensional (or 2.5-dimensional) model.

Figure 3 presents results where a homogeneous half-space ground is assumed having the properties of the lower material. Material 2 has been assumed as it is this that exists at the depth of the trench and controls the low-frequency asymptote of vibration response. The figure shows that a much lower absolute level of response would be expected. However, no significant insertion gain emerges until a frequency of 8 Hz rather than 5 Hz in the case of the layered-ground model. The frequency corresponding to $d/\lambda_R = 0.6$ in this case is about 14 Hz. At this frequency, for the specific railway and trench geometry, only about 8 dB of the expected 12 dB insertion gain [3] is actually predicted by the model. At 14 Hz, and indeed for the important 9 to 12 Hz peak, Fig. 2 for the layered-ground assumption actually shows a more promising insertion gain albeit without it continuing to increase with frequency as the insertion gain for the homogeneous half-space does.

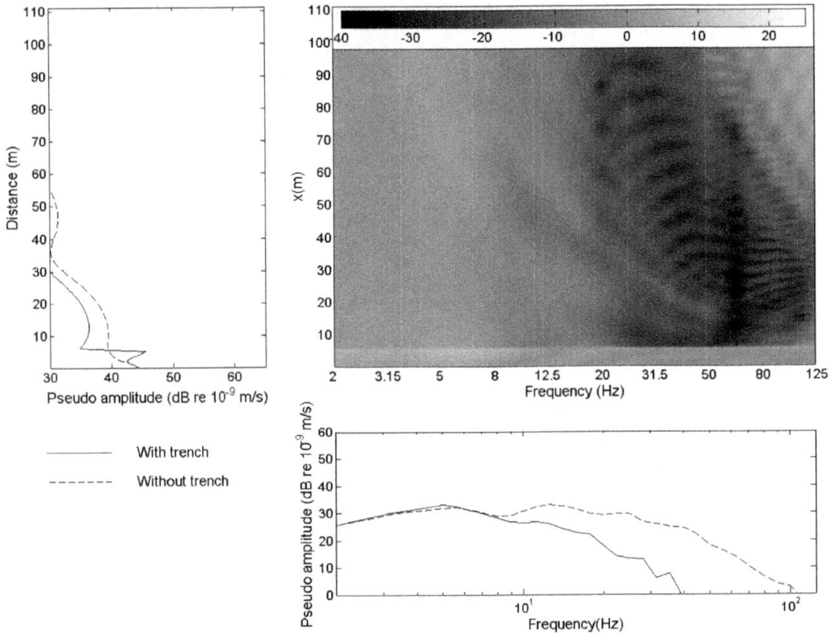

Fig. 3. The summary of results with and without a 10 m deep trench at 5.75 m in a half-space as layer 2, Table 2.1. Top left: variation with distance at 10 Hz; top right: insertion gain; bottom right: variation with frequency at 40 m

The trench depth of 10 m is on the very limit of what could be feasible to construct. Figure 4 therefore presents responses versus distance at 10 and 14 Hz for a range of trench depths in order to test whether worthwhile benefits are available for more reasonable depths of trench. At 5 Hz (not shown) the differences in response are small. The 10 m trench is even predicted to exacerbate vibration at short distances up to 30 m (as much as 4 dB at around 15 m).

At 10 Hz (Fig. 4), the 10 m trench is predicted to give at least 12 dB of mitigation from the trench to about 70 m. However, halving the depth of the trench to a more achievable 5 m depth lowers this to about 4 dB from 20 m outward although there are greater reductions from the trench to this point. Trenches of 2 and 3 m depth produce almost as much benefit as the 5 m case.

At 14 Hz, insertion losses rise more consistently with trench depth and a worthwhile 8 dB is predicted for a reasonably feasible 5 m depth. At this frequency, even a 2 or 3 m trench is predicted to provide a reduction of vibration amplitude of more than 6 dB.

Figure 5 presents the prediction summary for the 5 m trench case. This reveals that, although useful reductions are predicted around the peak of response, at 20 Hz and for higher frequencies, there is no consistent mitigation, with amplification at some frequencies. However, this may be acceptable as the response is more than 20 dB below the peak.

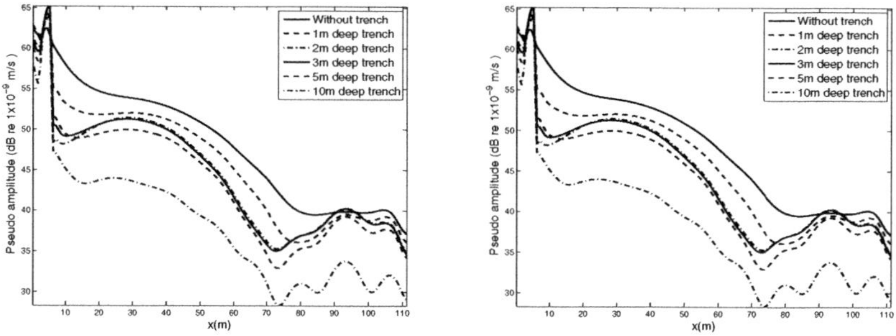

Fig. 4. Pseudo-resultant response as a function of distance for different trench depths at (a) 10 Hz and (b) 14 Hz.

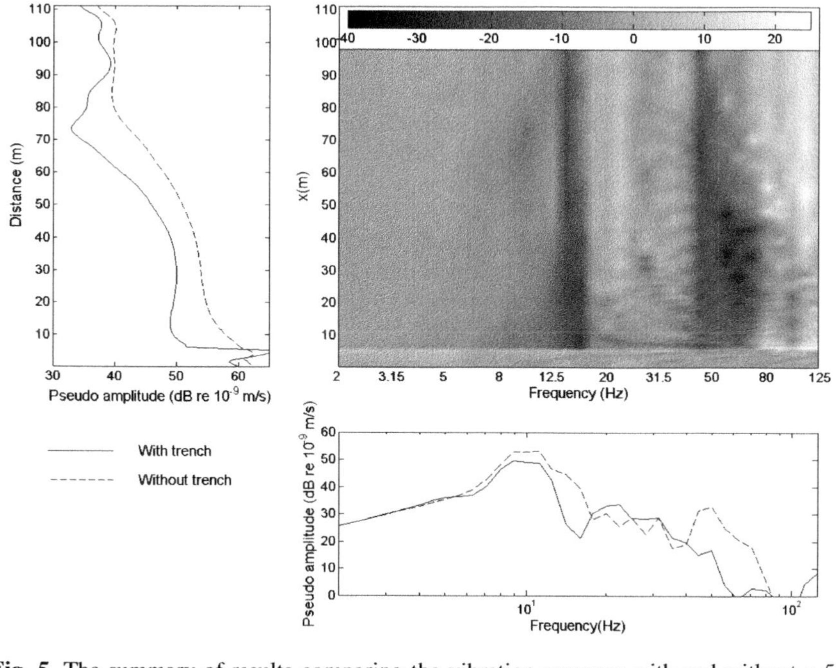

Fig. 5. The summary of results comparing the vibration response with and without a 5 m deep trench at 5.75 m from the track centre-line. Top left: variation with distance at 10 Hz; top right: insertion gain; bottom right: variation with frequency at 40 m

4 Conclusions

A preliminary study to indicate whether 'ground barriers' of feasible engineering depth might produce worthwhile reductions in surface propagating vibration from trains has been carried out. For the alluvial soil site studied, the results indicate that a reduction of 4 dB at 10 Hz and 8 dB at 14 Hz could be expected from a trench 5 m deep. Even a two or three metre trench provides significant benefits at 14 Hz over useful line-side distances. More specific studies with 2.5- or three-dimensional models for more specific engineering barrier structures and other soils are warranted.

References

[1] Woodroof, H.J., Griffin, M.J.: A survey of the effect if railway-induced building vibration on the community. ISVR TR 160, University of Southampton (1987)

[2] Yokoshima, S.: A study on factors constituting annoyance due to Shinkansen railway vibration. Journal of Architecture, Planning and Environmental Engineering 526 (1999)

[3] Richart, F.E., Hall, J.R., Woods, R.D.: Vibration of soils and foundations. Prentice-Hall, Englewood Cliffs (1970)

[4] Jones, C.J.C.: Low frequency ground vibration. In: Railway Noise and Vibration: Mechanisms, Modelling and Means of Control, ch. 12. D.J. Thompson, Elsevier, Amsterdam (2009)

[5] Müller, R.: Mitigation measures for open lines against vibration and ground-borne noise: a Swiss overview. In: Schulte-Werning, B., et al. (eds.) Noise and vibration mitigation for rail transportation systems, NNFM 99, pp. 264–270 (2008)

[6] Lang, J.: Ground-borne vibrations caused by trams, and control measures. Journal of Sound and Vibration 120(2), 407–412 (1988)

[7] Beskos, D.E., Daskupta, B., Vardoulakis, I.G.: Vibration isolation using open or in filled trenches. Journal of Computational Mechanics 1(1), 43–63 (1986)

[8] Ahmad, S., Al-Hussaini, T.M.: Simplified design for vibration screening by open and in-filled trenches. Geotechnical Engineering 117(1), 67–88 (1991)

[9] Yang, Y., Hung, H.: A parametric study of wave barriers for reduction of train-induced vibrations. International Journal for Numerical Methods in Engineering 40(20), 3729–3747 (1997)

[10] Jones, C.J.C., Thompson, D.J., Petyt, M.: A model for ground vibration from railway tunnels. Proc. Instn. Civ. Engs, Transp. 153(2), 121–129 (2002)

[11] Kausel, E., Roesset, J.: Stiffness matrices for layered soil. Bulletin of the Seismological Society of America 71(6), 1743–1761 (1981)

How Does Noise Annoyance Relate to Traffic Intensity?

T. Jerson[1], M. Ögren[2], E. Öhrström[3], and A. Gidlöf Gunnarsson[3]

[1] WSP Acoustics
Rullagerg. 6, SE-415 26 Gothenburg, Sweden
Tel.: +46 31 7272 647
e-mail: tomas.jerson@wspgroup

[2] VTI the Swedish National Road and Transport Research Institute,
Box 8077, SE-402 78 Gothenburg, Sweden
Tel.: +46 31 750 26 04
e-mail: mikael.ogren@vti.se

[3] Sahlgrenska Academy at the University of Gothenburg
Box 414, SE-405 30 Gothenburg, Sweden
Tel.: +46 31 786 36 10
e-mail: evy.ohrstrom@amm.gu.se, anita.gidlof@amm.gu.se

Summary

This study is part of the Swedish research program TVANE, where a number of questionnaire surveys have been performed. A comparison between three areas demonstrates that the reported annoyance is higher where the traffic intensity is higher even if the equivalent noise level is the same. The traffic intensity varies with a factor of 7 between the areas, and the reported annoyance is increased 1.3 – 2.5 times.

1 Introduction

In an international context, Sweden has relatively moderate railway traffic intensity, even if you consider the main railway lines between the three largest cities (Stockholm, Göteborg and Malmö). However, traffic flows are increasing, and can be expected to increase even more rapidly in the future due to environmental concerns and pressures to reduce CO2 emissions.

Noise from train, road and air traffic causes annoyance and reactions of different kinds (Öhrström 2004). Among the most common health effects are general annoyance, speech disturbance, sleep disturbance and effects on rest and relaxation. Traffic noise can also cause psychological and physiological stress-related symptoms and for that reason affects the overall well-being.

T. Maeda et al. (Eds.): Noise and Vibration Mitigation for Rail Trans. Sys., NNFM 118, pp. 323–328.
springerlink.com

The parliament of Sweden have in the infrastructure proposal (1996/97:53, TU7) established guidelines for traffic noise. The equivalent noise level indoors for a 24-hour period shall not exceed $L_{Aeq,\,24h}$ 30 dB and the maximum noise level shall not exceed $L_{AFmax,\,22\text{-}06\,h}$ 45 dB. The equivalent noise level outdoors shall not exceed $L_{Aeq,\,24h}$ 55 dB (free field) and the maximum noise level in the outdoor environment close to a residence shall not exceed L_{AFmax} 70 dB (free field).

Noise from railway traffic has a "bonus" of 5 dB in the outdoor environment close to the residence. The equivalent noise level is then allowed to reach $L_{Aeq,24h}$ 60 dB. There is approximately a quarter million people exposed to noise levels from railway traffic that exceeds the long-term goal of the infrastructure proposal, $L_{Aeq,\,24h}$ 55 dB.

This paper describes the connection between train traffic intensity, noise levels and annoyance determined in questionnaire surveys.

2 Difference in Traffic Intensity

Within the research program TVANE (Train Vibration and Noise Effects) [1 - 3] part of the focus has been on residents close to railway lines that are affected by the increased traffic volumes. In the municipality Sollentuna outside Stockholm the current traffic is close to 500 train passages per day, which is the most intense railway traffic in Sweden. In Fig. 1 the current train traffic is shown as the number of train passages per hour during day and evening in a histogram.

The four tracks in Sollentuna have different kinds of passenger traffic. There are commuter trains (see Fig. 2), intercity trains and high-speed trains traveling on domestic lines at shorter or longer distances up to approximately 1000 kilometers. Speeds vary from 200 km/h for the high-speed trains and down to 160 km/h, but commuter traffic that stops in the area decelerate and accelerate so speeds can be lower. Freight traffic moves at approximately 90 km/h.

The traffic situation in Töreboda and Falköping is similar regarding train speeds but the overall traffic is a substantially lower, see Fig. 1. The total traffic during the day and the evening (06-22) is approximately a factor 4 lower than the traffic in Sollentuna, but with a larger proportion of freight traffic at night time. In both areas there are two parallel tracks, not four as in Sollentuna.

In Sollentuna there is a non-profit association for home owners who partly address the noise situation. The focus is more on a major highway, through the area and in our investigation we have avoided all areas exposed to road traffic noise. Apart from this there is no reason to expect any bias between our research areas.

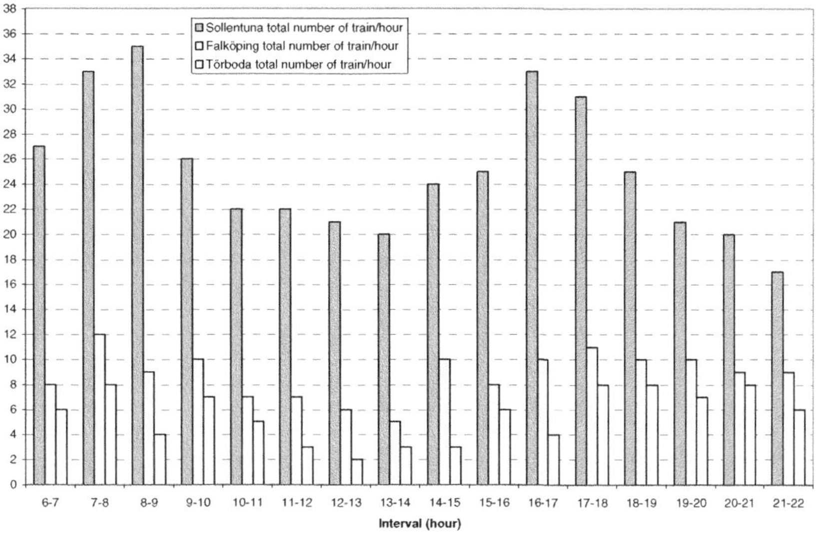

Fig. 1. Histogram showing train traffic in Sollentuna, Falköping and Töreboda during day and evening (color figure online)

Fig. 2. A modern commuter train X60 is departing from a station in Sollentuna. (color figure online)

3 Calculated Railway Noise Exposure

Theoretical calculations using the Nordic prediction method for railway noise [4] and measurements of noise exposure have been used to estimate how many times per hour different noise events occur. The diagram in Fig. 3 shows the predicted number of trains that exceed 70 dB maximum level for the different areas. Note that there is a small probability that a train passage is louder than 70 dB even if the maximum level is 70 dB, and with high-traffic intensities there may be a few trains that are exceptionally loud. This partly explains the different annoyance response with the same maximum level, but different traffic intensities, since more noisy events can be expected per hour where the traffic is high (Sollentuna).

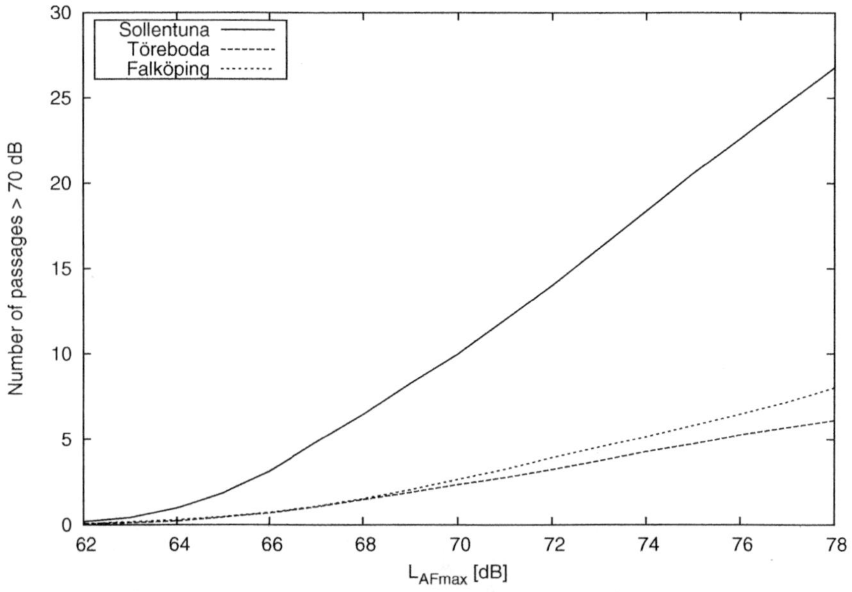

Fig. 3. Predicted number of train passages exceeding 70 dB maximum level during the peak hour as a function of the calculated maximum level L_{AFmax}.

4 Results from Questionnaires

A questionnaire was sent to inhabitants in Sollentuna, Töreboda and Falköping, and about half of the subjects answered the questionnaire. The total number of responding subjects were 724 (Sollentuna), 210 (Töreboda) and 296 (Falköping). The age of the participants ranged from 18 to 75 years. There were 50 main questions in five parts. The different parts in the questionnaire considered living environment, disturbance from different kinds of traffic, background questions and also possibility to write comments.

In Fig. 4 data on noise annoyance from the questionnaires and noise calculations from Sollentuna are compared to data from the two other areas, Töreboda and Falköping. In these areas the traffic intensity is around 125 passages per day, and they are all located along a major railway line. The noise annoyance is higher in the high traffic area Sollentuna compared to Falköping and Töreboda at the same equivalent level, see Fig. 4.

The results from the questionnaires show that 13 % of the inhabitants in Sollentuna are annoyed due to railway traffic with an outdoor noise level of $L_{\mathrm{Aeq,24h}}$ 46–50 dB. This is 5 percentage points more annoyed inhabitants compared to the answers from Töreboda and Falköping. In areas with an outdoor noise level of $L_{\mathrm{Aeq,24h}}$ 56–60 dB from railway traffic there are up to 49 % annoyed in Sollentuna. This is 31 percentage points more annoyed inhabitants compared to the answers from Töreboda and Falköping. All differences are statistically significant.

In areas with higher outdoor noise levels from train traffic, $L_{\mathrm{Aeq,24h}}$ 61–65 dB the difference between annoyed inhabitants in Sollentuna and Töreboda, Falköping decreases with 16 percentage points. This probably is because the noise levels from the railway traffic are so high that a more intensive traffic no longer increases the annoyance.

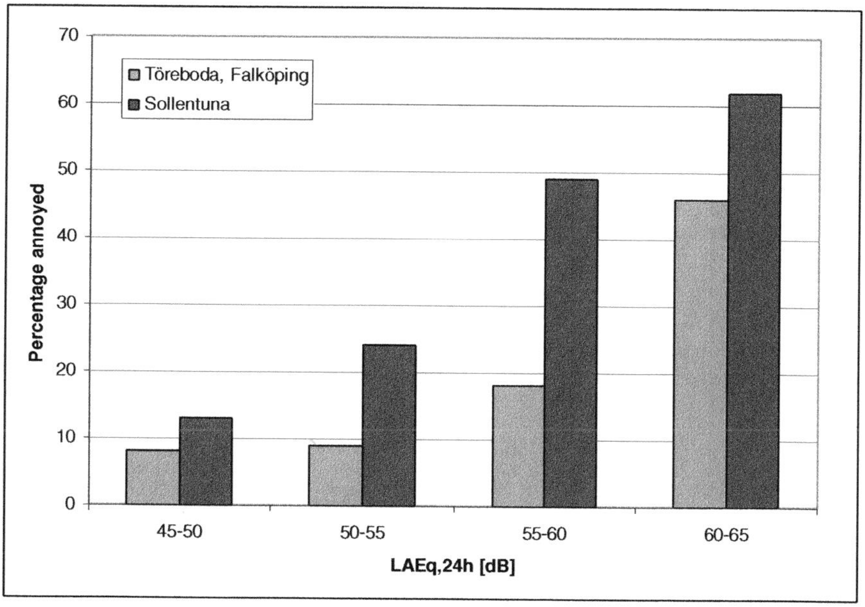

Fig. 4. Percentage of noise annoyed respondents in 5 dB intervals of the equivalent noise level. (color figure online)

Noise disturbed communication and noise disturbed listening to radio and television are extensive when the noise levels are over $L_{Aeq,\ 24h}$ 55 dB. More inhabitants are annoyed due to railway noise during rest, relaxation and sleep in Sollentuna compared to Töreboda and Falköping.

For example 43 % of the inhabitants in Sollentuna with outdoor noise levels L_{night} 55–59 dB reported disturbed sleep quality. At corresponding levels in Töreboda and Falköping 12 % of the inhabitants reported disturbed sleep quality.

The place where the balcony / patio is situated has great importance when it comes to the disturbance. When the balcony/patio is oriented toward the railway, noise annoyance and disturbed activities are higher.

The answers in the questionnaire from Sollentuna and Töreboda, Falköping show that it is not only the equivalent noise level that affects the number of disturbed, but also the number of trains per 24 hours which has a large influence on the noise disturbance. During the day time hours (06–22) an average of 26 trains per hour pass by in Sollentuna. This corresponds to 4 times more railway traffic than in Töreboda and Falköping (6 trains per hour).

One way of summarizing the results is to conclude that intense railway traffic behaves more like road traffic, which gives higher annoyance than "ordinary" railway traffic, but this is overly simplified. Many other factors such as sleep disturbance, speech interference and so on should be included if such a conclusion should be made in future research.

References

[1] Jerson, T., Ögren, M., Öhrström, E.: Combined effects of noise and vibration from train and road traffic, Paper P08 - IWRN 9, Munich (2007)

[2] `http://www.tvane.se/` (accessed June 14, 2010)

[3] Öhrström, E., Gidlöf-Gunnarsson, A., Ögren, M., Jerson, T.: Effekter av tågbuller vid en starkt trafikerad järnväg. Rapport 2:2010, Arbets- och miljömedicin, Samhällsmedicin och Folkhälsa, University of Gothenburg (2010)

[4] Naturvårdsverket. Buller från spårburen trafik – Nordisk beräkningsmodell. Rapport 4935, Naturvårdsverket (Swedish Environmental Protection Agency), Stockholm

Internal Occupational Noise and Vibration Environments of Intermodal Freight Trains

L. Zoontjens

Norman Disney & Young, Level 10, 200 St Georges Terrace, Perth, Australia
Tel.: +61 8 9281 6800; Fax: +61 8 9281 6888
e-mail: l.zoontjens@ndy.com

Summary

The occupational environment of inter-modal freight trains is of particular interest to operators in Australia due to the large distances and extended shift lengths undertaken by drivers; and the subsequent risks associated with fatigue and stress induced by noise and vibration. The management of driver fatigue is identified as critical to operational and occupational safety.

'Crew vans' are specially modified railcars providing sleeping and rest facilities on long distance freight forwarding operations. In this paper, physical assessments of the noise and vibration environments of crew vans are presented and compared to relevant guidelines and a limited amount of driver feedback. Measurements were undertaken onboard for four different crew van variants on major freight rail line services in southern regional areas of Australia.

None of the sleeping quarters assessed had acoustic environments recommended for preventing sleep disturbance. Measured internal noise levels significantly exceeded recommended limits, and intermittent 'rattles and squeaks' added annoyance and interference to sleep pattern. Vibration levels were found to be in general compliance with set criteria except for 'buffing and pulling' vibration modes due to modulating tension in the various couplings along the train.

1 Introduction

Train drivers undertake shift work to cover the large distances and extended time of travel across Australia. When off-shift, the drivers will have approximately 5 hours for sleeping and making use of kitchen, lounge and shower amenities onboard the train. It is important that these areas have a reasonable degree of amenity, as driver fatigue is identified as critical to operational and occupational safety outcomes [1].

'Crew vans' are specially modified railcars providing the sleeping and rest facilities on long distance freight forwarding operations. Crew vans typically

T. Maeda et al. (Eds.): Noise and Vibration Mitigation for Rail Trans. Sys., NNFM 118, pp. 329–338.
springerlink.com

provide a common kitchen and lounge area with individual sleeping quarters (berths) for drivers to use when not in the forward engine. The crew van vehicle is generally located between the locomotives and the load, and it is subjected to significant structural dynamic forces and airborne noise ingress.

The intent of this study is to consider the quality of rest and sleeping areas onboard selected crew vans in terms of airborne noise and whole-body vibration. Figure 1 presents an image of significant external noise and vibration-ingress pathways into each crew van, annotated as follows: (a) emissions from plant and equipment onboard the forward locomotive; (b) dynamic loading via couplings between the crew van and attached vehicles; (c) wheel/track interactions; (d) ancillary plant such as generators and condenser units; and (e) internal activities, fixtures and fittings.

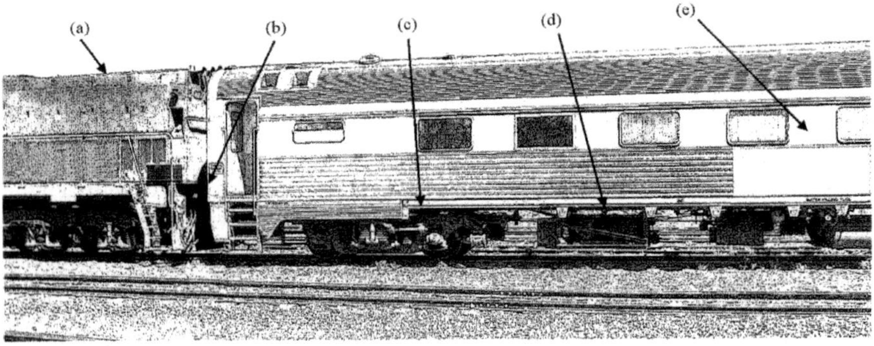

Fig. 1. Annotated image indicating significant rail vehicle noise and vibration aspects in regard to sleeping berth amenity.

2 Approach

There appears to be few scientific publications or standards regarding acoustic amenity *within* rail-bound freight vehicles. The following subsections discuss the relevant literature in the context of noise, vibration and general ergonomics.

2.1 On Assessing the Impact of Noise upon the Sleeping Environment

Our review did not find objective criteria specific to sleeping areas on rail- bound vehicles; published criteria for reducing the rate of sleep disturbance usually assume the sleeping environment to be within a building. However, it is acknowledged that

(a) environmental noise is linked to sleep degradation effects [2] such as difficulty in obtaining sleep, sleep duration, reductions in REM stages of sleep, induced physiological effects such as increased blood pressure and heart rate, and secondary effects such as increased fatigue, increased stress, reduced mood and alertness;

(b) in general, noise levels exceeding A-weighted Equivalent Continuous Sound Pressure Level (L_{Aeq}) values of 30dB(A) or A-weighted Maximum Sound Pressure Level (L_{Amax}) values of 45dB(A) induce adverse sleep disturbance effects [2]; a European study group on Night-Time Noise Guidelines (NNG) [3] indicated that a L_{Amax} value of 42dB(A) is a threshold above which transportation noise is expected to induce adverse effects (it was stated that a L_{Amax} of 60dB(A) within sleeping areas *"should never be exceeded in any case."*); and

(c) shift-workers and those sleeping in temporary accommodation are among the most sensitive to environmental noise.

Description of the internal rail noise vehicle environment is dependent on many dynamic variables such as train speed, track condition and external environment, notwithstanding the structural considerations, noise sources onboardor even the internal fit-out and furnishings. In this study, internal noise levels were assessed using L_{Aeq}, L_{Amax} and Room Criterion Combined (*RCC*) [4] values within sleeping areas over representative time periods of at least 10 minutes duration, in which the vehicle is at relatively constant speed in open country. The *RCC* value is based upon the Preferred Speech Interference Level (PSIL) and Room Criterion (RC) measures; it is defined by Hardy [4] as equal to $0.188\ PSIL + 0.822\ RC$.

2.2 On Assessing the Impact of Vibration upon the Sleeping Environment

ISO 2631.1:1997 [6] outlines general requirements for the evaluation of human exposure to whole-body vibration. 'Ride quality' or 'ride comfort' is used to describe the subjective response to vibration in moving vehicles. Acceptable weighted equivalent RMS vibration ($a_{w,e}$) limits compatible with the Basic Evaluation Method were developed from ISO 2631.1:1997 with relevant frequency weightings taken from ISO 2631.4:2001 [7]. In accordance with this standard, the applicability of the Basic Evaluation Method was assessed on the basis of ratios formed using Maximum Transient Vibration Velocity (MTVV), Vibration Dose Value (VDV) and $a_{w,e}$ terms. As the ratio $MTVV\ /\ a_w$ often exceeded 1.5, and the ratio $VDV\ /\ (a_w\ T^{0.25})$ occasionally exceeded 1.75, the fourth-power Vibration Dose method was used in addition to the Basic Evaluation Method ($a_{w,e}$) as it is more sensitive to transient vibration.

A review of other rail-specific international standards [8] did not yield in the clear specification of acceptable vibration comfort limits using vibration dose values (*VDV*), especially environments associated with freight trains. Quoting ISO 2631.1, *"There are not sufficient data to show a quantitative relationship between vibration exposure and risk of health effects"*. Despite this, we have taken 'Night-time' residential 'preferred' and 'maximum' limits for intermittent vibration from BS6472 [9] to provide an indication of comfort using the *VDV*.

2.3 Ergonomics and the Sleeping Environment

Noise, vibration and light reduction are critical to improve the sleeping environment however, there appears to be no objective human-factor criteria which may be applied to define a particularly ideal or acceptable rail vehicle sleeping environment. Some studies [10, 11, 12] indicate that the physical properties of the mattress have been found to be relatively unimportant, with the exception of the mattress width and thickness. The mattress width, thickness and supporting construction influence the results of the whole-body vibration measurements, and the risk of ejection under sudden vehicle movement.

3 Criteria

Table 1 lists set criteria for noise and vibration within sleeping areas onboard each variant of crew van.

Table 1. Recommended criteria for sleeping areas onboard rail-bound vehicles.

Aspect	Descriptor	Recommended	Maximum
Average ambient noise	L_{Aeq}	45dB(A)	50dB(A)
	L_{Amax}	60dB(A)	75dB(A)
	RCC	50	55
Average ambient vibration	$a_{w,e}$	0.315m/s^2	0.5m/s^2
	VDV	0.13m/s$^{1.75}$	0.26m/s$^{1.75}$

The feedback received from stakeholders and drivers indicated that the expected quality of accommodation is less than that experienced at home. On this basis, set criteria were developed using relevant standards for residential areas with penalties applied.

3.1 Method

The following subsections describe the methods employed to obtain data compatible with the set criteria. Table 2 lists the four crew van variants assessed in this study. Noise and vibration levels within crew vans were obtained over the same service route between Cook and Port Augusta, South Australia. Results were obtained over the same service measured for conditions where the train was at idle (stationary), at speed between 60 to 80km/hr and at speed between 105 to 115km/hr, as estimated using a GPS receiver. Results have been excluded outside these speeds and where track segments were not considered straight, of good condition and level. The locomotives used on all runs were of same class and specification.

Table 2. Crew van variants assessed in this study.

Type Variant	Service	Date	Indic. Tonnage
A	Cook to Port Augusta	October 23, 2007	~4,500t
B	Port Augusta to Cook	October 24, 2007	~2,200t
C	Cook to Port Augusta	October 17, 2007	~2,200t
D	Port Augusta to Cook	October 16, 2007	~4,300t

Airborne noise levels were measured using a calibrated sound analyzer at head positions of each sleeping area over periods not less than 10 minutes.

Vibration levels were assessed using a triaxial whole-body vibration pad. The weighted acceleration was assessed over a measurement period of typically one hour in several berths per van. Vibration dose values used a four hour exposure time, which is considered typical of the sleeping time allocated on relay operations (5 hours total off-duty typical).

4 Results and Discussion

The following subsections present and discuss the results of acoustic and vibratory measurements taken onboard the four crew van variants.

4.1 Vibration

Figure 2 and Figure 3 present average RMS vibration levels and *VDV* ,respectively for sleeping berths in each crew van variant and speed condition with annotations. From comparison of *VDV* results with BS6472, adverse comment is possible for Type B and probable with Type C.

With the train stationary, the vibration impact from onboard sources such as the generator set and cooling fans are well below that considered to adversely affect the sleeping environment. At speed, the Type A and Type D variants were assessed to have the lowest vibration levels despite tonnages over twice that of the others. In traveling westward opposed to eastward, these variants encountered headwinds and increased aerodynamic drag which served to minimize slackness. Slackness in the train couplings leads to 'buffing and pulling' (also referred to as 'run-ins' and 'longitudinal slack action') which adds high-energy low-frequency transient vibration movements. The acceleration levels are readily perceptible and the cause of the high levels of measured vibration within the Type C van. It can be reduced through increasing the freight tonnage or active braking.

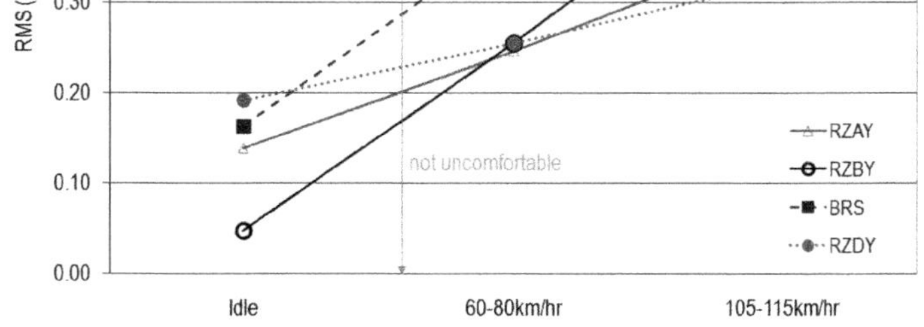

Fig. 2. Average aw,e values versus crew van variant and speed (color figure online)

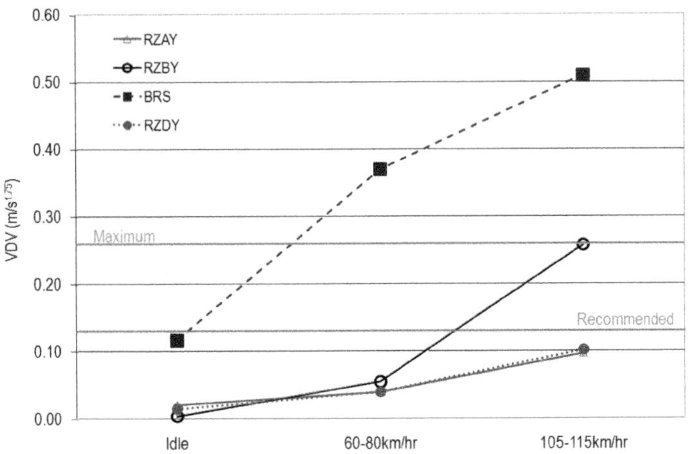

Fig. 3. Average VDV versus crew van variant and speed. (color figure online)

4.2 Noise

Figure 4 presents indicative noise level values for sleeping berths in each crew van variant and speed condition. At speed, all measurements exceeded the RC 'Region B' zone [13] and approximately 17% exceeded the RC 'Region A' zone. This indicates a moderate, but in some cases high probability, of *"noise-induced vibration in lightweight structures, induced audible rattle in light fixtures, doors, windows etc"*. These indications proved reasonably accurate.

All noise criteria recommended for sleeping areas were exceeded over the course of each journey. Time-averaged noise levels measured within each of the crew vans were typically 25 to 35dB above limits recommended for residential applications and 10 to 15dB above the criteria adopted in this study. It follows then that none of the crew van variants assessed are expected to have internal acoustic environments conducive to obtaining good quality sleep. Drivers reported that variants Type A and Type B are the least preferred of the four classes, and that most staff interviewed noted that they cannot reliably obtain sleep at all in Type A or Type B vans.

With average RCC values of the order of 60, the Type C and Type D crew van variants are identified as more comfortable than the Type A and Type B variants with RCC values closer to 65. According to Griefahn [14], the typical probability of awakening reactions at the speeds assessed is between 13 and 30 per cent.

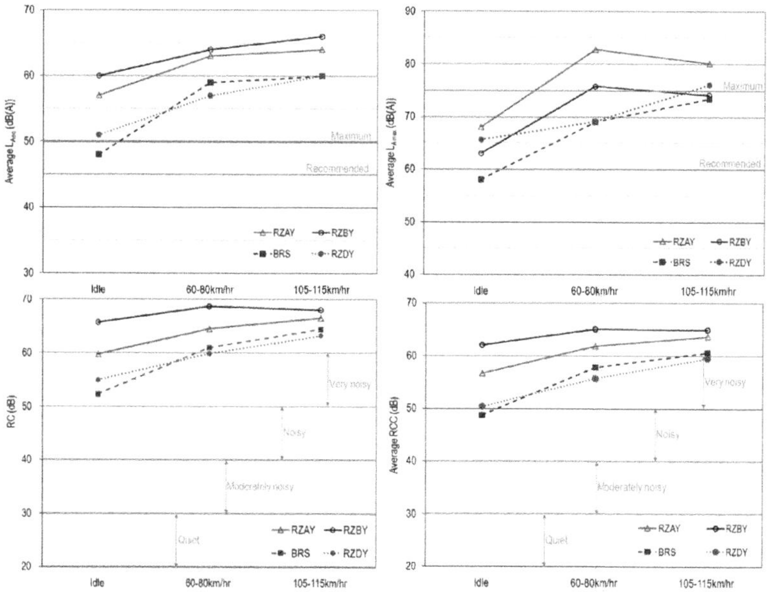

Fig. 4. Average internal noise levels versus crew van variant at various speeds (color figure online)

4.2.1 Influence of Speed upon the Sleeping Environments

Figure 5 presents A-weighted third-octave band data typical of sound levels within sleeping berths onboard the Type D crew van variant at each speed.

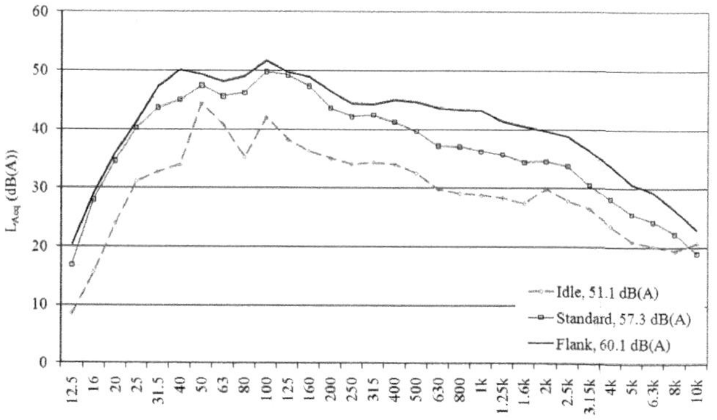

Fig. 5. Typical internal noise levels for Type D sleeping berths versus third octave band spectra at various speeds (color figure online)

At idle condition, noise ingress corresponding to generator sets is dominant for all four van variants, particularly for the Type B variant. At speed, impacts from the ancillaries are masked by other noise sources as described in **Figure 1** (with the exception of berths located directly above the generator). The bulk of sound energy within sleeping areas is now concentrated in the 31.5Hz, 125Hz and 250Hz octaves (100Hz to 315Hz third octaves), which may be linked not only to wheel-track interactions but also to cooling fan and pump systems onboard the locomotives. Noise levels under each condition were generally 59dB lower within the Type C and D variants at medium-to-high frequencies (>250Hz). This can be attributed to the fact that these two variants utilize carpet on the walls, which at these frequencies is more acoustically absorptive than the standard internal trim.

4.2.2 Intermittent 'Rattles and Squeaks'

Intermittent rattles and squeaks were present in at least one sleeping berth onboard each crew van. Annoyance in this regard was highest in the older crew vans (Type A and Type B), from rattling ceiling panels and creaking joints. Within the newer Type C and D vans, door and light fixtures had greater impact. 'Creaking', noise from structural joint flex, was particularly noticeable in the older Type A and B vans.

Audible 'rattling' or 'squeaking' of small fixtures and panels were reported as more intrusive than overall loudness. Most of the hardware used in crew van fitouts suit residential housing and are not designed for seismic loading. **Figure 6** indicates various examples of intermittent noise sources found, annotated as follows: (a) fixtures and securings left propped rather than loosely engaged or free to swing; (b) paper leaflet jammed into an edge crevice of a ceiling panel to minimize rattle; (c) hand towel jammed into door frame to remove slack in the door latching mechanism; (d) most of the installed light fittings have had the clear plastic inserts removed, which over time loosen and rattle.

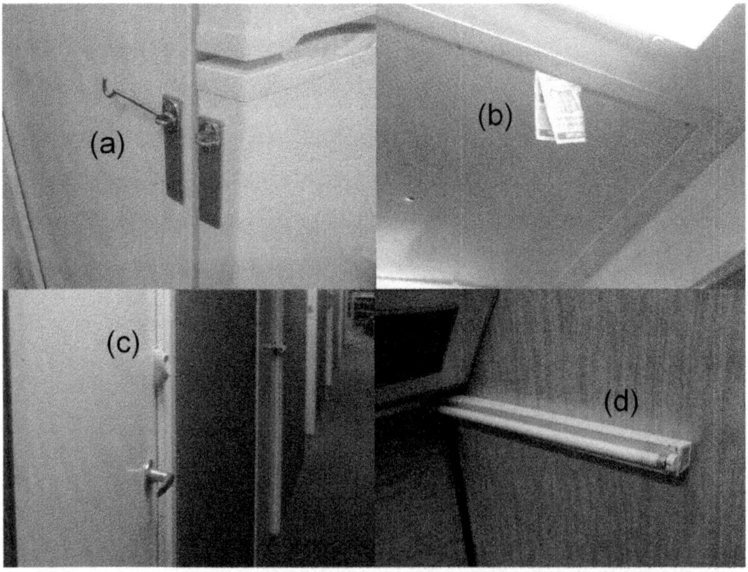

Fig. 6. Examples of potential intermittent noise sources onboard crew vans.

5 Conclusions

None of the sleeping quarters assessed were found to have continuous noise environments recommended for preventing sleep disturbance. Furthermore, the presence of intermittent 'rattles and squeaks' invoked additional annoyance to drivers sleeping in selected berths. Difficulty exists in tracking each rattling source, as it may only be practicably identified while the vehicle is at a certain range of speeds. Vibration levels measured align with modest criteria with the exception of 'buffing and pulling' actions which arose from excess slackness in the various couplings along the train.

6 Recommendations

From this study it is recommended that the primary noise control strategy includes ensuring ample control of reverberation within each berth first, then addressing intermittent and impulsive sources, such as minimization of rattles and squeaks. Gaining subjective feedback from regular users of the crew vans should be given highest priority, as they are best positioned to identify and record each rattle and squeak for maintenance or treatment at a later date.

It is also recommended that the vibration control strategy includes detailed consideration of likely service load conditions and associated transient event risks.

References

[1] AS 4292.1-2006 – Railway safety management – General requirements (2006)

[2] Berglund, B., Lindvall, T., Schwela, D., Goh, K.-T.: Guidelines for Community Noise (World Health Organisation (2000)

[3] World Health Organisation. Report on the third meeting on night noise guidelines. WHO European Centre for Environment and Health, Lisbon, Portugal (2005)

[4] Hardy, A.E.J.: Measurement and Assessment of Noise within Passenger Trains. Journal of Sound and Vibration 231(3), 819–829 (2000)

[5] AS/NZS 2107:2000 – Acoustics – Recommended design sound levels and reverberation times for building interiors (2000)

[6] ISO 2631.1:1997 – Mechanical vibration and shock – Evaluation of human exposure to whole-body vibration – Part 1: General requirements (1997)

[7] ISO 2631.4:2001 – Mechanical vibration and shock – Evaluation of human exposure to whole-body vibration – Part 4: Guidelines for the evaluation of the effects of vibration and rotational motion on passenger and crew comfort in fixed guideway transport systems (2001)

[8] IS ENV 12299:1999 – Railway applications – Ride comfort for passengers – Measurement and evaluation (1999)

[9] BS 6472:1992 Guide to evaluation of human exposure to vibration in buildings (1 Hz to 80Hz) (1992)

[10] Karwowski, W.: International encyclopedia of ergonomics and human factors. Taylor & Francis, New York (2001)

[11] Haex, B.: Back and Bed: Ergonomic Aspects of Sleeping. CRC Press, Florida (2004)
[12] Buckle, P., Fernandes, A.: Mattress Evaluation - Assessment of Contact Pressure, Comfort and Discomfort. Applied Ergonomics 29(1), 35–39 (1997)
[13] Bies, D.A., Hansen, C.H.: Engineering noise control: theory and practice. E & FN Spon, London (1997)
[14] Griefahn, B.: Noise control during the night – proposals for continuous and intermittent noise. Acoustics Australia 20(2), 43–47

Noise Reduction of Steel Bridges with Non Ballast Tracks

H. Venghaus[1], P. Balmer[2], M. Muncke[3], and F. Poisson[4]

[1] Schrey&Veit GmbH, 55576 Sprendlingen Zotzenheimerstrasse 42, Germany
[2] Schweizerische Bundesbahnen SBB Bereichsleiter Umwelt SBB Projekt Management Luzern, Bahnhofplatz, 6003 Luzern, Switzerland
[3] OeBB-Infrastruktur AG, Forschung, Entwicklung u. Systemtechnik, 1070 Wien, Zieglergasse 6, Austria
[4] Direction de l'Innovation et de la Recherche, 75379 Paris, 45 rue de Londres, France

Summary

Steel bridges without ballast are very often sources of annoying railway noise. While the insertion of a trough filled with ballast will lead to a noise reduction of about 6 dB(A), it causes an increase of the static mass of the bridge construction of about 30%, which then requires additional measures of reinforcement.

In this paper 4 different systems of noise mitigation are presented, which are based on decoupling and damping effects. A noise reduction of 4.5 dB(A) to 10 dB(A) was achieved with an additional mass of less than 15% of the weight of the bridge and therefore no reinforcement was needed.

1 Introduction

Bridges in urban areas are a necessary part of railway infrastructure to carry tracks over roads, rivers and channels and in some cases over other railway tracks. The main disadvantage of bridges is the increased noise radiation in comparison to plain tracks.

In particular, steel bridges with rails fastened directly to the bridge structure are known to be very noisy. Modern railway bridges which are built of concrete with ballasted tracks are often as quiet as the plain track.

For noise mitigation this will lead to the idea to replace the old steel bridges by modern concrete bridges, but there are some practical and economical reasons which deter from doing so:

- The number of old steel bridges in use is high, so replacement will lead to high costs
- Steel bridges provide a smaller construction depth under the rails, which is a critical requirement in urban areas

T. Maeda et al. (Eds.): Noise and Vibration Mitigation for Rail Trans. Sys., NNFM 118, pp. 339–347.
springerlink.com

- Bridges with a large span are more easily constructed from steel and ballast-less track reduces weight, requiring less structure and foundations.

Steel bridges therefore will remain a part of railway infrastructure in future and techniques are needed to reduce the radiated noise. This paper will discuss some measures which were implemented at steel bridges by increasing the damping of the rails and/or the bridge structure and by decoupling the tracks from the bridge structure.

2 Theoretical Aspects

On plain tracks the noise of a passing train is radiated to the upper half-space while the surface has characteristic absorbing effects. This changes when the train is crossing a bridge. From an elevated point it may radiate over a street or water, and both are highly reflective. This will lead to broader noise propagation especially at low frequencies.

The structure of a steel bridge adds further sound sources to the noise radiated by a train, including:

- steel deck
- supporting structure
- plate girders, box beams lattice girders.

All these structures consist of large surface areas, so that even low vibration energy can result in high-noise radiation.

The impedance of the rail will change from plain track to bridge track. Where rails are directly fastened to bridge structures, timber sleepers or transoms are generally used. The introduction of an isolation component as a decoupling element ("impedance mismatch") between rail foot and bridge causes an energy reflection at the rail foot which will lead to a higher movement of the rail and for a certain amount of the wheel. This effect will lead to higher-noise radiation from wheel and track.

The effect of the "impedance mismatch" is already known on plain tracks when soft rail pads are in use. With these rail pads the mobility of the rails is increased and because of the higher decoupling from the sleepers the rails will have less energy loss. This energy will remain inside the rails and will travel over a longer distance with a significant noise radiation. This effect is described by the Track-Decay-Rate (TDR), with high-TDR values significant for low-noise radiation by rails.

The mission of reducing the noise radiation of steel bridges is to

- reduce the vibration energy inside the bridge structure to prevent deck plates, box beams and other large surfaces from noise radiation by inserting a maximum of "impedance mismatch" between rail and bridge
- equip vibrating parts of the bridge with damping material

- equip rails with damping material to keep the TDR high
- grind rail heads to keep excitation induced by wheel - rail contact at the lowest level.

3 Noise Mitigation at 4 Steel Bridges

The following subchapters will describe successfully performed measures for noise reduction at steel bridges of different construction. Some main results are presented and will be discussed.

3.1 Wasserpark-Bridge Vienna

The Wasserpark-bridge is located in the north of Vienna as part of the line Vienna-North – Bernhardsthal crossing the old Danube River. The length of the bridge is nearly 98m long carrying 2 tracks. Two end-to-end plate girders with a height of 1.95m are cross connected by bracing plates with a section length of 3.89m (see Fig.1).

After the first tests in 1998 it was shown that replacing the rail expansion joints away from the bridge and grinding of the rails was necessary. With this a noise reduction of about 12 dB(A) was achieved [1].

Fig. 1. Wasserpark-bridge Vienna

Fig. 2. Absorbers at rail and bridge

After these maintenance works the primary effort to reduce noise from the bridge was started by mounting specially designed absorbers to the rail (see Fig.2). This was followed by mounting absorbers to the bridge structure (see Fig. 3, 4) along the plate girders and each bracing plate. In the last step all rail absorbers were temporarily dismantled, to determine the efficiency of the bridge absorbers alone.

Fig.3 shows the achieved noise reduction for the three different stages.

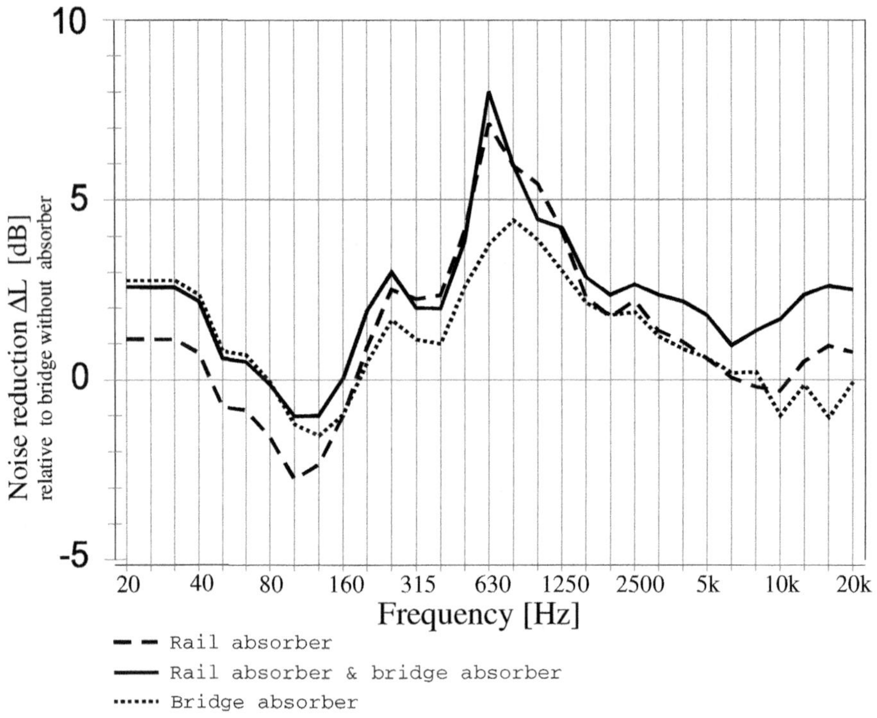

Fig. 3. Wasserpark-bridge Vienna Noise reduction

The observed overall noise reduction by rail and bridge absorbers is 6 dB(lin) achieved by 16t additional masses in relation to 250t weight of the bridge [1].

3.2 Gavignot Bridge – Bridge Across Kleine Emme Bridge

The Gavignot bridge nearby Paris (F) and the bridge across the river Kleine Emme nearby Emmenbruecke (CH) are of the same construction (see Fig 4, 5). The lattice girder span of the Gavignot bridge of 21m and the span of the Kleine Emme bridge is about 40m.

Fig. 4. Gavignot bridge

Fig. 5. Bridge across Kleine Emme

While the Gavignot bridge has a closed steel deck plate, the accessible area of the Kleine Emme bridge is generated by filling sleepers. The first step of noise reduction for both bridges was executed by very soft rail fasteners designed by Pandrol (Gavignot-bridge, see Fig. 6) and Delkor (Kleine Emme bridge, Fig.7). With this decoupling of rail and bridge structure a noise reduction of about 4 dB(A) was achieved. In the second step rail absorbers of the Schrey&Veit Company were mounted, specially tuned to the remaining noise radiation of the bridges (Kleine Emme bridge see Fig. 9).

Fig. 6. Soft rail fastening system VIPA by Pandrol

Fig. 7. Rail absorber between two Delkor rail fasteners

The efficiency of the rail absorbers is different; at the Kleine Emme bridge an additional noise reduction of 3.7 dB(A) was achieved [5], while the overall noise reduction of decoupling rail fasteners and rail absorbers at the Gavignot bridge is about 10 dB(A) (see Fig. 8) [3]. In Fig. 8 it is shown that the achieved noise reduction changes with the types of trains passing the bridge. This is mainly depending on the speed of the trains and to a certain extent to the unsprung masses of wagons and locomotives. A comparison of the achieved noise reduction for the GTW with the engine directly fixed to the bogey and the FLIRT - a new type of these commuter trains – shows this.

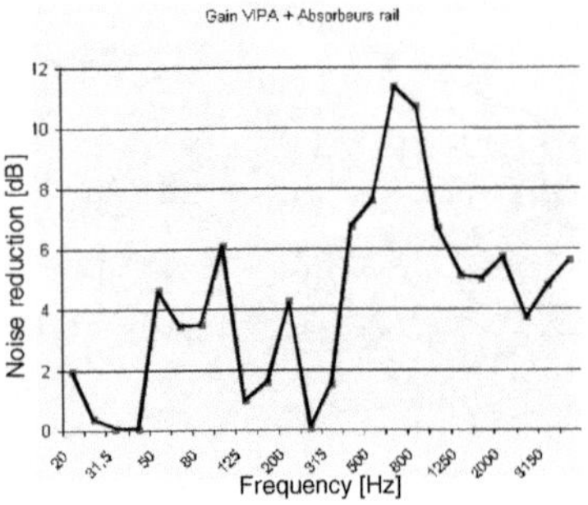

Fig. 8. Noise reduction at the Gavignot bridge by soft rail fasteners & rail absorbers [4] (color figure online)

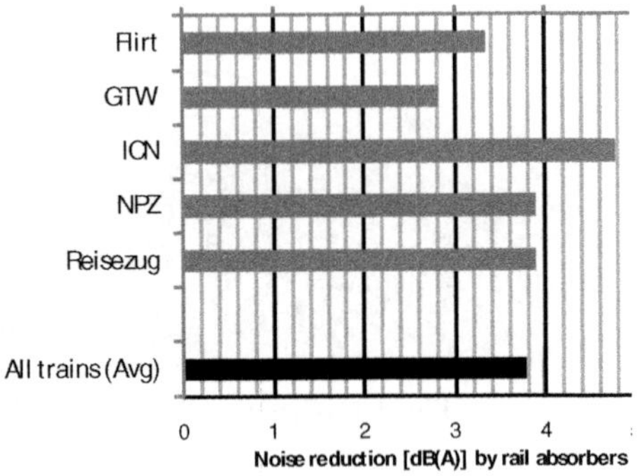

Fig. 9. Noise reduction depending on type of trains - Kleine Emme bridge

At the Gavignot bridge a special low-frequency noise was disturbing the neighborhoods [4]. Special absorbers mounted on the deck (see Fig 10) were therefore trialed to reduce this boom noise.

Fig. 10. Low-frequency absorber designed for Gavignot bridge

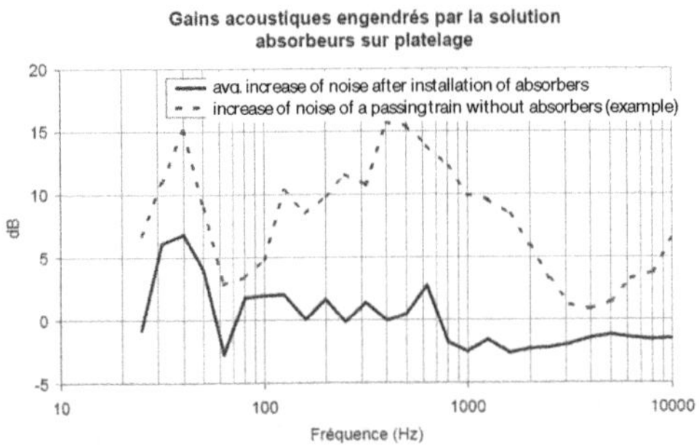

Fig. 11. Effectiveness of low-frequency absorber (color figure online)

Fig 11 shows the increased noise level of a train crossing the Gavignot bridge in relation to the normal track (dotted line). The mounted absorbers reduce the noise radiation in the requested frequency bands at 40 Hz and from 125 Hz to 2 kHz. The resulting noise level is relatively close to the normal track

At the Gavignot bridge the complete weight of the rail absorbers was about 1000kg while the weight of the low-frequency absorbers was about 780kg. At the Kleine Emme bridge the total mass of the rail absorbers was about 2200kg.

3.3 Hainburg Bridges

The Bridges in Hainburg are of a very different type, with several bridges with a length of about 12m separated by embankments, which have a length of about

10m (see Fig. 12). The track is directly fixed to two parallel plate girders which are stabilized by crossbars.

Due to the complaints from the nearby residents about boom noise radiated by the bridge, the decision here was to decouple the track from the bridge structure by inserting very soft elastomeric pads produced by Getzner underneath the wooden sleepers, to obtain a lower cut-off frequency for damping effects compared to the soft rail fastening systems (see Fig. 15), and to control the expected increase of the radiated noise by the wheel-rail-contact, rail absorbers designed by Schrey&Veit were mounted.

Fig. 12. Very old photograph of the Hainburg bridge system

Fig. 13. Getzner elastomeric pad with cutouts for the rivet heads

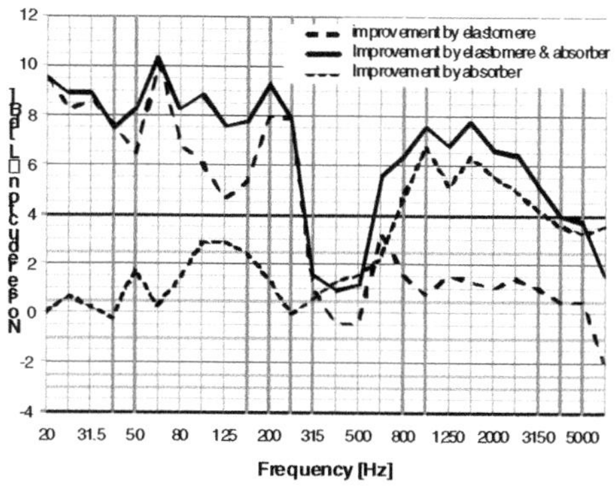

Fig. 14. Noise reduction by elastomeric pads (- -), rail absorbers (....) and both (___)

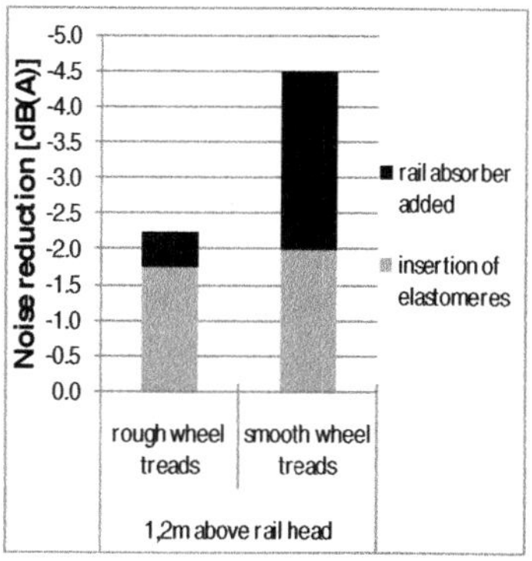

Fig. 15. A-weighted noise reduction for different wheel treads

The insertion of the soft elastomeric pads and rail absorbers was very successful except for a frequency band from 315 Hz to 500Hz (see Fig 14) where no noise reduction was achieved. Although there is a very high noise reduction in the lower frequency range – which was positively commented by the residents – the overall level was only reduced by 4.5 dB(A) for trains with smooth wheels and only about 2 dB(A) for trains with rough wheel treads [6] (see Fig. 15).

It is assumed that internal resonances of the elastomeric pads cause this unwanted quality fall-off over two-third-octave bands. If this gap could be avoided, the noise reduction would be around 6.5 dB(A). This can be achieved by adding tuned absorbers to the bridge structure as it was done at the Wasserpark bridge.

The additional mass by the rail absorbers was about 840kg per bridge.

4 Conclusions

At 4 different types of steel bridges it was shown that a significant noise reduction can be achieved by decoupling the track from the bridge structure with soft rail pads underneath the rails or soft pads underneath the sleepers, but the increased mobility of the rails due to this measure necessitates mounting rail absorbers to reduce the emitted noise of the wheel-rail contact. With this technique a noise reduction of steel bridges without ballasted tracks of more than 6 dB(A) is generally achievable in most of the cases. The additional masses by absorbers are about 56.5kg/m/track.

5 Final Remarks

While steel bridges with/without ballast tracks are radiating noise in the frequency bands from 40Hz to 630Hz, the A-weighted noise level suppresses the importance of this low-frequency range. This situation sometimes may lead to noise reduction efforts being concentrated in the mid-frequency range, although the boom noise is much more annoying because of less damping of the air at low frequencies.

When the absorbers at the bridge structure of the Wasserpark bridge were mounted it was possible to follow each bogie of the passing trains by ear. This is normally found only on steel bridges with ballast tracks. Both types of bridges have a higher damping of their structure, so that only local areas are excited by the vibration sources. Similar to the Track-Decay-Rate of rails it seems to be beneficial to implement a similar decay rate of bridges, as a means to reduce noise radiation from that component.

References

[1] Veit, G.: OeBB-Strecke Wien Nord – Bernhardtsthal, km 3.982 Eisenbahnbruecke ueber die Alte Donau (Wasserbruecke) Schalltechnische Untersuchung Abschlussbericht, Republik Oesterreich Bundesministerium fuer Verkehr, Innovation und Technologie MA 22 – 4533 (2000)

[2] Poisson, F., Dieleman, L.: Noise from railway steel bridges: a complete study. In: WCRR (2004)

[3] Poisson, F.: Performances des solutions de reduction du bruit des ponts metalliques. SNCF internal report (2006)

[4] Menoret, A.: Absorbeurs sur platelage Project "Reduction du bruit des ponts metalliques". SNCF internal report (2005)
[5] Hoin, R.: Pilotprojekt "Schwingungsabsorber bei Stahlbruecken" Vergleiche, Bericht an die SBB, internal paper (2009)
[6] Kalivoda, M., et al.: Laermsanierung HainburgbrueckeObjekt 19 – Massnahmenvergleich, Bericht an die OeBB Infrastruktur Bau AG (2009)

Field Verification of Curving Noise Mechanisms

J. Jiang[1], R. Dwight[1], and D. Anderson[2]

[1] University of Wollongong, Wollongong, NSW, Australia
Tel.: +61 42213183; Fax: +61 42213020
e-mail: Jiandong@uow.edu.au
[2] Railcorp, NSW, Australia
David.Anderson4@railcorp.nsw.gov.au

Summary

This paper deals with field observations of curving noise from revenue service trains on a mainline corridor. A specific trackside measurement methodology has been designed to verify the influences of the main parameters: wheel angle of attack, vertical and lateral rail forces, wheel lateral position, train speed and weather conditions. This paper focuses on the role of the wheel angle of attack on curving noise generation. A detailed method of curving noise identification, including the type of curving noise and its generating wheel, is presented. It is observed that all severe squeal noise events (>100dBA) were generated from high wheel angle of attack (>10mrad), while flanging noise has no obvious relationship with wheel angle of attack. A surprising finding from this measurement exercise is that squeal noise was generated from the outer wheel/high-rail interface, which is contradictory to the conventional knowledge.

1 Introduction

In the 1970s Rudd [1] postulated that lateral creepage between the wheel tread and the top of the rail was the main cause of squeal noise, while longitudinal creepage and flange contact were discounted. Rudd developed a linear frequency-domain model based on a lateral stick-slip mechanism to predict squeal noise as a function of train speed, curve radius and vehicle bogie wheelbase. In Rudd's model, a friction model with negative damping is assumed: the friction force decreases after saturation with increasing creep velocity. Various models have been subsequently developed [2-5] based on this concept. It is considered that these are proposed without sufficient proof.

There is increasing evidence that the lateral stick-slip instability due to negative friction damping may not be the only mechanism resulting in curve squeal. A recent squeal model [6] suggests that the negative friction slope only plays a minor role and is not a necessary condition for squeal noise to be generated, while the

T. Maeda et al. (Eds.): Noise and Vibration Mitigation for Rail Trans. Sys., NNFM 118, pp. 349–356.
springerlink.com

position of the wheel-rail contact point, the lateral displacement, the angle of attack and the level of friction are relevant to squeal generation. The absence of a negative friction slope when squeal noise is generated has been observed in laboratory tests [7]. The presence of longitudinal creepage has also been observed to eliminate squeal noise generation on a 1:3 scale twin-roller test rig [8]. Flange contact appears to generate a different form of curving noise termed 'flanging noise' [9-11], which is a high-frequency noise normally at a relatively lower level but more broadband and intermittent than squeal.

As part of a study to confirm curving noise models for revenue service trains, the study of the relationship between wheel angle of attack and curving noise is specifically reported in this paper. The study of the influences of other parameters, such as wheel/rail interface friction condition, wheel lateral position, train speed, humidity, and temperature may also be relevant and are the subject of a continuing investigation.

2 Trackside Measurement Setup

Trackside measurements were carried out on a curve with a 284m curve radius, in Sydney Australia. The curve is located on the main north rail line carrying a substantial proportion of the freight traffic traversing the city urban network. The track is ballasted, founded on a dry reactive soil formation, with concrete sleepers (ties) and an e-Clip fastening system. The super-elevation of the track is 130mm. Only the up direction traffic was monitored, which is traveling on a downhill grade of 1:52.

The data collection consisted of two stages:

Stage 1: A permanent angle of attack detector system (WID TBOGI) was installed at the inner side of the 284m curve (see Fig. 1) in 2007 for the purpose of conducting a curve squeal noise study. Following this installation, parameters including wheelset angle of attack and lateral position, train speed, humidity, ambient and rail temperature were recorded and stored in the permanent detector system continuously. Noise data collection was also carried out adjacent to the angle of attack detector during some isolated periods from August 2007 to February 2008. The microphone was located 2 meters from the inner rail.

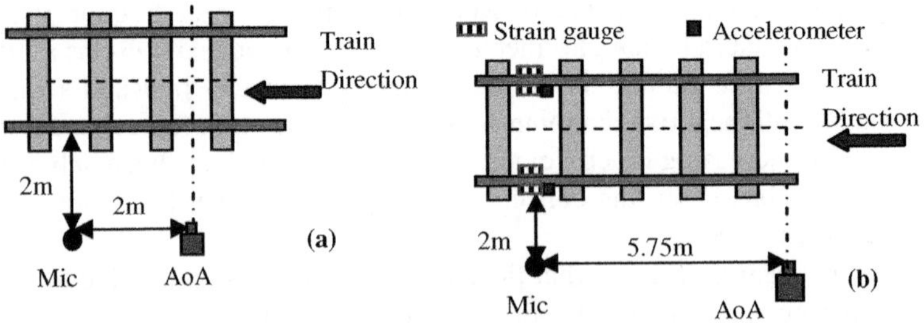

Fig. 1. Diagram of instrumentation at the measurement site: (a) Stage 1 and (b) Stage 2. (color figure online)

Stage 2: Strain gages and tri-axial accelerometers were installed on the rails near the angle of attack detector in March 2010. The vertical rail force measurement pattern and the base chevron pattern as identified in Ref. [12] were used to measure the vertical and lateral rail forces respectively. One sleeper was displaced to allow a span of 600mm to be achieved between sleeper edges in order to achieve a sufficient unsupported rail span. Vertical and lateral rail forces and three-dimensional accelerations at both low and high rail were collected along with the noise and angle of attack data. The data collection was conducted only during the daytime over 9 days from 22nd March 2010 to 6th May 2010. There were 32 freight train pass-bys recorded during this monitoring period in stage 2.

3 Curving Noise Identification and Analysis

In order to investigate the noise mechanisms, the type of noise and its origin has to be identified in a verified and rigorous manner. Noise identification involves three aspects: the type of noise, and the origin of the noise event, in particular which specific wheelset and which wheel (inner or outer). The method of noise identification employed is developed in the following paragraphs.

3.1 What Type of Curving Noise

Automated squeal and flanging noise identification has been developed by the authors [13] and is used to identify squeal and flanging noise events. The method is based on the different spectrum characteristics of these noises. For example, a typical one-thord Octave spectra of the squeal noise and flanging noise recorded from different wheelsets on the same freight train is shown in Fig. 2. This example shows that squeal noise is a pure tonal noise, with a dominant frequency component (~2.06kHz) and some harmonics; while flanging noise exhibits broadband frequency components, ranging from several hundred Hz up to 10kHz.

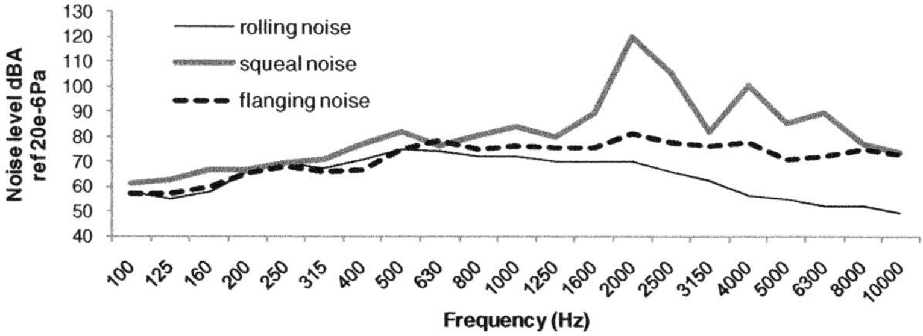

Fig. 2. 1/3 Octave spectra of rolling noise, squeal and flanging noise recorded from different wheelsets on the same freight train passing the test site. (color figure online)

3.2 Which Wheelset Generates Curving Noise

A key aspect of noise identification is to associate the identified noise event with a specific axle. For a single microphone-based system the identification of the axle that is generating squeal or flanging noise is based on the temporal characteristics of that noise event. The noise level increases as the wheel emitting squeal noise approaches and then reduces as the wheel departs from the microphone. It would be expected that the maximum noise level is observed when the wheel is adjacent to the microphone. In a simple monitoring system then, the wheel responsible for the noise may be identified as the one closest when the noise event reaches a peak level. This method leads to a false detection of the noise generating wheel if the noise is intermittent and stops before that wheel arrives at the position opposite to the measurement microphone. This is more of an issue for flanging noise. It will also be imprecise where more than one adjacent wheelset is generating curving noise. The allocation of curving noise to a particular wheelset may sometimes be verified by the point of frequency shift due to the Doppler Effect.

3.3 Inner or Outer Wheel

It is also useful to identify which of the wheel-rail interfaces, outer wheel or inner wheel, is generating the noise in order to determine the noise source and take appropriate noise mitigation measures. For the test site, accelerometers were installed on both rails in the Stage 2 for this purpose. Examples of low- and high-rail squeal events are shown in Fig. 3(a) and (b) ,respectively. In the case of inner wheel squeal, the lateral rail vibration at the low rail is about 20 dB higher than the corresponding vibrations at the high rail. Similarly the high-rail vibration is approximately 20dB higher than the corresponding low-rail vibration when the outer wheel generates squeal.

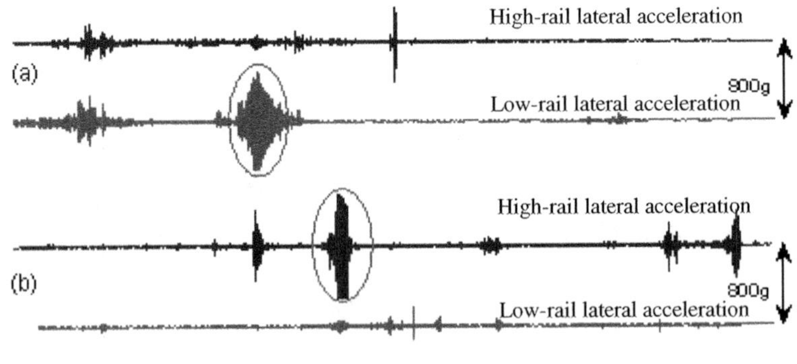

Fig. 3. Lateral rail accelerations indicating the origin of squeal events (circled area): an inner (a) and an outer (b) wheel squeal event recorded from different freight train pass-bys. (color figure online)

During the short monitoring period (9 days) in Stage 2, 28 squeal noise events were recorded from 32 freight train pass-bys. Unexpectedly, only 4 of those 28 squeal events were identified as being generated from the low-rail/inner-wheel interface. The pure tonal squeal noise is generally thought to be generated from the inner wheel, as reported from a number of field observations [6, 14-15]. It is unclear at this stage why squeal noise occurs predominantly from the high rail at this site.

4 The Influence of Angle of Attack on Noise

It is generally thought that wheel angle of attack is one of the key factors in the occurrence of curve squeal noise. Squeal noise is likely to occur if the angle of attack exceeds a certain threshold subject to other conditions, such as the wheel/rail interface friction condition and damping. The threshold of angle of attack is identified as 10mrad by some authors [15], and around 79mrad by others [15].

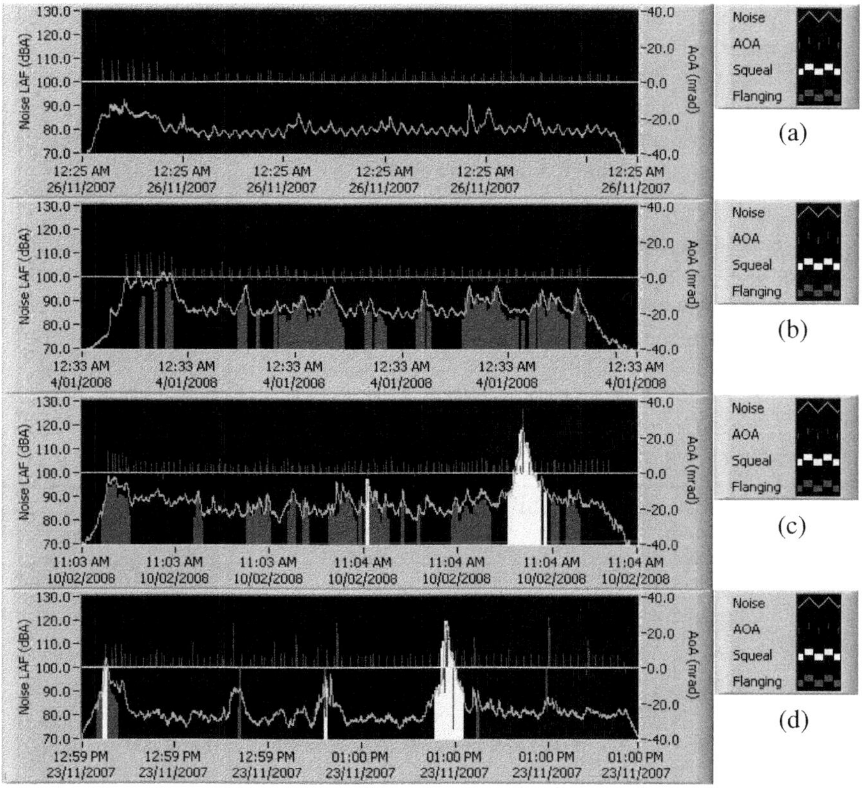

Fig. 4. Examples of noise versus angle of attack from different freight train pass-bys: (a) a freight train with regular angle of attack and no squeal or flanging noise emission; (b) one with regular angle of attack and flanging noise; (c) one with an abnormal angle of attack and one squeal generated from that wheel; (d) one with several abnormal wheel angles of attack but only one squeal is identified exceeding 100dBA. (color figure online)

The raw noise data recorded in stage 1 was analyzed by using the method previously described, embedded into an automated system. In total there were 131 squeal events detected with the noise level exceeding 100dBA recorded during a period (from August 2007 to February 2008). The relationship between the identified curving noise and wheel angle of attack were analyzed. Some examples are shown in Fig. 4 to illustrate the typical behavior of curving noise with wheel angle of attack.

A freight train with better curving behavior: with a 'normal' wheel angle of attack; can be relatively quiet as shown in Fig. 4 (a), or can generate significant flanging noise in some conditions as shown in Fig. 4 (b). It should be noted that flanging noise has also been observed to be generated from passing wheels with normal angle of attack. In contract, curve squeal is typically only associated with wheels with high angle of attack as in the example in Fig. 4 (c). Conversely, only a small percentage of wheels with high angle of attack generate high-level squeal (see Fig. 4 d).

A summary of the distribution of squeal events by noise level and angle of attack category is shown in Table 1. It appears that squeal noise rarely occurs when the wheel's angle of attack is less than 10mrad. The chance that squeal noise will be observed increases with the angle of attack, and the percentage of observed squeal noise occurrence reached 50~60% when the angle of attack exceeds 40mrad. In total, approximately 7.5% of wheels with angle of attack greater than 15mrad were observed to generate squeal noise above 100dBA. This observation allows some tentative conclusions to be drawn:

- Squeal noise is highly related to wheel angle of attack. The higher noise events were associated with higher angles of attack. Events exceeding 120dB were only observed when wheels with AoA exceeding 25mrad. Conversely, only a few wheels with irregular AoA were observed to generate squeal noise.
- The occurrence of flanging noise has no obvious relationship to wheel angle of attack.

Table 1. Summary of squeal events by the category of angle of attack and noise level measured from August 2007 to February 2008

Angle of attack		**Noise category (dBA)**							**Subtotal**	
(mrad)	**Num**	100 - 105	105 - 110	110 - 115	115 - 120	120 - 125	125 - 130	130 - 135	**Num**	**%**
<=5	27382	1	-	-	-	-	-	-	1	0.00%
5-10	13923	6	2	-	-	-	-	-	8	0.06%
10-15	1694	-	1	-	-	-	-	-	1	0.06%
15-20	575	1	-	1	-	-	-	-	2	0.35%
20-25	459	2	9	6	7	-	-	-	24	5.23%
25-26	360	3	11	11	9	3	-	-	37	10.28%
30-35	201	1	4	4	7	2	4	-	22	10.95%
35-40	84	-	2	5	4	4	1	-	16	19.05%
40-45	26	-	1	-	3	5	4	1	14	53.85%
>45	3	-	1	1	-	-	-	-	2	66.67%
Total	44707	14	31	28	30	14	9	1	127	0.28%

It should be noted that the result presented in Table 1 only represents those squeal events with the noise level exceeding 100dBA with a frequency between 0.5 and 10kHz. Squeal noise occurring in the ultrasonic range (greater than 20kHz) has been reported by Rudd [1]. The percentage of squealing axles will be higher if all squeal events with lower amplitude or with a frequency out of the specific range are counted. Nevertheless, it is expected that this will alter the observation that many wheels with a high angle of attack do not generate squeal noise. This is evidence of the absence of the other conditions required for the generation of squeal such as friction conditions, which according to Rudd's theory [1] influences the occurrence of squeal noise. Research is continuing in order to verify the influence of other parameters that play a role in squeal noise generation.

5 Conclusions

A detailed method of curving noise identification, including the type of curving noise and its generation source, has been devised and implemented. The pure tonal squeal noise and multi-tonal flanging noise can be identified by their temporal and frequency characteristics and can be identified with a particular wheel. The ability to differentiate inner or outer wheel generated curving noise using rail acceleration measurements taken at the low and high rail has been demonstrated.

Specific trackside measurements undertaken to study the relationship between curving noise and its influencing parameters indicate a relationship between wheel angle of attack and curving noise. The high-level squeal noise measured was associated with wheels that have an angle of attack >10mrad; and conversely, only a few (about 7.5%) of wheels with high angle of attack were observed to generate >100dBA squeal events in the 0.5~10kHz frequency range. It appears that both the noise level and the chance of squeal noise occurrence increase with increased wheel angle of attack. The occurrence of flanging noise was found to have no obvious relationship to wheel angle of attack.

A surprising finding from this field study is that predominantly outer wheel/high-rail squeal events were observed which is contradictory to observations reported by others.

Acknowledgements

This work is being conducted under CRC for Rail Innovations Project R1-105 Improved Rail Noise Management. The authors are grateful to the CRC for Rail Innovation (established and supported under the Australian Government's Cooperative Research Centres program) for the funding of this research.

References

[1] Rudd, M.: Wheel/rail noise—part 2: wheel squeal. Journal of Sound and Vibration 46, 381–394 (1976)

[2] Fingberg, U.: A model for wheel-rail squealing noise. Journal of Sound and Vibration 143, 365–377 (1990)

[3] Périard, F.: Wheel-Rail Noise Generation: Curve Squealing by Trams. PhD Thesis, Delft University of Technology (1998)

[4] Heckl, M.: A curve squeal of train wheels—part 1. Journal of Sound and Vibration 229, 669–693 (1999)

[5] de Beer, F.G., et al.: Squeal noise of rail- bound vehicles influenced by lateral contact position. Journal of Sound and Vibration 267(3), 497–507 (2003)

[6] Glocker, C.: Curve squealing of trains: Measurement, modelling and simulation. Journal of Sound and Vibration 324(1-2), 365–386 (2009)

[7] Koch, J.R., et al.: Curve squeal of urban rolling stock – Part 2: Parametric study on a 1/4 scale test rig. Journal of Sound and Vibration 293, 701–709 (2006)

[8] Monk-Steel, A.A., et al.: An investigation into the influence of longitudinal creepage on railway squeal noise due to lateral creepage. Journal of Sound and Vibration 293, 766–776 (2006)

[9] Thompson, D.J.: Curve Squeal Noise. In: Railway Noise and Vibration Mechanisms, Modelling and Means of Control, 1st edn., ch. 9. Elsevier Ltd, Amsterdam (2009)

[10] Eadie, D.T., Santoro, M.: Railway Noise and the Effect of Top of Rail Liquid Friction Modifiers: Changes In Sound And Vibration Spectral Distributions. In: 6th International Conference on Contact Mechanics and Wear of Rail/Wheel Systems, Gothenburg, Sweden (June 2003)

[11] Jiang, J., Dwight, R.: Determining Wheel-Rail Wear Conditions Using Wheel-Rail Noise. In: 7th International Conference on Contact Mechanics and Wear of Rail/Wheel Systems (2006)

[12] Harrison, H., Ahlbeck, D.: Development and evaluation of wayside wheel/rail load measurement techniques. In: Proceedings / International Conference on Wheel/Rail Load and Displacement Measurement Techniques, January 19-20 (1981)

[13] Jiang, J.D., Dwight, R.: On-board wheel wear and wheel-rail noise assessment for the direction of track maintenance, D4 - CRC Rail Project #36 Overall Report, University of Wollongong (July 2004)

[14] Stappenbeck, H.V.: Zeitschrif Vereinschaft Deutscher Ingenieuren 96, 171 - 175. Streetcar curve noise (1954)

[15] Vincent, N.: Curve squeal of urban rolling stock—Part 1: State of the art and field measurements. Journal of Sound and Vibration 293, 691–700 (2006)

[16] Remington, P.J.: Wheel/rail squeal and impact noise: what do we know? What don't we know? Where do we go from here? Journal of Sound and Vibration 116(2), 339–353 (1985)

Effectiveness of Using Elastomeric Pads to Mitigate Impact Vibration at an Urban Turnout Crossing

S. Kaewunruen

Rail Corporation – Track Engineering,
Level 13, 477 Pitt St., Sydney NSW 2000 Australia
Tel: +61 2 8922 1551, Fax: +61 2 8922 1154
e-mail: sak.kaewunruen@railcorp.nsw.gov.au

Summary

A turnout is a structural track system used to divert a train to other directions or other tracks. A turnout system includes rails, switches, crossings (special track components), steel plates, rubber pads, biscuits, fasteners, screw spikes, concrete bearers, ballast and formation. The wheel rail contact over the crossing transfer zone has a dip-like shape where the wheel trajectory is not smooth. The wheel/rail interaction with such imperfect contact transfer can cause detrimental impact loads on the railway track and its components. In addition, the large impact emits disturbing noises (either impact or ground-borne noise) to railway neighbors.

A methodology to reduce the wheel/rail impact at a turnout crossing is to adopt more resilient elastomeric pads. The functions of the turnout rubber pads are quite similar to those of plain-track rail pads. Nonetheless, it is important to note that the stiffness of a turnout structure is considerably higher than that of a plain track. As a result, it is desirable that the turnout pads should be softer than general rail pads, in order to moderate global track stiffness as well as to attenuate a higher impact force due to the wheel/rail interaction over the crossing transfer zone (dip-like trajectory). A trial of such methodology has been implemented at an urban turnout diamond in RailCorp's urban rail network and the effectiveness of such a method has then been evaluated using integrated numerical train/track simulations, axle box acceleration and ride quality data obtained from the calibrated track inspection vehicle "AK Car". The field trial demonstrates that using more resilient elastomeric pads in an urban turnout is effective in mitigating impact vibration and upgrading passenger ride comfort. However, it is important to note that an overly soft or hard pad is not recommended because it tends to reduce lives of other adjacent track components.

1 Introduction

A turnout is a structural track system used to divert a train to other directions or other tracks. A turnout system consists of rails, points (or called 'switches'),

T. Maeda et al. (Eds.): Noise and Vibration Mitigation for Rail Trans. Sys., NNFM 118, pp. 357–365.
springerlink.com

crossings (special track components), steel plates, rubber pads, biscuits, fasteners, screw spikes, concrete bearers, ballast and formation. The typical turnout geometry is shown in Fig. 1. In general, the turnout area contains no superelevation or has a flat surface installed on a canted formation. Although a new geometric design method provides a smooth transition from a tangent track, a wheel still has to travel over the wing rail to the crossing nose, or vice versa.

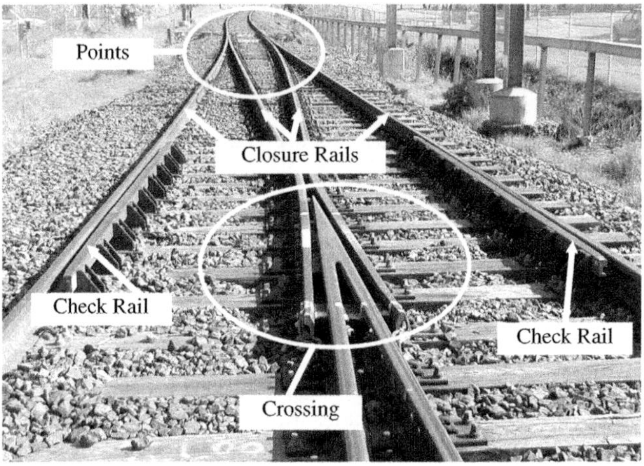

Fig. 1. Typical turnout geometry (color figure online)

A turnout is an inevitable structure in railway tracks whose crossing imparts a significant discontinuity in the rail running surface. The wheel/rail interaction on such imperfect contact transfer can cause detrimental impact loads on railway track and its components [1-4]. The transient vibration could also affect surrounding building structures. In addition, the large impact emits disturbing noises to railway neighbors [5]. The impact and ground-borne noises are additional to the normal rolling noise.

Many previous studies have predicted impact forces and noise using numerical models [5]. However, only a few have implemented impact mitigation strategies in the field and even fewer field trial reports are available in the literature [5]. The impact mitigation strategies at an urban turnout include wheel/rail transverse profiling and longitudinal profiling of crossings, increased turnout resilience and damping, changes to rolling stock, external noise/vibration controls, etc. This paper presents the implementation and the effectiveness of a methodology of using softer elastomeric pads in order to improve crossing impact at an urban turnout in Australia.

2 Interaction of Train and Turnout Structure

The wheel/rail contact over the crossing transfer zone has a dip-like shape where the wheel trajectory is not smooth. The accurate shape of the wheel trajectory (running top) and dip angle will depend on the wheel and running rail profiles. The associated dip angles, which are the acute angle between the tangents to the wheel trajectory at the point where it abruptly changes direction, can then be estimated from the wheel trajectories, as illustrated in Fig. 2.

a) wheel traversing a v-crossing (white paint showing the contact band)

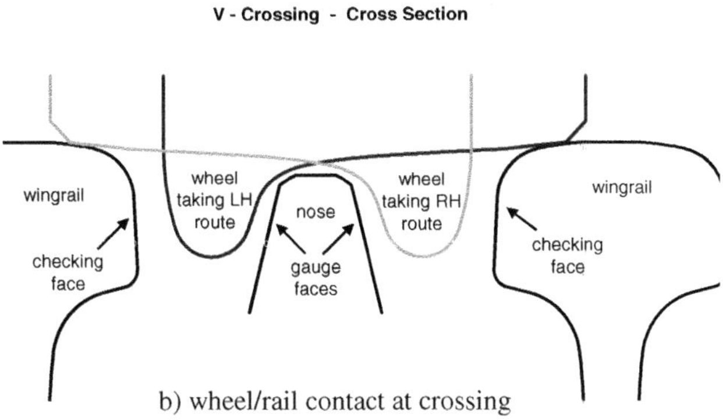

b) wheel/rail contact at crossing

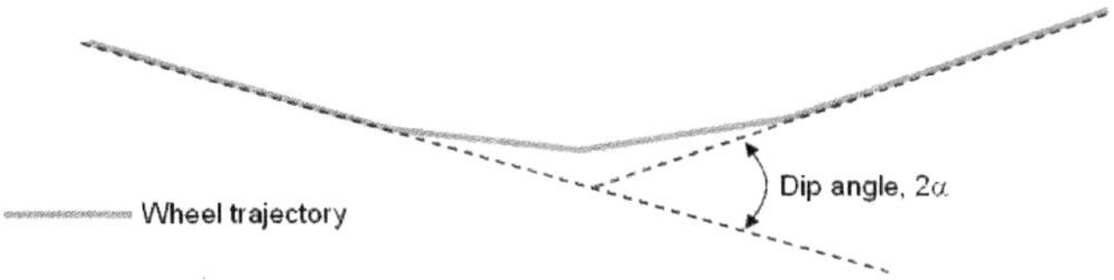

c) wheel trajectory in transfer zone

Fig. 2. Transfer zone at crossing (color figure online)

2.1 Turnout Force

It is generally assumed that the high-frequency impact force (P_1) that occurs either at a nose or at a wing rail has little effect on the rail foot [7]. On the other hand, the dynamic P_2 force (the second peak in the impact force history) has significant influence on the crossing components. The distance from the point of impact to the point of the peak impact force depends on a number of factors including train speed.

The common damage zones to be considered are the rail foot within 0.75 m of any joints to plain rail; and the base of the crossing in transfer zone (extending 0.75 m on both sides of the pick up point); and other components in vicinity of the crossings.

In a calculation of P_2 force, the track damping C_t is normally negligible. For plain tracks, it is commonly found that the track mass is relatively low in comparison with the wheel set mass and is then neglected. In contrast, for a turnout crossing, the track mass tends to be of significance and it cannot be neglected. Jenkins et al. [6] has proposed a formula for estimating a dynamic P_2 force as follows:

$$P_2 = P_0 + 2\alpha \cdot v \cdot \left[\frac{M_u}{M_u + M_t}\right]^{\frac{1}{2}} \cdot \left[1 - \frac{\pi \cdot C_t}{4\sqrt{K_t \cdot (M_u + M_t)}}\right] \cdot [K_t \cdot M_u]^{\frac{1}{2}}$$

where P_2 = Dynamic vertical force (kN)
P_0 = Vehicle static wheel load (kN)
M_u = Vehicle unstrung mass per wheel (kg)
2α = Total joint angle or equivalent dip angle (rad)
v = Vehicle velocity (m/s)
K_t = Equivalent track stiffness (MN/m)
C_t = Equivalent track damping (kNs/m)
M_t = Equivalent track mass (kg)

2.2 Track Stiffness

Track stiffness K_t can be estimated using a beam on elastic foundation model. The deflection under a point load can be written as:

$$K_t = \frac{P}{y} = 8\beta^3 EI$$

$$\beta = \left[\frac{k_f}{4EI}\right]^{\frac{1}{4}}$$

where E = Young's modulus
I = Moment of inertia of the rail or crossing section
k_f = foundation modulus (track modulus)
$k_f = 4.104P_o$ (axle load in tonnes) +14.61, MPa

2.3 Equivalent Track Mass

Considering the bearer spacing of 600 mm, a bearer mass of 600 kg (half can be associated with the crossing), a mass per unit length of the crossing is 120 kg/m, the track mass per unit rail length (m) can be estimated as 620 kg/m. The equivalent track mass (lumped) at crossing $M_t = 3m/2\beta$. Thus, M_t is varied from 705 kg to 756 kg.

2.4 Crossing Design Issues

It is evident that the track stiffness at crossing is much higher than normal plain tracks, resulting in a higher impact force. A concern in crossing design is that the foundation might be subjected to excessive bearing pressure and the track components will degrade dramatically due to the equivalent dip angle. The impact attenuation capacity might be compromised, leading to ballast pulverization and noise nuisance.

3 Impact Damage of Turnout

There was a problem in relation to significant degradation of turnout components due to large impact force from the Clyde 700 diamond (RailCorp's western main rail network). White ballast (ballast pulverization) could be observed near the crossings, as shown in Fig. 3. The ballast was pulverized due to the impact vibration of crossing structures and bearers, which is excited by the rail dip-like irregularity. Fig. 4 also shows a cracked bearer underneath the stockrail at Country end v-crossing. This negative bending failure is due to a wheel impact at the v-crossing [8-9].

Fig. 3. White ballast (color figure online)

Fig. 4. Broken concrete bearer (color figure online)

4 Field Trial

In turnout structures, pads are installed between the steel plate directly underneath the rail and the concrete bearer. Turnout pads used are selected based on the UIC-864-5 code, which specifies normal rail pads used in a plain track. Field experience shows that the specified properties of pads are not optimal for turnouts. Although the turnout pads have similar functions as the normal rail pads, the stiffness of the turnout components is much higher than that of plain tracks. Accordingly, it is desirable that the turnout pads should be softer than general rail pads, in order to moderate global track stiffness as well as to attenuate a higher impact force due to the wheel/rail interaction over the crossing transfer zone. A methodology to reduce the wheel/rail impact at a turnout crossing is to adopt more resilient elastomeric pads [10]. A trial of such methodology has been implemented at an urban turnout diamond. Currently, the existing pads used in a turnout are specified as SA-47 6 mm thick. In this study, the SA-47 12 mm thick pads have been chosen for trial (40% softer than the common turnout pads). This is the thickest pad with the maximum resiliency that could fit within the existing fastening systems. The pads selected are also commercially available.

To understand the role of the rubber pad stiffness on the load bearing area, a 3D finite-element model of 6m concrete bearer, 12 mm thick rubber pad, 20 mm thick steel plate has been developed [9]. In the simulation, only the load bearing transfer area has been considered via actual track and operational parameters [9]. The preloading of 40 kN was applied for each screw spike. The track support modulus used in this study is 0.4 MPa/mm. The validated FEM suggests that an excessively soft pad will cause bowing of the steel plate (see Fig. 5) and then break the screw spikes. Soft to moderate resilience of elastomeric pads is thus more desirable. It should be noted that the soft pads also help reduce the bending stress in the bearer; while the hard pad tends to transfer the flexure to the bearer.

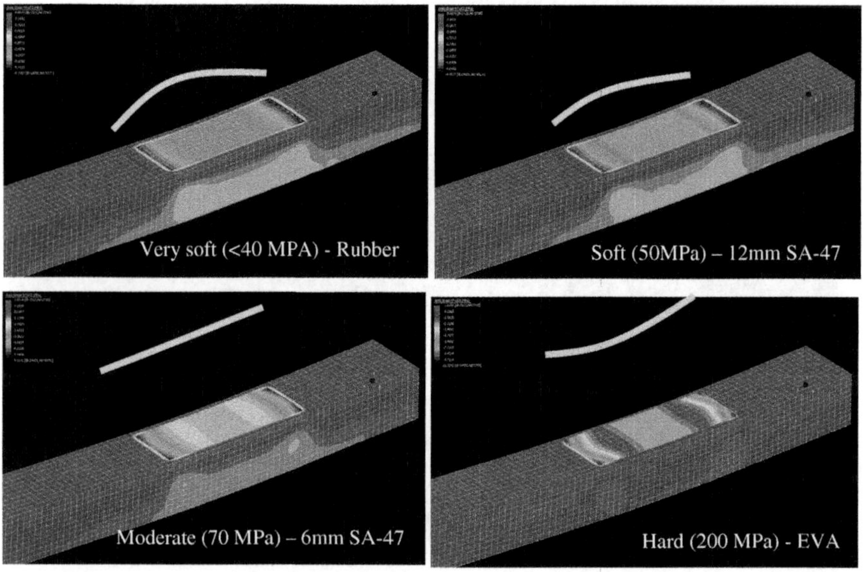

Fig. 5. Quasi-static stress redistribution on the elastomeric pads (color figure online)

After considering these issues, the SA-47 12 mm thick pad has been used for this field trial. The common turnout pads in the diamond in-fill have been replaced by the softer pads. Intermediate stiffness pads have been used to provide transition to the trail soft pads [9].

5 Results and Discussion

To evaluate the noise and vibration performance of the softer rubber pads in the turnout system, the axle box acceleration data was collected from AK Car archive for the runs between April 2005 and November 2008. It is noted that the data sampling rate is 3500 Hz, resulting in a maximum limit of frequency analysis of 1600 Hz. The previous numerical simulations [11] have been used to predict the impact forces from the historical axle box acceleration data. Filtering bands have been applied to the vibration data in order to distinguish between P1-related vibration and P2-related vibration. In relation to track dynamics, the frequency band used in this study is 70–200 Hz for P1 and 20–70 Hz for P2.

Fig. 6 shows the average summary data of the track dynamic forces, which are associated with the recorded wheel vibrations. The estimated linear trend line of track condition has been appended into the curves to project the level of vibration. This study focuses on the track deterioration before and after the track reconditioning (when the pads were replaced) in July 2006. It should be noted that the wheel vibration could vary because the AK car speeds were inconsistent (57 km/h – 11/04/2005; 38 km/h – 13/07/2005; 40 km/h – 19/12/2005; 56 km/h – 23/03/2006; 49km/h – 12/07/2006; 40 km/h – 22/03/2007; 49 km/h – 8/09/2008; and 52 km/h – 20/11/2008). Accordingly, a train/track simulation that has already taken the vehicle speeds into account has been used in order to calculate the equivalent dynamic dip angle [11]. Ride quality is shown in Fig. 7. It is found that the overall combined ride comfort has been significantly improved [12].

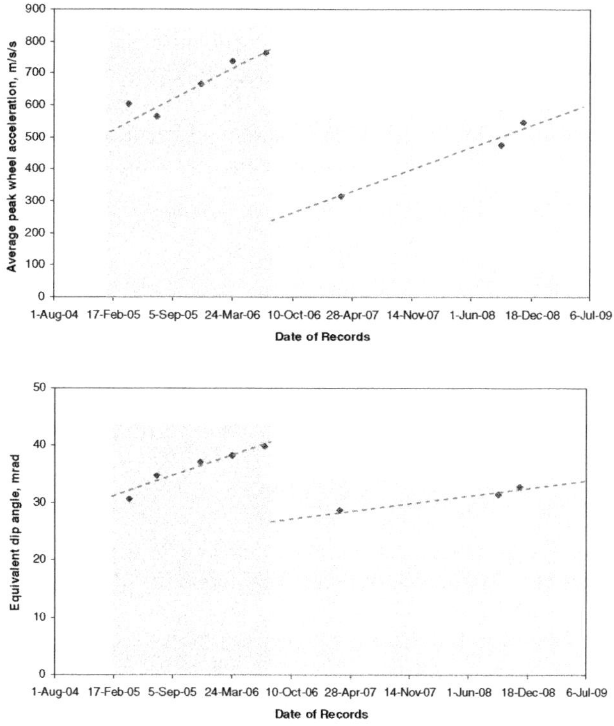

Fig. 6. Wheel peak impact forces (top – axle vibration; bottom – equivalent dip angle) (color figure online)

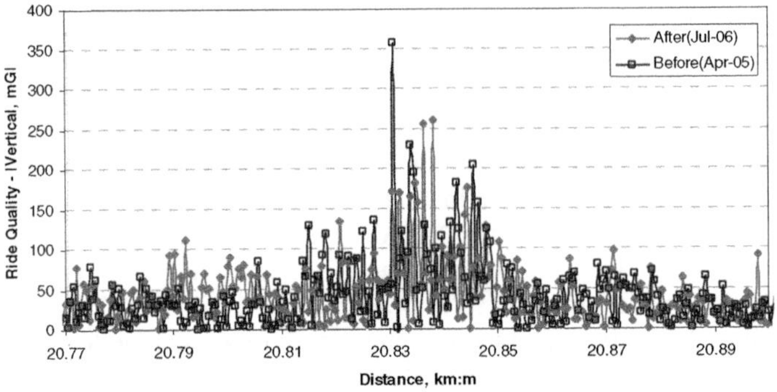

Fig. 7. Ride comfort (color figure online)

6 Remarks

This study shows the trial and the assessment for the effectiveness of using elastomeric pads in an urban turnout to mitigate impact vibration and improve passenger ride comfort. The effectiveness of the elastomeric pads can be assessed by considering the difference between the vibration after upgrading and the vibration before upgrading. Considering the condition before and after replacing the softer elastomeric pads, the unfiltered wheel vibrations reduce significantly. Taking into account the train speeds, the train/track simulation shows that the equivalent dip angle leading to the dynamic P2 force could be decreased as much as 27.5%. It is also found that the ride quality has significantly been improved. Note that although impact force is reduced by decreasing pad stiffness; an overly soft pad is not recommended because of the potential damage to other components.

Acknowledgement

The author is very grateful to RailCorp for the permission to publish this paper. The author would also like to sincerely thank Malcolm Kerr, Allan Churchill, Jeff Stead, Bob Fogarty, Andrew Wilson and RailCorp district staff for the assistance during the field investigation.

References

[1] Andersson, C., Dahlberg, T.: Wheel/rail impacts at a railway turnout crossing. Proceedings of the Institution of Mechanical Engineers, Part F: Journal of Rail and Rapid Transit. 212F, 135–146 (2003)

[2] Wu, T.X., Thompson, D.J.: On the impact noise generation due to a wheel passing over rail joints. Journal of Sound and Vibration 267(3), 485–496 (2003)

[3] Remennikov, A.M., Kaewunruen, S.: A review of loading conditions for railway track structures due to train and track vertical interaction. Structural Control and Health Monitoring 15(2), 207–234 (2007)

[4] Kaewunruen, S., Remennikov, A.M.: Dynamic properties of railway track and its components: recent findings and future research direction. Insight: Non-Destructive Testing and Condition Monitoring 52(1), 20–22 (2010)
[5] Thompson, D.J.: Railway Noise and Vibration. Elsevier, Amsterdam (2010)
[6] Jenkins, H.H., Stephenson, J.E., Clayton, G.A., Morland, J.W., Lyon, D.: The effect of track and vehicle parameters on wheel/rail vertical dynamic forces. Railway Eng. Journal, 2–16 (1974)
[7] Railtrack PLC: RT/CE/S/012 Specification of Cast Austentic Manganese Steel Crossings, Railtrack PLC (February 2002)
[8] Kaewunruen, S., Remennikov, A.M.: Dynamic flexural influence on a railway concrete sleeper in track system due to a single wheel impact. Eng. Failure Analysis 16(3), 705–712 (2009)
[9] Kaewunruen, S.: Performance Review of Elastomeric Pads in Turnout. RailCorp. Technical Report No TR159. 66 p, Sydney, Australia (2009)
[10] Standards Australia: AS1085.19 Railway fastening assemblies, Sydney, Australia (2001)
[11] Sun, Y.Q., Cole, C., McClanachan, M., Wilson, A., Kaewunruen, S., Kerr, M.B.: Rail short-wavelength irregularity identification based on wheel-rail impact response measurements and simulations. In: Proceedings of the 9th International Heavy Haul Conference, Shanghai, China, June 21-23 (2009)
[12] RailCorp Rolling Stock Engineering: Minimum Operating Standards for Rolling Stock, Sydney, Australia (2010)

A 2.5D Coupled FE-BE Methodology for the Prediction of Railway Induced Vibrations

S. François[1], P. Galvín[1,2], M. Schevenels[1], G. Lombaert[1], and G. Degrande[1]

[1] Department of Civil Engineering, K.U.Leuven,
Kasteelpark Arenberg 40, B-3001 Leuven, Belgium
Tel: +32 16 321668, Fax: +32 16 321988
e-mail: stijn.francois@bwk.kuleuven.be
[2] Escuela Técnica Superior de Ingenieros, Universidad de Sevilla,
Camino de los Descubrimientos, E-41092 Sevilla, Spain

Summary

Ground vibrations induced by railway traffic are often studied by means of two-and-half dimensional (2.5D) models that are based on the longitudinal invariance of the track geometry. In this paper, a 2.5D methodology is used where the finite-element method is combined with a boundary-element method, based on a regularized boundary integral equation. In the formulation of the boundary integral equation, the Green's functions of a layered elastic half space are used, so that no discretization of the free surface or the layer interfaces is required. The methodology is applied to two cases. In the first case, two alternative models for a ballasted track on an embankment are compared. In the first model, the ballast and the embankment are modeled as a continuum using 2.5D solid elements, whereas a simplified beam representation is adopted in the second model. A very large difference is found for the free field mobility of both models, which demonstrates the need for detailed 2.5D modeling of the embankment. In the second case, the efficiency of a vibration isolating screen is studied. A vibration isolating screen is a soft or stiff wave barrier. Due to the impedance contrast between the isolating screen and the soil, incident waves are reflected. As a result, the vibration levels behind the screen are effectively reduced. Both examples demonstrate that the use of detailed 2.5D models result in a better insight in the mechanical behavior of the coupled soil-track system.

1 Introduction

For the prediction of railway-induced vibrations, the geometry of the track-soil system is often assumed to be invariant in the longitudinal direction [5]. This allows for a Fourier transform of the coordinate in the longitudinal direction of the track and leads to a solution in the frequency-wave number domain where the

T. Maeda et al. (Eds.): Noise and Vibration Mitigation for Rail Trans. Sys., NNFM 118, pp. 367–374.
springerlink.com

original three-dimensional (3D) problem is replaced by a two-dimensional (2D) problem for each wave number. This two-and-half dimensional (2.5D) methodology results in a considerable reduction of the computation time [1, 5] compared to full 3D models.

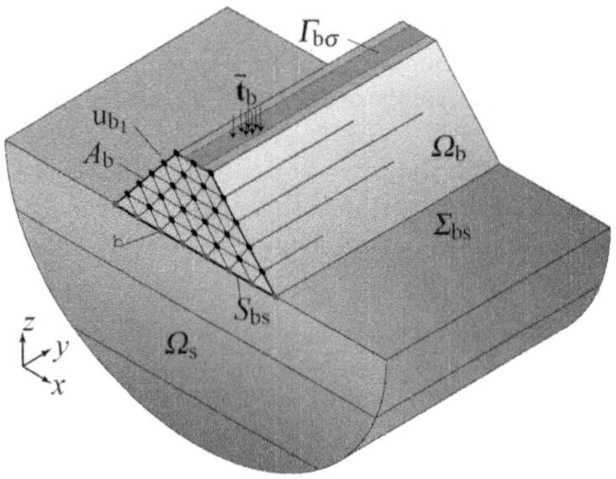

Fig. 1. The geometry of the coupled soil-structure system.

Recently, a novel 2.5D FE-BE methodology has been proposed by François et al. [1]. The 2.5D FE-BE methodology combines the classical 2.5D FE method with a novel 2.5D BE method, based on a regularized boundary integral equation in terms of the Green's functions of a layered half space.

In this paper, the 2.5D coupled FE-BE models are applied to the prediction of railway-traffic-induced vibrations. Two cases are considered. In the first case, two alternative models for a ballasted track on an embankment are compared. The 2.5D BE-FE method is used to model the ballast and the embankment as an elastic continuum, as an alternative to simplified ballast and embankment models used in the literature [3, 5]. In the second case, the efficiency of a vibration isolating screen is studied.

2 Methodology

The dynamic interaction between a railway track and the underlying soil is a problem of dynamic soil--structure interaction. A domain decomposition method is used to solve the problem, where the sub-domain Ω_b represents the structure and the sub-domain Ω_s the soil (Fig. 1).

It is assumed that the geometry of the track is invariant with respect to the coordinate y in the longitudinal direction. The soil is modeled as a horizontally layered half space and, therefore, invariant with respect to the y-direction as well. The dynamic track-soil interaction problem is assumed to be linear and all equations are elaborated using virtual work in the frequency domain.

The equilibrium equation for the dynamic soil-structure interaction problem is formulated in a variational form. For any virtual displacement field $\mathbf{v}_{\mathrm{b}}$ imposed on the structure Ω_{b}, the sum of the virtual work of the internal and the inertial forces is equal to the virtual work of the external loads:

$$-\omega^2 \int_{\Omega_{\mathrm{b}}} \mathbf{v}_{\mathrm{b}} \cdot \rho_{\mathrm{b}} \mathbf{u}_{\mathrm{b}} d\Omega + \int_{\Omega_{\mathrm{b}}} \boldsymbol{\epsilon}_{\mathrm{b}}(\mathbf{v}_{\mathrm{b}}) : \boldsymbol{\sigma}_{\mathrm{b}}(\mathbf{u}_{\mathrm{b}}) d\Omega$$

$$= \int_{\Omega_{\mathrm{b}}} \mathbf{v}_{\mathrm{b}} \cdot \rho_{\mathrm{b}} \mathbf{b}_{\mathrm{b}} d\Omega + \int_{\Gamma_{\mathrm{b}\sigma}} \mathbf{v}_{\mathrm{b}} \cdot \bar{\mathbf{t}}_{\mathrm{b}}^{\mathbf{n}_{\mathrm{b}}} d\Gamma + \int_{\Sigma_{\mathrm{bs}}} \mathbf{v}_{\mathrm{b}} \cdot \mathbf{t}_{\mathrm{b}}^{\mathbf{n}_{\mathrm{b}}}(\mathbf{u}_{\mathrm{b}}) d\Gamma$$

where $\mathbf{u}_{\mathrm{b}}$ is the displacement vector in the structure, $\rho_{\mathrm{b}}\mathbf{b}_{\mathrm{b}}$ denotes the body force in the domain Ω_{b} and $\mathbf{t}_{\mathrm{b}}^{\mathbf{n}_{\mathrm{b}}} = \sigma_{\mathrm{b}} \cdot \mathbf{n}_{\mathrm{b}}$ is the traction vector on a boundary with unit outward normal vector $\mathbf{n}_{\mathrm{b}}$ (Figure 1). Tractions $\bar{\mathbf{t}}_{\mathrm{b}}^{\mathbf{n}_{\mathrm{b}}}$ are imposed on the boundary $\Gamma_{\mathrm{b}\sigma}$.

Accounting for the equilibrium of stresses on the interface Σ_{bs} and using a 2.5D finite-element formulation for the interpolation of the displacement field with respect to the coordinates x and z, equation can be elaborated as [1]:

$$\left[-\omega^2 \mathbf{M}_{\mathrm{bb}} + \mathbf{K}_{\mathrm{bb}}^0 - ik_y \mathbf{K}_{\mathrm{bb}}^1 - k_y^2 \mathbf{K}_{\mathrm{bb}}^2 + ik_y^3 \mathbf{K}_{\mathrm{bb}}^3 + k_y^4 \mathbf{K}_{\mathrm{bb}}^4 + \mathbf{K}_{\mathrm{bb}}^{\mathrm{s}}(k_y, \omega)\right] \tilde{\underline{\mathbf{u}}}_{\mathrm{b}}(k_y, \omega)$$

$$= \tilde{\underline{\mathbf{f}}}_{\mathrm{b}}(k_y, \omega)$$

where $\mathbf{K}_{\mathrm{bb}}^0$, $\mathbf{K}_{\mathrm{bb}}^1$, $\mathbf{K}_{\mathrm{bb}}^2$, $\mathbf{K}_{\mathrm{bb}}^3$ and $\mathbf{K}_{\mathrm{bb}}^4$ are the stiffness matrices, $\mathbf{M}_{\mathrm{bb}}$ is the mass matrix, $\tilde{\underline{\mathbf{f}}}_{\mathrm{b}}(k_y, \omega)$ is the external load vector and $\mathbf{K}_{\mathrm{bb}}^{\mathrm{s}}(k_y, \omega)$ represents the dynamic soil- stiffness matrix. A tilde above a variable denotes its representation in the frequency-wave number domain. The finite element matrices $\mathbf{M}_{\mathrm{bb}}$ and $\mathbf{K}_{\mathrm{bb}}^0$ to $\mathbf{K}_{\mathrm{bb}}^4$ in equation are independent of the wave number k_y and the frequency ω and are only assembled once.

The dynamic soil stiffness matrix $\mathbf{K}_{\mathrm{bb}}^{\mathrm{s}}(k_y, \omega)$ is computed by means of a 2.5D boundary-element method, where a regularized version [1] of the 2.5D boundary integral equation is applied. In the formulation of the boundary integral equation, the Green's functions of a layered elastic half space [4] are used, so that no discretization of the free surface or the layer interfaces is required. Both the 2.5D finite-element and boundary-element methods have been implemented in MATLAB using a C++ mex interface.

3 A Ballasted Track on an Embankment

In this section, the proposed 2.5D FE-BE model is used to predict the transfer functions between the track and free field at a site in Reugny (France) situated along the high-speed railway line LGV Atlantique. The track in Reugny is a classical ballasted track, situated on top of an embankment. The continuously welded UIC60 rails are supported by rail pads and fixed with clips on twin block concrete sleepers with a spacing of $d = 0.60\,\mathrm{m}$. The concrete twin block sleepers have a length $l_{\mathrm{sl}} = 2.41\,\mathrm{m}$ and are composed of two tied concrete blocks with a length $l_{bl} = 0.84\,\mathrm{m}$. The total mass of the sleepers $m_{sl} = 250\,\mathrm{kg}$. The track is supported by a ballast layer with a thickness $h_b = 0.30\,\mathrm{m}$ and a density $\rho_b = 1400\,\mathrm{kg/m^3}$. The embankment has a width $w_{e1} = 6\,\mathrm{m}$ at the top supporting the railway track, a width $w_{e2} = 13\,\mathrm{m}$ at the soil's surface, and a height $h_e = 2\,\mathrm{m}$.

Two alternative track models are considered. In the first model (Fig. 2a), the ballast and the embankment are modeled as an elastic continuum using 2.5D solid elements. The second model (Figure 2b) is a simplified model where the ballast is represented by distributed springs and dampers while the embankment is modeled as an Euler--Bernoulli beam. Similar simplified models of the ballast and the embankment are frequently used for the prediction of railway-induced vibrations [3, 5].

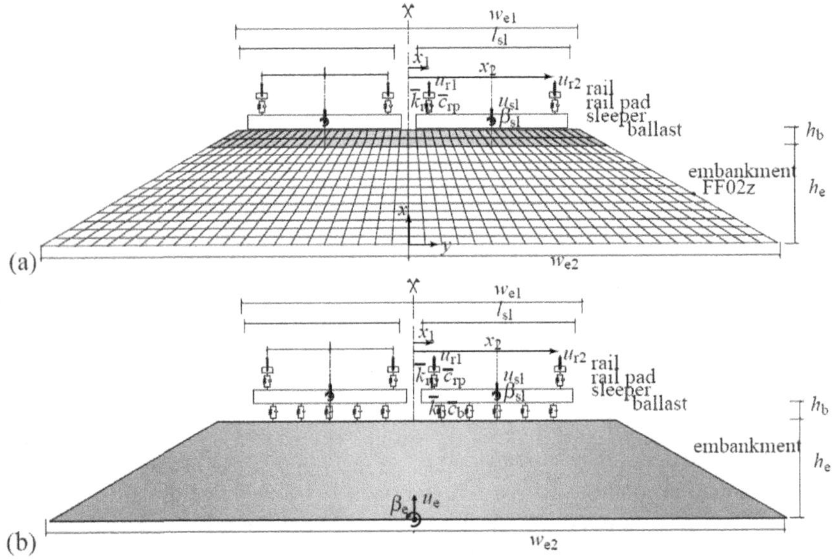

Fig. 2. Cross-section of (a) model 1 and (b) model 2 of the ballasted track on the embankment.

In both models, the rails are represented by Euler--Bernoulli beams. The positions of the rail are determined by $x_1 = 1.145\,\mathrm{m}$ and $x_2 = 2.580\,\mathrm{m}$, with $x_2 - x_1$ equal to the track gage r_d. The internal energy dissipation in the rail is modeled by a loss factor $\eta_r = 0.05$. The rail pads are modeled as continuous spring-damper connections. The rail pad stiffness k_{rp} of a single rail pad is used to calculate an equivalent stiffness $\bar{k}_{rp} = k_{rp}/d = 130 \times 10^6\,\mathrm{N/m^2}$. A loss factor $\eta_{rp} = 0.23$ is used to account for internal energy dissipation in the rail pad.

The concrete sleepers are assumed to be rigid in the plane of the track cross section and are modeled as a uniformly distributed mass $\bar{m}_{sl} = m_{sl}/d$ of

417 kg/m. The sleeper's rotational inertia $\rho_{sl}\overline{I}_{sl} = \rho_{sl}I_{sl}/d$ has been estimated as 298 kgm^2/m taking into account the eccentric position of the two blocks.

In model 1, the ballast bed is modeled as an elastic continuum, using 88 2.5D solid elements [1]. The ballast has a Young's modulus $E_b = 111 \times 10^6\,\mathrm{N/m^2}$, a Poisson's ratio $\nu_b = 0.36$, a density $\rho_b = 1550\,\mathrm{kg/m^3}$ and a loss factor $\eta_b = 1.00$. The embankment is modeled as an elastic continuum using 528 2.5D solid elements and has A Young's modulus $E_e = 170 \times 10^6\,\mathrm{N/m^2}$, Poisson's ratio $\nu_e = 0.36$ and a density $\rho_e = 1400\,\mathrm{kg/m^3}$.

In model 2, the ballast is represented by a set of distributed linear springs and dampers. The smeared ballast stiffness $\overline{k}_b$ is computed from the vertical spring stiffness k_b per sleeper [N/m] as k_b/d and equal to $300 \times 10^6\,\mathrm{N/m^2}$. The loss factor $\eta_b = 1.00$. The equivalent ballast mass $\overline{m}_b$ is computed from the ballast mass m_b situated under each sleeper as m_b/d. The ballast mass m_b is estimated from the height h_b of the ballast layer and a width $w_{b1} = l_{sl}$ and $w_{b2} = 3\,\mathrm{m}$ at the top and the bottom of the ballast layer, respectively, as $m_b = 0.5\rho_b h_b(w_{b1} + w_{b2})b_{bl}$. This leads to a value of $608\,\mathrm{kg/m}$ for the equivalent ballast mass $\overline{m}_b$. The embankment is represented by an Euler--Bernoulli beam, which implies that the cross-section of the embankment is assumed to be rigid.

The soil is modeled as a horizontally layered elastic half space, with a single layer with a thickness of $2.0\,\mathrm{m}$ and a shear wave velocity $C_s = 211\,\mathrm{m/s}$ on top of a half space with a shear wave velocity of $403\,\mathrm{m/s}$. These properties have been obtained from a Spectral Analysis of Surface Waves test on site. The density ρ is equal to $1400\,\mathrm{kg/m^3}$ for the top layer and equal to $2650\,\mathrm{kg/m^3}$ for the underlying half space. The Poisson's ratio ν is 0.36 for the top layer and 0.16 for the half space. The material damping ratio β in both deviatoric and volumetric deformation has a value of 0.05 and 0.06 for the top layer and the half space, respectively.

Fig. 3 compares the free field mobility at a distance of $2\,\mathrm{m}$, $12\,\mathrm{m}$, $32\,\mathrm{m}$ and $72\,\mathrm{m}$ from rail 2 as computed with both models. The point located at $2\,\mathrm{m}$ from the outer rail is situated on the embankment.

A very large difference is found between both results. The assumption of an Euler--Bernoulli model for the embankment leads to high tractions near the edges of the embankment, whereas in the case of the solid embankment model, the tractions are distributed more smoothly along the interface. As demonstrated by Galvín et al [2], model 1 leads to more accurate results as it allows for a better approximation of the stress distribution at the interface between the embankment and the soil.

At frequencies below $20\,\mathrm{Hz}$, the free field mobility is similar for both models, whereas at higher frequencies the free field mobility obtained by model 1 is larger. This might be due to the fact that at low frequencies, the wavelength in the soil is large with respect to the width of the interface between the embankment and the soil, so that the differences in the traction distribution do not significantly affect the radiated wave field. At higher frequencies, the wavelengths are smaller, so that the free field mobility is sensitive to the differences in the traction distribution. The smaller value of the free field mobility for model 2 is probably due to the assumption of a rigid cross-section for the embankment, which leads to very high values of the tractions near the edges of the embankment and, therefore, a substantial filtering effect when the wavelength in the soil is of the same order of magnitude as the width of the interface.

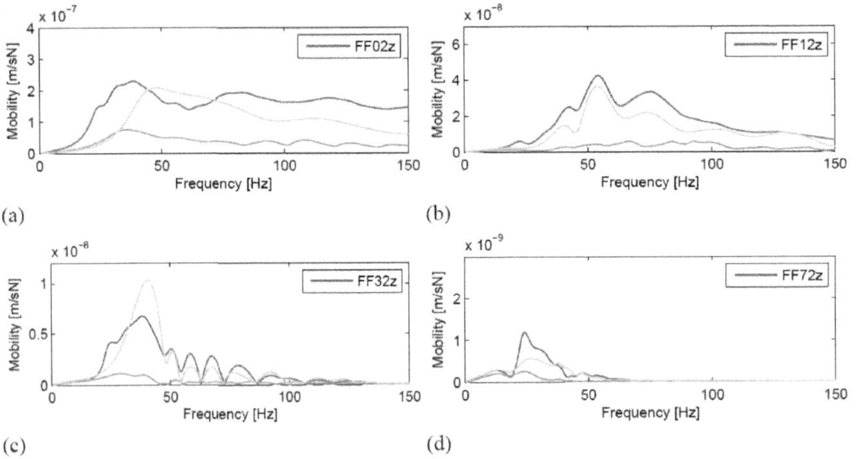

Fig. 3. Free field mobility at (a) 2 m, (b) 12 m, (c) 32 m and (d) 72 m from the outer rail computed from model 1 (dark grey line), model 2 (grey line) and model 2 without embankment (light grey line).

4 Vibration Isolating Screen

In this section, the 2.5D coupled FE-BE methodology is applied to study the efficiency of a vibration isolating screen in the soil. A vibration isolating screen is a soft or stiff wave barrier. Due to the impedance contrast between the isolating screen and the soil, waves are reflected, effectively reducing vibration levels behind the screen.

First, the reference case of a unit point load on the surface of a homogeneous half space is considered. The soil has a shear wave velocity $C_s = 150\,\mathrm{m/s}$, a dilatational wave velocity $C_p = 300\,\mathrm{m/s}$, a density $\rho = 1800\,\mathrm{kg/m^3}$ and a material damping ratio $\beta_s = \beta_p = 0.05$ in both deviatoric and volumetric deformation. The displacement in the soil is dominated by Rayleigh waves with cylindrical wave fronts and a velocity $C_R = 139.8\,\mathrm{m/s}$ which corresponds to a wavelength $\lambda_R = C_R/f = 6.95\,\mathrm{m}$ at the frequency $f = 20\,\mathrm{Hz}$ (Fig. 4a) and $\lambda_R = 3.48\,\mathrm{m}$ at a frequency $f = 40\,\mathrm{Hz}$ (Fig. 4b). The penetration depth of the Rayleigh waves is proportional to the wavelength: most of the wave energy is located above a depth of one wavelength. The embedment depth of the vibration isolating screen should therefore be larger than the penetration depth of the Rayleigh waves.

In the present case, a concrete vibration isolating screen with a depth of 8 m and a width of 0.6 m is considered, which is expected to isolate vibrations above a frequency of 20 Hz. The concrete has a Young's modulus $E_c = 30\,\text{GPa}$, a Poisson's ratio $\nu_c = 0.2$ and a density $\rho_c = 2600\,\text{kg/m}^3$. A vertical point source is considered at the surface of the half space at a distance of 5 m from the centerline of the screen. Figures Figure 4c and Figure 4d show the real part of the vertical displacement in the free field at frequencies of 20 Hz and 40 Hz, respectively. The incident waves are reflected on the vibration isolating screen, reducing vibration levels behind the screen. At 40 Hz, two lines of destructive interference between a direct and reflected Rayleigh waves are observed. As the concrete is much stiffer than the soil, it acts as a rigid wave barrier and a small displacement amplitude is observed along the soil-screen interface.

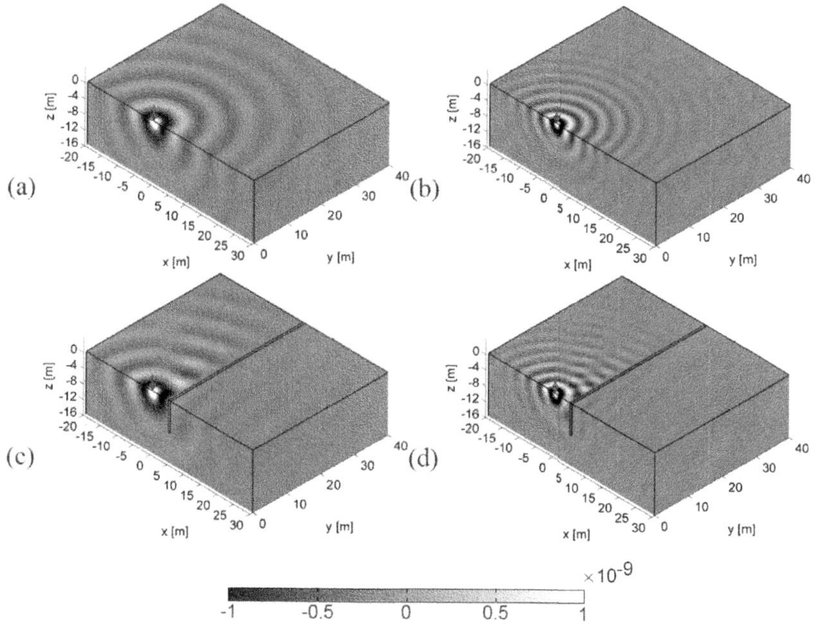

Fig. 4. Real part of the vertical displacement for a unit harmonic point load for a unit vertical point load at the surface of a homogeneous half space at a frequency of (a) 20 Hz and (b) 40 Hz and at distance of 5 m from the center of the concrete vibration isolating screen at a frequency of (c) 20 Hz and (d) 40 Hz.

5 Conclusion

In this paper, the prediction of vibrations from a track on an embankment and a vibration isolating screen in the soil have been computed using a 2.5D coupled FE-BE methodology.

Two alternative models for a ballasted track on an embankment have been considered. In the first model, 2.5D solid elements are used to model the ballast and the embankment as a continuum, whereas in the second model a simplified representation is used. A considerable difference is observed in the free field mobility due to a different tractions distribution at the embankment-soil interface.

In the second case, a vibration isolating screen has been studied. Due to the impedance contrast between the isolating screen and the soil, incident waves are reflected and the vibration levels behind the screen are effectively reduced.

These examples demonstrate that the use of detailed 2.5D models result in a better insight in the mechanical behavior of the coupled soil-track system

References

[1] François, S., Schevenels, M., Lombaert, G., Galvín, P., Degrande, G.: A 2.5D coupled FE-BE methodology for the dynamic interaction between longitudinally invariant structures and a layered halfspace. Computer Methods in Applied Mechanics and Engineering 199(23-24), 1536–1548 (2010)

[2] Galvín, P., François, S., Schevenels, M., Bongini, E., Degrande, G., Lombaert, G.: A 2.5D coupled FE-BE model for the prediction of railway induced vibrations. Soil Dynamics and Earthquake Engineering 30(12), 1500–1512 (2010)

[3] Lombaert, G., Degrande, G., Kogut, J., François, S.: The experimental validation of a numerical model for the prediction of railway induced vibrations. Journal of Sound and Vibration 297(3-5), 512–535 (2006)

[4] Schevenels, M., François, S., Degrande, G.: EDT: An ElastoDynamics Toolbox for MATLAB. Computers & Geosciences 35(8), 1752–1754 (2009)

[5] Sheng, X., Jones, C.J.C., Petyt, M.: Ground vibration generated by a harmonic load acting on a railway track. Journal of Sound and Vibration 225(1), 3–28 (1999)

Advantages of 3-D Vibration and Ground-Borne Noise Model

B.J. Murray

Wilkinson Murray (HK) Limited,
1101 Dannies House, 20 Luard Road, Wanchai, Hong Kong
Tel: +852 2861 2811, Fax: +852 2861 2887
e-mail: barrym@wilkinsonmurray.com.hk

1 Introduction

Airborne noise models are widely used for railway systems and most of these are based on 3-D calculations. In contrast, vibration and ground-borne noise levels are more often calculated using a 2-D model, often in the form of a spreadsheet.

This is probably because of the lack of commercially available 3-D vibration and ground-borne noise models and also the implication of widely accepted reports and guidelines that levels can be predicted accurately using 2-D models that is, not to say that the widely accepted reports and guidelines recommend 2-D models, but they are often written in a form that allows the conclusion that 2-D models will suffice.

Vibration and ground-borne noise predictions using 2-D models are not reliably accurate in all situations, particularly if the inputs are not adjusted to account for the short comings of such a model. The 2-D model calculation to one receiver is based on a single set of parameters usually relating to one point on the track and therefore assumes that the operational conditions applying at that point also apply at all other points on the track. However, the real situation is that there are generally varying conditions along the length of the track. The 2-D model therefore gives accurate results for a situation of tangent (straight) track, fixed speed and constant ground mobility conditions where no turnouts are located near the receiver location. However, it becomes inaccurate under the conditions where the speed is changing, the ground mobility changes over short distances and/or there is a turnout near the receiver location, unless the input parameters are manually adjusted to overcome the limitations.

A 3-D model can overcome the shortcomings of the 2-D model. This paper defines a 3-D model as one which takes into account all three dimensions with varying conditions along the length of the track, but not necessarily a finite-element model or other analytical methods.

This paper compares ground-borne noise levels predicted by a 2-D model with those predicted by a 3-D model, using the same prediction methodology. It shows that there are situations where the 2-D model inaccurately predicts the levels.

T. Maeda et al. (Eds.): Noise and Vibration Mitigation for Rail Trans. Sys., NNFM 118, pp. 375–382.
springerlink.com

2 2-D & 3-D Models

Most models, 2-D or 3-D, are based on a widely accepted prediction methodology. While there are a number of different methodologies used throughout the world, a very common methodology is that described in the Transit Noise & Vibration Impact Assessment report prepared for the US Federal Transit Administration [1].

This report describes, in general terms, methods for the prediction of vibration and ground-borne noise levels from rail systems, particularly underground rail systems. The prediction relies on a number of parameters which need to be measured or developed for the predictions.

It is not the purpose of this paper to discuss these parameters nor the accuracy of the FTA prediction procedure, but to use these parameters and the prediction procedure to predict and compare ground-borne noise levels. The 2-D and 3-D models discussed below have used identical parameters and identical prediction methodologies, based on the FTA report.

2.1 2-D Models

The most common 2-D model uses a spreadsheet for the prediction. The ground-borne noise level at the receiver is predicted using a series of parameters at the source point, which is commonly taken as the closest point (unless there are complications which do not allow this). The speed, track insertion loss, turnout factor and tunnel coupling loss are assigned in accordance with the factors applying at the source point. The ground mobility is then assigned in accordance with the ground between the source point and receiver point over the slant distance. Other factors relating to the building (building coupling loss, building amplification and conversion to noise) are assigned according to the building type.

Those parameters which are assigned according to the source point and the ground mobility parameter assigned according to the ground between the source point and receiver point are all based on one point on the track. It may be that the engineer chooses to base some of these factors on a different point on the track from the others (because of some practical reason), but only one value can generally be assigned to each parameter for each receiver and, therefore, the parameter must be based on a particular point on the track.

It is possible to use several source points, rather than one, spread out along the length of the track to take account of the factors which vary along the length of the track (possibly using point forces, rather than force density), but this then turns the 2-D model into a 3-D model. From my experience, this is not often done anyway.

2.2 3-D Models

A 3-D model can be used for the prediction of ground-borne noise levels and such a model would often take a form similar to commercially available airborne noise models for rail systems. Essentially, the movement of the train along the track is simulated in the model and this allows the speed of the train to vary along the length of the track and, accordingly, the force density level. It is also possible to assign a short length of track for each turnout whereby the force density level over that short length is increased in accordance with the assumed turnout adjustment (often taken as 10dB). Each section of track is assigned a ground mobility factor so that the ground mobility changes as the train moves along the track and also along the length of the train.

Wilkinson Murray has developed a 3-D vibration and ground-borne noise model called *MoleRat*. This model allows the track to be divided into a series of segments, typically 5 or 10m long, and also allows the train (of fixed length) to be moved along the track. For each train position the vibration and ground-borne noise level from each segment is calculated at each receiver point and the total level from the full train length is then calculated.

This procedure allows a noise profile to be calculated for the movement of the train along the full length of the track. From this, the maximum level and the L_{eq} level can readily be determined.

The important point about such a 3-D model is that the noise level at any point in time, and consequently the maximum and L_{eq} levels, is a function of the full track length, as opposed to one source point.

3 Factors Affecting 2-D Models

As indicated above, the 2-D model assumes all track and ground parameters are based on one source point, even though some of these parameters may change along the length of the track. Comparative predictions using 2-D and 3-D models are used here to demonstrate the weakness of the 2-D model, using identical methodologies based on the FTA report.

3.1 Speed

The speed of a train often varies along the length of the track, owing to things such as speed limits and the need to stop at stations. This is particularly relevant for receivers close to stations where the train speed is changing quickly. In these circumstances, it is possible for the maximum noise level to be generated when the train is not at the closest source point to the receiver point, but at a source point further away where the speed is higher. In addition, the speed of the train is commonly defined by the rail engineer at the front of the train. While the train speed is varying, the back of the train can be traveling past a receiver point at a significantly different speed from that of the front of the train.

Since the 3-D model divides the track into segments, it can assign a different speed for each segment for each part of the train, or at least the correct speed for each segment. The varying speed is therefore accounted for.

Using the 2-D model, it is necessary to make an assumption of speed applying to the calculation. Often the speed of the train at the closest source point is assigned for the calculation, and again this speed is often based on the speed of the front of the train.

By reference to Fig. 1, a low speed is often assigned near a station based on the nearest source point whereas the maximum noise level may be generated further from the station where the speed is substantially higher. The 2-D model in this case will not necessarily predict the correct maximum level.

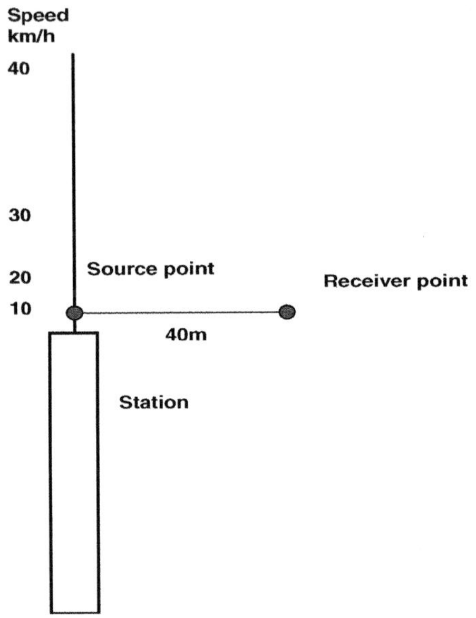

Fig. 1. Diagram showing 2-D calculation of speed factor (color figure online)

For the example in Fig. 2, it has been found that the 3-D model predicts higher ground-borne noise levels than the 2-D model. The predicted levels at the receiver point at a slant distance of 40m are as shown in Table 1.

Table 1. Calculated Ground-borne Noise Levels for Varying Speed at a Station

Model	Predicted Level (dBA)
3-D	24
2-D	21

Given that the 3-D model can allow for the varying speed and the 2-D model cannot, the 2-D model would under predict by 3dB because of this factor. This is not to say that the 3-D model is necessarily accurate because it must rely on the FTA prediction procedure and the available input information.

In some cases, the engineer may acknowledge the weakness of the 2-D model and may choose to compensate by assigning a speed which is not based on the nearest point, but is based on a close point where the speed is higher. While this approach has practical benefits, it requires engineers' judgment.

3.2 Turnouts

It is common practice to assume that a turnout will increase the vibration and ground-borne noise levels by 10dB, although recent measurements indicate that the increase may depend upon the type of turnout. Nevertheless, it is necessary to allow for the effect of such a turnout by applying an adjustment factor.

In the 3-D model, it is possible to assign a length, such as 20m, as the turnout and apply the adjustment factor over that length. If the turnout is at the closest part of the track to the receiver, this assumption may well increase the predicted ground-borne noise level at the receiver by an amount approaching the assumed adjustment. However, when the turnout is at some distance, say 50m, along the track from the nearest point, the increase in the maximum level is likely to be significantly less than the turnout adjustment. The 3-D model is able to reflect the real situation and predict the ground-borne noise level accordingly.

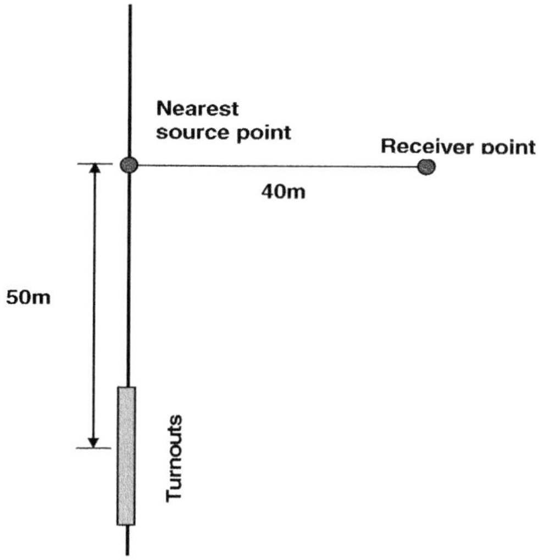

Fig. 2. Diagram showing 2-D calculation of turnout effect (color figure online)

In contrast, the 2-D model requires a relatively simplistic assumption about turnouts which will not necessarily result in the correct predicted level. If the turnout is close to the receiver, then the common approach would be to increase the ground-borne noise level predicted at the receiver by the turnout adjustment. Since the maximum noise level from a passing train is a function of the full train length and the turnout would be much shorter (say 20m), this method is equivalent to adding the turnout adjustment for the full train. This would clearly over estimate the overall effect of the turnout. When the turnout is some distance up the track (not at the closest point), the over estimation is increased.

Secondly, it is difficult to know in the 2-D model when to apply the turnout adjustment. Where the turnout is at the source point which is closest to the receiver point; then it would be logical to apply the adjustment. However, if the turnout were 50m along the track from the nearest source point, then should the adjustment be applied? What if the turnout were 100m along the track?

Figure 2 demonstrates this. Since the turnout is not at the closest source point to the receiver, should a turnout adjustment be applied to the predicted level? Is it correct to assume the full adjustment, given that the closest part of the train as it passes will not be on the turnout?

For the 2-D model, the difficulty in deciding on the assumption regarding the turnout is demonstrated by the prediction results based on Fig. 3. Three-dimensional and 2-D ground-borne noise predictions have been made for this situation. In the case of the 2-D prediction, a turnout adjustment has been both applied and not applied for comparison. The results in Table 2 show the significance of the decision.

Table 2. Calculated Ground-borne Noise Levels near Turnouts

Model	Turnout Assumption	Predicted Level (dBA)
3-D	20m turnouts	48
2-D	Turnouts ignored	46.5
2-D	Turnout included	56.5

The table also shows that the 2-D model will under predict the level if the turnout adjustment is not applied and substantially over predict the level if it is applied, for reasons of its inability to properly allow for the effect of turnouts.

3.3 Ground Mobility

The ground mobility depends upon the ground type and structure. While the mobility of rock is different from soil or sand, measurements also seem to indicate that the mobility depends upon the physical configuration, including the depth of the tunnel and the depth of the rock head. Accordingly, it is quite likely that the ground mobility along the length of any particular rail project will vary and it may change significantly over a very short length.

In preparing a 3-D model, mobility factors can be applied to each section of the track, as appropriate. Even so, there are some difficulties in establishing how to calculate ground-borne noise levels at a receiver where the vibration from a particular source point transmits through two ground types. However, overall the 3-D model seems to adequately account for the variations in ground mobility.

Accounting for variations in ground mobility in a 2-D model is more difficult. The common approach is to adopt the ground type between the nearest source point and the receiver point and apply this in the calculation of ground-borne noise levels. However, as demonstrated in Fig. 3, the highest ground-borne noise level may be generated by a section of the track not at the closest point because of a higher mobility of the ground between more distant source points and the receiver.

Figure 3 shows a situation where a receiver point may be affected by two ground mobility types. The 3-D model will take into account the two distinct ground types in the calculation and will assign each type for the relevant segments of the track. It is possible (if not likely) that the transmission through the rock to the receiver point will dominate the level at that point.

In contrast, the 2-D model will need to rely on only one assumption of the ground type, and in practice, the ground type between the nearest source point and the receiver point is commonly adopted. The 2-D calculation therefore would be based on soft ground and would ignore any affect from the rock.

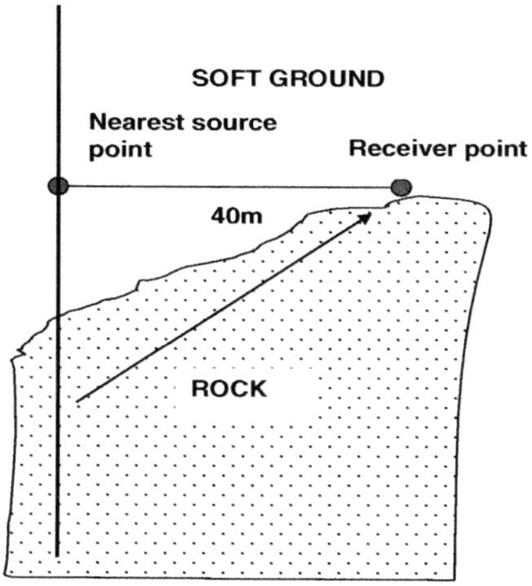

Fig. 3. Diagram showing 2-D calculation of the effect of ground mobility (color figure online)

Table 3 shows the results of calculations using the 3-D model and the 2-D model for the situation shown in Fig. 3.

Table 3. Calculated ground-borne noise levels for varying ground conditions.

Model	Predicted Level (dBA)
3-D	43
2-D	41.5

The table indicates that the 2-D model will under predict the ground-borne noise level by 1.5dB in the example in Fig. 4. This is because it may not be able to allow for the ground conditions along the relevant sections of track to each receiver. Nevertheless, a good engineer may be able to see this 2-D limitation and may choose to assign the rock mobility to the calculation so as not to underpredict.

To allow for the ground mobility for either the 2-D or the 3-D model, ground vibration propagation testing is often carried out. For the 3-D model, more testing may be required, or at least testing over larger distances may be required.

4 Conclusion

The comparative predictions of ground-borne noise levels using 3-D and 2-D models indicate that the 2-D model will not result in accurate ground-borne noise level predictions for all situations. The 2-D model can be incorrect for the following situations:

- Varying speed along the alignment, particularly near stations.
- Near turnouts.
- Near the interface between two ground types.

It may be possible to reduce the limitations of the 2-D model by carefully selecting the input parameters, rather than relying on the parameters at the nearest source point.

Any lack of accuracy of the 2-D model because of these factors needs to be considered in the light of the general inaccuracy of ground-borne noise prediction methodologies, such as the FTA procedure, primarily as a result of the limited availability of input information. Accordingly, it is difficult to demonstrate by measurement the problems of the 2-D model because of this basic inaccuracy of the noise-prediction methodologies.

Reference

[1] Federal Transit Administration: Transit Noise and Vibration Impact Assessment, USA (May 2006)

Development to Decrease Noise in Place along Railway-Tracks of Sanyo Shinkansen

Makoto Tsukanishi[1], Makoto Toyooka[2], Tomohiko Mori[1], and Hidemi Yano[1]

[1] Technical Research and Development Dept., West Japan Railway Company, 4-24, Shibata 2-chome, Kita-ku, Osaka-shi, Osaka, 530-8341, Japan
Tel.: (+81)6-6375-8819; Fax: (+81)6-6376-3193
[2] Hakata Car Maintenance Center (Okayama), West Japan Railway Company
e-mail: makoto-tsukanishi@westjr.co.jp, makoto-toyooka@westjr.co.jp, tomohiko-mori@westjr.co.jp, hidemi-yano@westjr.co.jp

Summary

Currently, Sanyo Shinkansen is in commercial operation at top speed 300km/h, and 16-car and 8-car trains running depending on the passenger volume. So far we have spent a great effort to achieve environmental standards of Japan, and the environment along railway tracks has been greatly improved.

However, because population continues to increase in developed areas along the railroad, it is essential to further reduce noise and vibration in order to continue to provide high-speed transport services in the future. Therefore, we are developing technologies that focus on differences in noise levels of 16-car and 8-car trains.

In this paper, we introduce two recent measures for noise reduction. One is the development of current collector applied to 8-car train, the second is the development to reduce the noise from the gaps between cars.

1 Development of Low Noise Current Collector

1.1 Introduction

Environmental standards of noise in the vicinity of railwaytracks are provided in Japan, and they are achieved by the 16-car train Shinkansen. However, there is concern that the closer distance between current collectors on the 8-car Sanyo Shinkansen will increase the total noise level. A new pantograph and insulator cover has been developed to decrease noise level. As a result, when the re-designed current collector is applied to the 8-car train, a noise level equal to the 16-car train will be achieved. The outline of development is described here.

T. Maeda et al. (Eds.): Noise and Vibration Mitigation for Rail Trans. Sys., NNFM 118, pp. 383–388.
springerlink.com

1.2 Decreasing Noise of Pantograph

The components of an existing pantograph are shown in Fig 1. To determine which parts should be improved to reduce noise, acoustic source localization measurements were conducted. Figure 2 shows the result.

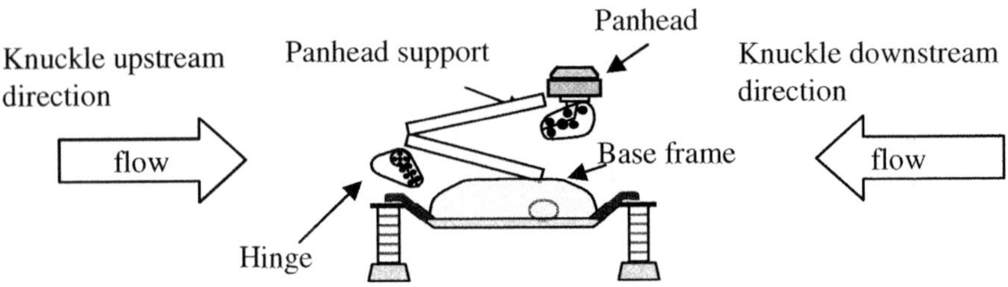

Fig. 1. Definition of pantograph (color figure online)

As a result, it is confirmed that the noise is chiefly generated by the following conditions.

① The sound of the high-frequency band is generated because the panhead is rectangular in shape.

② In the flow from the knuckle upstream direction, the noise is chiefly generated from the panhead support. In the flow from knuckle downstream direction, the noise is chiefly generated from the hinge.

Fig. 2. Acoustic source localization of pantograph (color figure online)

Noise reduction is therefore focused on the main panhead, panhead support and hinge of the collector.

The existing panhead is a rectangular shape as shown in Fig.3-(a). It is believed that the noise increases due to airflow separation from the corner. When the panhead is made to the shape of the wing type shown in Fig.3-(b), the decrease of noise and improvement of the aerodynamic upward force were projected.

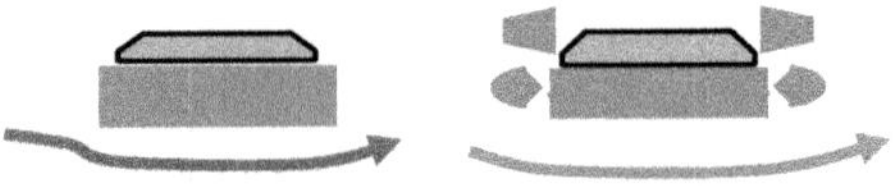

(a) rectangular geometry (b) wing type shape

Fig. 3. Shape of pantograph (color figure online) head

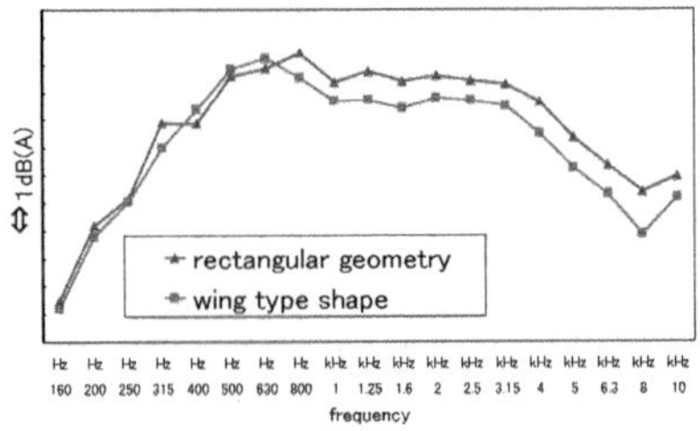

Fig. 4. Comparison by difference of panhead noise (color figure online)

As a result, the noise in the higher frequency bands has become reduced as shown in Fig.4.

In an effort to reduce the noise, a cover was put on the panhead support as shown in Fig.5. Moreover, a cover is put on the hinge and the base frame to attempt to reduce noise.

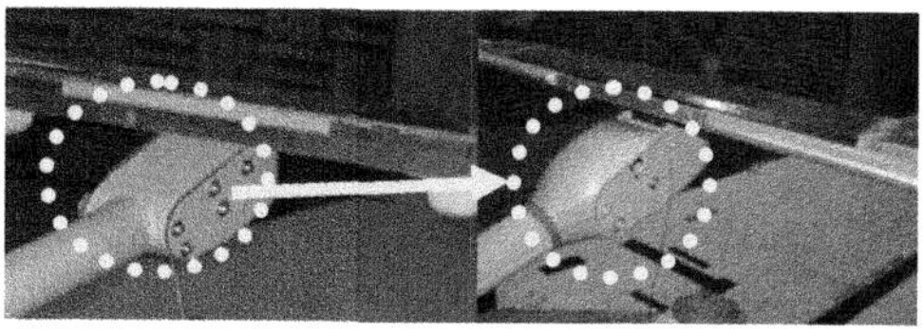

Fig. 5. Decreasing noise of panhead support (color figure online)

1.3 Decreasing Noise of Insulator Cover

A noise source survey of the insulator cover was performed by wind-tunnel tests. As a result, when the shape of the slope was adjusted to two dimensions, wind did not flow very much in the cavity, and the reduction of noise was confirmed. Moreover, it was confirmed that it is also effective to round the edge of the cavity.

Then the shape of the slope was adjusted to two dimensions as shown in Fig.6, and the edge of the cavity was rounded.

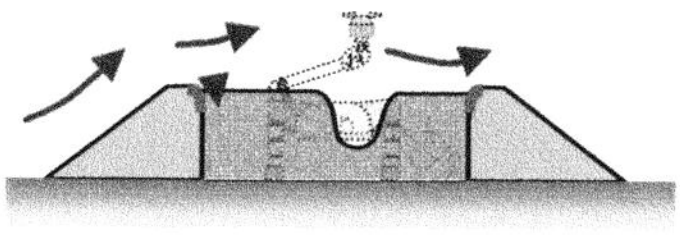

Fig. 6. Pattern diagrams of insulator cover (color figure online)

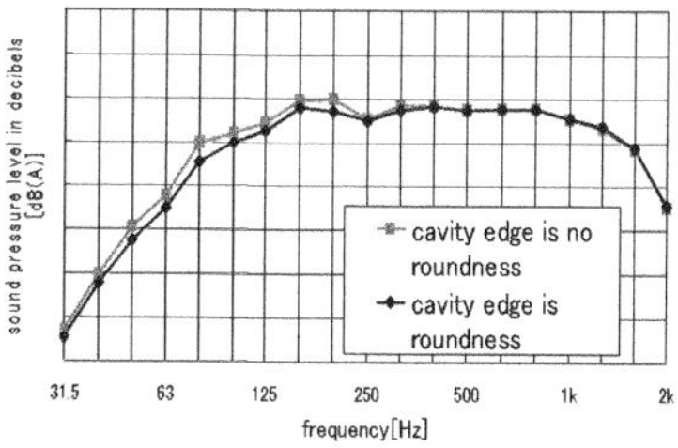

Fig. 7. Effect of putting roundness of cavity edge (color figure online)

Figure 7 shows the effect when roundness is put on the edge of the cavity.

1.4 Execution of Test Run

Measures that had been described up to now were applied, test run was executed and the effect was confirmed. As a result, it was confirmed that the noise decreased.

Fig. 8. Current collector installed in test run (color figure online)

1.5 Summary

The noise decrease was achieved by applying simple acoustical technology to the current collector. The developed device was adopted for the 8-car train that was the short configuration.

2 Reduction of Aeroacoustic Noise from the Gap Section of Shinkansen Train

2.1 Introduction

Aeroacoustic noise of the Shinkansen train increases in proportion to the sixth power of the train speed.

The control of aeroacoustic noise from the Shinkansen train has been successfully focused on decreasing noise from pantograph up to now. As a result, the relative contribution of aeroacoustic noise from the gap between cars has grown.

The aeroacoustic noise from the gap sections of a Shinkansen train includes impact sound generated by the wind around train which collides with the edge of the body and the resonance sound generated from the cavity between cars. (Fig.9)

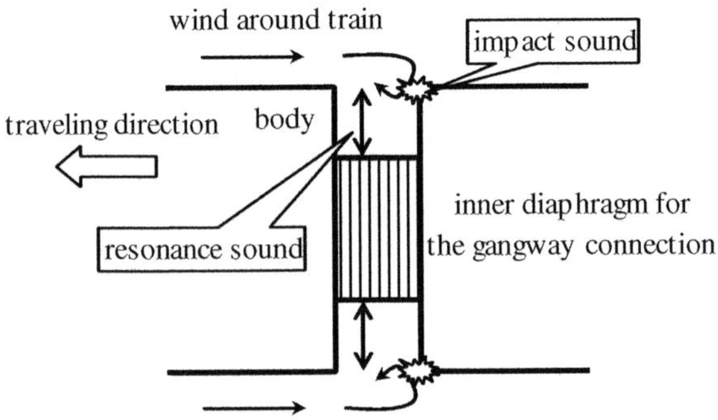

Fig. 9. The aeroacoustic noise from the gap section of Shinkansen train

We developed the gaps between adjacent cars that decreased both the impact sound and the resonance sound.

2.2 Outline of Development

It is understood that it is effective to round the edge of body to decrease the impact sound. [1]

This time, we developed the low-noise gaps between adjacent cars that rounded the edge of body to decrease the impact sound and closed the gap by a division hood to decrease the resonance sound. (Fig.10)

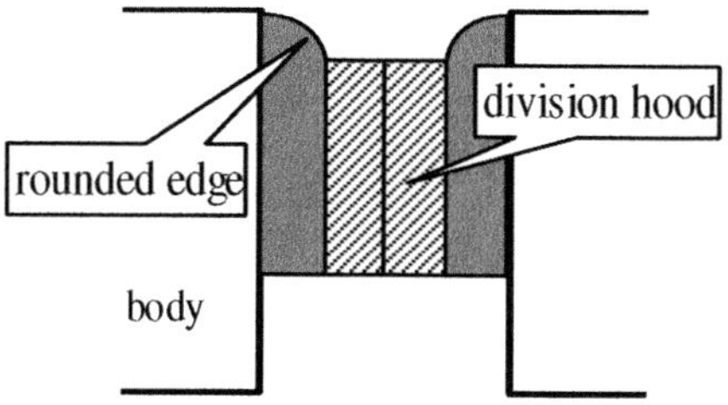

Fig. 10. Low noise gaps between adjacent cars

If the airflow blows in contacting part of division hood, noise is generated from it. Therefore the division hood is lowered with respect to body side and the edge of the body is rounded. (Fig.11)

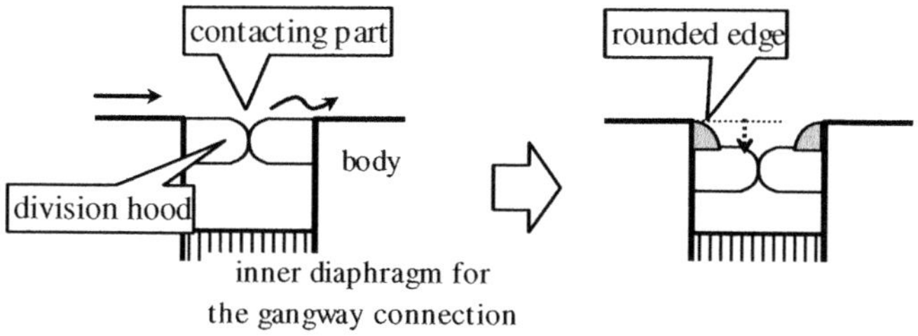

Fig. 11. Setting potion of division hood

If the roundness size is larger, the impact sound shows the tendency to become less. However, the noise from the contacting part of division hood increases because the airflow is directed down to the inside of body.

If the installed position of the division hood is lower, the noise from the contacting part of division hood is not generated. However, the resonance sound is generated because the cavity depth becomes deep.

We determined the best roundness size and installation position of the division hood to decrease both the impact sound and the resonance sound. It was an important point of this development. (Fig.12)

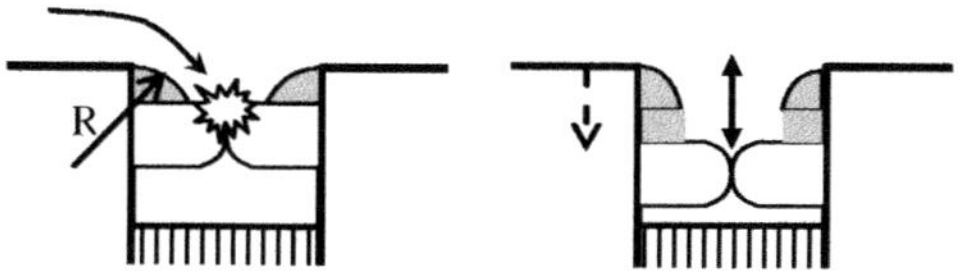

Fig. 12. Examination of shape

We tested to determine the low-noise gaps between adjacent cars in the large-scale wind tunnel, using a one-eighth scale model vehicle.

We were able to decrease the impact sound (400 Hz~) and the resonance sound (about 100 Hz) by the developed structure. (Fig.13)

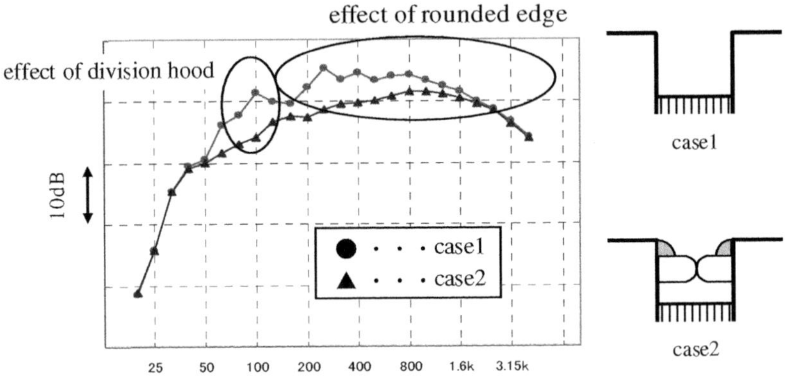

Fig. 13. Wind tunnel test (color figure online)

The noise-reduction effect by these techniques was verified in a field test.

Reference

[1] Takaishi, T., Yamazaki, N.: Wind Tunnel Tests on the Control of Aeroacoustic Noise from Gaps between Cars and Bogie Sections, vol. 21(3), pp. 12-16 (2007)

Acoustic Assessment of Micro-pressure Wave Emissions from High-Speed Railway Tunnels

Ch. Gerbig and K.G. Degen

Deutsche Bahn AG,
Voelckerstrasse 5, D-80939 Munich, Germany
Tel.: +49 89 1308 7393; Fax: +49 69 265 58006
e-mail: christian.gerbig@deutschebahn.com

Summary

With increasing train speeds, the emission of micro-pressure waves from tunnel portals, when high-speed trains enter the tunnel at the opposite side, can contribute significantly to the annoyance of residences near a high-speed railway line. In Europe at present no established assessment guidelines or standards exist explicitly devoted to this effect. Counter-measures can be applied to the tunnel portals or to the tunnel interior.

Appropriate consideration of measures to reduce the emission of micro-pressure waves at a very early stage of the planning process requires a prediction method and upper limits for noise levels. As micro-pressure waves are audible as "sonic boom", the generally used A-weighting leads to inappropriate results.

During the years 2007 to 2009 Deutsche Bahn performed a study how to predict the micro-pressure effects for new tunnel constructions as well as how to assess and to limit the micro-pressure effects. Main issues were the physical condition of humans in the close vicinity of the tunnel entrances (e.g. railway workers), the conspicuity in the portal region, the annoyance at residential areas in the vicinity of the railway line as well as micro-pressure effects on the structures of neighboring buildings and secondary acoustical effects by structural vibration excitation.

As a result, a set of two quantities and guidance levels were suggested. The following levels should not be exceeded: (I) At a distance of 25 m from the tunnel portal the C-weighted sound peak level $L_{pC,peak}$ of 115 dB(C) and (II) in dwellings in the neighborhood of tunnel portals the C-weighted sound exposure level L_{CE} of 75 dB(C). The method is currently applied for all tunnels on high-speed lines, which are currently in the planning phase in Germany.

1 Introduction

In December 2005, when first test runs were carried out on the new high-speed line Nuremberg-Ingolstadt, significant sonic boom incidents occurred at the portals of

T. Maeda et al. (Eds.): Noise and Vibration Mitigation for Rail Trans. Sys., NNFM 118, pp. 389–396.
springerlink.com

two 7700 m and 7260 m long tunnels, when trains with speeds higher than 280 km/h entered the tunnel at the opposite entrance. Both double-tracked tunnels with a cross-section of 92 m^2 are equipped with slab track. It was the first time that clearly audible effects, recognized as "sonic boom" by the residents, occurred during train operations under regular circumstances in Europe. In order not to delay the opening of the new high-speed line, Deutsche Bahn (DB) decided to take immediate countermeasures by retrofitting the two tunnels with sound absorbing elements covering track and tunnel floor. This led to a significant reduction of the micro-pressure wave emission and the commercial operation on the Nuremberg-Ingolstadt line started in May 2006 successfully in time and without operational restrictions [1, 2].

At present, new high-speed lines are in the planning stage or under construction in Germany. For safety reasons future tunnels will be single-tracked with a cross-section of about 60 m^2, thus leading to enhanced generation of micro pressure waves if no countermeasures are implemented. Consequently, most tunnels on high-speed lines are designed with portal hoods or specially designed portal sections to reduce the micro-pressure wave generation.

Proper design of measures to reduce the "sonic boom" requires a method to predict noise levels from micro-pressure waves not only close to the tunnel exit but also at dwellings, which are usually located at a certain distance from the tunnel.

As generally accepted physical quantities for assessing the effect of micro-pressure waves from high-speed lines do not exist in Germany, DB has developed in the recent years a prediction method for noise levels from micro-pressure waves and criteria for different situations, which are considered as upper limits for noise levels. This combination of prediction and limit values is essential for the proper choice of efficient measures to reduce the micro-pressure waves from tunnels as it guarantees the required protection on one hand while avoiding over-engineered solutions on the other hand.

This paper gives a brief overview of the acoustical properties of micro-pressure waves and describes the prediction method developed by DB. Particular attention is paid to the developed assessment criteria and their justification in view of the relevant noise legislation.

2 Characterization of Micro-pressure Waves

2.1 Generation Mechanism

When a high-speed train enters a tunnel, a compression wave is generated and propagates along the tunnel at the speed of sound. On its way through the tunnel the rising edge of the wave front becomes increasingly steeper (see Fig. 1), particularly because friction effects are much lower in tunnels with slab track than in tunnels with traditional ballasted track. At the tunnel exit the compression wave is partly reflected back into the tunnel and partly emitted outwards. The emitted wave is called a micro-pressure wave. Without countermeasures the micro-pressure wave can be clearly audible even at distances far away from the tunnel portal [1]. The perceptibility depends both on the spectrum and on the amplitude of the pressure pulse. The micro-pressure wave is particularly audible when the duration is short and the amplitude is large.

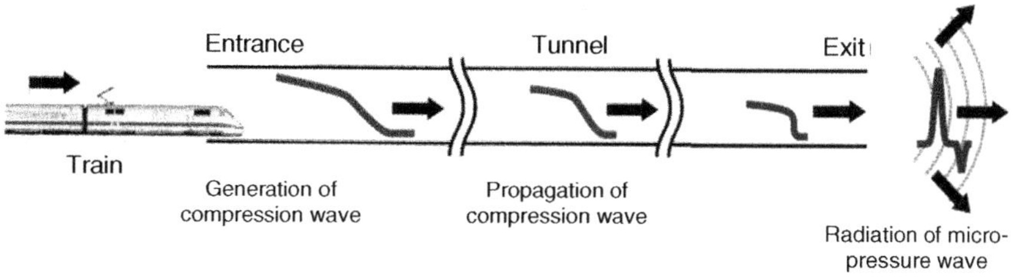

Fig. 1. Generation of a micro-pressure wave

2.2 Counter Measures

When a tunnel is designed in the planning stage, numerous parameters important for generation, transmission and emission of micro-pressure waves are fixed (e. g. design of the train nose, train speed, cross-sections of train and tunnel and track type), hence limiting the choices for the implementation of efficient reduction measures. However, an optimization of the portal design is possible in most cases when taken into account at an early stage of the planning. An optimized design reduces the gradient of the compression wave at its origin (i.e. when the train enters the tunnel) and is therefore the most efficient measure. The reduction of the pressure gradient can be achieved, for example, by sloped portals, expanded portal cross-sections and portal hoods with specially designed ventilation openings. In Japan a large variety of hood solutions are well known [3]. Fig. 2 shows some examples.

Fig. 2. Examples of portal hoods: Omata tunnel at Tohoku Shinkansen, Japan (left, photo by Mr. Kaltenbach, DB) and Katzenberg tunnel (under construction) at high-speed line Karlsruhe-Basel, Germany (right, photo by Mr. Zieger, DB)

Another possibility is to influence the propagation of the compression wave on its way through the tunnel. This can be achieved by the use of ballast, by porous absorber elements on a slab track, by baffles or by tunnel chambers, chimneys or ducts [1].

2.3 Acoustic Properties of Micro-pressure Waves

The acoustic properties of micro-pressure waves differ considerably from the other components of train noise (rolling noise, engine noise or aerodynamic noise). Fig. 3 shows time dependence and spectrum of a typical micro-pressure wave at a distance of 25 m from the tunnel exit without countermeasure (solid lines) and with a hood at the tunnel entrance as countermeasure (dashed line). Clearly visible is the influence of the shape of the entrance portal. The spectra of micro-pressure waves are characterized by pronounced contributions at low frequencies (f < 125 Hz), also including infra-sound below 20 Hz. In addition sound pressure levels at frequencies > 125 Hz may occur, which correspond to a hearing impression as sharp and bright bang when strongly pronounced.

The large sound emitting area (i.e. the full cross-section of the portal) causes a radiation of micro-pressure waves in the free field, which cannot be described by standard propagation models, especially not at small distances from the tunnel exit. Based on measurements of the C-weighted peak sound level $L_{pC,peak}$ at different distances d from the exit of DB´s Irlahuell tunnel, the following dependencies were derived for different distance domains:

$$\Delta L = 13.5 \log d_2/d_1 \text{ dB} \quad \text{for } 10 \text{ m} < d < 50 \text{ m} \quad (2.1)$$
$$\Delta L = 16.5 \log d_2/d_1 \text{ dB} \quad \text{for } 50 \text{ m} < d < 200 \text{ m} \quad (2.2)$$
$$\Delta L = 20 \log d_2/d_1 \text{ dB} \quad \text{for } d > 200 \text{ m} \quad (2.3)$$

where ΔL is the sound level difference between two positions with distances d_1 and d_2 from the portal. Eq. 2.1 to 2.3 corresponds to ΔL of 4 dB, 5 dB and 6 dB, respectively, when doubling the distance. At distances < 10 m, the distance dependency is estimated to be weaker than described in Eq. 2.1. At d >> 200 m atmospheric influences have to be taken into account.

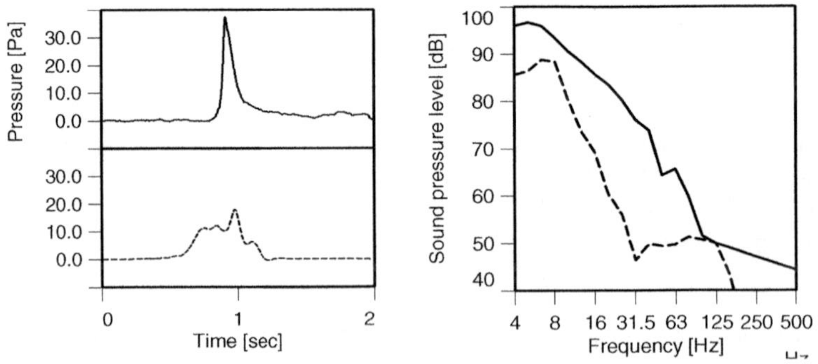

Fig. 3. Example for predicted pressure over time signal (left) and third octave band spectra (right) of a typical micro-pressure wave. Emission point located at 25 m distance from the exit portal of a tunnel without any countermeasure (—) and from the exit portal of a tunnel with entrance hood (---)

2.4 Prediction Process

In order to efficiently plan counter-measures for new tunnels on high-speed lines as well as to predict residual micro-pressure waves even after the implementation of countermeasures, a detailed and well-established prediction process was developed by DB. Using a three-stage model for (i) generation of the compressive wave, (ii) the steepening process during the propagation inside the tunnel and (iii) the emission of the micro-pressure wave at the opposite tunnel portal, the time dependent air pressure $p(t)$ is calculated at a distance of 25 m in front of the portal [4]. Based on $p(t)$, all relevant acoustical sound levels can easily be derived and compared with the suitable assessment quantities.

The prediction assumes the following reference conditions: The maximum regular train speed v_{max}, a reference train with disadvantageous front shape in accordance with the European technical specifications for high-speed rolling stock (TSI-HS). For double-tracked tunnels it is assumed that no train encounter happens when the train enters the tunnel. The metrological data are assumed to be $p = 1013$ hPa, air humidity 70%, air temperature 10 °C outside and 12 °C inside the tunnel and air velocity 0 m/s inside the tunnel.

3 Assessment Design Goals

Generally accepted regulations suitable for an assessment of the effects of micro-pressure waves from tunnels currently exist neither on a national level in Germany nor on a European level. In particular, the German traffic noise protection directive [5], which is dedicated to road and railway lines, does not mention such effects.

During the years 2007 to 2009 DB has performed a study how to assess and to limit the emission of micro-pressure waves. The purpose was to exclude negative health effects for people staying temporarily in the close vicinity of the tunnel entrances (e.g. railway workers), to avoid a strong audible boom close to the tunnel portal, to avoid annoyance of residents and to exclude effects on the structures of buildings and secondary acoustical effects due to vibrations excited by low frequencies in the spectrum of micro-pressure waves. The following boundary conditions were defined:

- Compliance with the German legislation comprising general requirements on prevention of harmful environmental effects.
- Harmonization with the German traffic noise protection directive [5] and its limit values for noise emission control.
- Compliance with the European directive 2003/10/EC considering health and safety requirements of workers and its derived national directive [6].
- Assessment based on physical quantities, which are well-accepted in the field of noise protection.

In addition, the German gun noise management directive [7] should be taken into account for orientation, especially concerning the quantitative guidance levels for impulsive sound incidents.

4 Results

Based on the above-stated requirements, a set of two quantities and guidance levels were suggested. An overview is given in Fig. 4.

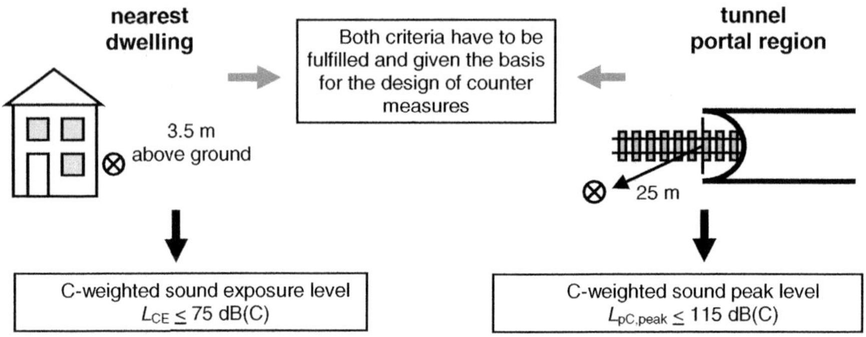

Fig. 4. Schematic design of locations and levels for the assessment quantities

Two representative points of emission were chosen. The first one is located at a fixed distance of 25 m from the portal and at a height of 1.2 m above rail head (representative for the area, where people might temporarily stay but outside the near field of the sound emission). The second point is chosen to be at the front of the nearest dwelling 3.5 m above the ground.

4.1 Noise Protection Close to the Tunnel Portal

Concerning the physical condition of humans, the European directive 2003/10/EC gives clear guidelines based on the C-weighted sound peak level $L_{pC,peak}$. Although the C-weighting of levels is not very common in railway noise assessment, it takes into account that most of the micro-pressure wave energy is located at low frequencies (< 100 Hz), even extending into in the infra sound region (< 20 Hz).

The directive defines a *lower exposure action value* at $L_{pC,peak}$=135 dB(C), which should not be exceeded. Based on this definition, the following guidance level was recommended:

- In a distance of 25 m from the tunnel portal a C-weighted sound peak level $L_{pC,peak}$ = 115 dB(C) should not be exceeded.

With this threshold value, the *lower exposure action value* $L_{pC,peak}$ = 135 dB(C) should not be exceeded at any point even very close to the tunnel portal. With this limitation, also secondary acoustical effects caused by the excitation of vibrations in buildings and unacceptable noise levels, e.g., in garden plots are avoided. Potential damages to nearby buildings are excluded as well.

4.2 Noise Protection in Residential Areas Close to Tunnels

Similarly, a guidance level based on the C-weighted sound exposure level L_{CE} was defined for residential areas in the vicinity of tunnels taking into account the acoustic properties of micro-pressure waves (see Sect. 2.3):

- At the front of dwellings in the vicinity of the tunnel exit a C-weighted sound exposure level L_{CE} of 75 dB(C) should not be exceeded during daytime and nighttime.

In order to keep the regulation as simple as possible, no distinction is made between dwellings in pure residential areas, in industrial areas or in mixed areas.

The German gun noise management directive [7] sets an upper limit of $L_{CE} = 80$ dB(C) for single sound incidents in residential areas during the nighttime. Hence, the value $L_{CE} = 75$ dB(C) is well below this limit.

Measurements in the vicinity of the south portal of the Euerwang tunnel (high-speed line Nuremberg-Ingolstadt) have shown that micro-pressure waves with $L_{CE} \leq 73$ dB(C) are hardly audible.

Within the framework of national German noise legislation, the assessment of the 'ordinary' pass-by noise is based on A-weighted sound pressure levels L_{eq}. Therefore the question had to be answered to what extend a micro-pressure wave with $L_{CE} \leq 75$ dB(C) contributes to the A-weighted level L_{eq}. In a calculation based on measurements of micro-pressure waves and typical pass-by noise, it could be shown that the contribution of the micro-pressure wave to the overall A-weighted level L_{eq} is negligible. In consequence, there is no need to take the micro-pressure wave into account for an assessment according to the German traffic noise protection directive [5].

5 Conclusions and Outlook

The above-presented assessment procedure is the basis of all tunnels either in the planning phase or under construction at present in Germany. It is applied successfully for dimensioning counter-measures to efficiently reduce micro-pressure waves. Up to now experience has shown that counter measures are strongly required for most of the tunnels on new high-speed lines. At the same time sufficient measures, e.g., portal hoods are feasible to meet the above-mentioned assessment criteria without going beyond of what is economically justifiable.

At present the coordination with the German national railway authority is in process but not finalized yet. Thus modifications of the assessment procedure in the next future are still possible.

DB expects to have with this assessment procedure a well-balanced protection level, which gives a practicable basis for the successful approval of new infrastructure projects. Acoustical problems occurring after a high-speed line goes into operation can thus be avoided.

Acknowledgement

The authors want to give their special thanks to the colleagues at the Japanese Railway Technical Research Institute (RTRI) for intensive discussions, the sharing of their experience on this topic of many decades and their hospitality on the occasion of a visit of high-speed tunnels in Japan in 2006.

References

[1] Tielkes, T., Kaltenbach, H.-J., Hieke, M., Deeg, P., Eisenlauer, M.: Measures to counteract micro-pressure waves radiating from tunnel exits of DB´s new Nuremburg-Ingolstadt high-speed line. In: Noise and Vibration Mitigation for Rail Transportation Systems, 9th IWRN, Munich, Germany, published in Notes on Numerical Fluid Mechanics and Multidisciplinary Design (NNFM) 99, September 2007, pp. 40–47. Springer, Heidelberg (2008)

[2] Degen, K.G., Gerbig, C., Onnich, J.: Acoustic assessment of micro-pressure waves radiating from tunnel exits of DB high-speed lines. In: Noise and Vibration Mitigation for Rail Transportation Systems, 9th IWRN, Munich, Germany, published in Notes on Numerical Fluid Mechanics and Multidisciplinary Design (NNFM) 99, September 2007, pp. 48–55. Springer, Heidelberg (2008)

[3] Maeda, T.: Protecting the Trackside Environment. Japan Railway & Transport Review 22, 48–57 (1999)

[4] Hieke, M., Kaltenbach, H.-J., Tielkes, T.: Prediction of micro-pressure wave emissions from high-speed railway tunnels. In: Proceedings of the 13th International Symposium on Aerodynamics and Ventilation of Vehicle Tunnels 2009, New Brunswick, USA, pp. 487–501. BHR Group Limited, UK (2009)

[5] German federal traffic noise protection directive: Sechzehnte Verordnung zur Durchführung des Bundes-Immissionsschutzgesetzes (Verkehrslärmschutz¬verordnung – 16. BImSchV) (1990)

[6] Directive 2003/10/EC of the European Parliament and of the Council, minimum health and safety requirements regarding the exposure of workers to the risks arising from physical agents (noise) (2003)

[7] German gun noise management directive: Richtlinie für das Lärmmanagement auf Schießplätzen (Lärmmanagementrichtlinie – LMR), German Ministry of Defence (2007)

Model Experiments on the Tunnel Compression Wave Using an Axisymmetric and Three-Dimensional Train Model

T. Fukuda, H. Saito, T. Miyachi, K. Kikuchi, and M. Iida

Railway Technical Research Institute,
2-8-38 Hikari-cho, Kokubunji-shi, Tokyo 185-8540, Japan
Tel.: +81 42 573 7318; Fax: +81 42 573 7329
e-mail: takashi@rtri.or.jp

Summary

Generation of the compression wave by a train entering a tunnel is investigated by model experiments. In the model experiments, the train and the tunnel are represented by an axisymmetric model, a three-dimensional mirror image model and a three-dimensional model. The experimental results indicate that the effect of the three-dimensionality of the train nose shape is approximately 2 % for the pressure gradient of the compression wavefront when the train nose is streamlined, hence without large flow separation around the train nose. Furthermore, the relationship between the pressure gradient of the compression wavefront and the train position in a cross-section at the tunnel portal is clarified.

1 Introduction

When a high-speed train enters a tunnel, a compression wave is generated in the tunnel. The compression wave propagates through the tunnel and arrives at the tunnel exit. At this moment, an impulsive pressure wave termed as "micro-pressure wave" is emitted out of the tunnel exit toward surroundings [1, 2]. For high-speed railways, e. g., Shinkansen in Japan, the micro-pressure wave causes an infra-sound problem. The strong micro-pressure wave is even accompanied an audible sound. The magnitude of the micro-pressure wave depends on the compression waveform arriving at the exit. The micro-pressure wave phenomenon can be divided into three stages: (a) generation of the compression wave at the tunnel entrance, (b) propagation of the compression wave through the tunnel and (c) emission of the micro-pressure wave from the tunnel exit.

The principle of countermeasures for the micro-pressure wave is to reduce the pressure gradient of the compression wavefront at the tunnel exit. In the Shinkansen, the most prevailing countermeasures are those at the stage (a), that is, installation of a tunnel entrance hood and an optimization of the train nose shape. Therefore, the prediction and the generation of the compression wave are essential

T. Maeda et al. (Eds.): Noise and Vibration Mitigation for Rail Trans. Sys., NNFM 118, pp. 397–404.
springerlink.com

for the estimation of the magnitude of the micro-pressure wave and the examination of the countermeasures thereto.

For the estimation of the compression waveform generated by the train entering the tunnel and the development of the countermeasures, a model experiment is one of the principal methods [2, 3, 4]. The model experiments using axisymmetric train and tunnel models have been conducted frequently, and effectively applied to clarification of the micro-pressure wave phenomena and the development of the countermeasures thereto. However, it is considered that clarification of the effects of the three-dimensionality of the train nose shape on the generation of the compression wave is necessary for improving the accuracy of the estimation of the compression waveform [5, 6]. In this study, we have conducted the model experiments using an axisymmetric model (AX model), a three-dimensional mirror-image model (3D-mirror model) and a three-dimensional model (3D model) for investigating the effects of the three-dimensionality of the train nose on the generation of the compression wave.

2 Model Experiments

2.1 Train Nose Shape

In the following model experiments, we assume a real-scale simplified train nose as shown in Fig. 1 [7]. The length of the nose, the width and the height of the train body are approximately equal to those of a general Shinkansen train. The train nose shape is a two-dimensional wedge with rounded edge for preventing a large flow separation. Figure 2 shows the cross-sectional area distribution of the train nose.

We have modeled the train and the tunnel using three kinds of methods as shown in Fig. 3. The three-dimensional train runs on the ground and enters the horseshoe-shaped

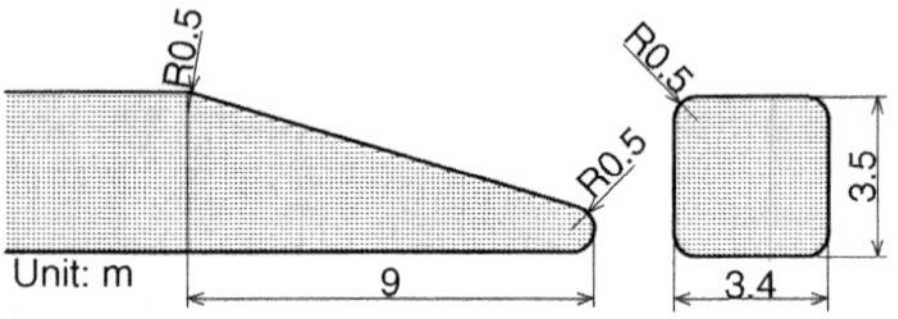

Fig. 1. Train nose shape

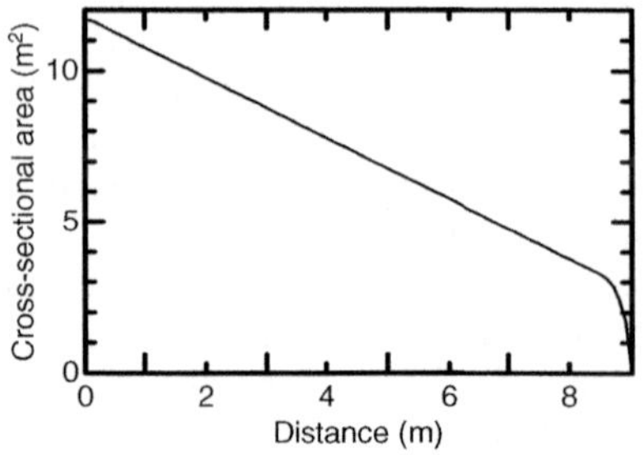

Fig. 2. Cross-sectional area distribution of train nose

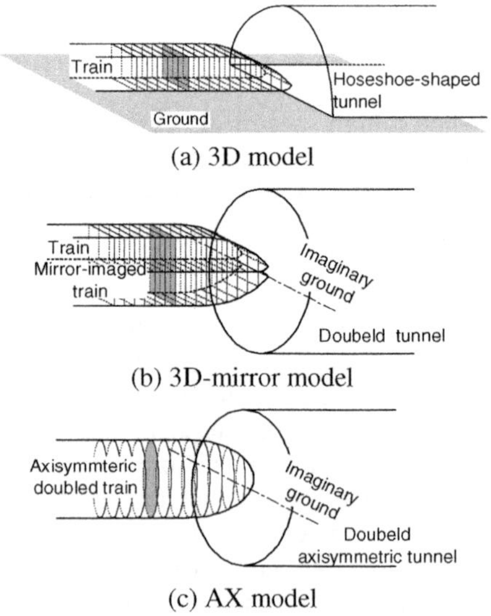

Fig. 3. Modeling of train and tunnel

tunnel (3D model, Fig. 3 (a)). Alternatively, the effect of the ground can be simulated using a mirror-image method (3D-mirror model, Fig. 3 (b)). The AX model uses a body of revolution as the train and its mirror-image and a circular pipe as the tunnel and its mirror-image as shown in Fig. 3 (c) [8, 9].

2.2 Three-Dimensional Model (3D Model)

Figures 4 (a) shows the train and tunnel of the 3D model. The front half of the train model (3D part) is an edge-rounded wedge-shaped train body similar to the real train (Fig. 1) fabricated from a duralumin, and the rear half (cylinder part) is a circular cylinder body fabricated from a nylon plastic material.

The tunnel model is fabricated from an aluminum alloy, and the cross-section of the tunnel model is horseshoe-shaped similar to the existing tunnel. The ground in front of the tunnel portal is simulated using a sheet of plywood. The model scale is 1/70 according to the ratio of the equivalent radius of the tunnel model to that of the real tunnel.

2.3 Three-Dimensional Mirror-Image Model (3D-Mirror Model)

Figure 4 (b) shows the 3D-mirror model. The front half of the train model (3D part) is a double edge-rounded wedge-shaped train body fabricated from a duralumin and the rear half (cylinder part) is a circular cylinder body fabricated from a nylon plastic material similar to the train of the 3D model.

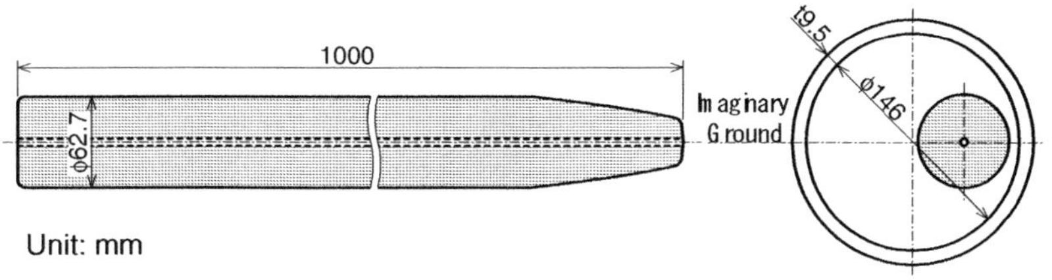

(a) 3D model (model scale: 1/70)

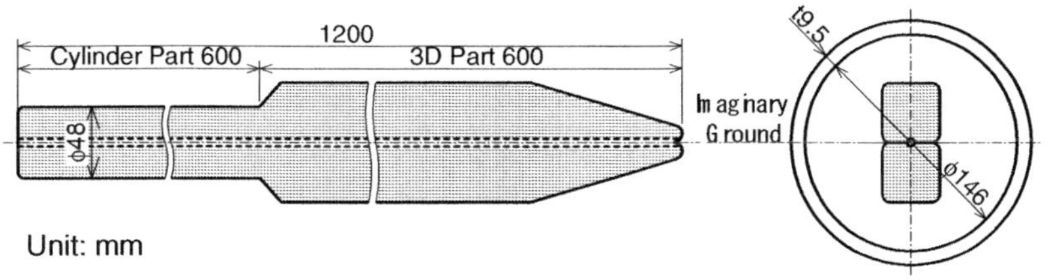

(b) 3D-mirror model (model scale: 1/87)

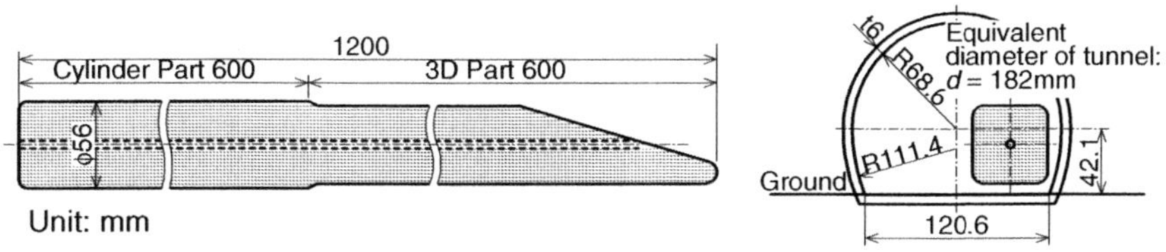

(c) AX model (model scale: 1/87)

Fig. 4. Train and tunnel models

The model scale is 1/87. The gap between the train and ground is omitted and the tunnel model consists of an 8 m long horizontal, circular cylindrical pipe made of vinyl chloride with inner diameter equal to 146 mm.

2.4 Axisymmetric Model (AX Model)

The AX train model is an axisymmetric body fabricated from a nylon plastic material of the size and shape specified in Fig. 4 (c). The train and its mirror image in the imaginary ground plane are modeled by one body of revolution together with the gap between the train and ground omitted. The tunnel model is the same as that for the 3D-mirror model.

The model scale is 1/87 according to the ratio of the radius of the tunnel model to that of the circle equivalent to the existing tunnel (the cross-sectional area: 63.4 m^2) combined with its mirror image in the ground. The blockage ratio of the train model-to-the tunnel model is identical to that of the real train-to-tunnel, and the cross-sectional area distribution along the longitudinal of the train model nose is similar to that of the real-scale train nose (Fig. 2). The double-tracked tunnel can be simulated by shifting the train center axis from the tunnel center axis [8, 9].

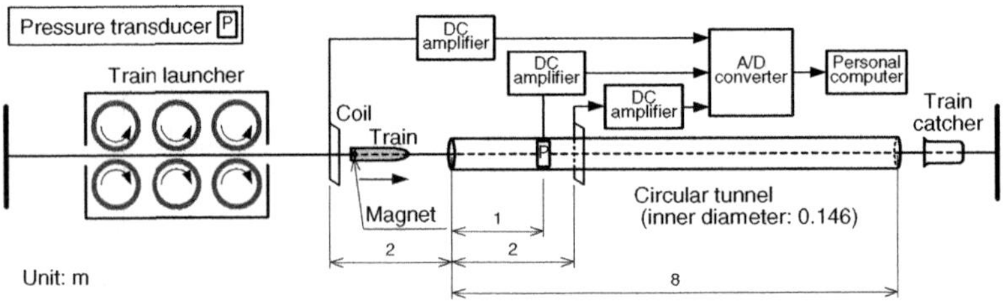

(a) AX model and 3D-mirror model (model scale: 1/87)

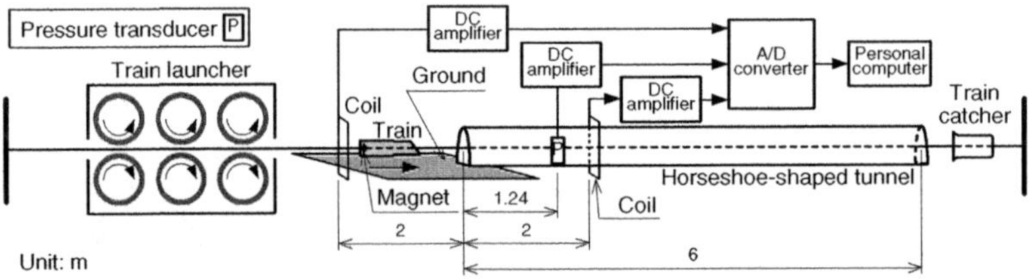

(b) 3D model (model scale: 1/70)

Fig. 5. Schematic of the model experiments

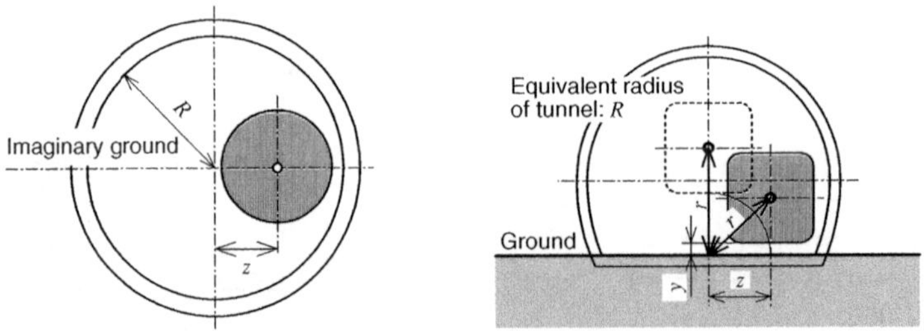

(a) AX model and 3D-mirror model (b) 3D model

Fig. 6. Cross-section of train and tunnel model

2.5 Experimental Method

We conducted model-scale experiments using the apparatus illustrated schematically in Fig. 5 [8, 10]. Each train model is projected into the corresponding tunnel model by means of a three-stage friction drive 'launcher' involving three-pairs of vertically aligned wheels; the train is guided along a 5 mm diameter taut steel wire extending along the tunnel. The cross-section of the wheels is concaved to fit the cylinder-shaped train model. The train speed is controlled by varying the rates of rotation of the drive wheels.

As mentioned in sections2.2 and 2.3, the front half and the rear half of the trains of the 3D model and the 3D-mirror model are the 3D part and the cylinder part, respectively. The rotation around the steel wire is prevented by the 3D part being concaved cross-section of the vertical aligned wheels of the launcher, and the train model can be launched at high speed because the cylinder part of the train fits the concaved cross-section of the wheels of the launcher. Furthermore, in the model experiment using the 3D-mirror model, we set the train model on its side because it can be between the vertically aligned wheels of a train launcher of the experimental apparatus.

Small permanent magnets made of neodymium are inserted in the train. The train speed is measured by detection of the magnetic field as the train passed through two wire coils placed 4 m apart, one in front and the other to the rear of the tunnel entrance.

We measured the pressure within the tunnel by two pressure transducers (Kulite Semiconductor Products XCS-190-5G), flush mounted in the tunnel wall at distances of 1 m (the AX model and the 3D-mirror model) or 1.24 m (the 3D model) from the tunnel entrance. In the analysis of the experimental results, the pressure and the pressure gradient of the compression wave are normalized as

$$p^* = \frac{p}{p_c}, \quad p_c = \frac{1}{2}\rho U^2 \frac{1-(1-R_T)^2}{(1-M)\{M+(1-R_T)^2\}}, \qquad (1)$$

$$t^* = \frac{t}{t_c}, \quad t_c = \frac{d}{U}, \qquad (2)$$

where p is the pressure, p^* is the normalized pressure, p_c is the pressure by Hara's formula [11], t is the time, t^* is the normalized time, t_c is the characteristic time of the pressure rise of compression wavefront, ρ is the air density, U is the train speed, M is the train Mach number, R_T is the blockage ratio (the cross-sectional ratio of train to tunnel) and d is the tunnel diameter. Furthermore, we define the running position of the train r, y and z in the cross-section as Fig. 6.

3 Results and Discussion

Figure 7 shows the normalized pressure p^* and pressure gradient dp^*/dt^* of the compression waveform when the train model enters along the center axis of the tunnel. In this example, the train position of the 3D model in the vertical direction of the train y corresponds to that of the actual one ($y/R = 0.07$). It can be seen that the compression waveforms of the 3D and 3D-mirror are very similar, and there is no significant difference in the maximum pressure gradient between the 3D and 3D-mirror model. In these model experiments, the under body of the train is smooth because bogies and equipments under the train body are omitted, and the ground surface is also smooth because the tracks on the ground is omitted. Therefore, it is considered that the effect of the flow between the train and the ground on the generation of the compression wave is smaller than that by the real train and tunnel.

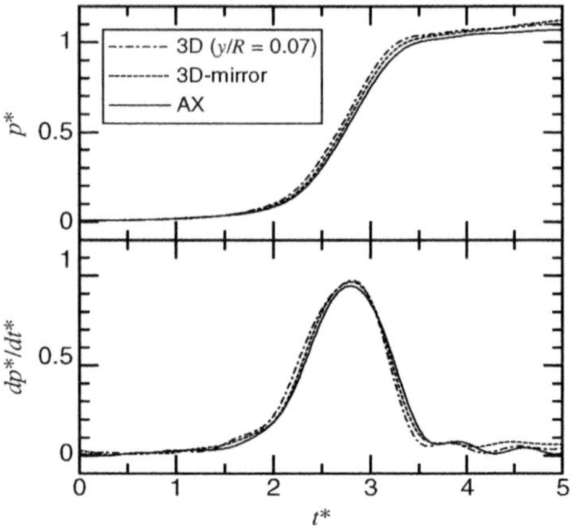

Fig. 7. Normalized pressure and pressure gradient of the compression waveform ($z/R = 0$, $U = 220$ km/h)

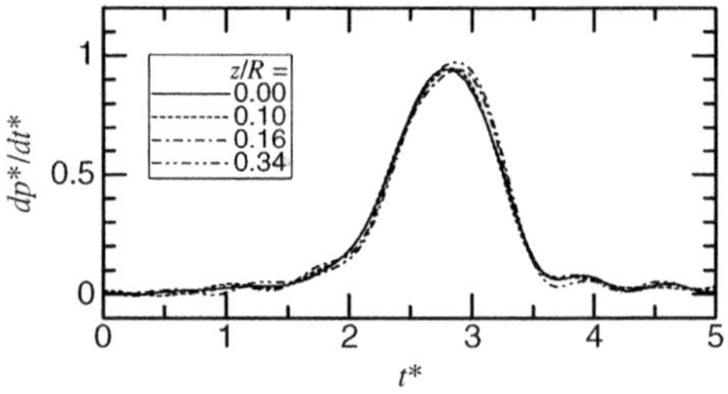

Fig. 8. Normalized pressure gradient of the compression waveform when the train is offset in horizontal direction (AX model, $U = 220$ km/h)

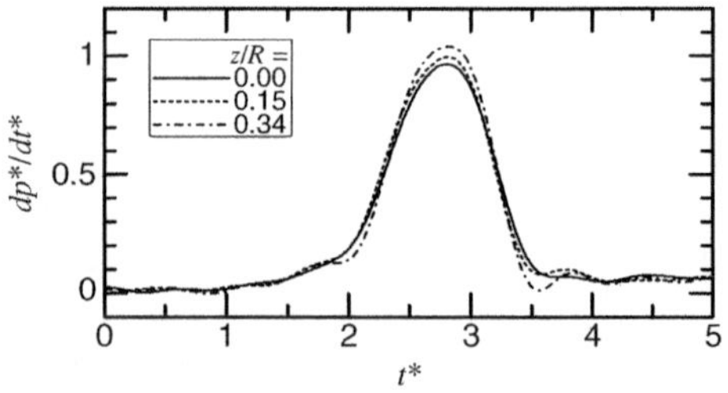

Fig. 9. Normalized pressure gradient of the compression waveform when the train is offset in horizontal direction (3D model, $y/R = 0.07$, $U = 220$ km/h)

Furthermore, it can be seen from Fig. 7 that the values of the pressure gradient of the 3D and 3D-mirror are larger than that of AX. The difference between the maximum pressure gradient of 3D-mirror and that of AX is approximately 2 %. The AX train model does not incorporate the three-dimensional shape of the train nose, though the cross-sectional area distribution of the train nose is similar to that of the real train. In these model experiments, the train nose shape is streamlined and there is no large-flow separation around it. Therefore, it is considered that the effect of the three-dimensionality on the generation of the compression wave in the model experiments is smaller than that by the real train and tunnel, where separation of flow occurs to some extent.

Figures 8 and 9 show the normalized pressure gradient of the compression waveform dp^*/dt^* when the train is offset in horizontal direction from the tunnel center axis. It can be confirmed that the pressure gradient of the compression wavefront becomes larger with the train offset in horizontal direction z.

Figure 10 shows the pressure gradient of the compression waveform when the train is offset in a vertical direction from the ground obtained by the model experiment using the 3D model and the 3D-mirror model. The pressure gradient of the compression wavefront becomes larger with the train position in the vertical direction y.

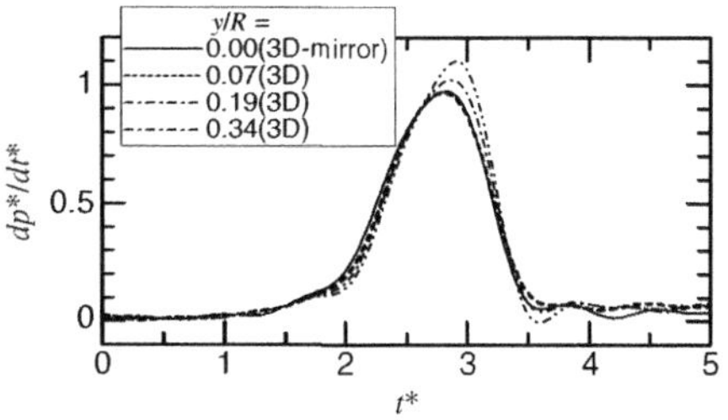

Fig. 10. Normalized pressure gradient of the compression waveform when the train is offset in vertical direction (3D-mirror model and 3D model, $z/R = 0$, $U = 220$ km/h)

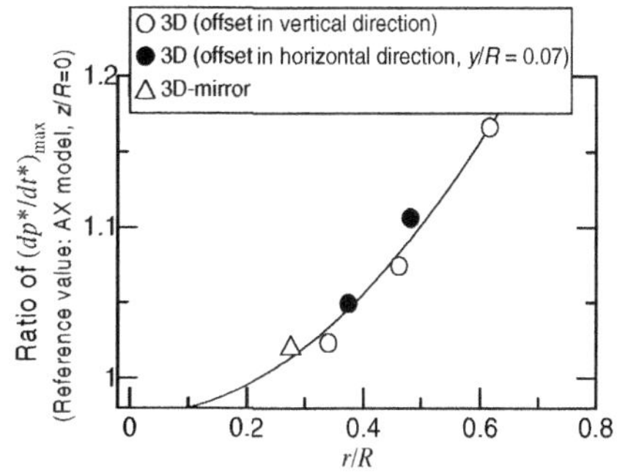

Fig. 11. Relation between the maximum value of the pressure gradient of the compression wavefront $(dp^*/dt^*)_{max}$ and the distance between the train center and the base center of the tunnel r (3D-mirror model and 3D model)

Figure 11 shows the relation between the maximum pressure gradient of the compression wavefront $(dp^*/dt^*)_{max}$ and the distance between the train center and the center of the tunnel base r. From Fig. 11, the maximum pressure gradient of the compression wavefront $(dp^*/dt^*)_{max}$ increases with the distance r equally for the horizontal and vertical directions. Therefore, it is considered that the distance r is a significant parameter for the estimation of the compression wave in the tunnel generated by the train entry.

4 Conclusions

We performed model experiments using an axisymmetric model, a three-dimensional mirror image model and a three-dimensional model to investigate the generation of the compression wave when a train enters a tunnel. The experimental results indicate that the effect of the three-dimensionality of the train nose shape is approximately 2 % for the pressure gradient of the compression wavefront when the train nose is streamlined. Furthermore, it is considered that the distance between the train center and the center of the tunnel base, r, is a significant parameter for the estimation of the compression wave in the tunnel generated by the train entry.

References

[1] Yamamoto, A.: Micro-pressure wave radiated from tunnel exit. Preprint of the Spring Meeting of Physical Society of Japan(1977) (in Japanese)

[2] Ozawa, S., Uchida, T., Maeda, T.: Reduction of micro-pressure wave radiated from tunnel exit by hood at tunnel entrance. Quarterly Report of RTRI, Japanese National Railways 19(2), 77–83 (1978)

[3] Ozawa, S., Maeda, T.: Model experiment on reduction of micro-pressure wave radiated from tunnel exit. In: Proceedings of the International Symposium on Scale Modeling, pp. 33–37 (1988)

[4] Iida, M., Matsumura, T., Nakatani, K., Fukuda, T., Maeda, T.: Optimum nose shape for reducing tunnel sonic boom. In: International Conference on Better Journey Time - Better Business, IMechE Conference Transactions, pp. 271–280 (1996)

[5] Bellenoue, M., Kageyama, T.: Reduced scale simulation of the compression wave generated by the entry of a high-speed train into tunnel. TRANSAERO - A European Initiative on Transient Aerodynamics for Railway System Optimisation, Notes on Numerical Fluid Mechanics and Multidisciplinary Design 79, 206–216 (2002)

[6] Doi, T., Kaku, J., Masubuchi, T., Ogawa, T.: Development of an experimental apparatus for measuring a pressure wave generated by a model train of actual shape. In: Proceedings of the INTER-NOISE (2006)

[7] Saito, H., Fukuda, T.: Three-dimensional model experiment on the compression wave generated by a train entering a tunnel. In: Proceedings of the 16th Joined Railway Technology Symposium, vol. 09(65), pp. 367–370 (2009) (in Japanese)

[8] Howe, M.S., Iida, M., Fukuda, T.: Theoretical and experimental investigation of the compression wave generated by a train entering a tunnel with a flared portal. Journal of Fluid Mechanics 425, 111–132 (2000)

[9] Tanaka, Y., Iida, M., Kikuchi, K.: Method to simulate generation of compression wave inside a tunnel at train entry with a simple geometry model (influence of the offset of train axis and the ground plane). Transaction of the Japan Society of Mechanical Engineers, B 69(683), 1607–1614 (2003) (in Japanese)

[10] Fukuda, T., Iida, M.: Model experiments on aerodynamics in train-tunnel system. The Journal of the Acoustic Society of Japan 63(9), 543–548 (2007) (in Japanese)

[11] Hara, T.: Aerodynamic force acting on a high speed train at tunnel entrance. Quarterly Report of RTRI, Japanese National Railways 2(2), 5–11 (1961)

Micro-pressure Wave Calculation Using Three Dimensional Compressive Fluid Analysis

K. Morita, Y. Abe, and Y. Takano

Hitachi Research Laboratory, Hitachi, Ltd.,
832-2 Horiguchi, Hitachinaka, Ibaraki, 302-0034, Japan
Tel: +81-292- 353-3111, Fax: +81-292- 353-3869
e-mail: kiyoshi.morita.ph@hitachi.com

Summary

We describe three-dimensional analysis of compressive fluid and three-dimensional scale tests to evaluate and verify the pressure gradient of compression waves that are generated when high-speed trains enter tunnels. We introduce an unstructured mesh and a total variation diminishing (TVD) method to improve the accuracy of calculations. A 1/30 scaled model was adopted in both the calculations and experiment. The traveling velocities of the train model ranged between 270 and 320 km/h. The pressure changes in the tunnel during the experiment were measured to verify the calculation results. The experimental results and calculations agreed well with an accuracy of approximately 10 % in both cases with and without a portal hood. We also confirmed that the installation of a portal hood reduced the pressure gradient by approximately 35% because of the increased duration of elevated pressure.

1 Introduction

Compression waves, which are generated when trains enter a tunnel are often called "micro-pressure waves", and may induce explosive noise at the tunnel exit. Because the noise level is proportional to the pressure gradient of the compression wave, it is necessary to reduce the pressure gradient to avoid noise issues.

There are two solutions to reduce the pressure gradient: the first is to optimize the shape of the nose including extending the nose length of the train, and the second is to install a hood at the tunnel portal [1]. In the past, simplified model tests and numerical simulations using a two-dimensional axisymmetric scale model were adopted to evaluate the pressure gradient because of limitations in experimental facilities and computational power. It was also obviously insufficient to adopt a two-dimensional model to evaluate the effects of nose shapes and hoods.

T. Maeda et al. (Eds.): Noise and Vibration Mitigation for Rail Trans. Sys., NNFM 118, pp. 405–410.
springerlink.com

Therefore, we introduce a three-dimensional analysis of compressive fluid and our calculations of the behavior of fluids around the nose of a train we examined closely. We also tested a three-dimensional scale model to verify the accuracy of calculations.

2 Method of Calculation and Experimental Setup

2.1 Method of Calculation

We used an inviscid program to analyze three-dimensional fluid especially to evaluate the compression waves generated by a train in this study [2]. The program has two main features; the first is that it adapts an unstructured mesh to precisely form the complicated nose shape of a train, and the second is that it adapts the total variation diminishing (TVD) method. The TVD method is known to be especially advantageous for precisely evaluating compression waves.

We also chose Euler equations as governing equations in the program, which means that the viscous effects of fluid were neglected. No boundary layer, which is governed by viscosity, usually develops around the nose of the train. Thus, this assumption is reasonable under constraints of restricted computational power.

The calculations were carried out using a high-speed technical parallel computer, the HITACHI SR11000. There were a total of about 4 million meshes, and the total time for calculation was about 24 hr.

2.2 Experimental Setup

A three-dimensional scale model was tested using the Transient Railway Aerodynamic Investigation Rig (TRAIR) made by DeltaRail Group Ltd., which is located in the UK.

Figure 1 has a photograph of the train model we examined in this study. The fabricated model was 1/30 scale and its total length was 1.4 m, which were determined so that the facilities at TRAIR could be adopted. Figure 2 plots variations in the cross-sectional area of the nose of the model, which were determined based on two-dimensional fluid analysis. The cross-sectional area ratio between train and tunnel was 0.17. The train model at TRAIR is powered by catapult-like mechanical propulsion and travels into the center of a miniature tunnel. The traveling velocities of the train model ranged between 270 and 320 km/h in the experiment.

The time history of static pressure at the center of the roof of the tunnel was measured to evaluate the magnitude of compression waves. A semiconductor pressure sensor was installed 1.33 m from the tunnel portal.

Figure 3 has two illustrations of the experimental conditions at the tunnel portal. We carried out the experiment both with and without the portal hood to clarify what effect it had on reducing the pressure gradient. The length of the portal hood was 0.83m, and the cross-sectional area ratio between portal hood and tunnel was 1.4 [3]. And, three side openings were installed at the one of side walls of hood. The height of side openings was 43 mm and total width was 426 mm. Thus total area of side openings was 18500 mm^2.

Fig. 1. Photograph of 1/30 scale train model (color figure online)

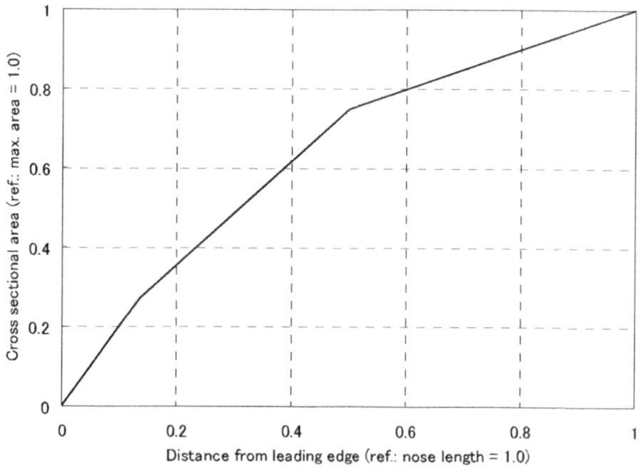

Fig. 2. Cross-sectional variations in nose of train model

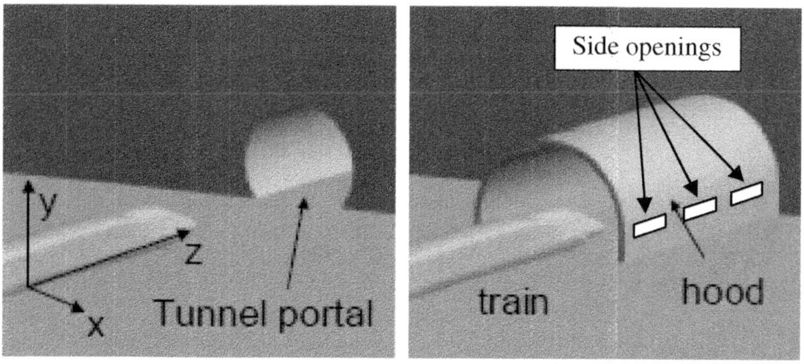

(a) Without portal hood (b) With portal hood

Fig. 3. Schematic of tunnel portal and portal hood (color figure online)

3 Results and Discussion

3.1 Pressure Distribution and Gradient

Figure 4 shows pressure distributions on the surface of a train and a tunnel at the moment of entry without a portal hood. The white areas in this figure indicate higher-pressure regions and the black areas indicate lower-pressure regions. We found that there were high pressures on the surface close to the tunnel portal. This indicates that compression waves in the tunnel are generated at the moment of entry.

Figure 5 compares calculated and experimental results for the time history of static pressure and the pressure gradient on the tunnel surface. The two results agree reasonably well in the duration from 0 to 0.04 s. The static pressure increases 0.03 s after a train enters the tunnel, and the pressure gradient reaches maximum at a time of 0.035 s. The maximum pressure gradients obtained from calculations and the experiment were 286 kPa/s and 312 kPa/s for the latter with an accuracy of 8%.

After the gradient had reached maximum, the error in calculations compared to that in the experiment gradually increased because of the lack of viscous effects in the calculations.

From these considerations, we confirmed that the proposed calculations were suitable to precisely evaluate the maximum pressure gradient of compression waves.

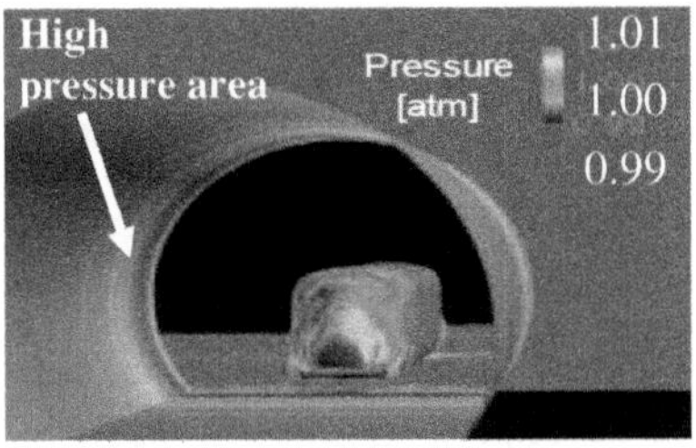

Fig. 4. Surface pressure distribution

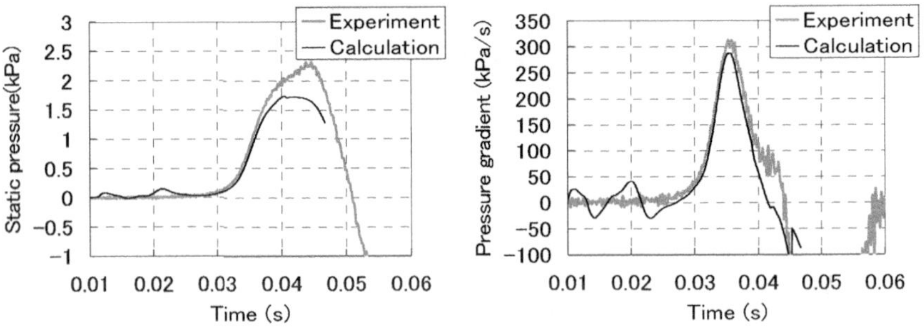

(a) Static pressure (b) Pressure gradient

Fig. 5. History of static pressure/pressure gradient (without portal hood and train speed of 315 km/h)

3.2 Effect of Portal Hood

Figure 6 compares the calculated and experimental results for the time history of static pressure and the pressure gradient on the tunnel surface when a train enters the tunnel with a portal hood. The nose of the train model first enters the portal hood firstly at 0.021 s and then enters the tunnel at 0.03 s.

We found that the increases in static pressure in Fig. 6 (with the portal hood) were slower than those in Fig. 5 (without the portal hood) and the calculated results indicate two strong peaks during the pressure increase. These results indicate that two compression waves were generated; the first occurred at the moment the train entered the portal hood and the second occurred at the moment the train entered the tunnel. Both the experimental and calculated results demonstrate that the second peak is larger than that of the first peak and the maximum pressure gradients in the calculation and experiment are 184 kPa/s for the former and 210 kPa/s for the latter with an accuracy of 12%.

Figure 7 compares the maximum pressure gradients with and without a portal hood. This indicates that the installation of a portal hood reduced the maximum pressure gradient by 33% in the experiment and 36% in the calculations.

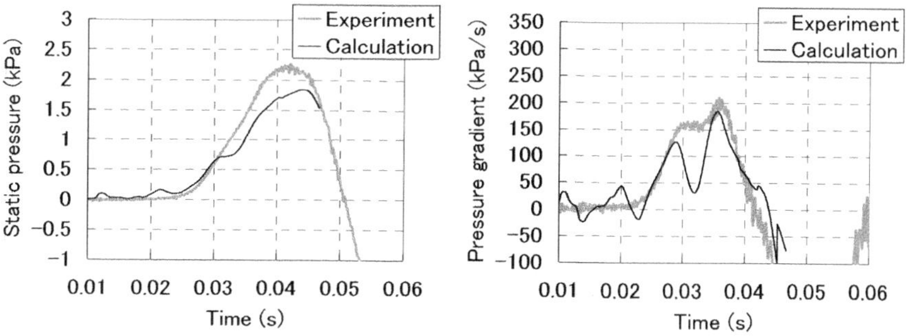

(a)Static pressure (b)Pressure gradient

Fig. 6. History of static pressure/pressure gradient (with portal hood and train speed of 315 km/h)

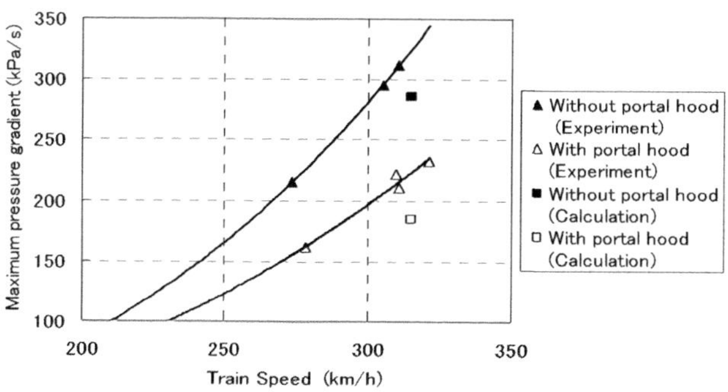

Fig. 7. Comparison of maximum pressure gradients

4 Conclusions

We carried out 1/30 scale three-dimensional compressive fluid analysis and a three-dimensional scale test to evaluate and verify the pressure gradients of compression waves that are generated when high-speed trains enter tunnels.

An unstructured mesh and the TVD method were introduced to the calculations to improve their accuracy. The experimental results and those from calculations agreed well with an accuracy of approximately 10 % in both cases with and without a portal hood. We also confirmed that the installation of a portal hood reduced the pressure gradient by approximately 35% because of the lengthened duration of pressure increase.

Acknowledgments

We would like to express our gratitude and appreciation to Mr. T. Johnson, and Ms. G. Figura-Hardy of DeltaRail Group Ltd., for their generous support throughout the experiment.

References

[1] Howe, M.S., Iida, M., Fukuda, T., Maeda, T.: Theoretical and experimental investigation of the compression wave generated by train entering a tunnel with a flared portal. J. Fluid Mech. 425, 111–132 (2000)

[2] Abe, Y., Morita, K., Terashima, O.: Evaluation of Micro-Pressure Wave Performance in Tunnels by Scaled Model Experiments and Numerical Calculations. In: 17th Transportation Logistics, Japan Society of Mechanical Engineers (2008) (in Japanese)

[3] Miyachi, T., Iida, M., Ozawa, S.: Optimization of tunnel entarance hood for reducing micro pressure wave from long slab-track tunnel portal. In: 17th Environmental Engineering Symposium, Japan Society of Mechanical Engineers (2007) ((in Japanese)

Recent Studies on Aerodynamic Noise Reduction at RTRI

T. Takaishi, N. Yamazaki, T. Sueki, and T. Uda

Noise Analysis Group, Railway Technical Research Institute
Umegahara, Maibara, Shiga, Japan
Tel: +81 749 52 4681, Fax: +81 749 52 4683
e-mail: takaishi@rtri.or.jp

Summary

As the maximum speed of high-speed trains increases, the effect of aerodynamic noise on the sound level at the wayside of the track becomes important. When the surface of bluff bodies such as pantographs is covered with porous materials, the aerodynamic noise generated by unsteady motion of vortices is significantly reduced. Experimental evaluation techniques of instantaneous flow fields using time-resolved PIV enable prediction of sound in the far field based on the theory of vortex sound. Aerodynamic noise emitted from a partial model of a pantograph is predicted numerically by coupling the calculation of unsteady flow with the evaluation of acoustical behaviour. The simulation succeeds in giving detailed information on the structure of aerodynamic sound sources.

1 Introduction

Noise generated aerodynamically has become a dominant noise source for today's high-speed trains, which run at the maximum speed of 300 km/h (190 mph) in Japan. The aerodynamic noise is produced mainly by projecting objects such as pantographs and cable heads on the roof, gaps between cars and cavities at the bogie sections. This noise is proportional to the 6th–8th power of the wind (vehicle) speed, and at high speeds it exceeds other noises such as the rolling noise from wheels and rails, which is normally proportional to about 3rd power of the vehicle speed. Therefore, it is crucial to reduce the aerodynamic noise caused by trains.

Railway Technical Research Institute (RTRI) established a large-scale low-noise wind tunnel in Maibara in 1996. Since then, many studies on aeroacoustic noise have been carried out. In this paper, we introduce some of the recent studies on noise-reduction measures, experimental techniques for clarifying the cross-correlation relationship between flow and emitted sound and a computational simulation method to predict the detailed structure of aerodynamic sound sources.

T. Maeda et al. (Eds.): Noise and Vibration Mitigation for Rail Trans. Sys., NNFM 118, pp. 411–418.
springerlink.com

2 Aerodynamic Noise Reduction by Applying Porous Materials to High-Speed Pantographs

In order to reduce aerodynamic noise from bluff bodies such as pantographs on the roof, we have proposed a new method that controls a flow around them by covering their surfaces with open-cell type porous materials [1]. The porosity of the applied materials is more than 97 percent and approximately 13 cells are included per inch (25.4mm).

Table 1. Parameters of cylinder specimens

Type	Cylinder diameter d	Porous material	Outside diameter d_0
A	25mm	None (bare cylinder)	—
B	45mm	None (bare cylinder)	—
C	25mm	Porous metal	45mm

To compare characteristics of aerodynamic noise and flow fields in the wake, we tested three types of cylinder specimens as listed in Table 1 in a small-scale low-noise wind tunnel of RTRI in Tokyo. The wind tunnel is a single return type with an open test section located inside an anechoic room, and its cross-section is 720mm wide × 600mm high. The thickness of the porous metal can be freely selectable and fixed as 10mm in the present study. The diameter of the Type B corresponds to the outside diameter of the cylinder covered with porous metal (Type C). Wind speed is set to 150km/h.

Figure 1 shows a comparison of aerodynamic sound radiated from each cylinder. Types A and B generate intense narrow-band sound, which correspond to the frequencies of vortices shedding from each cylinder. On the other hand, noise from the Type C which consists of a core cylinder and porous metal is reduced.

Then we applied this method to a pantograph of high-speed trains and executed tests at our low-noise wind tunnel in Maibara whose cross- section is 3000mm wide ×2500mm high. We installed porous metal on four fiberglass reinforced plastic (FRP) covers (the base- frame cover, the main shaft cover, the hinge cover and the panhead support cover) and the

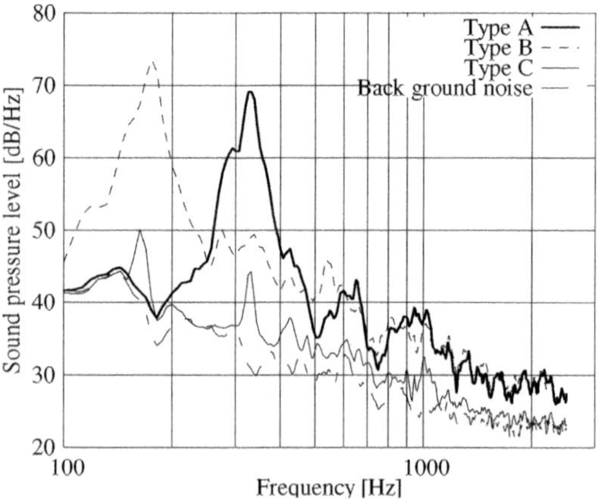

Fig. 1. Comparison of aerodynamic sound from each cylinder

bottom side of the base frame of the pantograph as illustrated in Fig. 2. We selected these areas to minimize influence on current collection performance and for ease of installation and removal. The thickness of the porous metal is 10mm (in application to the base frame cover and the bottom side of the base frame) and 5mm (in application to the main shaft cover, the hinge cover and the panhead support cover). Wind speed is set to 360km/h.

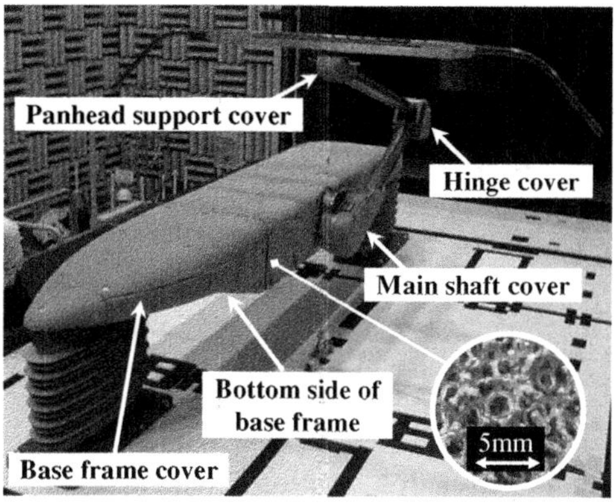

Fig. 2. Pantograph of high speed trains with porous metal

Figure 3 shows comparison of aerodynamic noise level measured with the omni-directional microphone placed at 5m beside the pantograph. The spectrum indicates that noise level of 315Hz band is reduced remarkably due to the porous metal. This noise is mainly generated from insulators. The porous metal might affect the flow around insulators, as a result, the noise generated from insulators becomes lower. The overall noise level from the pantograph with porous metal is 0.8dB lower compared to that without it.

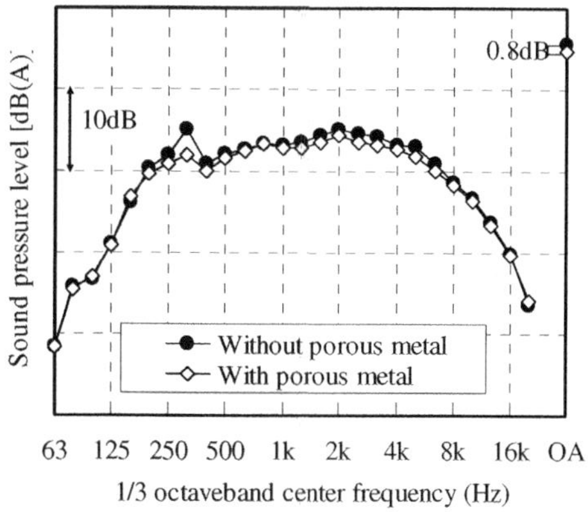

Fig. 3. Comparison of aerodynamic sound from pantographs

Figure 4 shows a comparison of noise source distribution (sound pressure levels) measured with the microphone-array system. In the case of the porous metal, noises emitted from the panhead and the main shaft covers are reduced.

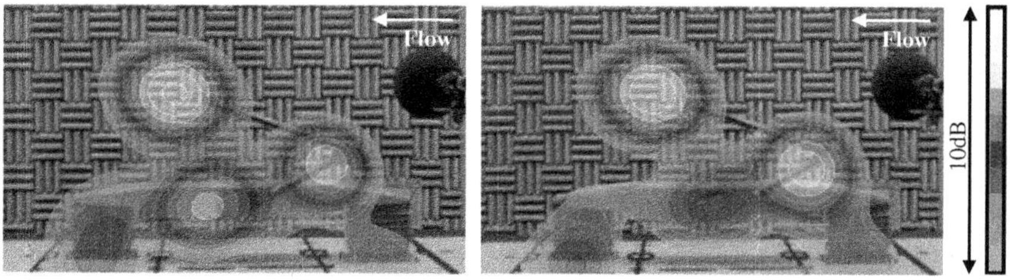

(a) Without porous metal (b) With porous metal

Fig. 4. Comparison of noise source distribution around a pantograph at 360km/h (1k-1.25kHz band)

From these results, the application of porous materials is proved to be effective in reducing aerodynamic noise caused by the pantograph of high-speed trains.

3 Experimental Evaluation Techniques of Aerodynamic Noise Using Time-Resolved PIV

Although the measurement with microphone array system gives profitable information regarding sound source locations, the relation between flow and emitted sound is not necessarily clear. Many prior references on aeroacoustics have pointed out that unsteady motion of turbulent flow, especially unsteady motion of vortices, generates aerodynamic noise. Therefore, we tried to clarify the cross-correlation relationship between instantaneous velocity fields and sounds measured outside the flow [2].

We assume that the Mach number is low and that the source region is small compared to the wavelength of the emitted sound; in other words, we assume that the source region is "acoustically compact". In order to relate the flow with the radiated sound, the equation of Howe's vortex sound theory [3], which is based on Lighthill's acoustic analogy, is applied and transformed in consideration of the finite analysis region as follows.

$$p_a(\mathbf{x},t)=\frac{\rho_0 x_i}{4\pi c_0|\mathbf{x}|^2}\int\frac{\partial}{\partial t}(\boldsymbol{\omega}\times\mathbf{u})(\mathbf{y},t-|\mathbf{x}|/c_0)\cdot\nabla\varphi_i d\mathbf{y} \quad (3.1)$$

where, p_a is the estimated sound pressure, t :time, $\mathbf{x}$:the position vector of the observer, $\mathbf{y}$:the position of sound sources, ρ_0 :the density of air, c_0 :the speed of sound, $\boldsymbol{\omega}$:vorticity vector, $\mathbf{u}$:velocity vector and φ_i :the acoustic effect of the body.

Test facilities for this study were set up at the University of Tokyo. The test section of the wind tunnel used is 75mm wide × 150mm high, and the wind velocity is fixed to 15.0m/s. A glass cylinder with a diameter of 6.0mm is used as the specimen in this experiment; this cylinder is bounded on both sides by end plates made of acrylic. The radiated sound is measured using an omni-directional microphone placed 300mm above the cylinder. We applied the time-resolved particle image velocimetry (PIV) technique to acquire the temporal evolution of the flow field around the cylinder, and the aeroacoustic sound in the far field is predicted by Eq.(3.1). A high-repetition rate Nd:YLF laser, which provides irradiation in the double-pulsed mode, is used as a light source to illuminate tracer particles added to follow the flow. The interval of the double-pulsed laser is 9μs, and the repetition rate per one cavity is set to 5 kHz. The displacement of the tracer particles between the laser pulses are determined by image-evaluation methods for PIV, and then velocity vectors at each interrogation area are calculated.

Figure 5 shows a comparison between the spectral analysis results of the sound pressure predicted by Eq.(3.1) with PIV results and that measured by the microphone. Strouhal number (*St*) in the lateral axis represents a dimensionless frequency; $St = 0.2$ corresponds with the peak frequency of vortices shedding from the cylinder. The power spectrum density predicted using PIV is scaled, assuming that the sound source structure is perfectly identical in the axial direction of a cylinder. Both the actually measured and PIV-predicted sound pressures are in good agreement in the low frequency, especially around $St = 0.2$. The discrepancy between the two in the higher frequency might result from the noise included in the PIV measurement and data processing.

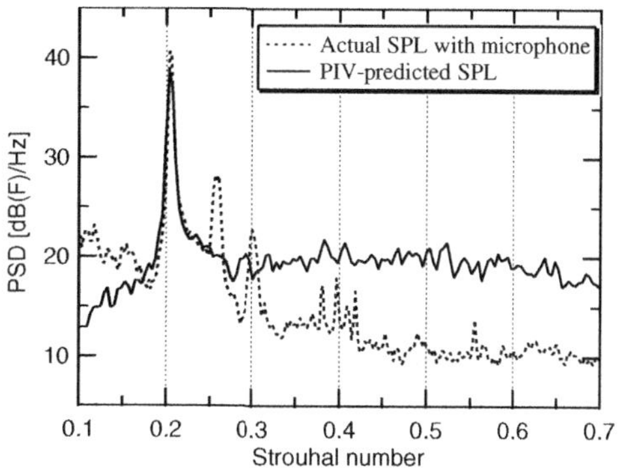

Fig. 5. Comparison of aerodynamic sound from a cylinder

Figure 6 shows the coherence and phase distribution between the actual sound pressure fluctuation measured by the microphone and the sound sources evaluated by integrant of Eq.(3.1) with PIV results. A significant cross-correlation is observed between the PIV-predicted sound sources and the measured sound pressure at $St = 0.2$. In addition to the coherence, the phase distribution at $St = 0.2$ reveals that the phase structure is well-ordered behind the cylinder and even in the immediate vicinity of the cylinder.

These results confirm that our experimental method based on Eq.(3.1) gives good prediction of the aeroacoustic sound radiated at the far field and that it also clarifies the detailed structure of sound sources.

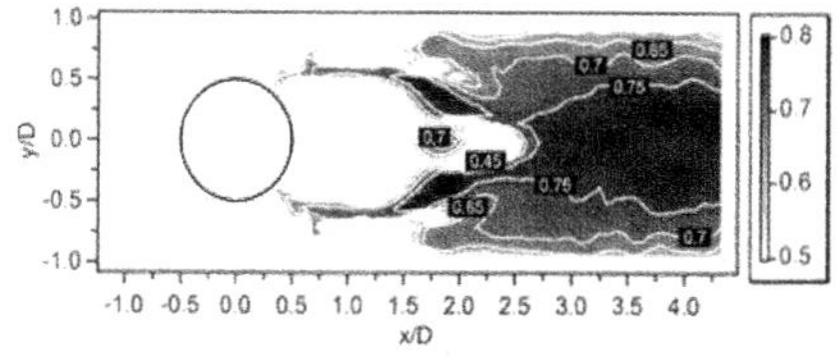

(a) Coherence at St=0.2

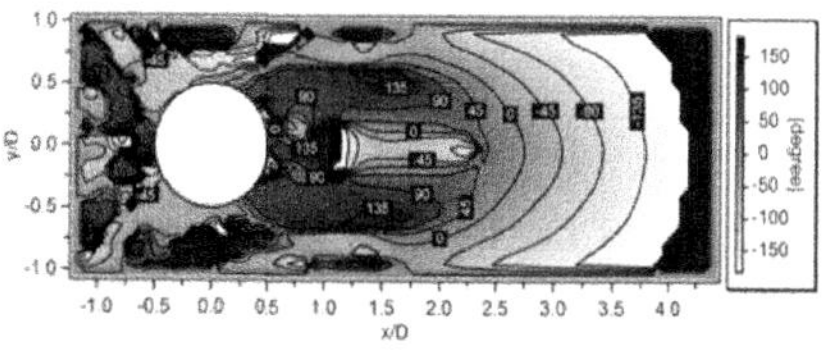

(b) Phase at St=0.2

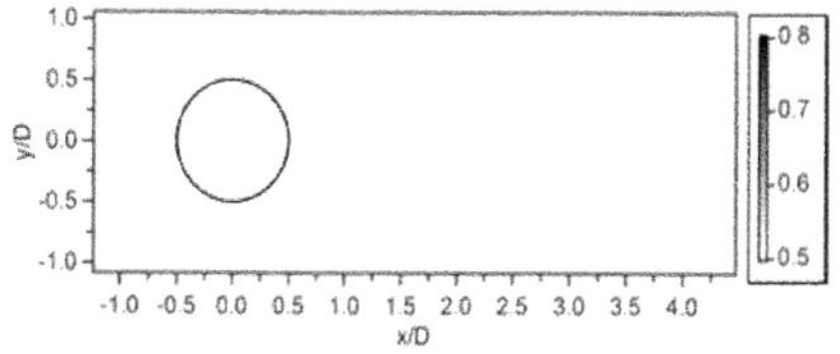

(c) Coherence at St=0.4

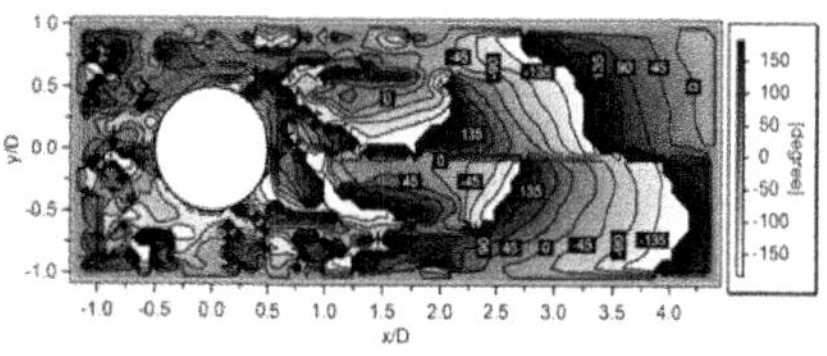

(d) Phase at St=0.4

Fig. 6. Coherence (left) and phase (right) between the sound pressure predicted using PIV and the actual sound pressure measured by the microphone

4 Computational Simulation Method of Predicting Aerodynamic Sound Sources

Experiments in a low-noise wind tunnel are effective ways of examining preventative measures against aeroacoustic problems, but it is difficult to investigate the complex 3-D structure of sound sources from experiments. Many previous numerical simulations calculated the pressure fluctuations at the body surface and then estimated the acoustic pressure fluctuations in the far field using Curle's equation. However, they could not directly identify the structure of sound sources in the flow, and the relationship between flow and radiated sound was unclear.

Then we have developed a numerical method based on the theory of vortex sound, which suggests that the unsteady motion of vortices produce aerodynamic sound. We combined the calculation of unsteady flow and the evaluation of acoustical performance numerically [4]. First, the unsteady flow around the body is obtained using the large-eddy simulation (LES) technique, which solves the spatially filtered equations of continuity and momentum for incompressible flow. Second, the acoustical performance by the presence of the body is estimated by solving a Green's function adapted to the body shape. Finally, by coupling the instantaneous flow properties with the Green's function, the distribution of aerodynamic sound sources around the body is obtained. We also proposed a new idea by which the sudden termination of vortices at the outer boundary of a finite -computational domain is avoided. At a low Mach number, since the radiation efficiency of sound scattered at the body surface is larger than the direct sound, the net contribution from so-called dipole sound sources is extracted. The dipole sound pressure p_a^s at the observer $\mathbf{x}$ becomes

$$p_a^s(\mathbf{x},\omega) = -\rho_0 \int (\boldsymbol{\omega}\times\mathbf{u})(\mathbf{y},\omega)\cdot\nabla_y G^s(\mathbf{x},\mathbf{y},\omega)d\mathbf{y} \tag{4.1}$$

where ω is the angular frequency and G^s is the scattered part of the shape adapted Green's function. Then the integrand of Eq.(4.1) can be regarded as the dipole sound source. Equation (3.1) is a simplified form of Eq.(4.1) under the assumption of acoustical compactness.

As an example of computational simulation, the aerodynamic noise from a partial model of a pantograph of high-speed trains is predicted numerically. In order to validate the numerical results, an experiment with a partial model of a pantograph is performed in the large-scale low-noise wind tunnel of RTRI. The model consists of a panhead of 0.9m×0.1m×0.055m, an upper frame made of a pipe with a diameter of 70mm that inclines from the vertical line by 70 degrees and a panhead support that connects the panhead and the upper frame. The pantograph model is set on a 4m×8m support table in the open type test section of the wind tunnel. The center of the lower side of the panhead is placed at the position 3m downstream of the nozzle exit. Wind speed is set to 300km/h.

Figure 7 shows instantaneous stream lines around a pantograph model. The flow separates at the leading edges of the upper and lower surfaces of the panhead, and rolls up strongly behind the panhead and its support. Longitudinal vortices are formed along the joint and cause strong disturbance in flow.

Figure 8 shows numerical results of the distribution of aerodynamic sound sources defined by the integrant of Eq.(4.1) when the observer is fixed 5m above the panhead. White indicates isosurface of sound source strength with positive phase, and black indicates those with negative phase. The influence of sound reflection at the support table (train roof) is also included in the calculation of G^s by Boundary-Element Method (BEM). In the low frequency range, large separations occurring at the upper and lower sides of a pantograph head, at lower sides of the pantograph head support and at an upper frame are main sources of scattered sound propagating upward. In the high- frequency range, sound sources are located behind the panhead and its support. Large-scale vortices shed from the body surface become small gradually as they are thrown downstream, and unsteady motion of these small vortices seems to produce aerodynamic sound of short periods, namely high frequencies.

Figure 9 shows a comparison of sound pressure level at the observer 5m above the panhead. Sound pressure levels calculated by Eq.(4.1) roughly agree with the experimental results in the frequency range of 800Hz or less. Although the accuracy of computational result is not enough for quantitative evaluation of sound pressure level, the qualitative tendency seems to be captured satisfactorily.

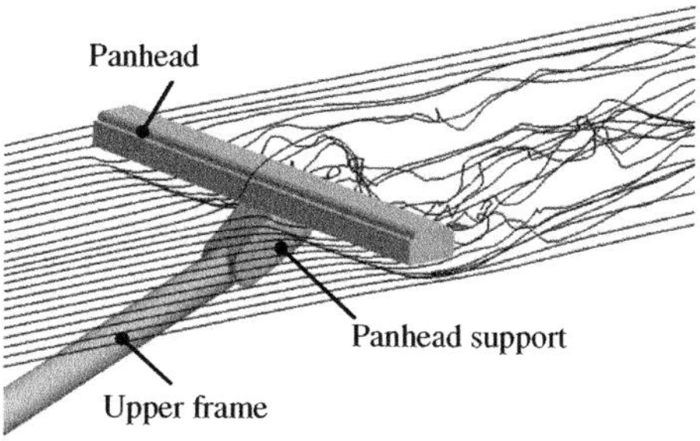

Fig. 7. Instantaneous stream lines around a pantograph model

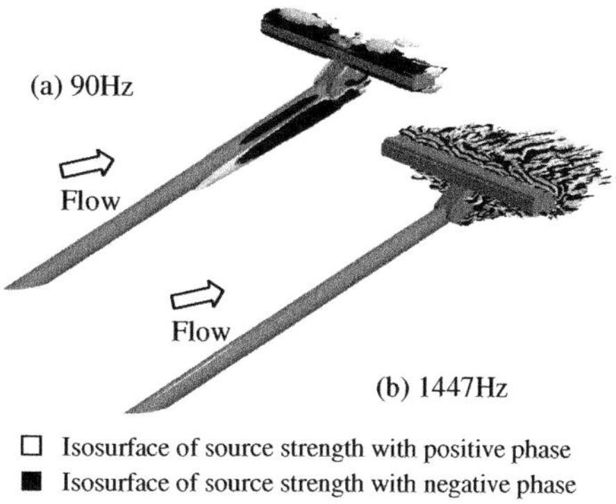

Fig. 8. Numerical results of the aerodynamic sound sources around a pantograph model

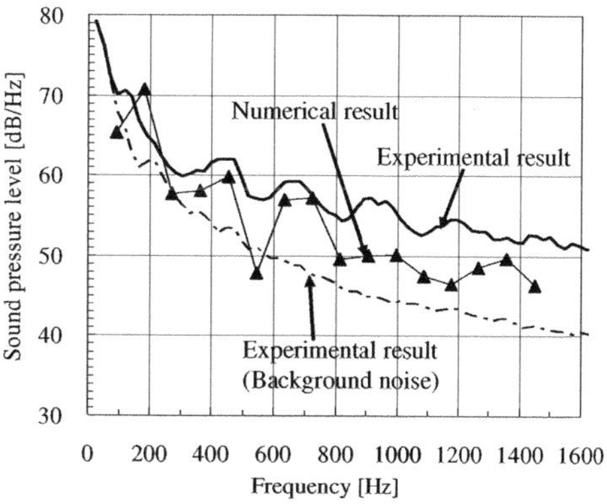

Fig. 9. Comparison of aerodynamic sound from a pantograph model

5 Conclusions

Reduction of aerodynamic noise generated from high-speed trains is a crucial issue in Japan, and we summarized the results of some recent studies at Railway Technical Research Institute in this paper.

—When the surface of a cylinder is covered with open-cell type porous materials, the aerodynamic noise generated by unsteady motion of vortices is significantly reduced. Then this method is applied to a pantograph of high-speed trains. Experimental results at the wind tunnel show that the overall noise level from the pantograph with porous metal is 0.8dB lower compared to that without it.

—The temporal evolution of the flow field around a cylinder is determined using a time-resolved PIV technique; these data are then applied to Howe's vortex sound theory in order to estimate the aeroacoustic sound in the far field. A significant cross-correlation is observed between the PIV-predicted and the measured sound pressure at the frequency of the Aeolian tone of a cylinder.

—Aerodynamic noise emitted from a pantograph model is predicted numerically by coupling the calculation of unsteady flow with the evaluation of acoustical performance. Distribution of aerodynamic sound sources indicates that large separations at each component produce low-frequency sound and that unsteady motion of small vortices produce high-frequency sound. Although the accuracy of computational result is not enough for quantitative evaluation of sound pressure level, the qualitative tendency seems to be captured satisfactorily.

Acknowledgements

The authors would like to express their sincere gratitude to Mr. T. Kurita and Mr. H. Yamada at the East Japan Railway Company for their cooperation in the wind tunnel testing of the pantograph of high-speed trains.

References

[1] Sueki, T., et al.: Application of porous material to reduce aerodynamic noise caused by a high-speed pantograph. In: Proceedings of Inter Noise 2008, Shanghai, China (2008)

[2] Uda, T., et al.: Cross-correlation analysis of aeroacoustic sound and flow field using time-resolved PIV. In: Proceedings of 15th Int. Symp. on Application of Laser Techniques to Fluid Mechanics, Lisbon, Portugal (2010)

[3] Howe, M.S.: Theory of vortex sound. Cambridge University Press, Cambridge (2003)

[4] Takaishi, T., et al.: A computational method of evaluating noncompact sound based on vortex sound theory. Journal of the Acoustical Society of America 121(3), 1353–1361 (2007)

Aerodynamic Noise Reduction of a Pantograph by Shape-Smoothing of Panhead and Its Support and by the Surface Covering with Porous Material

M. Ikeda[1], T. Mitsumoji[1], T. Sueki[2], and T. Takaishi[2]

[1] Railway Technical Research Institute, 2-8-38, Hikari-cho, Kokugunji-shi, Tokyo, Japan
Tel: +81-42-573-7288, Fax: +81-425-73-7332
e-mail: mikeda@rtri.or.jp, mitumoji@rtri.or.jp
[2] Railway Technical Research Institute, Umegahara, Maibara-shi, Shiga, Japan
Tel: +81-749-52-4681, Fax: +81-749-52-4683
e-mail: sueki@rtri.or.jp, takaishi@rtri.or.jp

Summary

To reduce aerodynamic noise generated by a pantograph, which is one of the dominant noise sources of high-speed trains, the authors proposed some noise reduction techniques, that is, shape-optimization of a panhead, relaxation of aerodynamic interference between panhead and articulated frame and surface covering with porous material. To evaluate total noise reduction effect of them, wind tunnel tests were performed with a prototype pantograph to which these techniques were applied. The test results show that noise level of the prototype pantograph is lower than that of the currently-used pantograph by about 4 dB. Furthermore, it was also confirmed that the prototype pantograph has enough aerodynamic stability against change of attack angle.

1 Introduction

Reduction of wayside noise caused by a high-speed train is important to preserve surrounding environments along railway lines. The acoustic energy of aerodynamic noise is proportional to the 6th to 8th power of the train's velocity, while the energy of other kinds of noises such as the rolling noise and structure-borne noise is governed by a lower-power law. Accordingly, as the operational train velocity increases, the aerodynamic noise becomes predominant as the cause of the wayside noise of high-speed railways. Therefore, the reduction of the aerodynamic noise is an important subject for the high-speed railways.

T. Maeda et al. (Eds.): Noise and Vibration Mitigation for Rail Trans. Sys., NNFM 118, pp. 419–426.
springerlink.com

A pantograph installed on train roof is one of the major aerodynamic noise sources of the high-speed train; hence a low-noise pantograph has been used for Shinkansen trains. As shown in Fig. 1, the low noise pantograph takes a relatively simple configuration to reduce aerodynamic noise, consisting mainly of a panhead, an articulated frame and a base frame. The panhead is mounted on the top of the articulated frame by a panhead support. The panhead is known to be one of the major aerodynamic noise sources of the low-noise pantograph, suggesting that aerodynamic noise caused by the panhead must be reduced in order to lower overall pantograph noise. Hence, we proposed a unique panhead design procedure combining computational fluid dynamics (CFD) analysis with an optimization method to obtain a suitable panhead shape that fulfils two requirements, i.e., low aerodynamic noise performance and stable lift force characteristics [1,2].

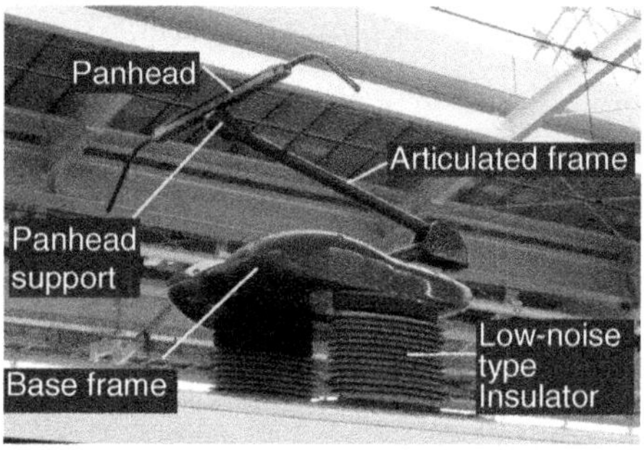

Fig. 1. Example of a low-noise pantograph (color figure online)

However, just small noise reduction effect could be achieved by only the panhead shape optimization. This is because distinct aerodynamic interference between the panhead and the articulated frame induces additional aerodynamic noise sources; therefore, relaxation of the interference is urgently required. We found out that shape modification of the panhead support is very important to reduce aerodynamic noise of the pantograph with a smoothed-shape panhead [3].

These design techniques are very effective for aerodynamic noise reduction of high-speed pantographs. However, shape modifications are not readily achievable for some members of the pantograph because strict restrictions are imposed on pantograph design due to securing of internal space or mechanical movements. To reduce aerodynamic noise generated by such members, we proposed surface covering with porous materials [4]. We have found out that this method is also very effective for the high-speed pantograph [5].

We produced a prototype pantograph along these aerodynamic noise reduction strategies by remodeling a currently-used low-noise pantograph which was mounted on two low-noise insulators with ellipsoidal cross-section as shown in Fig.1. By using this prototype pantograph, noise reduction efficiency was evaluated by wind tunnel testing.

2 Prototype Pantograph

Because the panhead has large projection area and it is exposed directly to high-speed oncoming flow, it is one of the main aeroacoustic noise sources of the pantograph. Moreover, it has a great influence on lift force characteristics of the pantograph. To secure contact quality, the pantograph must have steady lift characteristics, in other words, angle-of-attack dependence of the lift force must be small. For these reasons, design of the panhead is important work for the high-speed pantograph. But, to balance aerodynamic noise reduction with the steady lift force characteristics is difficult to accomplish since both are contradictory features. Therefore, we developed new design procedure to obtain an adequate shape of the panhead by combination of CFD analysis and optimization method. This technique provides a cross-sectional shape of the panhead with consideration for contact strips wear and changes of the angle of attack in operational conditions.

Figure 2(a) shows the optimized panhead shape for the prototype pantograph under restriction condition of its width of 120 mm, allowing anterior–posterior asymmetry of the shape. For comparison, the example of the currently-used panhead shape is shown in Fig. 2(b). In addition, to prevent the panhead from generating intense Aeolian tone, 20 through holes with diameter of 6 mm penetrate the panhead at intervals of 20 mm along the span axis as shown in Fig. 3. This panhead is mounted on the panhead support to operate in knee-upward direction.

It has been confirmed that the optimized panhead could achieve drastic aerodynamic noise reduction when noise generated by only the panhead without an articulated frame was evaluated. However, the aerodynamic noise generated by the optimized panhead mounted on the articulated frame is not effectively-reducing because strong aerodynamic interference between the smoothed-shaped panhead and the articulated frame induces intense aerodynamic noise in the relatively high-frequency range. To resolve this, we developed a new-type panhead support shown in Fig. 3 and Fig. 4(a). It is composed of a solid cylinder of 22 mm in diameter to separate the panhead from the articulated frame spatially. The top of the cylinder is connected to a cross bar which is placed inside of the panhead; the panhead is elastically connected to the crossbar by springs. The panhead is always supported at the level position by the linkage structure.

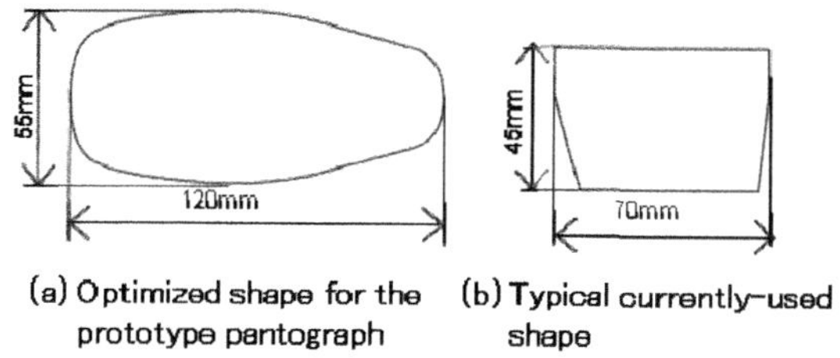

Fig. 2. Cross-sectional shape of panheads

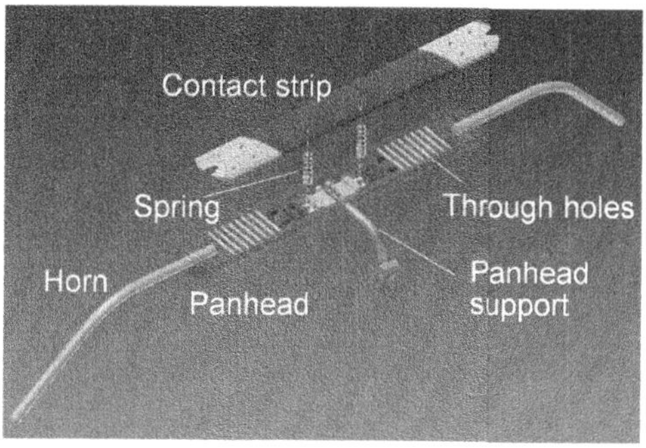

Fig. 3. Configuration of the panhead of the prototype pantograph (color figure online)

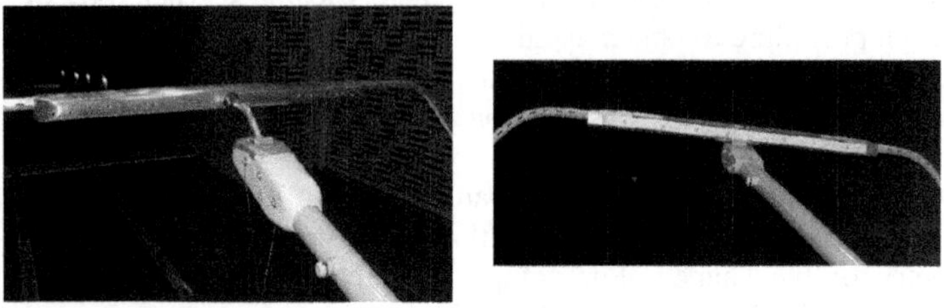

(a) New-type panhead support (b) Currently-used panhead support

Fig. 4. Configuration of the panhead support for the prototype pantograph (color figure online)

As described above, shape smoothing of a members is very effective for aerodynamic noise reduction. However, functionality requirements often restrict shape design of pantograph members. In case of a fairing of the knee of the articulated frame, shape modification can be permitted only to a limited extent; therefore, it is difficult to reduce the aerodynamic noise by smoothing its shape. The fairing of the base frame can be formed as a smoothed shape; but such modification let the fairing increase in size and weight to secure space for driving the pantograph inside the fairing.

To reduce aerodynamic noise caused by such fairings, we developed a noise reduction technique by surface covering with porous materials. Although porous materials have various configurations, we found that open-cell type porous materials, which have continuous pore, are effective for the noise reduction. For the prototype pantograph, we used open-cell type porous material with porosity of 97% having a three-dimensional skeleton structure with 13 cells per inch and thickness of 5–10 mm as shown in Fig. 5. The porous material applied to the high-speed pantograph must be high-strength metal. However, covering curved surface of the fairings with such hard metal needs many process with high cost. Hence, we redesigned the fairing of the base frame for ease in covering with hard porous metal. After considering noise reduction by the porous material, the fairing was modified to simple boxy shape shown in Fig. 5. However, the fairing of the knee of the articulated frame was not

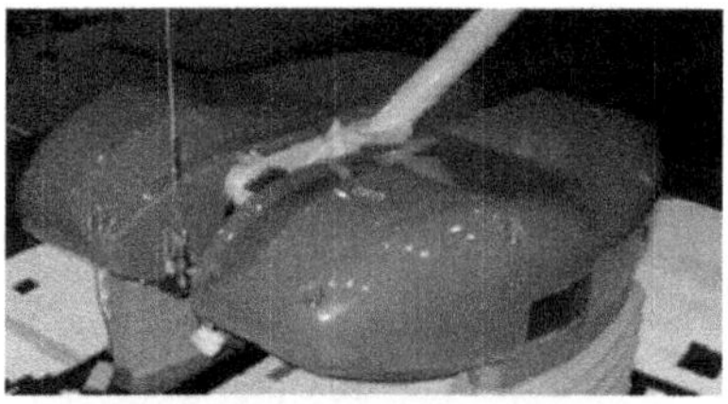

(a) Currently-used base frame fairing

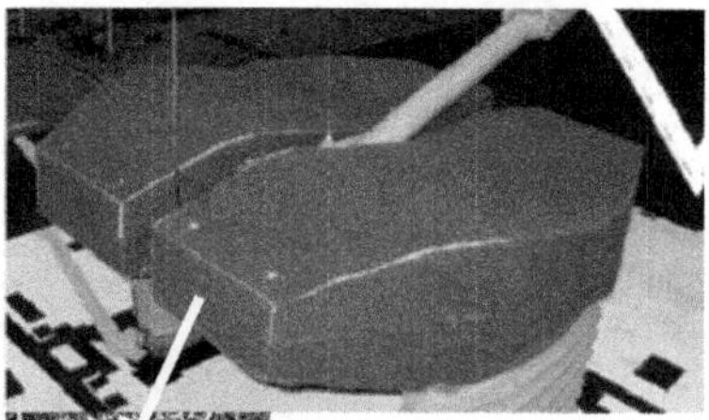

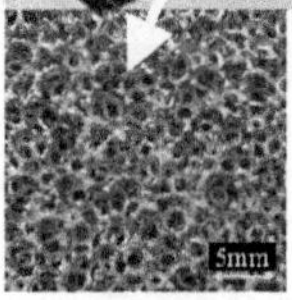

(b) Base frame covered with porous material

Fig. 5. Surface covering with porous material (fairing of base frame) (color figure online)

modified for covering with hard porous metal because its shape is sensitive to the lift force characteristics of the pantograph.

Results of our wind tunnel experiments indicate that an aerodynamic noise reduction effect is independent of rigidity of the porous material [4]. We chose porous urethane for the prototype pantograph which was tested only in the laboratory because the porous urethane is easier to handle than porous metals.

3 Wind Tunnel Test of Prototype Pantograph

3.1 Aerodynamic Noise Reduction Effect by Proposed Methods

To evaluate aerodynamic noise of the prototype pantograph, wind tunnel tests were conducted at Large-scale Anechoic Wind-tunnel of Railway Technical Research Institute (RTRI). The pantograph was set in open type test section with the nozzle size of 2.5 x 3 m. In this condition, the maximum flow velocity is 400 km/h (111 m/s). The pantograph was set on the ground plate using two low-noise insulators in the knee-upstream position. Aerodynamic noise was measured by an omni-directional microphone at the distance of 5 m sideways from the pantograph as shown in Fig. 6.

Figure 7 shows measurement results of aerodynamic noise at 300 km/h. This figure compares noise level of four conditions, that is, the currently-used pantograph, the pantograph only with modified panhead and its support, the pantograph only with fairings covered with the porous material, and the fully modified pantograph (the prototype pantograph). Figure 7 indicates that the noise level of the prototype pantograph is lower than the currently-used pantograph by about 3.7 dB.

To distinguish contribution of aerodynamic noise reduction effect of shape modification of the panhead and its support and surface covered fairings with porous material, noise source localization was carried out at 300 km/h by using a wheel-type microphone array system with the diameter of 4 m. Figures 8 and 9

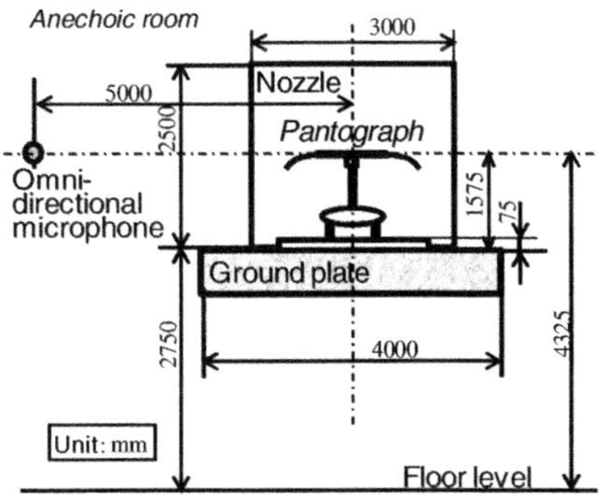

Fig. 6. Set-up of the pantograph in the wind tunnel

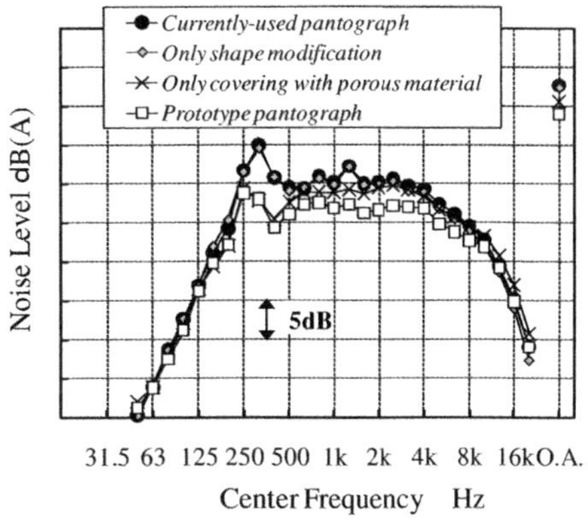

Fig. 7. Noise reduction effect of the prototype pantograph

show noise source distribution maps at 315 Hz band, 800 Hz band, and 2.5 kHz band. We can understand that contribution of noise generated by the base frame is large in the low-frequency range. In contrast, noise generated by the panhead and its support also contributes significantly in the high-frequency range. In case of the prototype pantograph, strength of noise sources in the vicinity of the pahnead and its support diminishes in high-frequency range and that around the base frame diminishes in low-frequency range compared to the currently-used pantograph.

For a more quantitative comparison, we estimated power averages of localized noise levels over region A and region B, respectively. The region A is the vicinity of the panhead and its support, and the region B is vicinity of the base frame as shown in Fig. 10. Figure 11 indicates comparison of the power averaged noise level over the region A and region B between the currently-used pantograph and the prototype pantograph. Although directivity of the array microphone depends on frequency, any compensation is not corrected. Aerodynamic noise emitted from two regions does not interact with each other; exact compensation is not needed only to estimate noise reduction effect due to the proposed methods.

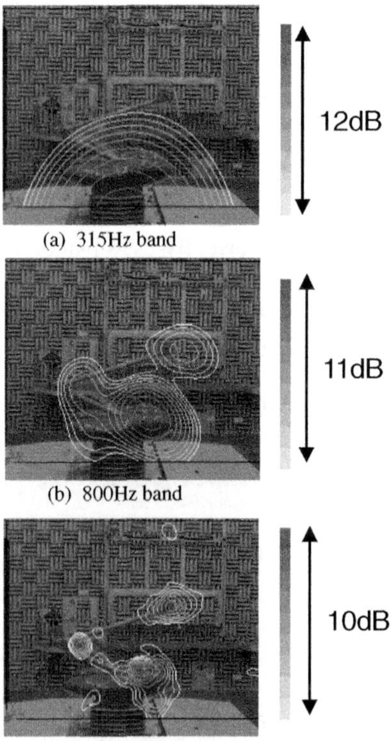

(a) 315Hz band

(b) 800Hz band

(c) 2.5kHz band

Fig. 8. Noise source distribution map (Currently-used pantograph)

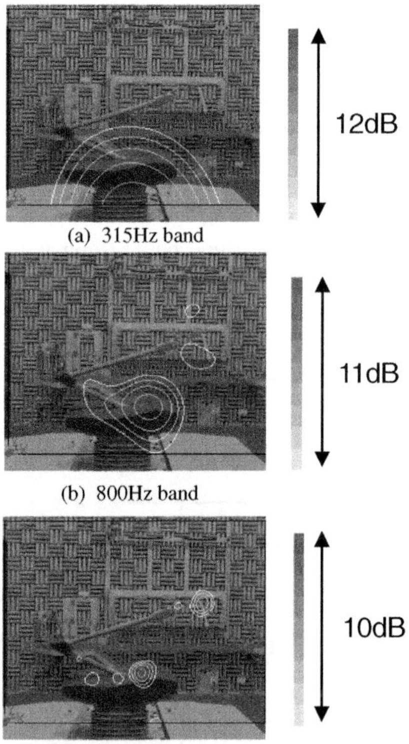

(a) 315Hz band

(b) 800Hz band

(c) 2.5kHz band

Fig. 9. Noise source distribution map (Prototype pantograph)

Figure 11 indicates that the aerodynamic noise sources of the prototype pantograph distributed both over the region A and over the region B reduce by about 4 dB, respectively compared with that of the prototype pantograph. To say that, the shape modification of the panhead and its support and covering the fairings with the porous material have almost the same noise reduction efficiency, and therefore result in a total aerodynamic noise reduction of about 4 dB from the currently-used pantograph.

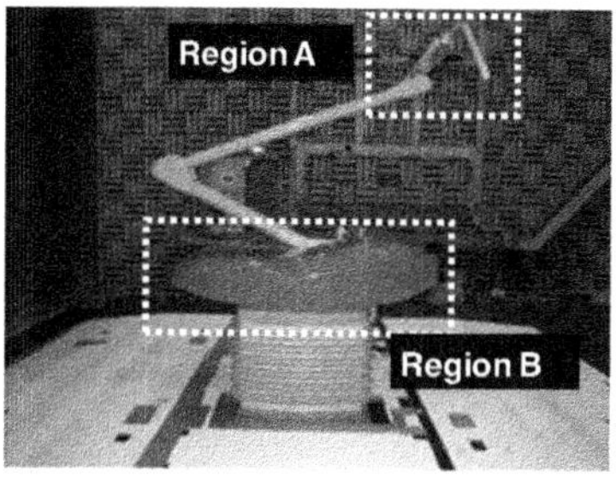

Fig. 10. Definition of noise source regions

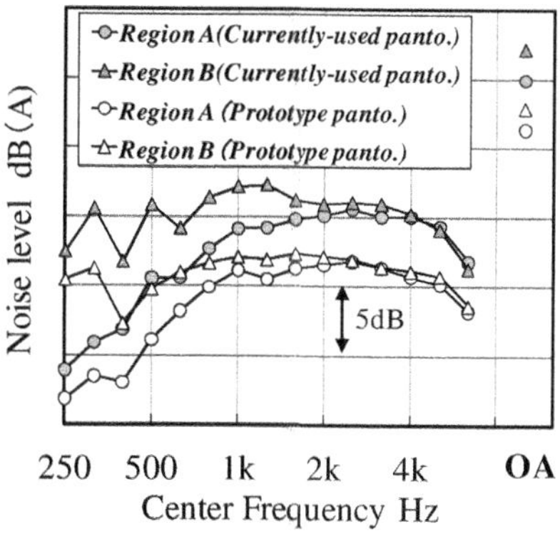

Fig. 11. Power averaged noise level over region A and region B

3.2 Lift Force Characteristics of the Prototype Pantograph

The shape modification of the panhead and its support can influence lift force characteristics of the pantograph; therefore, we carried out a wind tunnel test at the Large-scale Anechoic Wind-tunnel of RTRI to evaluate lift force characteristics of the prototype pantograph in comparison with that of the currently-used pantograph. To simulate change of flow angle near the panhead, mounting angle of the panhead was changed within ±2 degrees. Measurement results of lift force characteristics shown in Fig. 12 indicate that the lift curve slope of the prototype pantograph is almost same as that of the currently-used pantograph. In other words, shape modification of the panhead and its support does not affect aerodynamic stability.

Although the lift force of the currently-used pantograph at the mounting angle of 0 degree increases with flow velocity that of the prototype pantograph is nearly zero at any flow velocity. This is mainly due to relaxation of aerodynamic interference between the panhead and its

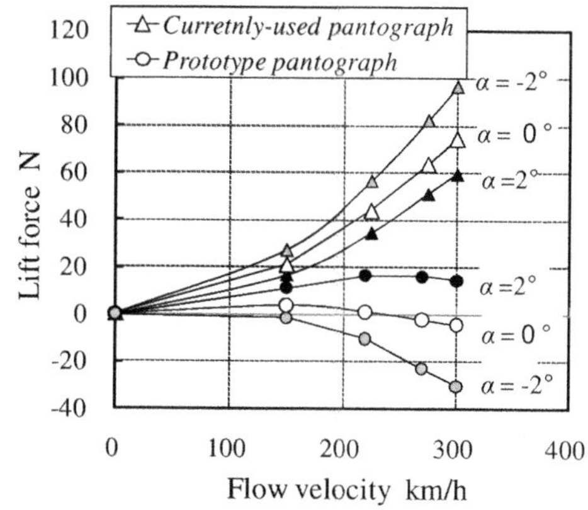

Fig. 12. Comparison of lift force characteristics

support. In the case of the currently-used pantograph, distinct aerodynamic interference arises in the wake region of the panhead and its support, causing positive lift force characteristics. In a different viewpoint, if we can evaluate the lift force characteristics of the panhead alone and that of the articulated frame without a panhead, the total lift force of the prototype pantograph can be easily estimated in disregard of effect of such aerodynamic interference; lift force control can be easily achieved.

4 Concluding Remarks

In this paper, effects of aerodynamic noise reduction due to panhead shape optimization, shape modification of the panhead support for relaxation of aerodynamic interference and surface covering with porous material to the fairings are described. The prototype pantograph employing these techniques reduces aerodynamic noise by about 4 dB from the currently-used pantograph. Furthermore, the lift curve slope of the prototype pantograph is almost same as the currently-used one. We will continue efforts aimed at practical realization of these methods.

References

[1] Suzuki, M., Ikeda, M., Yoshida, K.: Study on numerical optimization of cross-sectional panhead shape for high-speed train. J. of Mechanical Systems for Transportation and Logistics 1(1) (2008)

[2] Yoshida, K., Suzuki, M., Ikeda, M.: Optimization of 2-Dimensional Shape of Pantograph Panhead for High Speed Trains. In: Proceeding of Inter Noise 2004, Prague, Czech (2004)

[3] Ikeda, M., Mitsumoji, T.: Numerical Estimation of Aerodynamic Interference between Panhead and Articulated Frame. Quarterly Report of RTRI 50(4), 227–232 (2009)

[4] Sueki, T., Takaishi, T., Ikeda, M., Arai, N.: Application of porous material to reduce aerodynamic sound from bluff bodies. Fluid Dynamics Research, Res. 42, 015004 (2010)

[5] Sueki, T., Ikeda, M., Takaishi, T.: Aerodynamic Noise Reduction using Porous Materials and their Application to High-speed Pantographs. Quarterly Report of RTRI 50(1), 26–31 (2009)

Application of a Component-Based Approach to Modelling the Aerodynamic Noise from High-Speed Trains

D.J. Thompson, M.G. Smith, and F. Coudret

Institute of Sound and Vibration Research,
University of Southampton, Southampton SO17 1BJ, UK,
Tel.: +44 23 8059 2510; Fax: +44 23 8059 3190
e-mail: djt@isvr.soton.ac.uk

Summary

At sufficiently high speeds, aerodynamic noise from trains becomes more important than rolling noise. However, there remains considerable uncertainty concerning the level of aerodynamic noise from trains, the relative contributions of the various sources and the parameters of major influence. In this paper, a component-based model is used to predict the aerodynamic noise from high-speed trains, and in particular the noise from the bogie region and the pantograph. The predicted noise has been compared with measurements of overall noise spectra from high-speed trains at various speeds and the approach appears promising.

1 Introduction

High-speed trains are increasingly seen as a more sustainable means of transport than short-haul air travel. However, as train speeds increase so does the noise produced. Rolling noise increases at a rate of around 30 log V, where V is the speed, whereas aerodynamic noise increases at a higher rate, typically 60 log V.

Fig. 1 shows measured noise levels from French high-speed trains [1]. Above about 300 km/h, the noise level rises above the 30 log V curve, representative of rolling noise. This is due to the dominance of aerodynamic noise at high speeds [2]. Moreover, the aerodynamic noise generated by the power car exceeds that generated by the coaches [3]. Clearly, the noise from high-speed trains cannot be reduced significantly without a better understanding of the aerodynamic noise.

Aerodynamic noise is caused by the flow of air over the components of the train. This flow is very complex and includes a turbulent boundary layer (TBL) that generates some surface noise. However, the main noise sources are caused by the flow over smaller components, discontinuities and cavities [1]. Some computational fluid dynamics (CFD) modelling has been carried out in recent years.

T. Maeda et al. (Eds.): Noise and Vibration Mitigation for Rail Trans. Sys., NNFM 118, pp. 427–435.
springerlink.com

However, this involves considerable computational effort and does not readily allow parametric study.

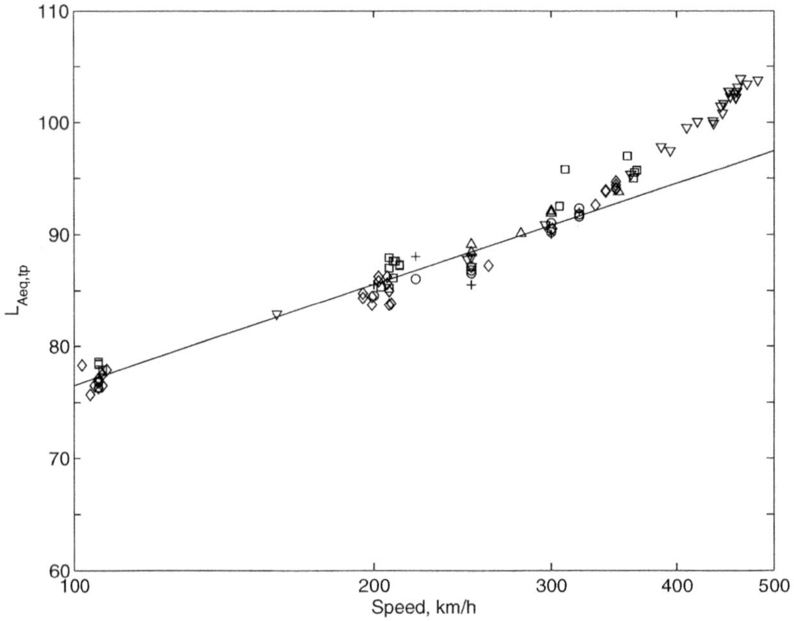

Fig. 1. $L_{Aeq,tp}$ measured at 25 m from TGV-A (∇, □, ◇), TGV-Duplex (O, Δ) and Thalys (+). Straight line has a slope of 30 $\log_{10}$V to represent rolling noise [1].

Simpler 'component-based' models have been developed by Smith [4,5] and more recently by Guo [6] for the prediction of aircraft landing gear noise. In this technique, a landing gear is divided into a number of components (wheels, main struts and small details) whose contributions to the overall noise are predicted by a semi-empirical model. This model is based on simple non-dimensional source spectra and directivities. This approach has shown encouraging results for landing gears when compared with measured data.

The current work investigates the feasibility of applying the same approach to the case of high-speed trains. The pantograph or the bogie can be divided into various components in a similar way to the components of an aircraft landing gear. These components are modelled by a source term based on Guo's model [6].

2 Component-Based Model

2.1 Guo's Model

Guo [6] decomposed the landing gear noise into three components corresponding to low, mid and high frequencies. These were associated with the wheels, the main struts and the small details. The three groups have significantly different sizes, which makes their dominant noise well separated from each other in frequency.

For each of these components, a source spectrum was defined, which is based on the aerodynamic sound produced by the flow over a cylinder. The overall level and frequency distribution were defined in terms of the dimensions of the component and the flow velocity. Guo also included directivity terms associated with the position of the landing gear under the aircraft, and a 'complexity' term used to represent different types of landing gear, but these are not considered relevant in the present work. Atmospheric absorption is also neglected as the receivers are closer to the train than for the aircraft case.

2.2 Application to Rail Vehicles

In the form used here, the mean square pressure is given by

$$\langle p^2 \rangle = \frac{(\rho_0 c_0^2)^2 M^6}{D^2 (1 + M \sin\theta)^4} \sum_{components} \beta_{0i} S_i F_i(St_i) \tag{2.1}$$

where $M = V/c_0$ is the Mach number, D is the distance from source to receiver and θ is the angle as shown in Fig. 2, ρ_0 is the density of air and c_0 is the speed of sound. The terms for each component include an amplitude, β_0, the surface area S_i and a normalised spectrum F_i that depends on the Strouhal number of component i,

$$St_i = \frac{fd_i}{c_0 M} \tag{2.2}$$

where f is frequency and d_i is a typical dimension for component i, e.g. strut diameter.

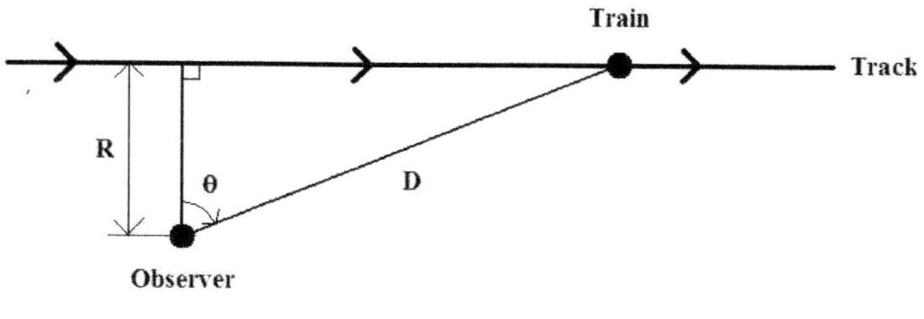

Fig. 2. Assumed source/receiver geometry

The normalised spectra in equation (2.1) have the form:

$$F(St) = A \frac{St^{\sigma}}{(B + St^{\mu})^{q}} \tag{2.3}$$

where σ, μ and q are empirical constants given in Table 2.1, and A and B are given by:

$$A=\left(\frac{q\mu}{\sigma}\right)^{q} St_0^{q\mu-\sigma},\quad B=\left\{\left(\frac{q\mu}{\sigma}\right)-1\right\}St_0^{\mu} \tag{2.4}$$

where St_0 is an empirical value of the Strouhal number also given in Table 1.

$F(St)$ is normalised to have a maximum value, of unity, at Strouhal number St_0. The resulting spectrum is converted to one-third octave bands.

Table 1. Empirical parameters used for the three landing gear noise components [6]. d is wheel diameter, w is wheel width, N is number of wheels in the landing gear assembly, a is the average strut diameter, L is total length of struts.

	Low	Mid	High
Amplitude, β_0	2.3×10^{-8}	7.5×10^{-9}	1.6×10^{-5}
Typical length used in Strouhal number, d_i	d	a	$l = 0.15a$
Surface area, S_i	πNwd	πaL	ηl^2
St_0	1.0	0.3	0.1
σ	4.0	3.0	2.0
μ	2.5	2.5	1.1
Q	2.6	4.2	4.2

2.3 Pantograph

The pantograph is considered first as it resembles a landing gear most closely; it is located in relatively clean flow and consists of readily identifiable struts. The pantograph is divided into main struts, control struts, pantograph head and 'details'. The latter are modelled using the 'high frequency' source terms while the remainder are modelled using the 'mid frequency' source terms with suitable dimensions. The number of components included in the 'details' source is determined by the empirical constant β_0 in Guo's model.

Fig. 3 presents an example prediction of pantograph noise at 25 m for a train speed of 300 km/h. The results show a fairly good agreement with measured data in [1,7] in terms of spectrum shape although no comparisons with sound level are available. However, the predicted spectrum contains no tonal components and seems to roll off too slowly at high frequency. Unlike for the landing gear, the various components have similar sizes and therefore have similar peak frequencies.

2.4 Bogie Region

Although the flow around the bogie is known to be very turbulent, this is initially ignored. The bogie is divided into transverse members, side frames, wheels, dampers, brake discs, small details and large details. The wheels and brake discs are considered as 'mid frequency' components, because it has been assumed that they resemble a strut more than a landing gear wheel. The small and large details are treated in the same way as for the pantograph although with different lengths.

2.5 Other Components

The nose of the train is modelled as a single large strut. Due to the large size of the source, its noise spectrum is concentrated at very low frequencies.

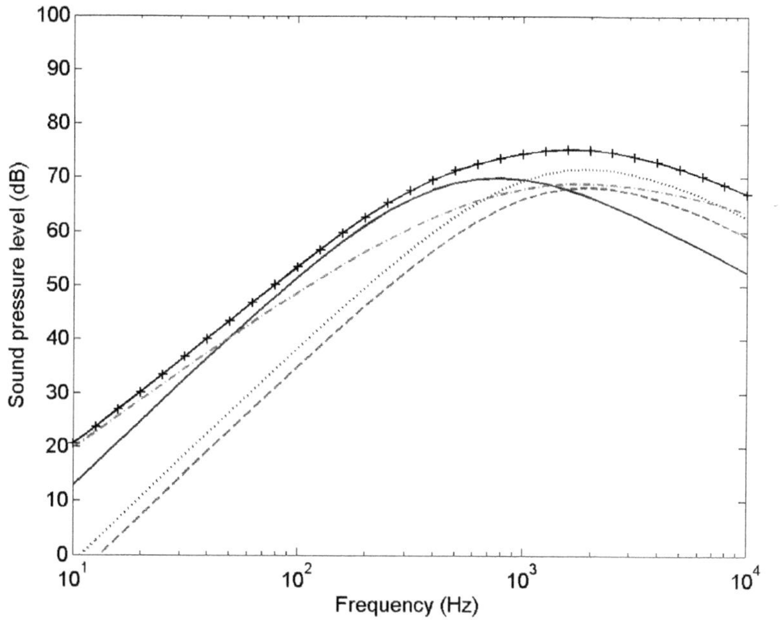

Fig. 3. Spectral decomposition of total noise for the pantograph at 25 m and 300 km/h. +—+, total noise; —, main struts; – – –, control struts; ······, pantograph head; – · – ·, details. (colour figure online)

3 Total Train Model

The above results are for stationary sources in a moving air flow. In this section, account is taken of the presence of multiple bogies and the movement of the train.

3.1 Turbulence

The turbulent flow incident on the bogie is likely to create some additional noise. On the other hand, the flow speed incident on the bogie will be less than the train speed due to boundary layer effects. Reduced flow speeds are therefore adopted for the bogie, as listed in Table 2. The flow speed over the pantograph is assumed to be 0.85 V.

An amplification due to incident turbulence is included in the model for all except the front bogie. This amounts to an amplification spectrum with a peak of approximately 2 dB at 50 Hz reducing to less than 0.5 dB above 500 Hz.

Table 2. Assumed flow speeds under and beside bogies. V is train speed.

	First bogie	Other bogies
Under the bogie	0.75 *V*	0.55 *V*
Beside the bogie	0.85 *V*	0.75 *V*

3.2 Motion of the Train

The effect of source motion is calculated, including the Doppler effect, time delay due to propagation and convective amplification. Each bogie is treated as a single source assembled from the various components. The overall A-weighted level calculated for a TGV at 300 km/h is shown in Fig. 4. The leading bogie can be seen to produce more noise than other bogies, due to the higher assumed incident flow speed, and the pantograph contributes to the peak at the rear of the train.

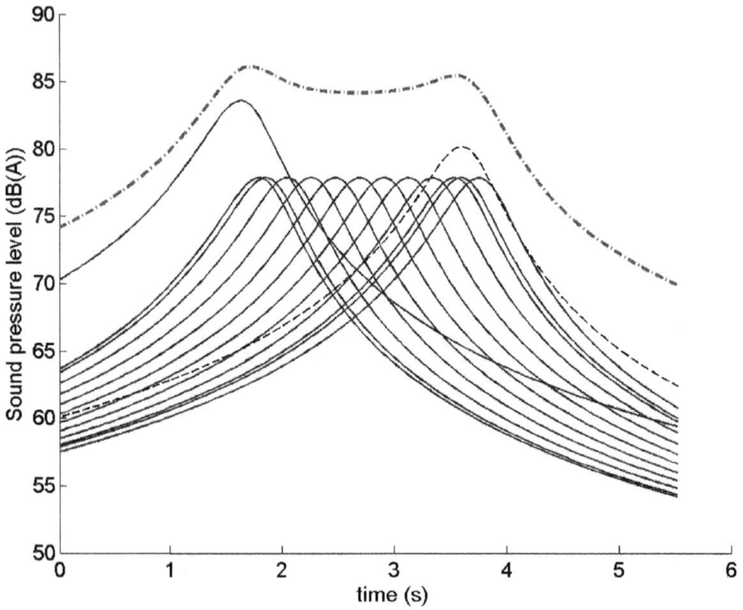

Fig. 4. Time signal of the whole train at 300 km/h for receiver at 25 m. —, noise from each bogie; – – –, noise from the pantograph; – · – ·, overall noise. (colour figure online)

Fig. 5 shows spectra of noise at a receiver 25 m from the track during a train pass-by. Here, the pantograph, bogies and inter-coach cavity are all included. Measured results for a TGV are shown for comparison. Although it is clear that some low frequency contributions are missing, the results at mid frequencies show a good match, especially at higher speeds. The estimate of aerodynamic noise at high frequencies appears to roll off too gradually. At lower speeds, the predictions are lower than the measurements but this is expected as rolling noise will be dominant at most frequencies at these speeds.

The contributions of the various components to the pass-by level at 300 km/h are listed in Table 3. In Table 4, the predictions are compared with results from an empirical model, MAT2S, developed in a French/German collaboration (see [1]). The results for the bogie agree well although the pantograph noise appears a little too high in the present model. An estimate of the noise due to the inter-coach gap is also derived from measured data from [8]. This is not included in Fig. 5 but partly accounts for the discrepancies at low frequencies.

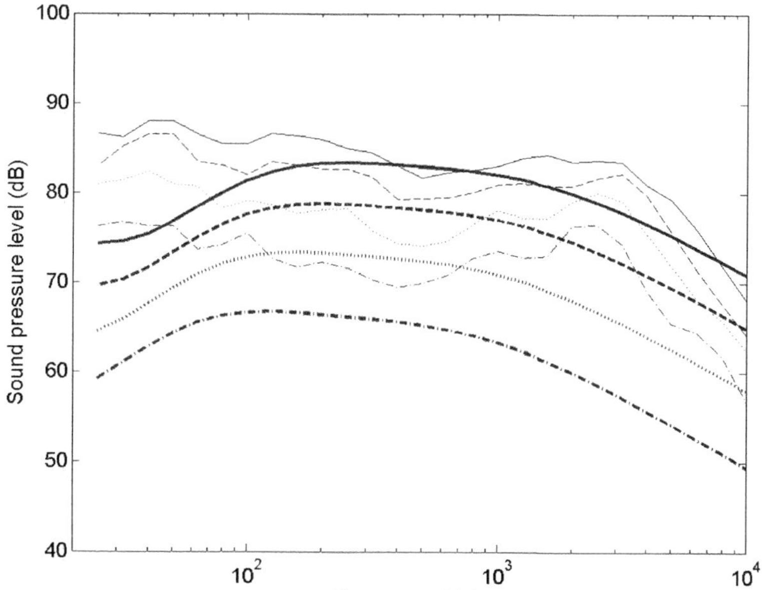

Fig. 5. Noise spectra at 25 m from a train running at different speeds. —, 350 km/h; – – –, 300 km/h; ······, 250 km/h; – · – ·, 200 km/h. Thick: predicted; thin: measured from [1]. (colour figure online)

Table 3. Predictions of A-weighted sound pressure levels at 25 m

Component	dB(A)	Component	dB(A)
Transverse members	63.2	Dampers	78.0
Side frames	74.9	Small details	78.2
Wheels	70.2	Large details	81.7
Brake discs	79.0	Nose	50.0
Inter-coach cavities	60.5	Pantograph	76.4

Table 4. Comparison of A-weighted sound pressure levels at 25 m with MAT2S [1]

Component	Present model	MAT2S
Bogies	86.0	86.6
Pantograph	76.4	69.2
Total aerodynamic	86.5	87.7

4 Conclusions

The component-based method is a promising approach for determining the relative importance of various sources of aerodynamic noise. In a very short computation time, it is possible to see the influence of the main parameters on the overall noise.

Further work is needed to refine the source terms and to validate the modelling of individual components. In particular, the flow speeds acting on the various components should be determined more precisely, and the influence of turbulence should be further investigated. At low frequency, the model under-predicts the noise. Here, models for sources of cavity noise, such as the inter-coach spacing, pantograph recess, louvers, etc, are required. A model for TBL noise and an improved model for the noise produced by the train nose are also required. At high frequencies, the model over-predicts the levels partly due to over-predicting the pantograph noise. Particularly for the 'details', a method of specifying the number of such sources and the statistical distribution of their sizes is required.

Although reasonable agreement is achieved using the empirical constants derived for aircraft landing gear, these could be improved by using wind tunnel experiments, for example on a scale model of a bogie and other components of a high-speed train. The speed dependence of the various sources also needs further investigation.

Despite its limitations, the model presented here can be seen as a useful framework, allowing source terms to be added or improved from field experiments, wind tunnel experiments or computational fluid dynamics (CFD) calculations.

References

[1] Thompson, D.J.: Railway noise and vibration: mechanisms, modelling and means of control. Elsevier, Oxford (2008)

[2] Talotte, C.: Aerodynamic noise: a critical survey. Journal of Sound and Vibration 231(3), 549–562 (2000)

[3] Mellet, C., et al.: High speed train noise emission: Latest investigation of the aerodynamic/rolling noise contribution. Journal of Sound and Vibration 293, 535–546 (2006)

[4] Smith, M.G., Chow, L.C.: Prediction method for aerodynamic noise from aircraft landing gear, AIAA Paper 98-2228 (June 1998)

[5] Smith, M.G., Chow, L.C.: Validation of prediction model for aerodynamic noise from aircraft landing gear, AIAA Paper 2002-2581 (June 2002)

[6] Guo, Y.: A component-based model for aircraft landing gear noise prediction. Journal of Sound and Vibration 312, 801–820 (2008)

[7] Lölgen, T.: Wind tunnel noise measurements on full-scale pantograph models. In: Proceedings Joint ASA/EAA meeting, Berlin, Germany (1999)

[8] Frémion, N., et al.: Aerodynamic noise radiated by the intercoach spacing and the bogie of a high-speed train. Journal of Sound and Vibration 231(3), 577–593 (2000)

The Numerical Prediction of the Aerodynamic Noise of the TGV POS High-Speed Train Power Car

E. Masson[1], N. Paradot[2], and E. Allain[2]

[1] SNCF Research Department,
45 Rue de Londres 75008 Paris, France
[2] SNCF Rolling Stock Engineering Center
e-mail: elisa.masson@netcourrier.com,
nicolas.paradot@sncf.fr,
eliane.allain@sncf.fr

Summary

For a high-speed train operating at 300km/h or more, the aerodynamic noise, which is induced by the pressure fluctuations of the air, reaches similar levels to the rolling noise, which is due to contact between the rail and wheel. For this reason, the study of the noise generated by the flow around the train is of great interest for the constructors and operators of high-speed trains. Although the main noise sources are now well identified, their localization is so far mainly based on experimental techniques using acoustic array measurements, applied either on small-scale models in a wind tunnel or on full-scale passing trains. Nonetheless, the numerical prediction of the aerodynamic noise is becoming more and more of a practical possibility, as the need is emerging for proper integration of technical solutions and optimizations at an early stage of design of new rolling stock.

In this study, the aerodynamic noise generated by the power car of the TGV POS high-speed train is considered. This generation of TGV is named POS for Paris Ost-Frankreich Süd-Deutschland as it has been dedicated to connect Paris to the East of France (Strasbourg) and the South of Germany (Stuttgart, Munich). It has been introduced on the French and German high-speed network in 2007, where the operational speed has been increased from 300km/h to 320km/h (on French East-European line).

For the two main reasons of the introduction of new rolling stock and the increase in speed, the aerodynamic noise is less well known and is expected to be more significant. A numerical study has been carried out to identify the sources of the noise and to quantify their level. Some results will be presented here.

In order to obtain results of a high level of quality, a very detailed geometry was used, and the unsteady simulations were carried out at full scale. The complexity of the real geometry and the consideration of all parts and elements of bogie or

T. Maeda et al. (Eds.): Noise and Vibration Mitigation for Rail Trans. Sys., NNFM 118, pp. 437–444.
springerlink.com

pantograph frames enabled the influence to be determined of several shape modifications or of the integration of fairings on the level of aerodynamic noise. This gives some clues for solutions in terms of noise reduction. Among all the solutions tested, two examples have been selected and are presented in this paper.

1 Introduction

Aerodynamic and aeroacoustic engineers and researchers have at their disposal three means to study flow and noise around a vehicle: measurements on track (i.e. on a real vehicle running in real operational conditions), measurements in a wind tunnel (i.e. considering a reduced-scale model) and calculations with a CFD (computational fluid dynamics) code. All these approaches have their advantages and disadvantages and are used depending on the needs and the means of the study [1,2].

As with the two experimental methods, the numerical method has some drawbacks. For instance, a detailed calculation on a representative configuration with a fine resolution is often limited by computational resources, and only some parts of the train can be treated in as much detail as wished. Nevertheless, as the interest here was only in the sources on the first power car, this method has been preferred. Indeed, this method has some advantages: it gives the access to understanding of the phenomenon through the knowledge of the flow, whereas the wind tunnel or the on-line measurements provide only partial information on speed and pressure fields. This was part of the specifications of the study as it was necessary to understand the phenomenon in order to propose solutions to reduce the noise generated by the flow.

In this study, aerodynamic and aeroacoustic calculations have been performed with the PowerFLOW code based on the Lattice–Boltzmann method. The acoustic propagation has been determined by a Ffowcs-Williams Hawking approach.

2 The Basic Configuration

2.1 Geometrical and Physical Description

First calculations have been performed on the original configuration of the TGV POS high-speed train, at full scale. The model is composed of a very detailed power car at the front, followed by three trailer cars and terminated by a simplified rear power car. A full-scale model has been considered where the bogie and pantograph cavities have been modelled in great detail as has the nose of the train (Fig. 1). The intercar gap between the leading power car and the first trailer car is also well represented.

Ten levels of refinement were considered for the mesh of the model, with a smallest cell dimension of 2.2 mm at the wall, leading to a grid of 22 million surface fluid cells (or surfels) at the wall and about 210 million volume fluid cells (or voxels) in the whole domain. The parameters of the mesh are listed in Table 1. Different areas of refinement, named voxel resolution (VR) zones, are considered.

The VR10 zone corresponds to the smallest refinement, whereas VR1 is the mesh at the border of the calculation domain. The velocity of the train is 320 km/h, which corresponds in atmospheric conditions to a Mach number of 0.259. The reference length considered here is the height of the train, giving a Reynolds number of 25×10^6. The boundary conditions used are a fixed-velocity condition at the entrance of the computational domain, a fixed static pressure at the outlet, a slip condition on side walls and a non-slip condition on the floor. The simulation parameters are also listed in Table 1.

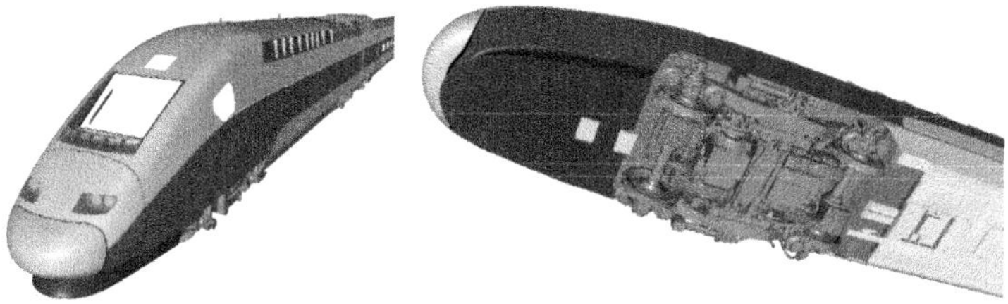

Fig. 1. TGV POS geometry considered for the simulations

Table 1. Parameters used for the mesh and the simulations

Train dimensions (l / h / w)	106.7m / 4.1m / 2.9m
Domain dimensions (l / h / w)	300m / 160m / 180m
Number of surfels	22.4×10^6
Number of voxels	209.3×10^6
Flow velocity	88.89 m/s
Mach number	0.259
Reference length L_{ref} (height of vehicle)	4.128 m
Viscosity	1.49×10^{-5} m^2/s
Reynolds number (based on L_{ref})	24.6×10^6
Temperature	20° C

2.2 Numerical Conditions

The Lattice–Boltzmann method (also referred to as LBM) has been used for the calculations with the PowerFLOW software, where turbulence is modelled with the k-ε model considered under its RNG (ReNormalization Group) form [3].

The calculations were processed in two steps. In the first step, a simulation was run from scratch with a coarser grid (with eight levels of refinement considered, from VR0 to VR7, corresponding to 67×10^6 voxels) during 1900 CPU hours. Then, the solution of the first simulation was used for initializing the calculation on the most refined grid. This technique allowed the time for reaching the 'stabilization' to the unsteady solution to be reduced by a factor of 2. A further 17,240 CPU hours were used to obtain 1.1-second (real time) record. An adapted small time-step was used to perform high-frequency (23 kHz) recordings in specific zones.

2.3 Aerodynamic and Aeroacoustic Results

2.3.1 The Mean Flow

A way to analyse the mean flow is to represent the iso-surfaces of zero pressure coefficient (noted Cp total and calculated as Cp total=0) in order to determine the separation zones of the flow. Fig. 2 shows these zones for the power car. The most important are the protection of the wiper on the windscreen, the zone between the motorcar and the passenger car, the pantograph cavity; concerning the under part, the flow is separated in front of the snowplough and just after the snowplough in the bogie cavity.

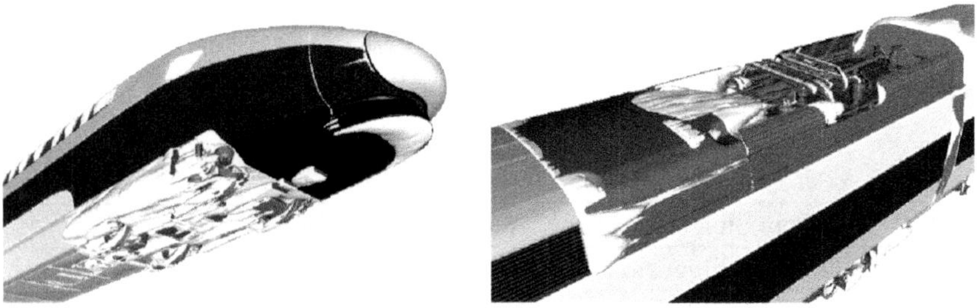

Fig. 2. Visualization of Cp = 0 iso-surfaces

This study has been carried out to understand and reduce the aeroacoustic phenomenon due to the flow around the train at high speed. However, the aerodynamic aspects are also of interest as the noise reduction solutions that will be proposed should not damage the aerodynamic performance and will be considered as better if they improve the drag coefficient (Cx) of the power car. So the drag of the train set has been calculated. It has been also defined for each element of the power car with the effect on the flow of its passage over the nose, the bogie, the pantograph and the intercar gap.

For each of the parts where large recirculation occurs, which are probably the source of noise, i.e. the train nose, the bogie elements, the pantograph cavity and the intercar gap, the unsteady flow has been analysed.

2.3.2 The Unsteady Flow

An analysis of the unsteady flow consists in visualizing the iso-surfaces of the $\lambda 2$ coefficient, which is defined as the second greatest eigenvalue of the quantity $S^2+\Omega^2$, where S is the symmetric part and Ω the antisymmetric part of the tensor of the velocity gradient. The term S is also called the deformation tensor and corresponds to a deformation of a fluid element as the term Ω is linked to the vorticity and corresponds to a simple rotation. With these two terms, the apparition and the distortion (deformation plus rotation) of the vortices inside the flow are located. In other words, the iso-surfaces of $\lambda 2$ allow the visualization the heart of the vortices, their quantity and their size. The quantification of this criterion in dB adds information on their intensity in terms of pressure fluctuation.

Fig. 3 represents this coefficient for the whole power car, together with the unsteady velocity in the mid-plane. It highlights the most important turbulent zones, which are the windscreen, the first bogie and then the second (with a smaller contribution), the roof at the pantograph location and the zone between the power car and the first trailer car.

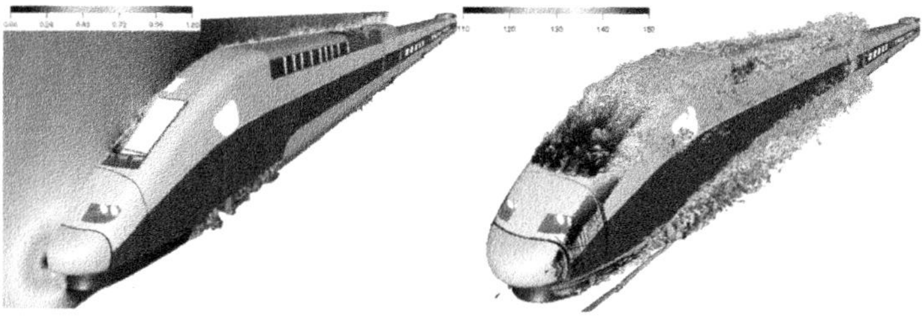

Fig. 3. Unsteady velocity (left) and $\lambda 2$ iso-surfaces (right)

Once the noise sources had been determined, they were propagated in order to evaluate their acoustic radiation (Fig 4.). Indeed, one of the advantages of LBM is the possibility to compute the acoustic propagation. Several methods have been developed. In this study, we have considered the Ffowcs-Williams Hawkings (or FWH) approach, which consists in a generalization of the wave equation taking into account all the aeroacoustic sources. More precisely, the aerodynamics results obtained with the PowerFLOW code have been processed with the implementation of Casalino method [4].

Through this numerical evaluation, the acoustic power radiated by the front part of the train, which is composed of the power car and half of the first trailer car with all their components plus the rails and the ground, is 130.6 dB(A).

The maximum value of the Lw of the front part of the train for receivers located at 7.5 m and 25 m is, respectively, 99.6 dB(A) and 90 dB(A). These figures take into account only the aerodynamic noise, as the rolling noise is not considered in the calculation.

The analysis of the reference case has highlighted and confirmed the main sources of aerodynamic noise: the snowplough, the windscreen, the bogie, the pantograph and the intercar gap. Now they are identified and evaluated, and the aim is to find a way to decrease their intensity in order to reduce the overall noise of the train.

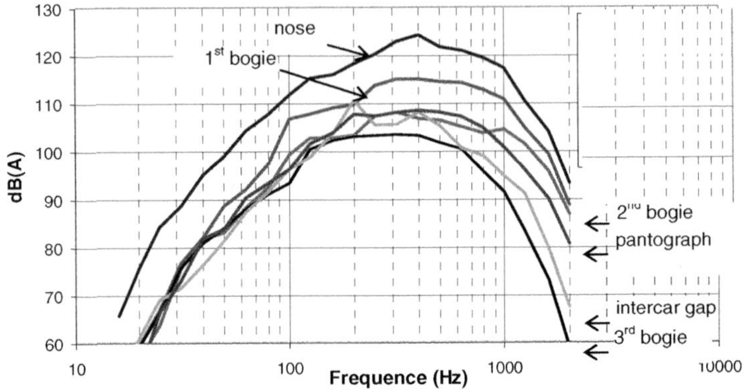

Fig. 4. Acoustic power in dB(A)

3 The Optimizations

In a second stage, the reduction in the aerodynamic noise of the TGV POS has been studied. Here, the study has been focused on the use of geometrical deflectors on the original structure. Some modifications have been locally added, and new calculations have been performed in order to estimate the benefits of these modifications.

Looking at the results of the first calculation on the source locations, two particular places have been considered: one on the underbelly zone around the bogie cavity and the other one on the roof with the pantograph cavity and all the roof equipment.

In order to be the more efficient, two criteria have been considered: (i) the origin of the generated noise (resonance in cavities, vortex due to protuberant parts …) and (ii) the feasibility of the proposed optimizations without a big influence on maintenance and exploitation of rolling stock.

The optimization of the leading bogie of the power car is handled in a first part; the optimization of the pantograph cavity, which has been treated independently of the bogie cavity, is presented in a second part.

3.1 Concerning the Leading Bogie of the Power Car

According to the results of the reference calculation, the main cause of the noise generated by the bogie cavity is the impact of the flow on its components. No resonance phenomena have been observed.

The flow is relatively undisturbed upstream of the bogie cavity; consequently, the mixing layer in front of the snowplough is very coherent. The bogie and the equipment, which are the first obstacle encountered by the flow, are impacted by a high flow speed. A new shape of snowplough has been imagined. An opening is located at the front, allowing the entrance of the flow into a V-shaped duct, and its exit is in two parts blowing towards the wheels of the bogie (see Fig. 5). A reduction in the mixing layer impinging on the bogie is expected.

In Fig. 6 is presented the evolution of structures, represented by iso-surfaces of Q-criterion, based on the velocity gradient tensor invariant. This indicator illustrates the coherency of flow vortices. Here, it clearly shows the predominant role of small structures, which are amplified with the optimized configuration.

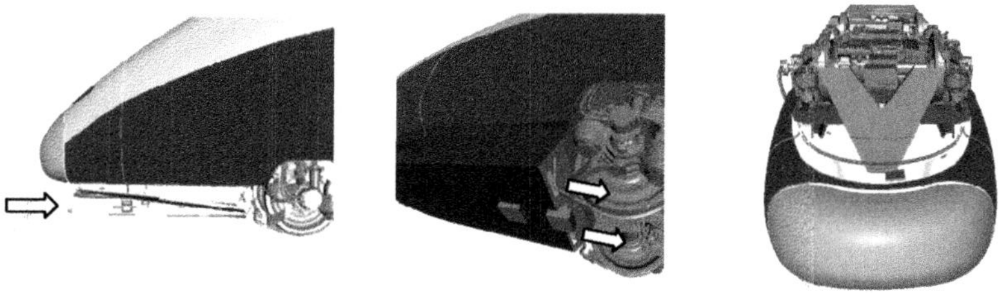

Fig. 5. First bogie with optimization (opening)

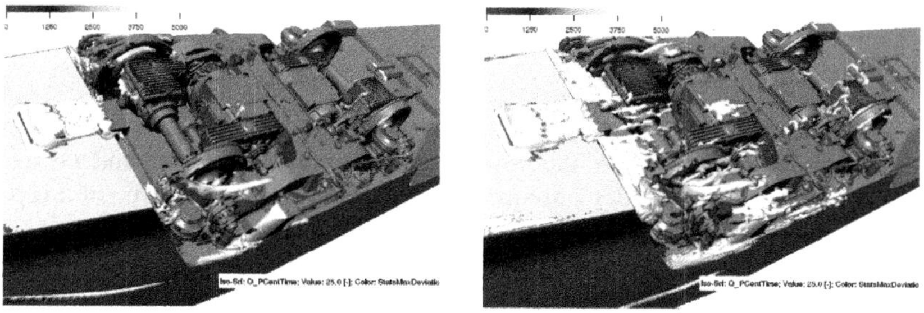

Fig. 6. Q-criterion iso-surfaces in bogie cavity: original (left) and optimized (right) configurations

For the FWH analysis, only the acoustic power of the elements of the bogie has been propagated (bogie, antenna and fairings). No peak can be noticed (Fig 7.), but it is noticeable that large structures are lowered; hence, a lower level is obtained at low frequency, whereas at high frequency, a slightly higher level is obtained. The spectrum is globally worse, with an overall level increased to about 1dB(A).

The SCx of the bogie has been slightly reduced in comparison with the reference configuration, with a decrease of almost 2%.

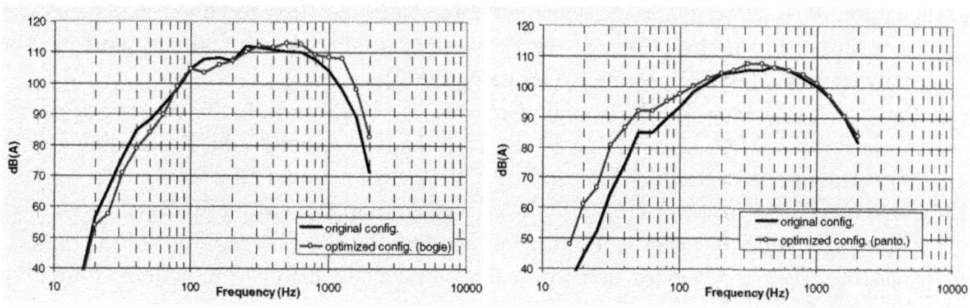

Fig. 7. Acoustic power for optimizations: bogie area (left) and pantograph cavity (right)

3.2 Concerning the Pantograph Cavity

The idea behind the optimization of the pantograph zone is to close the cavity. However, an opening is needed for the deployment of the two pantographs (which are used alternatively on classic or high-speed lines). Its shape has been optimized in order to limit the flow separation on the roof. The geometry of the protection is illustrated in Fig. 8.

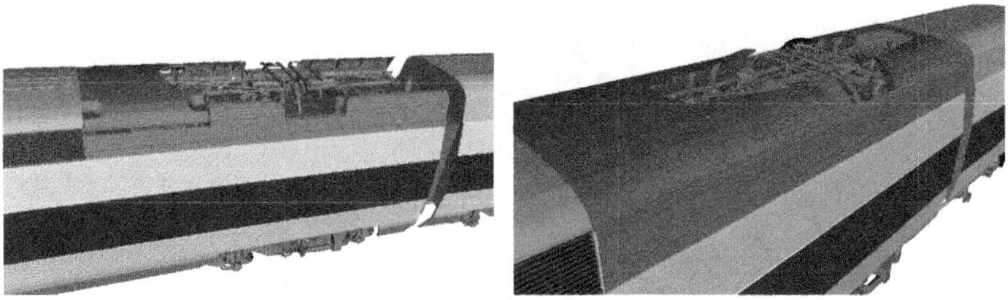

Fig. 8. Pantograph cavity, without fairings (left) and with fairings (right)

The opening for the deployment of pantograph is the source of the bad results obtained as a recirculation of the flow is observed inside the cavity and a resonance phenomenon has appeared. Consequently, the global level of acoustic of this part of the power car is downgraded, especially at low frequency.

4 Conclusion

Numerical simulations based on the Lattice–Boltzmann method have been carried out to study the aeroacoustic performance of the front shape of a high-speed train at full scale. The calculations on the basic configuration confirmed the main sources of aerodynamic noise, and the influence of several geometrical optimizations on both drag and aerodynamic noise has been evaluated. This numerical evaluation, once validated against experimental results, should give some useful clues towards industrial solutions to reduce the aerodynamic noise of future railway rolling stock.

References

[1] Gautier, P.-E., Poisson, F., Letourneaux, F.: Noise sources for high speed trains: a review of results in the TGV case. In: Proceedings of the 9th International Workshop on Railway Noise (IWRN), Munich, Germany (2007)

[2] Paradot, N., Masson, E., Poisson, F., Grégoire, R., Guilloteau, E., Touil, H., Sagaut, P.: Aeroacoustics Methods for High Speed Train Noise Prediction. In: Proceedings of the World Congress in Railway Research (WCRR), Seoul, Korea (2008)

[3] Teixeira, C.M.: Incorporating turbulence models into the Lattice Boltzmann Method, Exa Corporation. International Journal of Modern Physics C 9(8), 1159–1175 (1998)

[4] Casalino, D.: An advanced time approach for acoustic analogy predictions. Journal of Sound and Vibrations 261, 583–612 (2003)

Interior Noise Prediction in High-Speed Rolling Stock Driver's Cab: Focus on Structure-Borne Paths (Mechanical and Aero Sources)

J. Sapena[1], A. Tabbal[2], J. Jové[1], and F. Guerville[2]

[1] Alstom Transporte,
Santa Perpètua de Mogoda, Spain
Tel: +34 93 575 50 00, Fax: +34 93 574 29 83
e-mail: juan.sapena@transport.alstom.com
[2] Alstom Transport France

Summary

Interior noise in rolling stock is an important parameter in the evaluation of global comfort perceived by passengers and train crew. This paper describes the methods proposed for interior noise prediction in high-speed trains focusing on structure-borne paths that includes interior noise produced by the vibration of components (bogie, traction motor, gearbox…) and aero-acoustic sources due to the airflow at high speed around the driver's cab.

An experimental analysis of the different noise transfer paths and contributing sources is applied following panel contribution and transfer path analysis techniques. In order to obtain a model covering a wide frequency range, hybrid FE/SEA methodologies are chosen. Different loads measured in dynamic conditions are applied at the interfaces between equipment and structure to evaluate the contribution of each equipment to interior noise. Finally, the contributions are compared with the ones measured experimentally. The model obtained is also used to compute the aero-acoustic contribution to interior noise, mainly considered as a structural excitation.

1 Introduction

Numerical methodologies have helped in solving acoustic issues in systems such as trains for two different frequency ranges. On the one hand, methods like Finite Elements (FE) or Boundary Elements (BE) have usually worked well at the low frequencies, where the modal density is low and the system exhibits a global modal behaviour. On the other hand, energy methods such as Statistical Energy Analysis (SEA) have predicted with sufficient accuracy at high frequencies, where the modal densities of their components are high. However, the mid-frequency range

T. Maeda et al. (Eds.): Noise and Vibration Mitigation for Rail Trans. Sys., NNFM 118, pp. 445–452.
springerlink.com

remained unpredictable. This mid-frequency range is fairly large and covers an essential part of the acoustic spectrum.

Hybrid methods, introduced during the last decade, can address the whole frequency range. Nevertheless, even if they have been applied successfully in the automotive sector, these methods need to be further investigated in order to be applied to railway sector. The project presented in this paper deals with the investigation process carried out to set up the hybrid FE-SEA methodology described by Langley et al. [1] to the Duplex TGV driver's cab, with the goal of predicting the structure-borne contribution to the internal noise.

Furthermore, several new numerical approaches that appeared recently have given rise to the possibility of evaluating the different turbulent fields to a certain degree of accuracy that allows accounting for the aero-acoustic terms. Thus, the current Computational Fluid Dynamics (CFD) codes have low dissipation, being able to account for the physics of the interaction of the airflow with solid bodies. A methodology to couple the CFD outputs with the hybrid models is shown for the case studied.

2 Paths and Sources Contribution Analysis in a Driver's Cab of a High-Speed Train

2.1 Description of the Experimental Tests

The work presented is part of a research project, and its objective is not the structure-borne noise prediction itself, but the study of the capabilities of the hybrid methodology, the process and parameters necessary for future predictions. Indeed, the product on which the hybrid method is applied is a current and well-known product. Before starting the modelling process, an experimental quantification of the structure-borne contribution to the total interior noise in the area of study is done. Details of the methodology applied to the driver's cab can be found in the initial work of this project [2,3] where an internal developed analysis tool called META-X based on the Global Transfer – Direct Transfer (GTDT) techniques [4,5,6] is used. Several automotive manufacturers have applied this technique during the 90s [7,8]. Its main advantage is its applicability to rolling stock projects, where usually no prototype is available for testing, and testing availability in trains is limited.

2.2 Results of the Tests

Fig. 1 shows the overall noise levels (L_{Aeq}) measured inside the driver's cab at the height of the driver's ear. Measurements are done at a constant speed of 300 km/h. The main part of the acoustic energy is concentrated between 63 and 630 Hz. Fig. 2 shows the contributions of all the structure-borne paths where it can be seen that the main contributor for bogie structure-borne paths is the anti-yaw damper link.

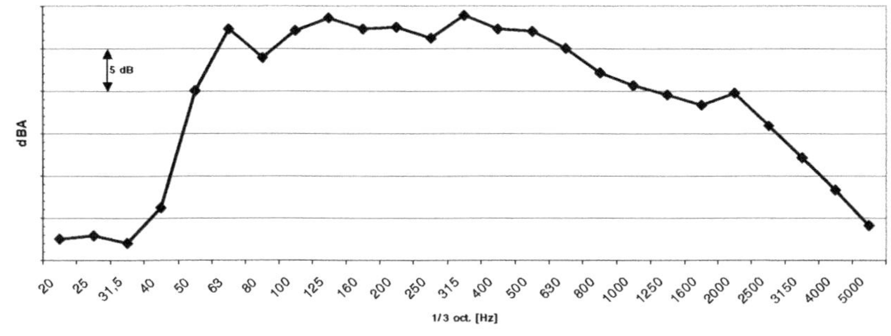

Fig. 1. Average of 6 different measurements of SPL at driver's right ear position at 300 km/h (colour figure online)

From this analysis, it can be concluded that the structure-borne noise in the driver's cab is quite significant and it covers a wide frequency range. Its influence can be up to 50% of the overall noise energy [2]. Furthermore, aero-acoustic excitation of the driver's cab is transmitted mainly through structure-borne paths (panel excited by turbulent flow).

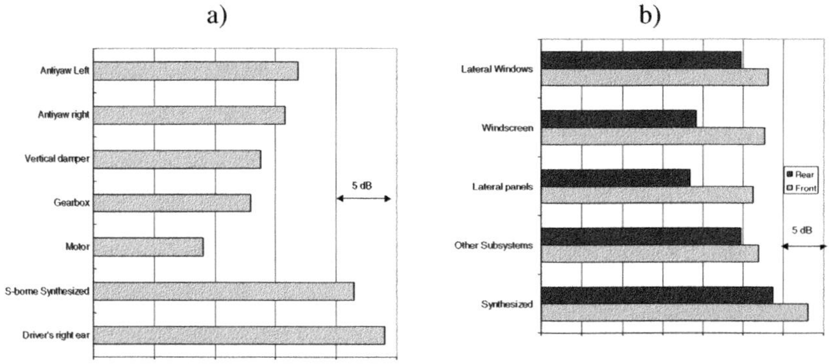

Fig. 2. a) Structure-borne path contributions in terms of overall levels b) Overall dBA panel contributions and synthesized level.

In Fig. 2, the overall interior noise panel contributions in dBA are presented [9]. Measurement results are obtained with the train running in both forward and backwards directions at 300 km/h. TSI noise levels are fulfilled for the case studied. However, future increases in commercial speed suggest the need of a prediction model covering a frequency range from 80 Hz up to 630 Hz focused on structure-borne transmission from the bogie and turbulent flow excitation.

3 Structure-Borne Interior Noise Model of a Driver's Cab

3.1 General Overview of the Hybrid FE-SEA Method

A detailed description of the hybrid FE-SEA methodology applied for the study is out of the scope of the present text. This is given in detail in [10]. A train cabin is formed by beams and stiffeners covered by thin and light plates and panels (see Fig. 4). The hybrid method models the former with a FE description and the latter by a SEA formulation. The hybrid method models the influence of the SEA subsystems on the FE ones and vice versa. The complete modelling process for structure-borne noise including vibratory sources and aero-acoustic sources is shown in Fig. 3.

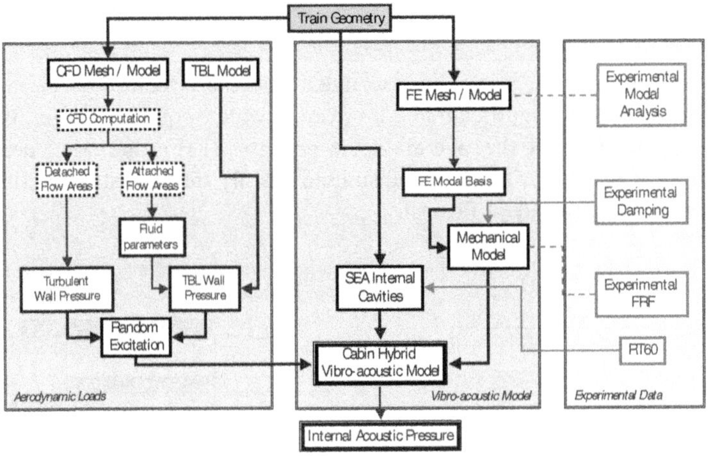

Fig. 3. Scheme of the different modelling steps (colour figure online)

3.2 Modelling Process

The modelled cabin is a complex structure formed by several layers and components (see Fig. 4). The main structure is made of beams and a heavy and rigid substructure designed for crash resistance. On the sides and the roof, several light metal panels are welded to the main structure closing both sides of the cabin. The windscreen and side windows are multilayered glass panels. The floor is made of a metal panel with an additional multilayered panel elastically suspended. Finally, the external side is covered with a polyester nose, while the internal side is covered by several kinds of trimmings and components.

Different complexity levels have been achieved, from the initial driver's cab body-in-white (BIW) up to the final model. The final model includes the windscreen, side windows and doors, as well as the polyester nose and the interior trim. Details of the different steps in the modelling are presented in [2,3] except for the final configuration discussed in this paper. The interior trim has been modelled as noise control treatments applied to the FE and SEA elements of the structural model. The acoustic cavity of the driver's cab is modelled as an SEA acoustic cavity, connected to FE and SEA subsystems. For each configuration, a complete FE model and a hybrid FE-SEA (by means of the commercial VA One software) are performed. The results of both models are compared and correlated with experimental data. The frequency range of interest is 100 to 1000 Hz. In order to apply correctly the SEA theory to any subsystem, a minimum modal density is required. In this case, the criteria to select an SEA subsystem are a number of modes between 3 and 5 per $1/3^{rd}$ octave band.

To apply the external aerodynamic flow contribution in the cabin, the procedure shown in Fig. 3 is applied [9]. Two major contributions of flow-induced noise are identified: turbulent boundary layer noise and pressure fluctuation noise [11]. An unsteady time-accurate simulation based on a Lattice-Boltzmann approach is performed on the complete train-set. A high level of geometry representation of the nose external surface, including all offsets and local details such as widescreen wiper and rain gutter, is included in the model. The unsteady pressure field on the whole train surface is obtained. In the flow-separated areas, the fluctuating pressure represents the major source of aerodynamic noise. Therefore, the average pressure spectra are computed on these surfaces and then applied as a random excitation model in the hybrid vibro-acoustic model.

The turbulent boundary layer on the surface cannot be directly computed at these very high Reynolds numbers even using the best state-of-the-art methods. In the CFD simulation, the turbulent boundary layer is modelled by a pressure gradient sensitive wall function. The results in the boundary layer cannot be used directly to compute the turbulent boundary layer noise. A TBL noise model is thus required to provide the noise contribution of the attached boundary layer. The Cockburn model [12] is applied to obtain the TBL pressure, but in order to improve the accuracy of this contribution, the TBL model parameters, namely the correlation lengths, the convection velocity and the averaged auto-spectrum pressure, are obtained from the CFD simulation.

3.3 Vibration Prediction in the Structure: Final Configuration

Several excitation points are computed, but to simplify the discussion only results due to bogie anti-yaw excitation in the vertical direction are shown, see Fig. 5. The results represent the acceleration averaged over the panel area for the full FE model, the hybrid model and measurements.

From Fig. 5, it can be concluded that a good correlation is achieved depending on the frequency range and the position of the subsystem analysed. Main problems for correlation arise at frequencies higher than 500 Hz. The main difficulty is to obtain a good FE correlation of the input power to the structure.

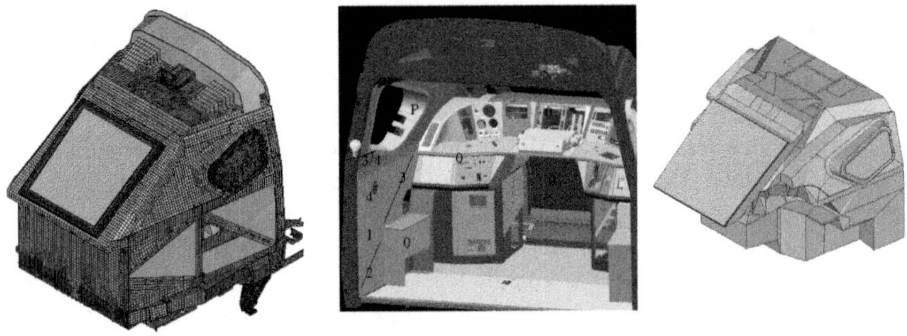

a) b) c)

Fig. 4. a) Hybrid structure FE-SEA model, b) Interior trim c) SEA acoustic cavity in the cab (colour figure online)

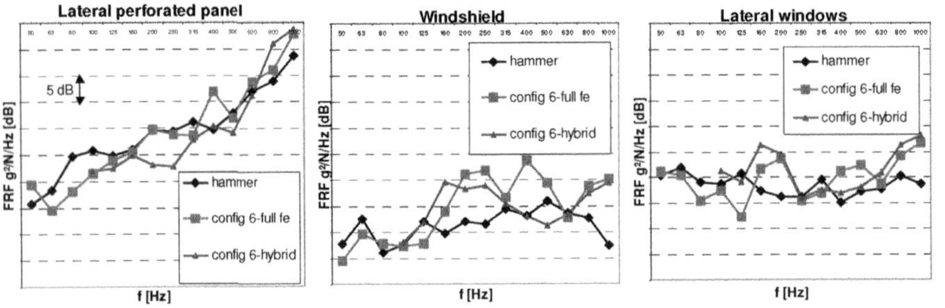

Fig. 5. Comparison for the averaged cross-inertance (measured with hammer, calculated with full structure FE model and hybrid model) at three panels between 100 and 1000 Hz. (5 dB per main division in the horizontal axis). (colour figure online)

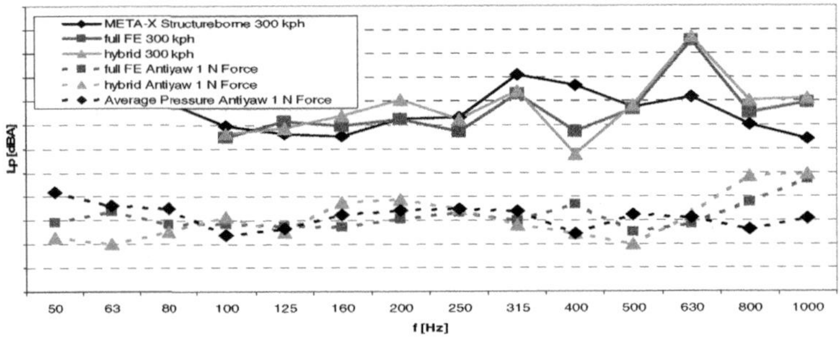

Fig. 6. Acoustic response in driver's cab due to a unit force at anti-yaw damper in vertical direction (dashed line) and to all the structure-borne sources at 300 km/h (solid line). 5 dB/division in horizontal axis. Levels not calibrated. (colour figure online)

3.4 Acoustic Prediction in the Cab: Final Configuration

Two different types of results are shown in Fig. 6 due to a unit input force and a dynamic operating force, calculated through an indirect force estimation method [13] at 300 km/h. The results show a mean deviation of the models of 3 dB in the range of 100-500 Hz for a unit force at the anti-yaw bogie position. For input forces at 300 km/h, the mean deviation is 2 dB in the range of 100–300 Hz. For frequencies higher than 500 Hz, the mean deviation increases with frequency, showing difficulties to obtain a good correlation. However, the contribution of higher frequencies to overall noise levels is not important, for this case.

3.5 Acoustic Prediction in the Cab: Aerodynamic Sources Excitation

The results of the aero-acoustic windshield contribution are summarized in Fig.7. The contribution from the turbulent boundary layer computed using the Cockburn model and the fluctuating pressure obtained directly from the CFD simulation are shown separately. The total aero-acoustic contribution is also shown on the same graph and compared with the calculated contribution of the windscreen (META-X). It can be seen that the TBL contribution is quite significant above 300 Hz, while the fluctuating pressure contribution is significant for the low-frequency range, with an over-estimation below 160Hz. This discrepancy can be explained by the fact that the pressure fluctuation contribution is computed explicitly and thus is favoured in comparison with the TBL contribution, which is only applicable to the attached flow regions. Also for highly separated flows such as the ones over the windscreen, aero-acoustic energy is concentrated at the lower frequencies. The experimental windscreen contribution (META-X) includes the aero-acoustic sources contribution, considered predominant at low frequencies and all other significant sources such as the rolling noise, motor and other equipment contribution.

Overall, it can be seen that the CFD solution with the current resolution gives acceptable resolution up to 500 Hz. The resolution is reduced above this frequency with a relatively sharp decay from 500 Hz to 1000 Hz.

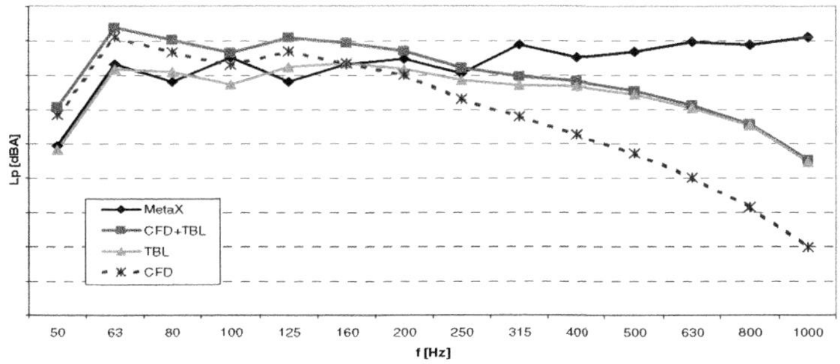

Fig. 7. Averaged sound pressure level contribution from TBL, CFD and CFD+TBL comparison with measurements. 5 dB/division in horizontal axis. (colour figure online)

4 Conclusions and Further Steps

A good correlation for acoustic predictions has been achieved up to 500 Hz (2-3 dB) for structural excitations, and a methodology for aero-acoustic contributions has been developed. The results and process allow the application of hybrid modelling to new developments but a significant effort is needed to correlate the models for medium and high frequencies. Adaptation of the numerical models for crash and static calculations is a challenge in order to reduce modelling times. In the domain of aero-acoustics, an effort should be done to characterize experimentally the sources to perform a better correlation. An experimental separation technique of the aero-acoustic contribution from panels, including structural and aerial part, is needed.

References

[1] Langley, R.S., Shorter, P.J., Cotoni, V.: Hybrid FE-SEA method for the analysis of complex vibroacoustic systems. In: Proceedings of NOVEM 2005. Saint Raphaël, France (2005)

[2] Sapena, J., Blanchet, D.: Interior noise structure-borne path prediction in a high speed train using "FE/SEA" hybrid modeling methodologies. In: Proceedings of the NOVEM 2009. "In CD-ROM", Oxford, April 5-8 (2009)

[3] Jové, J., Sapena, J., Guerville, F.: Interior structure borne noise prediction of the cabin of a high speed train using FE-SEA hybrid methods. In: Proceedings of Internoise 2009 Congress, Ottawa,

[4] Magrans, F.X.: Method of measuring transfer paths. J. Sound Vibration 74(3), 321–330 (1981)

[5] Magrans, F.X.: Definition and calculation of transmission paths within an S.E.A. framework. Journal of Sound and Vibration 165(2), 277–283 (1993)

[6] Guasch, O.: Direct transfer functions and path blocking in a discrete mechanical system. Journal of Sound and Vibration 321(3-5), 854–874 (2009)

[7] Fingberg, U., Ahlersmeyer, T.: Noise path analysis. A new approach based on practical experience. In: Proceedings of the 3rd Automotive Acoustics Conference, Zurich (1991)

[8] Tschudi, H.R.: The force transmission path method: an interesting alternative concerning demounting tests. In: Unikeller Conference 1991, Zurich (1991)

[9] Jové, J., Guerville, F., Vallespin, A.: Study of the Aerodynamic Contribution to the High-speed Train Cabin Internal Noise by means of Hybrid FE-SEA Modeling. In: Proceedings of Internoise 2010 congress, Lisboa (2010)

[10] Shorter, P.J., Langley, R.S.: Vibro-acoustic analysis of complex systems. Journal of Sound and Vibration 288(3), 669–699 (2005)

[11] Chen, S., Doolen, G.D.: Lattice Boltzmann Method for fluid flows. Annu. Rev. Fluid Mech. 30, 329–364 (1998)

[12] Cockburn, J.A., Jolly, A.C.: Structural-acoustic response, Noise transmission losses and interior noise levels of an aircraft fuselage excited by random pressure fields. AFFDL-TR-68-2 (1968)

[13] Van der Auweraer, H.: Wyckaert. K., Hendricx, W., van der Linden, P.: Noise and vibration transfer path analysis. ISAAC 7, KU Leuven (1996)

Noise Reduction of Trains Using the Operational Transfer Path Analysis – Demonstration of the Method and Evaluation by Case Study

Alex Sievi, Otto Martner, and Stefan Lutzenberger

Müller-BBM GmbH,
Robert-Koch-Str. 11, 82152 Planegg, Germany
Tel: +49 89 85602 208, Fax: +49 89 85602 111
e-mail: Alex.Sievi@MuellerBBM.de

Summary

Despite the fact that the traditional transfer path analysis (TPA) is a very well-known method, it has not yet become a standard tool for the railway industry. One reason is certainly the time-consuming effort required for determining the TFs and the source quantities such as the operational forces and volume velocities.

The operational transfer path analysis (OTPA) is a method, which derives the transfer functions (TF) with a so-called "cross-talk-cancellation" (CTC) technique directly from the operational data. Therefore, it is much faster and more effective than the traditional TPA.

The application of the OTPA is illustrated by an investigation on a meter-gauge multiple-unit from Stadler Rail. A dominant sound path could be identified.

As the CTC technique derives the TFs from the operational data, the captured operational conditions and the chosen reference signals become crucial. In order to assess the influence of the correlation between the reference signals, that effect is studied in this paper by means of several test cases.

For this purpose, reference signals of test data were filtered with randomly chosen TFs and added to a virtual total sound pressure. Then, the generated test cases were analyzed using the CTC technique to reversely calculate the TFs from the reference signals and the virtual sound pressure. As the test cases were generated without cross-talk, the CTC results have to be identical to the TF used generating the test case. By comparing the TFs, the applicability of the CTC technique on the generated test data was examined.

The first two test cases deal with 100 % uncorrelated and 100 % correlated reference signals. The cases show that the CTC technique can estimate the TFs precisely if the reference signals are uncorrelated. If the reference signals are 100 % correlated, the TFs become less accurate.

A further test was conducted using the experimental data from the presented OTPA example to assess the effect of typically correlated reference signals on the

T. Maeda et al. (Eds.): Noise and Vibration Mitigation for Rail Trans. Sys., NNFM 118, pp. 453–461.
springerlink.com

estimated TFs. For the investigated example, it could be shown that for realistic correlations, that are in the range from 0% up to about 90 %, the CTC technique could exactly estimate the TFs.

1 Introduction

The sound pressure inside railway vehicles is generated by various sound sources. Each sound source generates forces, leading to structure-borne sound, and volume velocities, leading to exterior airborne sound.

The structure-borne sound is transmitted via several structural connections into the train and generates vibrations of the interior panels, which are radiating the sound pressure. The exterior airborne sound excites the train's shell and according to the transmission losses of the train carriage floor, side walls, etc. certain sound pressure contributions will be generated in the train's interior.

Even for one sound source, the quantification of all structure-borne and airborne sound contributions to the overall sound pressure level inside the train can be a real challenge due to the complexity of the sound transmission. The TPA is an experimental technique to deal with this kind of issues.

2 Transfer Path Analysis Methods

2.1 Traditional Transfer Path Analysis (TPA)

The basic idea of the traditional TPA is the separation of the vibro-acoustic system into sources and transfer paths. In the traditional TPA, the operational force (f_{op}) is chosen to quantify the sources. This approach has the additional benefit that the transfer path H_{j} can be easily derived experimentally by impact measurements.

For the transfer path synthesis (TPS), the sound pressure contribution p of each single point of excitation j is calculated separately by multiplying the TFs H_{j} with the operational force $f_{\mathrm{op.j}}$. Eventually, the sum of all contributions j will be calculated in order to get the synthesis p_{syn}:

$$p_{syn} = \sum_j p_j = \sum_j H_j \cdot f_{op.j} = \sum_j \left[\frac{p}{f}\right]_j \cdot f_{op.j}$$

This is a very rough description of the principle of the traditional TPA. The real procedure is much more complex due to various facts and it needs time-consuming experimental measurements. Therefore, this technique has only rarely been applied in the railway industry in some validation projects [1] and has never become a standard tool. A detailed description of the traditional TPA is given in [2].

2.2 Operational Transfer Path Analysis (OTPA)

The approach of the OTPA uses the vibro-acoustic quantities (x) like acceleration (at the mounting points) and sound pressure (near the sound source) in contrast to force (and volume velocity) to characterize the sound sources. The sound pressure synthesis p_{syn} is the sum of all contributions j:

$$p_{syn} = \sum_j p_j = \sum_j H_j \cdot x_{op.j} = \sum_j \left[\frac{p}{x}\right]_j \cdot x_{op.j}$$

The challenge of this approach is that the acceleration at one (reference) point is not only generated by the force acting at this point, but also by other points nearby. This phenomenon is called "cross-talk." To overcome this issue, a cross-talk-cancellation (CTC) technique based on the singular value decomposition was developed. A detailed description of the method can be found in [3] and [5].

The biggest advantage of the CTC technique is that the TFs can be derived directly from the operational data and in consequence no separate and time-consuming measurements of the TFs are needed. This approach is therefore called Operational Transfer Path Analysis (OTPA). Furthermore, the technique is able to cope with both structure-borne and airborne sound paths at the same time.

In [4], different approaches of the traditional TPA are compared with OTPA for an analysis of the interior sound pressure of a car. The paper shows that for low frequencies, the different methods show similar results, and the OTPA is more accurate for sound synthesis under actual load conditions. The paper [5] describes the theory of OTPA and illustrates the analysis of a classic car with the OTPA.

Due to the fact that the TFs in an OTPA are derived from the operational data, highly correlated reference signals may influence the accuracy of the estimated TFs. The correlation between the measurement signals (reference signals) depends on the positions of the reference sensors and especially on the chosen data sets and therefore on the difference of the captured operating condition. Therefore, the captured operational conditions and the chosen reference signals become crucial. In order to assess the influence of the correlation, this effect is studied in this paper by means of several test cases.

3 Example: RBe 4/8 from Stadler Rail

3.1 Task

The meter-gauge multiple-unit RBe 4/8 from Stadler Rail is designed and produced for the railway company LEB. Some sections of the LEB transport network are in a very bad condition, which leads to an increased noise level, especially in the passenger area above the non-driven bogies.

Although the train meets all acoustical requirements, Stadler Rail investigated the reason for the increased level and the dominating sound transmission effects to find an effective measure to improve the perceived acoustical performance of the train running on the bad track. In this context, Müller-BBM carried out experimental analyses including an OTPA.

3.2 Measurement Set-Up and Measurements

The investigation was conducted under strict limitations in time and effort. Besides the OTPA, further analyses had to be performed; therefore, a reduced measurement set-up with five uniaxial acceleration sensors near the attachment lugs of the ARB, VD, and LD (see Fig. 1) on the crossbeam was used. Two microphones were positioned in the vicinity of the bogie in order to evaluate the contributions from the airborne sound path. One microphone was positioned inside the train.

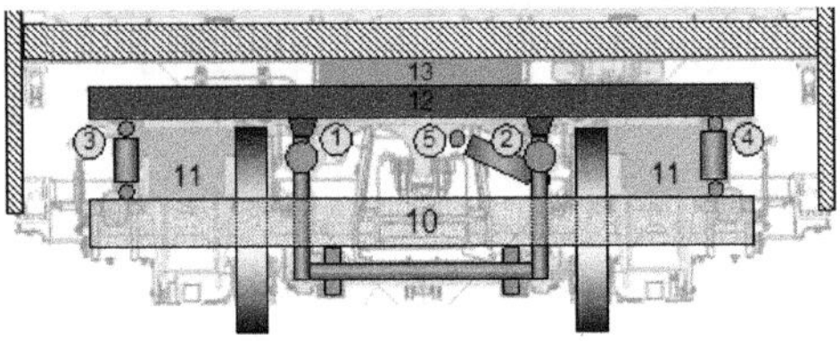

Fig. 1. Principle design of the bogie: 1,2: anti-roll bar (ARB); 3,4: vertical damper (VD); 5: lateral damper (LD); 10: bogie frame; 11: air-springs; 12: cross beam; 13: ring mount (color figure online)

3.3 Measurements, Analysis and Results

Several run-ups and runs with constant speed on different track sections were performed. The measurement data were analyzed with the CTC module of the PAK measurement system to achieve the TFs from the reference signals to the response signal (sound pressure in the train). The derived TFs were used to make a TPS for a measurement at a constant speed of about 75 km/h.

The results are shown in Fig. 2 as third octave band spectra. Between 50 Hz and 125 Hz, the synthesized SPL represents the measured SPL quite well. In this frequency range, the SPL is mainly caused by the structure-borne sound contribution. Obviously the chosen reference signals cover all relevant coupling points.

Above 160 Hz, the synthesized SPL is below the measured SPL, indicating that not all transfer paths and contributions were covered with the limited set of reference signals. This deviation is due to the fact that the airborne sound contribution normally becomes relevant above 250 - 500 Hz and only two microphones were used in this investigation. To cover all relevant airborne contributions, more microphones would have to be applied in the vicinity of the bogie, under the gangway and the entrance doors, which were nearby the investigated coach section.

Despite the limited number of sensors, the following detailed analysis was derived: The transmission via the attachment lugs of the vertical dampers are dominating the SPL in the frequency range of 31.5 Hz to 200 Hz. Between 250 Hz and 630 Hz, the attachment lugs of the ARB are the major contributors. The contributions of the bogie's airborne sound and the contribution from the lateral damper are marginal.

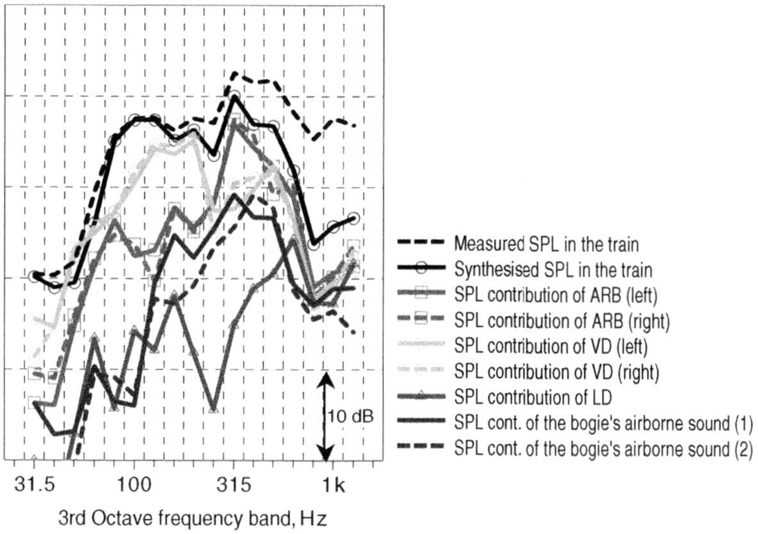

Fig. 2. The measured and synthesized SPL in the train; contributions from the attachment lugs of ARB, VD, and LD and airborne sound contributions (color figure online)

4 Evaluation of the Method by Means of Test Cases

The effect of signal correlation is demonstrated by two test cases using uncorrelated and correlated reference signals (see section 4.1). To assess the effect of the correlation on the above-derived results, the measurements of the presented example were used to conduct a further test case (see section 4.2).

4.1 Uncorrelated and Correlated Reference Signals

The first two test cases deal with uncorrelated and correlated reference signals. For the demonstration, test signals (accelerations of axle-boxes x_j from different measurements at constant speed, duration 20 s) and two arbitrary vibro-acoustic TFs p_j/x_j (H_1 and H_2) with magnitude and phase were taken from our database.

The two signals were filtered according to the two defined TFs (target TFs) and then summed up to a virtual response signal (sound pressure). This was done in the time domain using the TPS module of the PAK measurement system. The principle of the data processing is shown in Fig. 3.

Then the CTC of the PAK measurement system was used to reversely calculate the TFs using the signals 1 and 2 as reference signals and the sum of the filtered signals (result) as response signal. As the generated test examples do not contain any cross-talk, the TFs, which were calculated using the CTC method, have to be identical to the specified TFs (target).

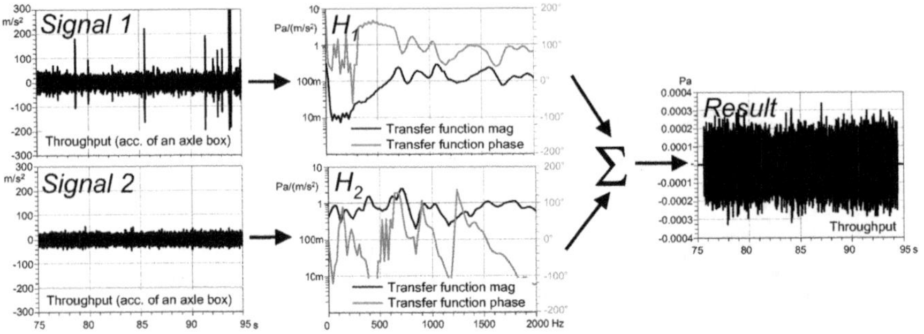

Fig. 3. Signal processing for generating the test case

4.1.1 Case 1: Two Uncorrelated Reference Signals

For the first test case, the signals 1 and 2 were uncorrelated. The reversely calculated TFs are displayed in Fig. 4 as magnitude and phase. The figure also shows the target TFs (TF, which were used generating the test case). It shows clearly that the CTC is suited for the calculation of the TFs from the input signals 1 and 2 and the sum of the filtered signals (results).

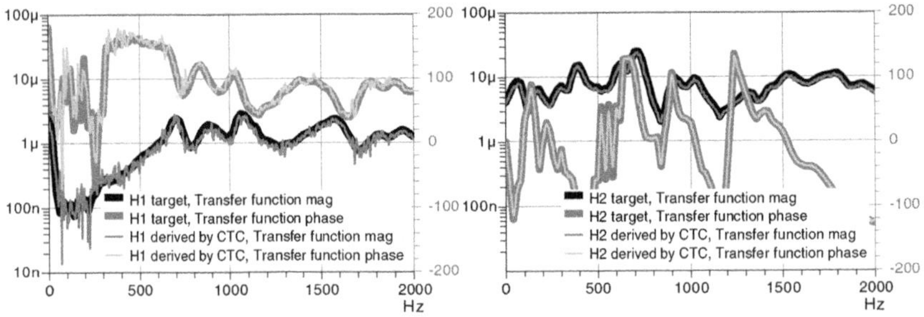

Fig. 4. Comparison of the target and the derived TFs H1 and H2 (magnitude in $Pa/(m/s^2)$ and phase in °); signal 1 and signal 2 are uncorrelated

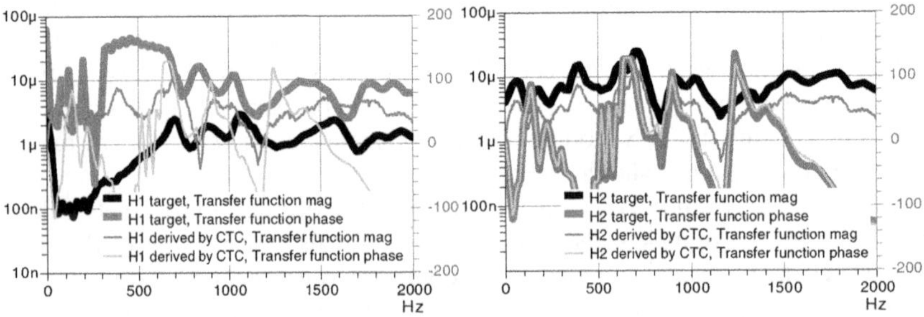

Fig. 5. Comparison of the target and the derived TFs H1 and H2 (magnitude in Pa/(m/s2) and phase in °); signal 1 and signal 2 are 100% correlated

4.1.2 Case 2: Two Correlated Reference Signals

The test case 2 for correlated input signals was generated by setting signal 2 equal to signal 1. That, of course, leads to 100 % correlation between signal 1 and signal 2. The signals were processed as described in the previous case. Afterwards the CTC was used to calculate the TFs from the reference signals 1, 2 (which is equal to 1) and the sum of the filtered signals.

The calculated TFs (derived by CTC) are displayed in Fig. 5 in comparison with the target TFs (TFs, which were used to generate the test case). In this example, the 100 % correlation of the reference signals leads to the following results: The calculated phase of the TF H_2 is absolutely correct and the curve progression of the magnitude is right, but 6 dB lower than specified (target). The estimated TF H_1 is equal to H_2. To summarize: Due to 100% correlation between the reference signals (1,2), the CTC technique derives two identical TFs, which are, ignoring the offset, equal to H_2, because the contribution via H_1 is marginal. In order to match the total sound pressure, the derived TFs have to be 6 dB lower than the TF H_2. Other deviations will occur if there are amplitude and phase differences between the reference signals 1 and 2 or if the TFs H_1 and H_2 are different.

This second case demonstrates that the estimation of the TF becomes less accurate if the reference signals are 100 % correlated. This is a quite artificial test case, because normally, real measurement signals are not 100 % correlated over the total frequency range.

4.2 Measurements of the Example RBe 4/8

The previous section shows that 100 % correlated reference signals lead to a less accurate estimation of the TFs. The correlation between the reference signals (expressed as coherence) on the swing beam lies between 90 % and 0 % (see Fig. 6).

To assess the effect of the typical signal correlation on the TF estimation, a third test case was generated using the data of the example RBe 4/8.

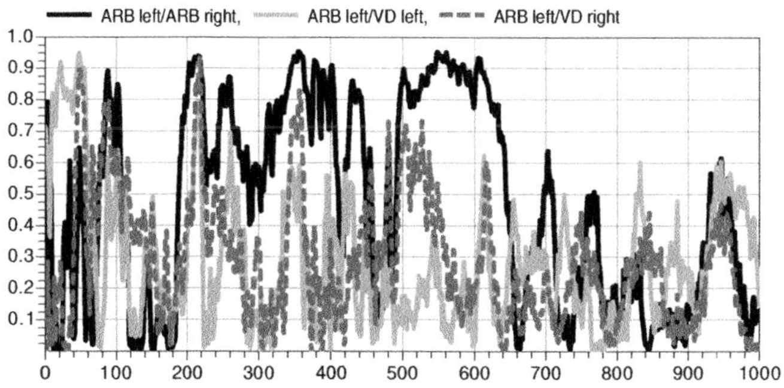

Fig. 6. Coherence over frequency between the measured accelerations on the cross beam at the attachment lug of the ARB left and the ARB right, the VD left and right; total duration 215 s with several run-ups and downs, drive with constant speeds on different tracks and in curves (color figure online)

4.2.1 Case 3: Four Typically Correlated Reference Signals

The time raw data of the four reference signals (ARB le, ARB ri, VD le, and VD ri) were used to generate the third test case. The signals were filtered according to the directly measured TFs (p/a) and then summed up to a total (virtual) sound pressure. For this test example, the directly measured TFs were used, but this is not relevant for the test case itself. Any other transfer function could have been chosen for the test example.

Then the CTC module of the PAK measurement system was used for a reverse calculation of the TFs from the input signals (references) and the resulting total sound pressure (response).

Fig. 7 shows exemplary the derived TF for the left VD in comparison with the target TF (which was used to generate the test case). The results for the other TFs are similar.

The result is obvious: Despite the quite high correlation of up to 90 %, in our example, it turns out that the CTC technique is able to estimate all TFs correctly in magnitude and phase.

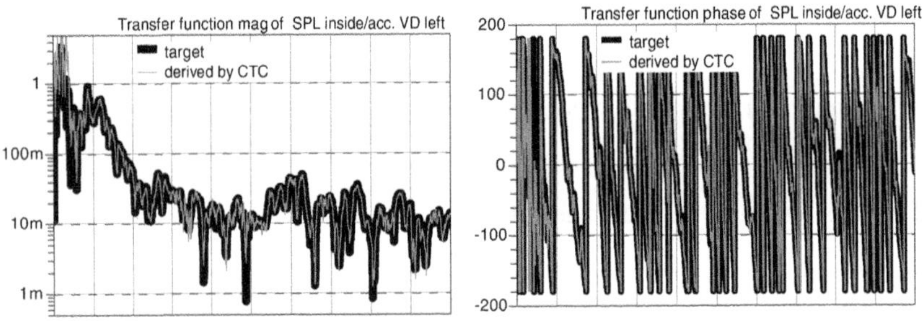

Fig. 7. Comparison of target and the derived TF for the left VD (color figure online)

5 Conclusions

In several test cases for the CTC technique of the OTPA, which is implemented in the PAK measurement system, it has been shown that this technique is suitable for a reverse determination of the TFs from the reference signals and the response signals, even if the reference signals are typically correlated.

The way of assessing the influence of the correlation between the reference signals, as presented here, can be applied in case of any doubts on the applicability of the CTC technique due to the high correlation between the reference signals. It should be helpful to assess the applicability of the CTC technique for the chosen test data.

References

[1] Arndt, R., Jalics, K., Schleinzer, G., Priebsch, H.-H.: TPA Schienenfahrzeuge – Körperschallübertragung Drehgestell. In: Proceedings of DAGA 2008 (2008)

[2] Vecchio, A., Van der Auweraer, H., Kindt, P., Houthoofdt, T., Sas, P.: Experimental Transfer Path Analysis of a Hybrid Bus. Paper presented at SAE 2005, Noise and Vibration Conference and Exhibition (2005)

[3] Noumura, Yoshida: Method of transfer path analysis for vehi-cle interior sound with no excitation experiment. FISTA, F2006D183
[4] Lohrmann, M., Hohenberger, T.: Methodenvergleich zur Ermittlung von Übertragungskoeffizienten: Linear/kraftbasiert und mittels Hauptkomponen- tenanalyse. In: Proceeding of DAGA 2008 (2008)
[5] De Klerk, D., Lohrmann, M., Quickert, M., Foken, W.: Application of Operational Transfer Path Analysis on a Classic Car. In: Proceeding of NAG/ DAGA 2009 (2009)

In-Car Noise Control for Epping to Chatswood Rail Link

C.M. Weber[1], J. Haling[1], G.P. Brown[2], and D. Hanson[2]

[1] SLR Consulting Australia Pty Ltd,
2 Lincoln Street, Lane Cove, NSW 2066, Australia
Tel.: +612 9427 8100; Fax: +612 9427 8200
e-mail: cweber@slrconsulting.com, jhaling@slrconsulting.com
[2] Sinclair Knight Merz,
100 Christie Street, St Leonards, NSW 2065, Australia
Tel.: +612 9928 2100; Fax: +612 9928 2500
e-mail: GPBrown@skm.com.au, DHanson@skm.com.au

Summary

Sydney's Epping to Chatswood Rail Link (ECRL) is a landmark rail tunnel project (12.5 km long) which opened to the public in 2009. During commissioning, in-car passenger noise levels within older style non-air-conditioned trains were found to exceed 100dBA at some locations.

Several noise mitigation strategies were optimised and implemented within a short timeframe to reduce in-car noise levels by an average of 10dBA. These included a new rail grinding technique, installation of additional sound absorptive panels on the walls and between the rails, and rail dampers (in a tunnel environment).

This paper presents a summary of the works undertaken to mitigate in-car noise, calculation methods and achieved noise reductions.

1 Introduction

Sydney's Epping to Chatswood Rail Link (ECRL) was designed to increase the capacity and reliability of the existing network. As such, the line experiences a wide range of electric multiple unit trains [1]. The older generation fleet is a mix of non-air-conditioned trains with openable windows, and air-conditioned trains with sealed windows (both with sliding doors). The later generation fleet includes design features such as plug doors, sealed windows and high transmission loss components which aid in reducing in-car noise levels within rail tunnels.

During commissioning, attended measurements were undertaken within two older generation trains. In the most exposed position on the non-air-conditioned R-Set train (standing in the vestibule area between the doors), $L_{Aeq(1second)}$ noise levels were typically around 95dBA to 100dBA with some short-term events up to

T. Maeda et al. (Eds.): Noise and Vibration Mitigation for Rail Trans. Sys., NNFM 118, pp. 463–470.
springerlink.com

105dBA. The high noise levels resulted from general rolling noise, rather than corrugated track or curving noise, and were up to 15dBA higher compared with other Sydney rail tunnels.

Whilst the design and construction of the ECRL incorporated two runs of sound absorptive panels on one wall to reduce in-car noise levels, the initial testing identified a need to incorporate additional mitigation measures prior to the commencement of fully integrated operations.

This paper explores the options that were considered and subsequently installed to achieve a satisfactory in-car noise environment for passengers and train staff.

2 Initial Investigations

2.1 In-Car Noise Level Variation for Different Rolling Stock

The rolling stock design and its ability to attenuate outside noise intrusion is a significant factor influencing in-car noise levels in a tunnel environment.

Initial testing identified that in-car noise levels in older generation trains (with windows closed) were typically 10dBA to 15dBA higher compared with a later generation train, and up to 20dBA higher with windows open.

This paper presents results for the vestibule area of older generation air-conditioned K-Set trains. This was on the basis that non-air-conditioned trains will shortly be replaced with newer generation trains.

2.2 Noise Levels in Existing Rail Tunnels

Detailed measurements were undertaken in four Sydney rail tunnels in order to quantify the existing noise levels, understand the reasons why noise levels were higher in the ECRL tunnels and identify potential mitigation measures.

Table 1. Noise Levels in K-set Vestibule - Sydney Rail Tunnels

Tunnel	20% Exceedance $L_{Aeq(5second)}$ Noise Level (dBA)	Typical Maximum $L_{Aeq(5second)}$ Noise level (dBA)
City Circle Outer	82	87
NSR Down	78	81
NSR Up	76	80
ESR Down	79	84
ESR Up	82	86
ECRL Down	91	95
ECRL Up	91	95

Note: The 20% exceedance noise levels represent the noise level exceeded for 20% of the measurement period in each tunnel. For 80% of the journey time, the $L_{Aeq(5second)}$ noise level is below the 20% exceedance level.

In accordance with the requirements of ISO 3381:2005 [2], the measurements were undertaken at a height of 1.6 m above floor level, representative of the standing position. The measurements were carried out under controlled test conditions with the trains running empty, except for essential staff and test personnel. The trains were operated at the normal track design speeds.

Overall and 1/3-octave L_{Aeq} noise levels were measured in consecutive 1-second intervals. This allowed sufficient detail to identify specific noise sources and simple calculation of other noise parameters such as the $L_{Aeq(5second)}$ which was adopted for the ECRL noise goal (refer Section 2.3).

A summary of the measured in-car noise levels is provided in Table 1. The measurement results indicate that the 20% exceedance noise levels are 9dBA to 15dBA higher in the ECRL tunnels compared with other Sydney rail tunnels.

The four rail tunnels have a range of design, maintenance and operating conditions which significantly influence in-car noise levels. Several key factors were found which contribute to increased noise levels in ECRL.

- In-car noise levels in the ECRL tunnels were variable (at constant speed), suggesting that the rail roughness associated with the initial rail grinding was not constant. This was verified by the presence of high axle box acceleration levels where high in-car noise levels were also apparent and by the results of corrugation analysis trolley (CAT) measurements.
- The shape of the single bored ECRL tunnels and inadequate sound absorption result in higher in-car noise levels for a given sound power level compared with the other rail tunnels - these comprise either ballast track or absorptive panels placed close to the wheel-rail interface to provide effective sound absorption.
- The low-stiffness Delkor Egg rail fasteners and increased spacing present for most of the ECRL tunnels result in a reduced track decay rate and additional rail radiated noise compared with the other rail tunnels.
- The train speeds and overall trip duration in the ECRL tunnels is higher than for the other rail tunnels. Whilst the train speed for the ECRL and NSR tunnels is the same (80 km/h), the increased cross-sectional area, ballast track and stiff rail pads present in the NSR tunnel results in significantly lower noise levels.

2.3 In-Car Noise Design Goals for ECRL

The passenger amenity noise goals (relating to speed intelligibility and mobile phone usage) were more stringent than the occupational health and safety (OHS) noise goals (relating to potential hearing loss).

In order to establish noise levels that are acceptable for rail passengers, criteria used locally and by international rail operators were reviewed. Consideration was also given to the existing noise levels in other Sydney rail tunnels.

Local design criteria for new rolling stock are defined on the basis of the maximum allowable L_{Amax} noise level in open country at maximum speed (typically 70dBA), and also the maximum allowable L_{Aeq} noise level measured over a defined section of existing track (typically 70dBA to 75dBA including rail tunnels and surface track locations).

For the design of new rail tunnels (such as ECRL), local design criteria have been based on achieving frequency-dependent reverberation time (RT60) levels.

International guidance [3, 4] indicated that the preferred in-car noise levels for passenger comfort are typically between 75dBA and 80dBA for trains operating within a tunnel environment. Whilst some studies indicate that the A-weighted noise level may not be the most appropriate parameter to assess human response [5, 6], our review concluded that the A-weighted noise level remains the most widely used, and that the L_{Aeq} assessment parameter (evaluated over a 5-second period) is preferred for noise mitigation design purposes. This finding is supported by Thrane [7].

The following provides a summary of the noise mitigation design goals adopted for ECRL.

The $L_{Aeq(5second)}$ noise level in the vestibule of an average 8-car K-set train shall not exceed 80dBA for more than 20% of the journey through the tunnel, and shall not exceed 85dBA at any location.

$L_{Aeq(5second)}$ noise levels below 80dBA are considered acceptable for passenger amenity on the basis that the specified levels will allow normal conversation to continue. Very short occasional exceedances (up to 85dBA) can be tolerated during a conversation without causing undue problems with communication.

For future rail tunnel projects in Sydney, it is recommended that in-car noise design goals are established at the outset (in lieu of reverberation time criteria) and that the tunnel design, track maintenance, rolling stock and operating conditions are taken into account as part of a complete noise mitigation design.

2.4 Potential Mitigation Measures

At the conclusion of the initial investigations, the ECRL In-car Noise Taskforce made the following recommendations:

1. Explore all rail grinding options until a consistently smooth running surface can be reliably delivered after grinding.
2. Install additional absorptive material and conduct field trials to optimise their performance and placement locations. It was recommended that materials be trialled on the tunnel walls and within the four foot (between the rails).
3. Trial two types of rail dampers developed in Europe to control the emission of rail radiated noise. The use of rail dampers is likely to be most beneficial at locations where the rail is directly fixed with low-stiffness baseplates.

3 Mitigation Measures

3.1 Rail Grinding

An initial rail grinding trial was undertaken on a 500-m section of tangent track within one of the ECRL tunnels using two methods. The first 250-m section was re-ground using standard course stones, and the second 250-m section was re-ground using finer grinding stones (after an initial pass with coarse stones).

On the basis of in-car noise measurements before and after grinding, a 3.5dBA change was observed over both test sections.

Following the initial trial, the entire length of both tunnels (except cross-overs) was re-ground using a revised rail polishing technique (which included several passes at a higher than normal train speed, but removing less material from the rail during each pass). Overall, the 20% exceedance $L_{Aeq(5second)}$ noise levels were reduced by approximately 3dBA and the entire trip L_{Aeq} noise levels were reduced by approximately 2.5dBA. The maximum noise level reductions exceeded 5dBA at several locations.

Following the rail polishing, baseline in-car noise levels were re-established for both tunnels for the purpose of determining the additional mitigation measures required to achieve the design goals.

3.2 Additional Sound Absorption

Room acoustic modelling techniques were initially utilised to determine the likely benefit of installing additional sound absorptive panels within the tunnels, and several field trials were undertaken to confirm the likely noise reductions.

The initial trial involved the installation of three additional 1.2 m high x 75 mm thick runs of polyester batts over a 500-m test section. These provided a measured in-car noise reduction of approximately 4dBA (with and without).

A more comprehensive test was undertaken which involved the temporary installation of five additional runs of absorption. Testing was initially undertaken with all five runs of additional absorption. One run of absorption was removed in each subsequent test. Whilst the order in which the absorption is removed is important, the logistical constraints of the test environment meant that not every permutation could be tested (room acoustics theory was used to evaluate alternative configurations). The test sequence in Table 2 was chosen to yield the greatest test certainty for the most significant treatment combinations.

Table 2. Noise reductions for additional absorptive panel runs

Additional runs of absorption	5	4	3	2	1	0
A - Under walkway	X	X	X	X	X	
B - Under existing runs	X	X	X	X		
C - Above walkway	X	X	X			
D - Four Foot	X	X				
E - Drains	X					
Noise Reduction (dBA)	9.3	8.6	6.2	5.0	2.5	0.0

On the basis of the encouraging results for the four foot treatment option, a separate trial was undertaken with full-width and half-width absorption (to mimic the absorption achievable with a realistic installation). The noise reductions were 3.5dBA and 2.5dBA for the full-width and half-width treatments respectively.

3.3 Rail Dampers

Two rail damper products (Schrey & Veit and Corus) were installed within the tunnel to investigate the likely noise reductions. The rail damper performance was investigated by direct noise measurement inside and outside the train, track decay rate measurements and rail vibration measurements during train passbys.

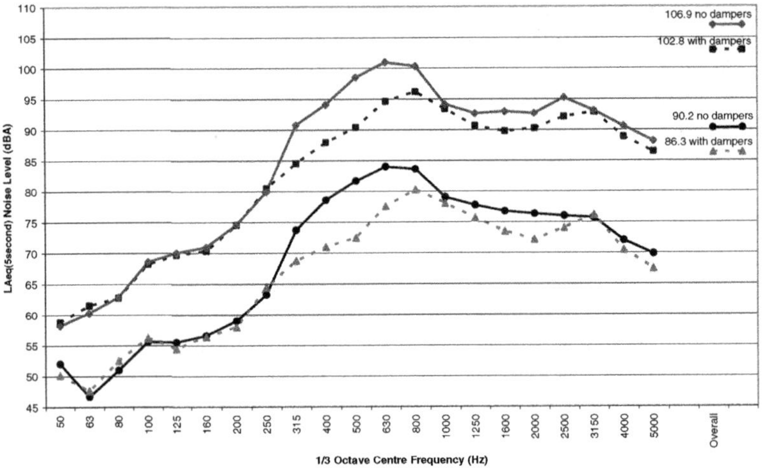

Fig. 1. Noise reduction due to rail dampers Higher noise spectra represent the measured levels outside the train Lower noise spectra represent the measured noise levels in the vestibule of the K-Set (colour figure online)

Schrey & Veit provided a sufficient quantity of rail dampers to undertake an in-car noise test over a 90-m section of tunnel with direct-fix rail fasteners (Delkor Egg fasteners with dynamic stiffness of 17 MN/m at 700 mm centres). The results of in-car noise testing showed that the rail dampers provided an overall noise reduction of 4dBA inside the vestibule of a K-Set. The noise levels outside the train were also reduced by 4dBA.

The attenuation properties of the rail dampers in terms of frequency are shown in the two pairs of spectra compared in Figure 1. This figure shows that the performance of the rail dampers is concentrated over the frequency range 315 Hz to 800 Hz, which also corresponds to the peak in the in-car noise spectrum.

Track decay rate and rail vibration measurements undertaken on smaller test sections approximately 20 m long indicated similar performance between the Schrey & Veit and Corus products. For reasons other than acoustic performance (e.g. maintenance and inspection), Corus rail dampers were selected.

4 Final Predictions and Measurements

A spreadsheet noise model was developed to assist in optimising cost-effective placement locations for each mitigation option. The baseline noise levels (after rail polishing) were adopted as the starting point. These were normalised to a design speed profile using a 30 log (speed) relationship to ensure that future noise level comparisons were based on a common speed profile.

1/3-octave L_{Aeq} noise spectra were derived from the baseline measurements for each 100-m track segment, with finer resolution in the vicinity of stations and changes in track form. For each absorption option and placement location, the noise reduction performance was evaluated on a 1/3-octave basis using a Sabine method, with a correction factor included to account for losses along the tunnel, and a weighting applied to reflect the fact that absorptive treatments are more effective (per unit area of material) in some locations than in others (i.e. close to the wheel-rail interface). The simple spreadsheet tool provided a means to evaluate several combinations of treatment options, noise mitigation strategies and overall cost. The tool provided a predicted in-car L_{Aeq} noise profile for each tunnel.

Table 3. Design and measured noise levels in K-set Vestibule

Tunnel	20% Exceedance $L_{Aeq(5second)}$ Noise Level (dBA)			Maximum $L_{Aeq(5second)}$ Noise level (dBA)		
	Design goal	Original level	Final level	Design goal	Original level	Final level[1]
Down	80	91	81	85	95	84
Up	80	91	80	85	95	83

Note 1: The maximum $L_{Aeq(5second)}$ noise levels exceeded 85dBA at some turnout locations where it was not feasible to install adequate sound absorption or undertake rail polishing. The reported levels represent the maximum levels at locations away from turnouts.

The final mitigation design incorporated 16.7 track kilometres of rail dampers (70% coverage), 21.3 track kilometres of four foot sound absorptive panels (89% coverage) and the equivalent of 35.3 track kilometres of additional sound absorption (in the form of absorptive wall panels and cloth-wrapped material fixed to the underside of cable trays).

Table 3 provides a summary of the design goals, original noise levels and the results of in-car noise compliance measurements undertaken in December 2009 following the installation of all mitigation measures. The results show that the final noise levels were within 1dBA of the design goals and provided an average noise level reduction of 10dBA.

5 Lessons Learned

The issue of high in-car noise levels in Sydney rail tunnels is not new and was recognised as a significant design issue in the early 1980s, following the opening of the Eastern Suburbs Rail (ESR) line [8]. Despite the fact that trains running on the ESR were the same type as used in the existing City Circle line tunnels, noise levels on the ESR were found to be significantly higher, and at times exceeded 100dBA inside carriages. In the mid 1980s, the ESR tunnels were retrofitted with absorptive panels to address reverberant noise. This work was highly successful, reducing the in-car noise levels by as much as 7dBA.

Experience on the ECRL project has identified that the RT60 specifications are not ideal for controlling in-car noise levels. It is recommended that future rail tunnels are designed on the basis of a comprehensive approach dealing with the tunnel design features, rolling stock, operating conditions and in-car noise goals.

Acknowledgements

The work associated with this project was jointly funded by the Transport Construction Authority and RailCorp.

Special thanks are extended to key members of the ECRL In-car Noise Task Force including David Coker and Graham Watson (TCA), David Anderson, Noel Snedden, Serge Vegh, Dave Morton, Malcolm Kerr and Nicholas Turton (RailCorp), and Richard Heggie, Antony Williams, Dominic Sburlati, Andrew Parker, David Geiger and James Hong (SLR Consulting Australia).

References

[1] CityRail Website. CityRail Fleet Information, `http://www.cityrail.info/about/fleet/`

[2] International Standards Organisation: ISO 3381:2005 Railway applications – Acoustics – Measurement of noise inside railbound vehicles (2005)

[3] Frommer, G.H.: Handbook of Noise and Vibration, ch. 96 (2007)

[4] American Public Transit Association (APTA) Noise and Vibration Guidelines (1981)

[5] Hardy, A.E.J.: Measurement and Assessment of Noise within Passenger Trains. Journal of Sound and Vibration 231(3), 819–829 (2000)

[6] Poisson, F., Boullet, I.: The loudness: a new indicator to specify the acoustic comfort inside trains. In: Proceedings of the 8th World Congress on Railway Research (WCRR 2008), Seoul, Korea (2008)

[7] Thrane, H.W.: Handbook of Noise and Vibration, ch. 97 (2007)

[8] Anderson, D., Weber, C.: Recent developments in operational rail noise and vibration in NSW, Australia. In: Proceedings of the 9th International Workshop on Railway Noise (IWRN9), Munich, Germany (2007)

Development of a System Using Piezoelectric Materials to Reduce the Interior Noise of a Railway Vehicle*

K. Yamamoto, D. Yamamoto, S. Mamada, and D. Sato

Railway Technical Research Institute,
2-8-38, Hikari-cho, Kokubunji-shi, Tokyo, Japan
Tel.: +42 573 7312; Fax: +42 573 7313,
Tel.: +42 573 7339; Fax: +42 573 7354
e-mail: katsu@rtri.or.jp, d-yama@rtri.or.jp,
mamada@rtri.or.jp, dsatou@rtri.or.jp

Summary

With respect to the reduction of the interior noise of railway vehicles, most of the countermeasures taken have been mainly passive procedures based on the mass law. Although the application of "active noise control" to railway vehicles represents a challenge, only a few experimental tests were carried out so far. This paper proposes a system to reduce the interior noise using piezoelectric materials. The effectiveness of the system will become apparent through the evaluation of the proposed noise reduction system as applied to the ceiling panels under a pantograph of the hinkansen.

1 Introduction

As the speeds of railway vehicles are increased, ride comfort in cabins accordingly becomes essential. On the contrary, the ride comfort or the cabin interior noise has worsened as lightweight railway vehicle construction has been implemented [1]. The countermeasures to the interior noise of railway vehicles have mainly been passive procedures based on the mass law, implying the increase in the material weight, which will be difficult in view of the current effort to design lightweight vehicles. Applying active noise control as well as passive measures to the vehicle body has been reported for automobile and aircraft, but for a railway vehicle, only a few experiential tests have been performed up to now [2]-[5]. A reason why the railway vehicle has lagged behind in the practical application is considered that it has to take into account the special conditions, such as the capacity of the cabin or the vibration change in the interior panels as a target plate, which are not required by other systems.

* Awarded Best Paper of IWRN10.

T. Maeda et al. (Eds.): Noise and Vibration Mitigation for Rail Trans. Sys., NNFM 118, pp. 471–478.
springerlink.com

In this paper, we present a new noise reduction system applicable to a railway vehicle. The system achieves the suppression of the noise transmitted through the target plate without identifying the vibration or transmitted noise characteristics of the target plate by making the void between the target plate and the noise reduction panels. Moreover, a running test result, when the system can be applied to a ceiling panel under a pantograph of the Shinkansen, will be demonstrated.

2 Outline of the Noise Reduction System

In order to reduce the noise in the cabin that has a large volume, a method that suppresses the incident acoustic energy from the interior panels is considered suitable, as described in [1].

Then, a system was developed to reduce the noise transmitted through the interior panel by suppressing the incident acoustic power to the plate, in order for noise reduction not to be affected by changes in the vibrational state. Figure 1 shows an application image of the noise reduction system to the railway vehicle, and Fig. 2 shows a cross-sectional view of the proposed noise reduction system that is applied to the interior panel. Here, the noise reduction system is attached onto the incident side of the interior panel (hereinafter called "target plate" for generalization). Moreover, an air layer is used to isolate the vibration of the target plate.

The noise reduction system consisted of the panels (hereinafter called "noise reduction panel") on which a plate with piezoelectric material (hereinafter called "noise insulation panel") and a control circuit is placed. All piezoelectric elements on the noise insulation panel are connected to each other and wired to a control circuit. This controller is an

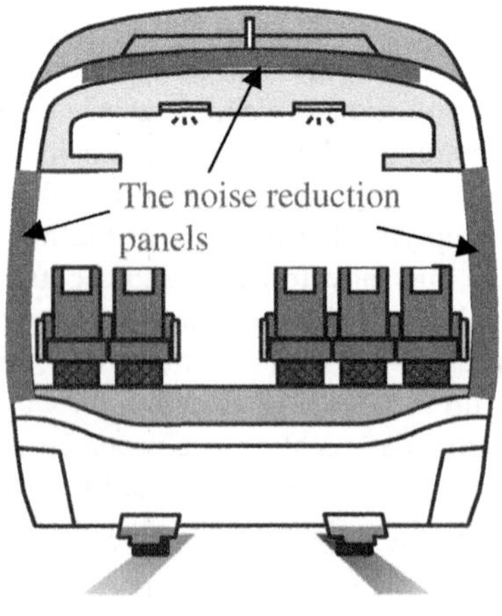

Fig. 1. Application of the noise reduction system to a railway vehicle (color figure online)

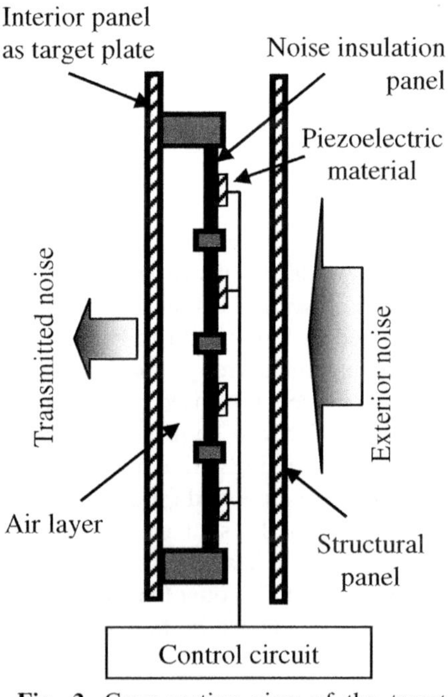

Fig. 2. Cross-section view of the target panel with the noise reduction system (color figure online)

equivalent inductance circuit of a simple analogue circuit using operational amplifiers. Because the piezoelectric material acts as a capacitor, a resonance circuit is made between the piezoelectric material and the control circuit.

When noise is incident on the noise reduction panel, the piezoelectric materials generate voltage. By making the resonance circuit, the antiphase voltage opposing the generated voltage is input to the piezoelectric material, and an antiphase force acts on the noise insulation panel. The antiphase force cancels the original vibration, and the vibration of the noise insulation panel can be suppressed. Consequently, the noise transmitted through the noise insulation panel reduces. The acoustic power in the air layer between the target plate and the noise reduction panel is suppressed, and the noise transmitted through the target plate becomes less.

Although Fig.2 suggests that the system reduces only the transmitted noise in as far this is airborne incident on the exterior side of the target plate, this system is capable of reducing the structure-borne noise on the panel installed on the transmitted noise side.

This noise reduction system has a number of advantages.

(1) Identifying the vibration condition of the target panel or the sound characteristics of the transmitted noise is not necessary.
(2) Thanks to the previous item, this system can reduce the transmitted noise without sensors such as microphones or acceleration pickups.
(3) Since the controller is composed of an electrical circuit with operational amplifiers and the piezoelectric material is driven according to the magnitude of the input voltage, the power consumption is small.
(4) As the noise reduction panel and the control circuit can be mass-produced, this system can be realized with small cost.

3 Design Approach of the Noise Reduction System

3.1 Size of the Noise Insulation Panel and the Inductance of the Control Circuit

The basic characteristics of the noise reduction system were described in detail, and the accuracy of the modeling was verified in references [6,7]. In these papers, it was concluded that the characteristics of the noise reduction ratio relate to the vibration condition of the noise insulation panels. That is to say, in order to reduce the noise at any frequency, it is better if the noise insulation panel has a size that vibrates in its (1,1) mode (hereinafter called "(1,1) vibrational mode"), so that it vibrates with an antinode around the center of the plate.

The characteristics of the noise insulation panel are defined by the material and support conditions. Each noise insulation panel is fixed on a frame, and the support condition is assumed to be clamped at all edges.

Suppose the noise insulation panel is square with side length a. The first natural frequency ω_1, corresponding to the (1,1) vibrational mode, is calculated as follows [8].

$$\omega_1 = 2\left(\frac{4}{3}\right)^2 \frac{\pi^2}{a^2}\sqrt{\frac{D}{\rho h}} \tag{1}$$

with

$$D = \frac{Eh^3}{12(1-\nu^2)} \tag{2}$$

where ρ , h , E, and ν denote the density, the thickness, Young's modulus, and Poisson's ratio of the noise insulation panel. Rearranging this, the length of the edge can be expressed as the following equation when the first natural frequency coincides with the target frequency f.

$$a = \frac{4}{3}\sqrt{\frac{\pi}{f}\sqrt{\frac{D}{\rho h}}} \tag{3}$$

Next, consider the tuning method of the inductance in the control circuit. When the piezoelectric materials with capacitance C are connected to each other, the resultant capacitance $\hat{C}$ becomes as follows.

$$\hat{C} = \frac{C}{N} \quad \text{(in series connection)} \tag{4}$$

$$\hat{C} = NC \quad \text{(in parallel connection)} \tag{5}$$

where N denotes the number of the piezoelectric elements. The resonance frequency ω_e of a resonant circuit (hereinafter called "electrical resonance frequency") between the inductance of the control circuit and the piezoelectric materials can be calculated using

$$L\omega_e = \frac{1}{\hat{C}\omega_e} \tag{6}$$

So, the inductance value can be set up such that this electrical resonance frequency ω_e corresponds with the (1,1) vibrational mode natural frequency of the noise insulation panel, which is equal to the target frequency. If the target frequency moves to another frequency, the electrical resonance frequency can be tuned by changing the value of a variable resistor in the control circuit.

4 The Noise Reduction Effect in Running Test of the Shinkansen

4.1 The Noise Reduction Panel and a Preliminary Bench Test Result

For the evaluation of the noise reduction system in a running vehicle, a test was carried out under the pantograph in a Shinkansen vehicle. The interior noise below the pantograph has a peak around 200Hz, which is assumed to be caused by noise transmitted through the ceiling panels from the pantograph. The target frequency of the test was set at approximately 200Hz from this characteristic of the noise that was measured previously.

The noise reduction panels for the test in the Shinkansen were manufactured with a size of 490×730mm. The base plate of the noise insulation panel was made of a thin aluminum plate, and thin piezoelectric material was bonded to it. The size of the noise insulation panels was determined by eq. (2). An array of twenty-four noise insulation panels with piezoelectric elements was arrayed in a frame. Table 1 shows a specification of the noise reduction panel and the noise insulation panel, and figure 3 is a photograph of a noise reduction panel.

Table 1. Specification of the noise insulation panel and the noise reduction panel

Noise insulation panel	
Size of vibrational area	100×100mm
Material of the base plate	Aluminum
Piezoelectric material	PZT
Capacitance at 200Hz	Ave. 100nH
Noise reduction panel	
Size	490×730×30mm
Number of the noise insulation panel	6×4=24 sheets
Resultant capacitance	144nH

Figure 4 shows the bench test result that is the reduction ratio of the transmitted noise to a noise reduction panel. The inductance value of the control circuit was set approximately to 4.4H according to eq.(6). Because the transmitted noise level had a peak around 190Hz, the (1,1) vibrational mode frequency was close to the target frequency. When the noise reduction panel was connected to the control circuit, the transmitted noise was reduced by up to approximate 15dB at 200Hz with control.

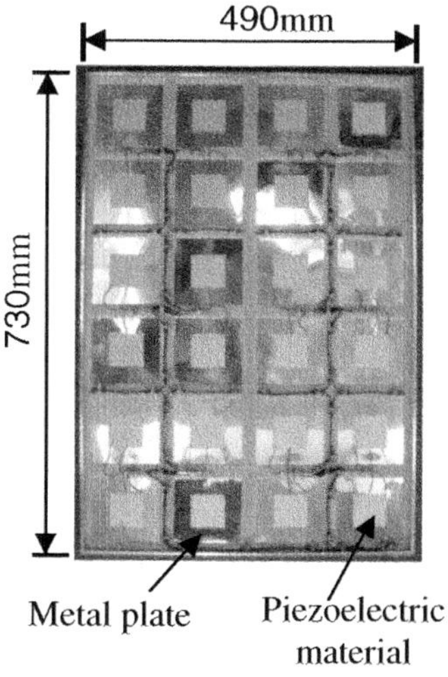

Fig. 3. Noise reduction panel (color figure online)

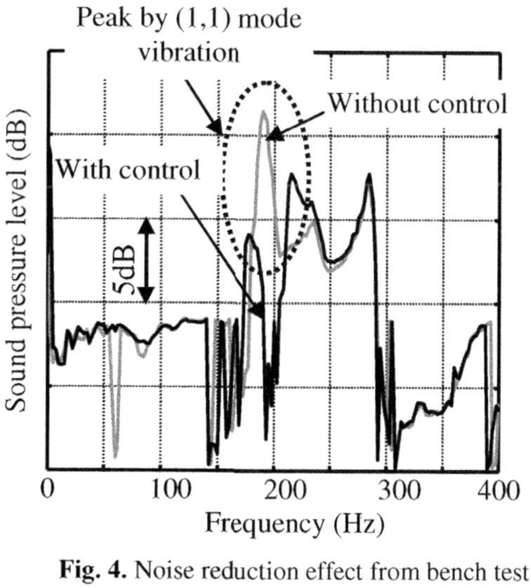

Fig. 4. Noise reduction effect from bench test

4.2 Test Result in Shinkansen Vehicle

In the Shinkansen running test, eight noise reduction panels were installed behind the ceiling panels. A pantograph of the Shinkansen is installed on the roof between shields that cover approximately 3m in length from the end of the vehicle. The noise reduction panels were applied in the range of the pantograph shields. Figure 5 shows the arrangement of the noise reduction panels, and figure 6 shows the area of the application that was as wide as possible without obstructing the service equipment such as lighting and public address loudspeakers. Sound level meters were placed just below the ceiling panel and at 1200mm above the floor under the test area. For the test, the Shinkansen train ran on the same track with almost the same velocity. A-weighted sound pressure levels (SPL) were calculated with and without control.

Figure 7 shows the interior noise spectra under the pantograph with and without control. A peak of the noise appeared around 210Hz without control under the pantograph of the vehicle. As a result of that, the electrical resonance frequency of each noise reduction panel was adjusted using the control circuit to match the frequency of the peak to the target. The noise level was reduced by approximately 5dB just below the ceiling panels and by up to approximately 4dB at 1200 mm above the floor at the peak.

Fig. 5. Arrangement of the noise reduction panels behind the ceiling panels (color figure online)

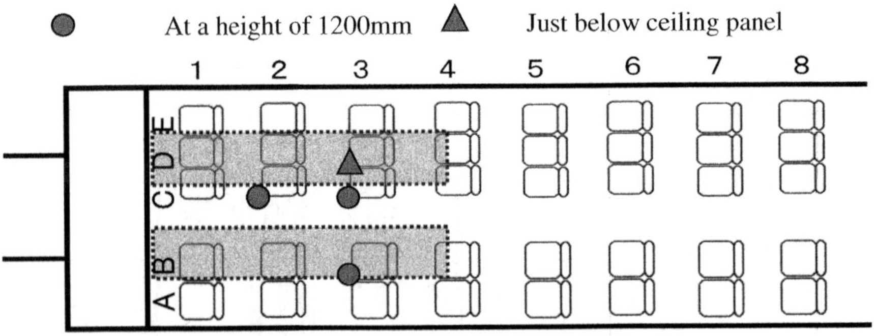

Fig. 6. Measurement points for the interior noise (color figure online)

This was an important result because the interior noise in the railway vehicle, which is difficult situation for a noise control system, was reduced on target without identifying the vibration condition of the ceiling panel or the characteristics of the interior noise. So, the effectiveness of this noise reduction system could be verified.

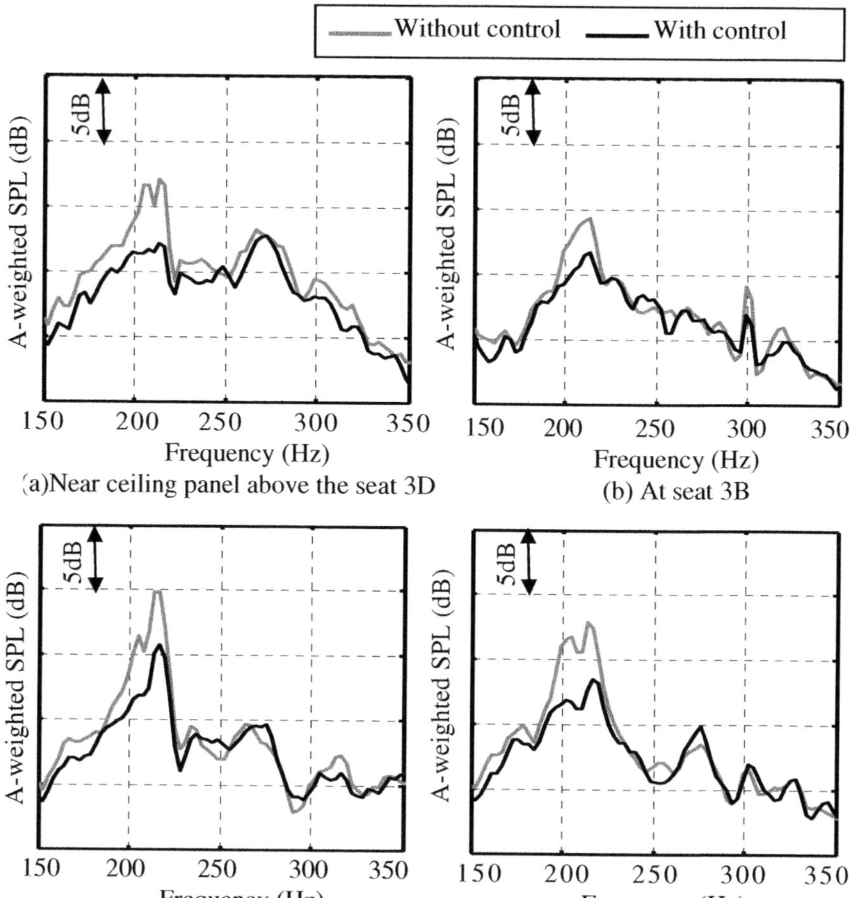

(a)Near ceiling panel above the seat 3D
(b) At seat 3B
(c) At seat 3C
(d) At seat 2C

Fig. 7. Noise reduction effect of the interior noise running in open section

5 Conclusion

This paper proposes a new noise reduction system with piezoelectric materials and demonstrates its effectiveness through the result of a test that was carried out for the interior noise below the pantograph of a Shinkansen vehicle, as given in the following.

(1) A new noise reduction system has been developed, and a method was proposed that suppresses the noise transmitted through the interior panel of the railway vehicle.

(2) For verification of the effectiveness of the noise reduction system when applied to the railway vehicle, the system was manufactured, and the test was carried out below the pantograph of a Shinkansen vehicle.
(3) The noise reduction system, which was installed actually on the ceiling panels in the area of a pantograph in a running Shinkansen vehicle, could reduce the noise by up to 4dB. From these results, the effectiveness of this noise reduction system for the railway vehicle has been verified.

References

[1] Yamamoto, K.: Interior Noise Reduction of Railway Vehicle. The Journal of the INCE of Japan 31(5), 368–373 (2007)
[2] Technical Report of IEEJ 762, 23-26 (1999)
[3] Nishimura, M.: Practical Use of ANC-Past and Future-. Journal of INCE/J 27(4), 223–225 (2003)
[4] Botto, M.A., Sousa, J.M.C., Sa da Costa, J.M.G.: Intelligent active noise control applied to a laboratory railway coach model. Control Engineering Practive 13(4), 473–484 (2005)
[5] Diaz, J., Egana, J.M., Vinolas, J.: A local active noise control system based on a virtual-microphone technique for railway sleeping vehicle applications. Mechanical Systems and Signal Processing 20(8), 2259–2276 (2006)
[6] Yamamoto, K., Tagawa, N., Tanaka, N.: Development of a system to control noise transmitted through interior panels of railway vehicles (1st report, Study on a method of controlling noise transmitted through a panel using a module-type noise control system). Transactions of the Japan Society of Mechanical Engineers, Series C 74(739), 610–618 (2008)
[7] Yamamoto, K., Tanaka, N., Asahina, M.: Development of a system to control noise transmitted through internal panels of railway vehicle (2nd report, Proposal for a Noise Control System Arrayed the Noise Insulation Panels). Transactions of the Japan Society of Mechanical Engineers, Series C 75(752), 890–891 (2009)
[8] Leissa, A.: Vibration of Plates. Acoustical Society of America through the American Institute of Physics, 58–65

Separation of Rolling Noise and Aerodynamic Noise by In-Service Measurement of Combined Roughness and Transfer Functions on a High Speed Slab Track

H.W. Jansen and M.G. Dittrich

TNO Technical Sciences
Oude Waalsdorperweg 63, 2597 AK The Hague, The Netherlands
Tel.: +31 (0)888 666 345; Fax: +31 (0)70 328 09 61
e-mail: erwin.jansen@tno.nl

Summary

Combined sound and vibration measurements during train pass-bys can be used to quantify contributions from the excitation and transmission of rolling noise. This is useful for the identification of sound sources and the assessment of the track contribution. In this paper, a practical application on a high-speed track is discussed.

Acoustic measurements were conducted on a high-speed slab track, with pass-bys of the Thalys and conventional intercity trains, at speeds of around 300 and 160 km/h, respectively. Both sound pressure and vertical rail vibrations were recorded simultaneously. These were used to derive an estimate for the vertical track decay rates, combined effective roughness and transfer functions, with the Pass-by Analysis Software (PBA).

The transfer function of the slab track is compared with a conventional concrete sleeper track, showing a distinct difference in the low and medium frequency ranges. The transfer function is characteristic of rolling noise, and if also measured at high speed, the contribution of aerodynamic noise can be quantified. For this train/track combination, aerodynamic sound sources are shown to contribute to the pass-by levels mainly in the lower frequency range. Based on the transfer functions and roughness levels, it is shown that rolling noise is still a significant noise source up to 300 km/h.

1 Introduction

In 2009, TNO Science and Industry conducted acoustic measurements on the Dutch high-speed track [1, 2]. The objective was to assess the noise emission and wheel roughness of the rolling stock that is currently operational on the high-speed track.

T. Maeda et al. (Eds.): Noise and Vibration Mitigation for Rail Trans. Sys., NNFM 118, pp. 479–486.
springerlink.com

In this paper, the focus is on wheel/rail roughness, noise emission of the slab track and separation of aerodynamic noise and rolling noise. For this purpose, the rail vibrations were also measured synchronously with the sound pressure levels. This enables characterisation of the railway rolling noise. In section 2, the main quantities to characterise railway rolling noise are briefly outlined. Until now, this method had never been applied at high speeds and on a slab track.

2 Definition of Combined Roughness and Transfer Functions

In order to separate the roughness excitation and the acoustic transfer of vehicle/track pass-by noise, a simple breakdown of pass-by noise into quantities determined by wheel and track was proposed in [3, 4]. Wheel roughness, rail roughness and train speed characterise the excitation of rolling noise at the wheel/rail contact. The ratios of the resulting trackside wheel and track noise components to the roughness excitation are expressed as transfer functions. These functions capture every aspect of vibro-acoustic transmission of the vehicle or track implicitly: the wheel and track vibration response to the excitation, its radiation and the sound transmission to the trackside. All transfer functions depend only on the respective track or vehicle design and are independent of the train speed and the roughness levels.

TNO implemented this approach into a software tool called '*Pass-by Analysis Software*' PBA [5]. This software allows the operational measurement of the wheel/rail rolling noise–related quantities, from a simple pass-by measurement of rail vibration and sound pressure.

2.1 Effective Roughness

The wheel and rail roughness amplitude levels are expressed as one-third octave spectra with respect to wavelength λ. The so-called *effective roughness* is used. This is the roughness spectrum as if the total contact patch was represented by a single point, which means that the effective roughness already includes the averaging and 'filtering' effects of the wheel/rail contact patch. The combined effective roughness of wheel and rail is derived directly from the measurement of vertical rail vibrations during train pass-bys [3]. In order to be able to conduct this conversion, the vertical decay rate of the track needs to be known. This can also be determined from the operational track vibration measurements during train pass-bys [3].

2.2 Transfer Functions

The combined transfer function $L_{H,tot}$ (summation of vehicle and track transfer function) is found by subtracting the total roughness from the measured sound pressure level:

$$L_{H,tot}(f_{to}) \equiv L_{p,tot}(V, f_{to}) - 10\lg\left(\frac{N_{axle}}{l_{wagon}}\right) - L_{R,tot}(\lambda(V, f_{to})) \tag{1}$$

where f_{to} denotes the one-third octave band centre frequency, $L_{p,tot}(V,f_{to})$ is the measured equivalent total sound pressure level at pass-by speed V taken over the time interval for the vehicle to pass the measurement location 'from buffer to buffer', N_{axle} is the number of axles per wagon, and l_{wagon} is the wagon length and $L_{R,tot}$ the total effective roughness.

3 In-Service Measurements

3.1 Measurement Set-Up

In November 2009, TNO performed a series of sound and vibration measurements on a Thalys-PBKA train, with disc brakes and auxiliary brake blocks, and a temporary intercity train (called *Fyra)* consisting of disc-braked Prio coaches with additional cast iron block brakes, hauled by a Traxx locomotive. Measurements were performed at two locations in the Netherlands, Schiebroek and De Wacht, both with slab track. Both vertical rail vibrations and pass-by levels were measured at 7.5 m from the centre of the track, as set out in figure 1. Accelerometers were installed under the rails while the track was out of service at night. Noise and vibrations were measured at 90 km/h, 160 km/h and 300 km/h for the Thalys, and only at 160 km/h for the Fyra train.

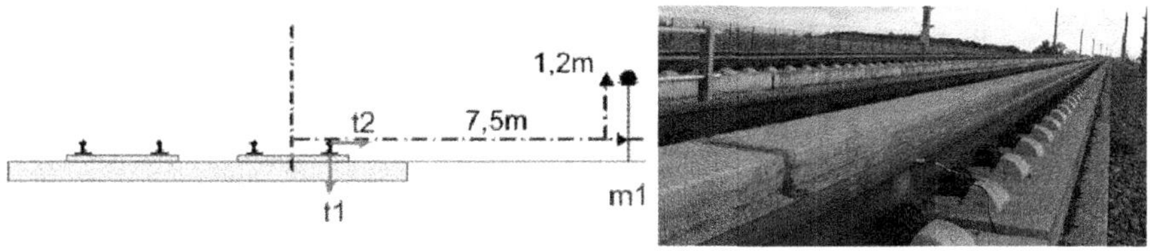

Fig. 1. Measurement positions of accelerometer and microphones in Schiebroek and De Wacht (left). The arrows indicate the accelerometers underneath the rail. The segmented slab track is shown on the right. (colour figure online)

3.2 Rheda 2000 Slab Track

About 90 km of the Dutch high-speed track is constructed from *Rheda 2000* slab track [6]. The main acoustic differences between the Rheda 2000 slab track and a conventional ballasted track are the massive concrete base, which acts as a large heavy sleeper system, the absence of acoustic absorption and a lower rail pad stiffness (a static stiffness of about 22.5 MN/m [6]). Usually, a Rheda track is constructed with a continuous slab. For the Dutch high-speed track, however, the slab has been divided into segments of 6.4 m each, see figure 1. This is required to optimise the dynamic behaviour of the track relative to the fixed substructure, supported by piles in the soft soil of the Netherlands [6]. The rail type is UIC60.

4 Results

4.1 Pass-by Levels

The measured pass-by levels ($L_{eq,tp}$) are shown in figure 2. The pass-by levels of the disc-braked Thalys are about 7 dB(A) lower than the Fyra train, with disc brakes with additional cast iron block brake blocks, at a speed of 160 km/h. This is consistent with the total roughness levels found for both trains, see section 4.2. In both cases, the additional noise emission between 500Hz and 1 kHz can be attributed to the slab track.

At the considered speed of 160 km/h, no aerodynamic noise sources contribute to the pass-by level, which is totally dominated by rolling noise. For a higher speed of 300 km/h, however, the low frequency increase in the pass-by level can be attributed to aerodynamic sources, such as turbulent boundary layers along the nose of the train and the bogies [7]. In section 5, the relative contribution of aerodynamic noise and rolling noise on the total pass-by level is studied.

The general 30lgV relationship for rolling noise [7] was not applicable for the pass-bys of the Thalys on the Rheda 2000 track, for which a 26lgV relationship was found to be more appropriate. To extrapolate the rolling noise emission of trains at higher train speeds, 30lgV is commonly used. It has also been found to be valid for pass-bys of TGVs [10] and is based on pass-by measurements on ballasted tracks. On such a track, the wheels are often the predominant sound source at higher speeds. Since the track contribution is relatively high for the slab track, sound pressure levels are not strictly 30lgV. The transfer functions that are applied in section 5 take this effect into account.

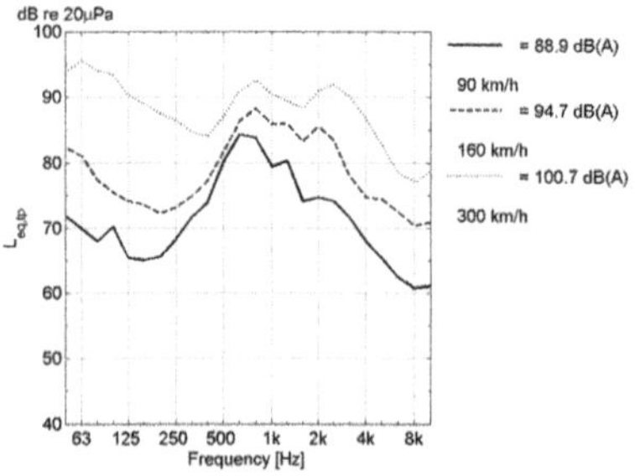

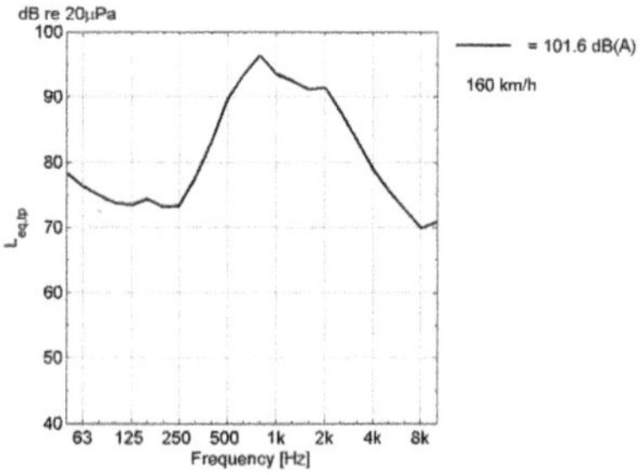

Fig. 2. Measured pass-by levels of the Thalys (left) and the Fyra (right) at 7.5 m from the Rheda 2000 track at various speeds as a function of frequency. (colour figure online)

4.2 Roughness Levels

As shown in figure 3, for the Fyra train, a typical block-braked wheel roughness spectrum is found. For the disc-braked Thalys, the roughness spectrum is much lower in the entire wavelength range. Due to the low wheel roughness of the Thalys, the total roughness level, which is the energy summation of wheel roughness and rail roughness, can be regarded as an upper limit for the rail roughness.

The total roughness levels are about equal for both considered speeds at both sites. This indicates that the indirect roughness method also works at high-speed pass-bys.

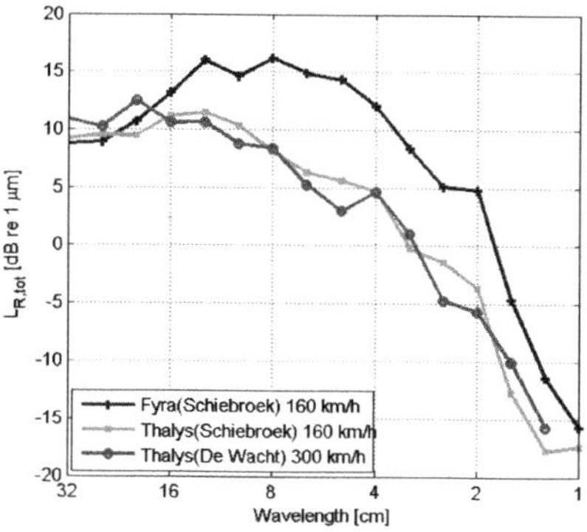

Fig. 3. Total roughness levels of the Fyra and the Thalys at various speeds as a function of wavelength. (colour figure online)

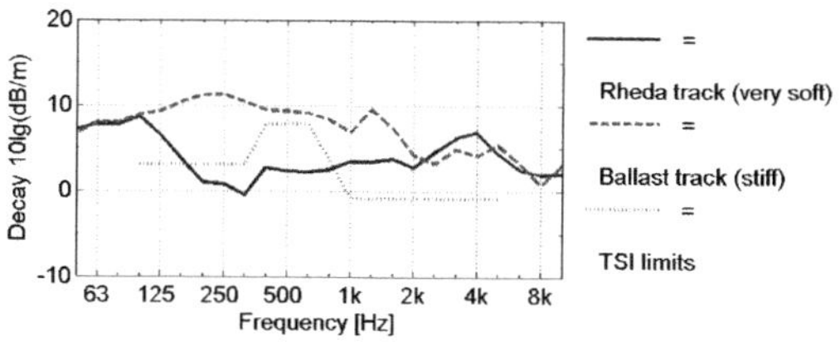

Fig. 4. Vertical spatial decay of the Rheda 2000 slab track and a conventional ballasted track with monoblock sleepers, as a function of frequency. (colour figure online)

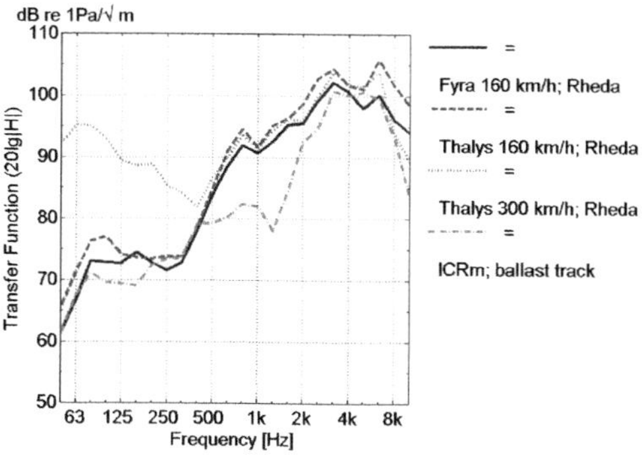

Fig. 5. Total transfer function levels near the Rheda track and a ballast track, measured with various train types, as a function of frequency. (colour figure online)

4.3 Decay Rates

The average vertical decay rate of the Rheda 2000 slab track, as derived from the rail vibrations [3], is shown in figure 4. For comparison purposes, also the decay rate of a ballast track with monoblock sleepers and stiff rail pads is shown in the figure, taken from [9]. In [8], it has been shown that the vertical decay rate is a good measure to quantify the rail pad stiffness. The vertical decay rate typically drops at the frequency at which the rail is decoupled from the sleeper, beyond the natural frequency of the rail on the sleeper [11]. For the considered sleepered track, this occurs at a frequency of about 1700 Hz. In this case, the pad stiffness would be qualified as stiff (about 1500 MN/m). For the slab track, however, the decoupling frequency is very low, about 100 Hz. This is caused by a combination of the heavy slab track segment, which can be treated as a heavy mass, and the very low static pad stiffness of about 22.5 MN/m [6]. The low-frequency decoupling of the rail results in a low track decay rate. Consequently, the attenuation of vibrations per unit rail length is small, and therefore, its sound radiation is high [11]. Also, TSI limits are shown in figure 4. Due to the low pad stiffness, the high-speed slab track is not TSI compliant.

4.4 Transfer Functions

The total transfer function is derived from the pass-by levels and the total roughness levels according to equation 1, see figure 5. This function is the ratio of the pass-by level to the total roughness, normalised to axle density (APL). The total transfer functions measured with the Fyra and Thalys trains on the Rheda track show similar behaviour, as is to be expected [3]. However, in the lower frequency range from 50 Hz to 400 Hz, at a pass-by speed of 300 km/h, the total transfer function of the Thalys is significantly higher than the transfer functions found at a lower speed of 160 km/h. This can be explained by aerodynamic sound sources that contribute to the noise level at this speed. The high noise levels in this frequency range are not generated by roughness excitation. Noise and roughness have no correlation in this frequency range.

As a comparison, also the transfer function of a ballasted track with monoblock sleepers is shown, taken from [9]. Clearly, in the mid-frequency range from 500 Hz to 2 kHz, the transfer function of the Rheda track is much higher than that of the ballasted track. This can be explained by the increased noise radiation of the Rheda track due to very low pad stiffness, as already discussed in section 4.3.

5 Separation of Aerodynamic Noise

The pass-by level of the Thalys on the Rheda track that is radiated by just rolling noise at a speed of 300 km/h can be assessed from the roughness levels and the transfer function at low speed:

$$L_{p,300km/h}(V, f_{to}) = L_{H,rheda}(f_{to}) + 10\lg\left(\frac{N_{axle}}{l_{wagon}}\right) + L_{R,160km/h}(\lambda(V, f_{to})) \quad (2)$$

The calculated spectrum of the pass-by level is shown in figure 6. When this calculated rolling noise spectrum is compared with the measured spectrum of the total pass-by level, including contributions of aerodynamic sources, the rolling noise can be seen to be dominant for the total A-weighed pass-by level. This effect was already shown in [7] and [10]. Low-frequency contributions from aerodynamic sources, in total 86 dB(A) in the frequency range up to 400 Hz, do not affect the total pass-by level of 101 dB(A).

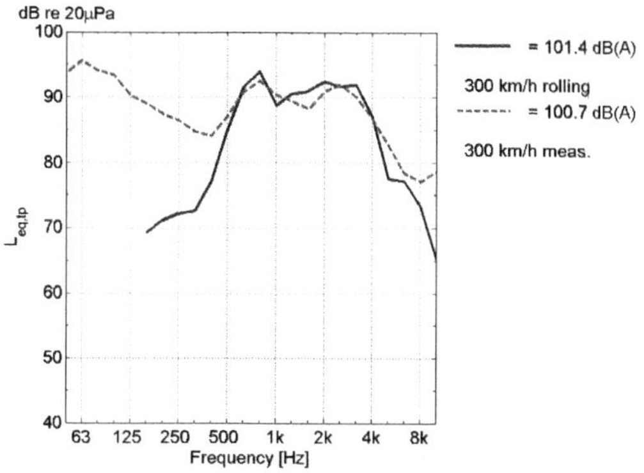

Fig. 6. Measured and calculated pass-by level of the Thalys on a Rheda high-speed track at 300 km/h as a function of frequency. (colour figure online)

6 Conclusions

The indirect roughness measurement was successfully applied at a high-speed slab track at high-speed pass-bys. Combined roughness and transfer function measurements allow an estimate of wheel/rail roughness and the different source contributions to be made.

For Intercity trains with cast iron brake blocks on the Rheda track, relatively high noise levels are found due to high wheel roughness and extra radiation from the slab track. A combination of low pad stiffness and hard reflecting surfaces of the Rheda 2000 track results in a high contribution of the rail to rolling noise.

For the Thalys at 160 km/h on the Rheda track, noise levels are higher than normal due to the slab track contribution, but this effect is much less at high speeds, due to the increased radiation of the wheels. Sound pressure levels are shown not to be strictly 30lgV in this case. Transfer functions take this effect into account.

Based on the transfer functions and roughness levels, the rolling noise can be separated from the aerodynamic noise. For the Thalys at 300 km/h on Rheda track, rolling noise still dominates the total noise level, but aerodynamic noise dominates the low frequency range.

Acknowledgements

The authors gratefully acknowledge the support of ProRail and the Netherlands Ministry of Transport.

References

[1] Eisses, A., Dittrich, M.G., et al.: Evaluatie geluidemissie van de HSL-Zuid. TNO report MON-RPT-033-DTS-2009-03096, Delft, the Netherlands (2009)

[2] Eisses, A., Dittrich, M.G., et al.: Evaluatie van de geluidemissie van hogesnelheidsmaterieel (Thalys) op de HSL-Zuid. TNO report MON-RPT-033-DTS-2010-00055, Delft, the Netherlands (2010)

[3] Janssens, M.H.A., et al.: Railway noise measurement method for pass-by noise, total effective roughness, transfer functions and track spatial decay. Journal of Sound and Vibration 293, 1007–1028 (2006)

[4] Jansen, H.W., Janssens, M.H.A., Dittrich, M.G.: Practical applications of total wheel/rail roughness measurements during train pass-bys. In: 9th IWRN, Munich, Germany (September 2007)

[5] Jansen, H.W.: User manual 'Pass-by Analysis Software V3.0'. TNO report IS-RPT-050074, Delft, the Netherland (2005)

[6] `http://www.railwaypeople.com/rail-projects/hsl-zuid-project-56.html`

[7] Thompson, D.J.: Railway noise and vibration: mechanisms, modelling and means of control. Elsevier, Great Britain (2009)

[8] Thompson, D.J., Jones, C.J.C.: Study on the sensitivity of the indirect roughness method to variations in track and wheel parameters (STAIRRS report) ISVR Contract Report 01/xx (April 2001)

[9] Jansen, H.W., Dittrich, M.G.: Geluid- en trillingsmetingen van ICRm-rijtuigen met LL-remblokken voor categorietoetsing volgens TR procedure. TNO report IS-RPT-033-DTS-2006-01162, Delft, the Netherlands (2006)

[10] Poisson, F., Gauttier, P.E., Letourneaux, F.: Noise sources for high speed trains: a review of results in the TGV case. In: Noise and Vibration Mitigation for Rail Transportation Systems. Notes on Numerical Fluid Mechanics and Multidisciplinary Design (NNFM 1999). Springer, Heidelberg (2008)

[11] Jones, C.J.C., et al.: The use of decay rates to analyse the performance of railway track in rolling noise generation. Journal of Sound and Vibration 293, 485–495 (2006)

Experimental Study on Flow Field under a High Speed Shinkansen Train

N. Yamazaki[1], A. Ido[2], T. Kurita[3], and M. Matsumoto[4]

[1] Railway Technical Research Institute (Former, East Japan Railway Company), Umegahara; Maibara, Japan
Tel: +81749524681, Fax: +81749524683
e-mail: nobuhiro@rtri.or.jp
[2] Railway Technical Research Institute
[3] East Japan Railway Company
[4] Dantec Dynamics K.K.

Summary

A flow velocity profile under a Shinkansen train car was measured to investigate the flow field, as it is essential in understanding the mechanisms of aerodynamic noise generation. Laser Doppler velocimetry (LDV) was used to measure the area outside of rolling stock gauge. The flow field near the underside of the car was also measured meticulously to clarify inlet conditions to the bogie cavity. It was found that mean flow velocity at a middle car (car #5 from the front car) decreases to 42 % of train speed at a point halfway between the underside of the car and the ground, whereas that of a leading car reaches approximately 78% at the same point. It was also found that, at car #5, mean velocity and turbulence level on ballast tracks are slightly higher than that of slab tracks.

1 Introduction

The maximum speed of typical high-speed trains in Japan, Shinkansen, has increased steadily since operations started in 1964. Increasing concern about environmental issues, particularly noise, has led to studies on effective noise reduction techniques. Aerodynamic noise has also been recognized as an important environmental issue in recent years due to being characteristically proportional to the 6th - 8th power of the train speed [1]. The difference in noise intensity at other sources distributed around cars has decreased, due to progress in noise reduction techniques for typical sources such as pantographs. Therefore, it is essential to focus on various noise sources in order to reduce the total noise from the high-speed trains. Among the noise sources, recent researches show that the bogie section is one of the main sources [2]. Since the aerodynamic noise closely relates to air flow, studies for noise reduction in the bogie section have already been widely conducted through wind tunnel tests and CFD [3, 4]. But precise inlet conditions to the bogie

T. Maeda et al. (Eds.): Noise and Vibration Mitigation for Rail Trans. Sys., NNFM 118, pp. 487–494.
springerlink.com

sections in these tests are still unknown. This is mainly due to the complex flow field formed between two planes, namely the underside of the car and the ground. Furthermore, characteristic long flow channels of a train with many cars make it difficult to predict the flow field under middle cars. In order to model the flow field around the bogie section more precisely, the flow profile at the inlet of bogie section should be clarified based on measurement of field tests. These results will be useful for the studies on noise reduction techniques in future work.

The main scope of this study is to clarify the flow field under train cars and investigate the influence of ground conditions in a field test. This paper describes the result of flow velocity measurement at the inlet of bogie sections of a high-speed train. Vertical flow velocity including the area near the underside of the car was obtained by non-intrusive velocimetry, that is Laser Doppler Velocimetry (LDV), the advantage of which is to measure the area outside of the rolling stock gauge [5].

2 Preliminary Test

Since the LDV system detects the velocity of seeding particles representing air velocity, the accuracy of the velocity measurement is ensured by the assumption that the velocity difference between the particle and the air is sufficiently small. In this study, water mist was chosen as the seeding instead of conventional oil mist to

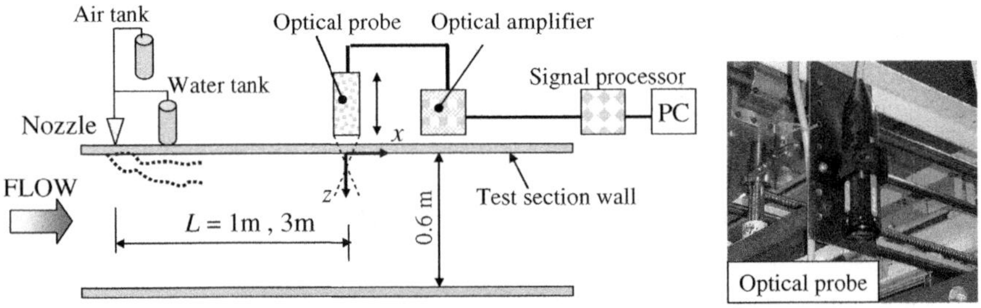

(a) Experimental setup

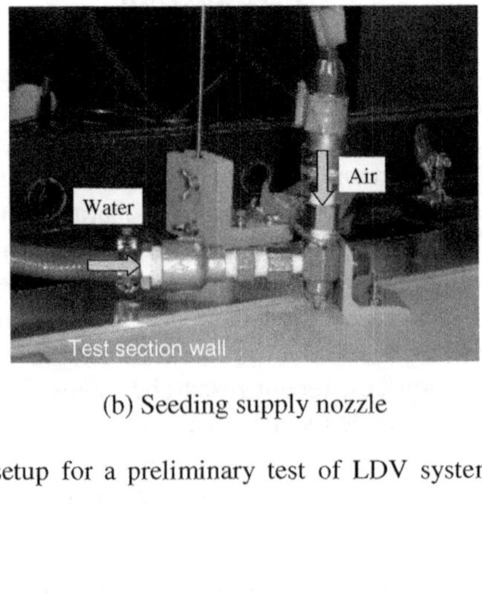

(b) Seeding supply nozzle

Fig. 1. Experimental setup for a preliminary test of LDV system in a wind tunnel test (color figure online)

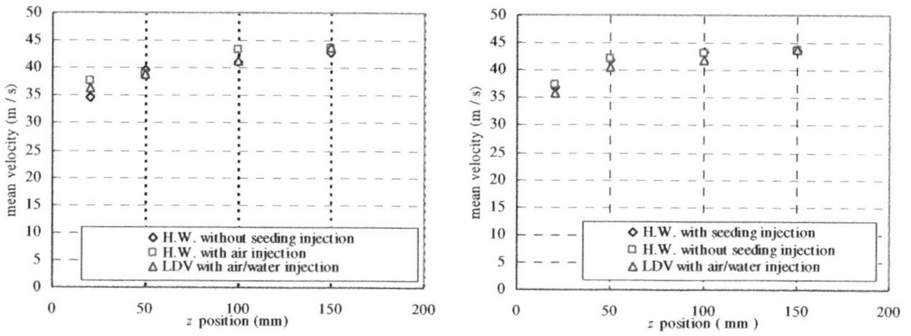

(a) $L = 1$ m (b) $L = 3$ m

Fig. 2. Accuracy of LDV measurement (color figure online)

prevent pollution for rail tracks. Before applying the LDV system to the field tests, the validity of the water mist was examined in a wind tunnel test with the same configuration.

The outline of the LDV system (FlowLite system, Dantec Dynamics K.K.) and experimental setup are illustrated in Fig.1, where the main flow and vertical directions are denoted by x and z, respectively. The LDV system is mainly composed of an optical probe for seeding particle detection, an optical amplifier, a signal processor, and a PC. Seeding particles were sprayed thorough a nozzle mounted upstream of a measurement section. This nozzle creates water mist by mixing air and water at the nozzle exit. The distance between the nozzle and the measurement section was set at 1 m and 3 m corresponding to the conditions in the field test. The result of LDV measurement, flow velocity in the x direction, was compared with that of hot wire anemometry (H.W.) that is independent of the seeding particle. Figure 2 shows distributions of mean velocity along the z direction. The result shows that the velocity difference between the measurement by LDV and H.W. is less than 5 %, and the water mist follows the main stream with sufficient accuracy.

3 Flow Velocity Measurement in Field Test

3.1 Measurement Methods

Flow velocity measurement in the field test was carried out on JR East's prototype Shinkansen train FASTECH360S. The measurement schedule is shown in Table 1. Based on a previous study showing that the flow field under train cars can be considered as fully developed turbulent flow after the second car [6], flow velocity in the direction of the train was measured at the inlet position of the rear bogie section of a leading car (final car for direction A) and a middle car (car #5 for direction B) as shown in Fig 3. The vertical coordinate with origin at the underside level of the car is denoted by h. The vertical measurement area was covered from h = 5 mm to 260 mm by traversing the optical probe. The schematic layout of the LDV system is shown in Fig. 4. All components except for the water and air tank were installed under the floor of the cabin. Water pressure and air pressure were adjusted to 0.175 MPa and 0.45 MPa, respectively. The optical probe was installed near upstream edge of the bogie cavity as shown in Fig. 5. Figures 6 and 7 show views of laser emission and water mist injection. A polycarbonate board was placed over the laser emission hole to protect the optical probe.

3.2 Measurement Condition

Since it is considered that the flow field under train cars depends on not only the floor shape but also ground conditions, measurements were conducted on slab and ballasted tracks. Although the LDV system is capable of data acquisition at a sampling rate up to 50 kHz, the sampling interval depends on the timing of incoming particles. Measurement duration time was then set to either 10 sec or the time in which 50,000 particles were detected. Since the train runs 890 m in 10 seconds at its maximum speed of 320 km/h, straight lines of approximately 1 km in length through an open section were chosen as measurement sections. In this paper, the vertical coordinate is non-dimensionalized by the distance $H = 400$ mm between underside of the car and the ground.

3.3 Results and Discussion

3.3.1 Particle Detection Rate

The number of seeding particles detected by the optical probe is assumed to affect measurement results. Change of the particle detection rate is plotted in Fig. 8 against the height for typical measurement days. The result shows that the water mist injected into the flow field covers a sufficiently wide area from the underside to $h / H = 0.5$. Note that the particle detection rates measured on the days 13 and 14 are less than that of day 12 in Fig. 8 (a). From Fig. 8 (b), it is also found that detection rates are restored on day 20 when the polycarbonate board was replaced with a new one and deteriorate again on day 25. One possible reason is that the protective window at the laser emission hole had been soiled during the test run.

Table 1. Measurement schedule

Date (May)	12th	13th	14th	20th	25th
Middle car	○	○	○		
Leading car				○	○

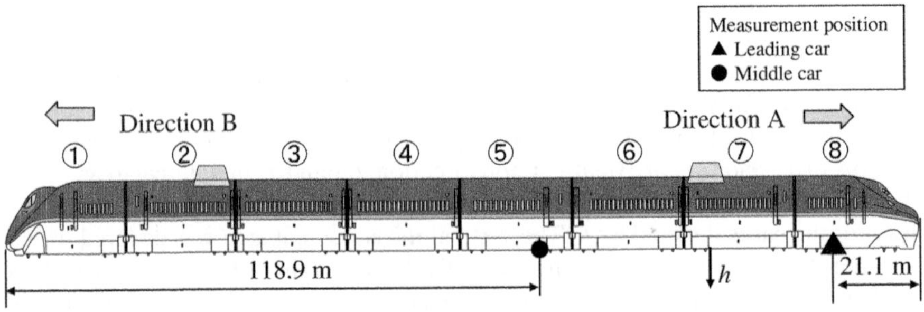

Fig. 3. Measurement points on Shinkansen train in the field test (color figure online)

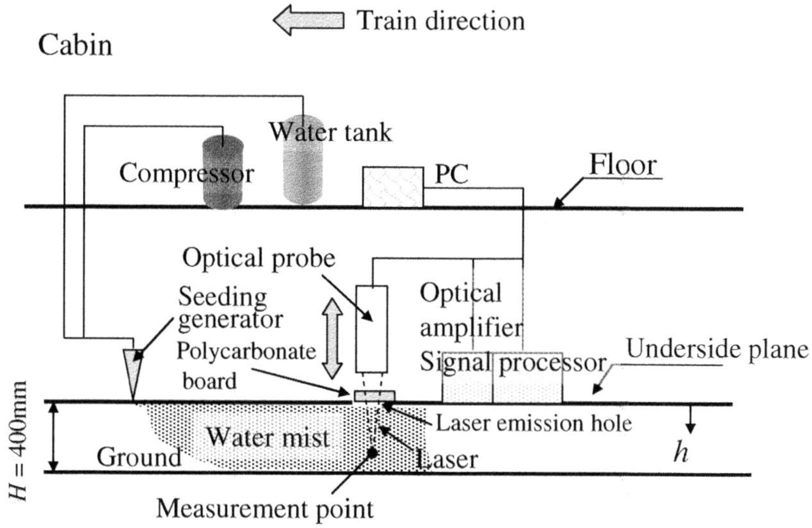

Fig. 4. Schematic layout of the LDV system (color figure online)

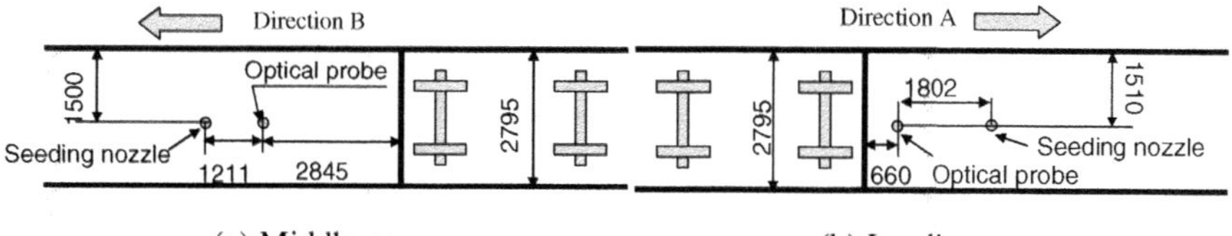

Fig. 5. Arrangement of the LDV system under floor of the train (color figure online)

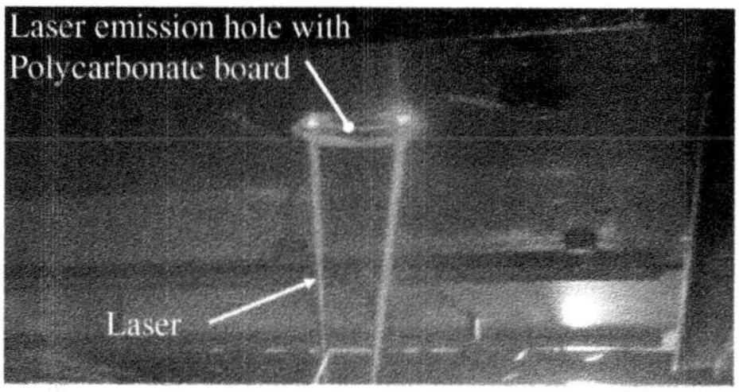

Fig. 6. Laser emission (color figure online)

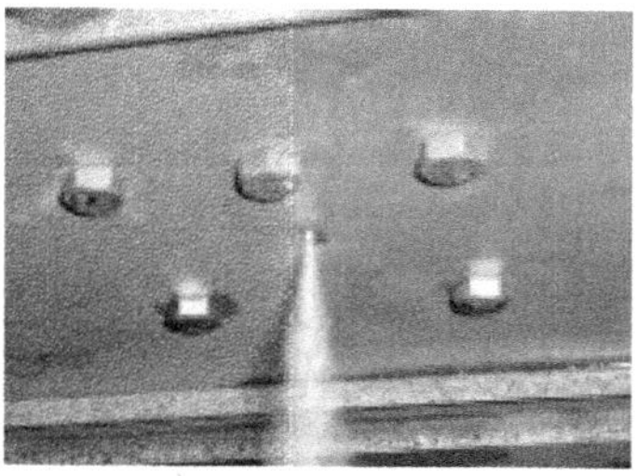

Fig. 7. Water mist injection (color figure online)

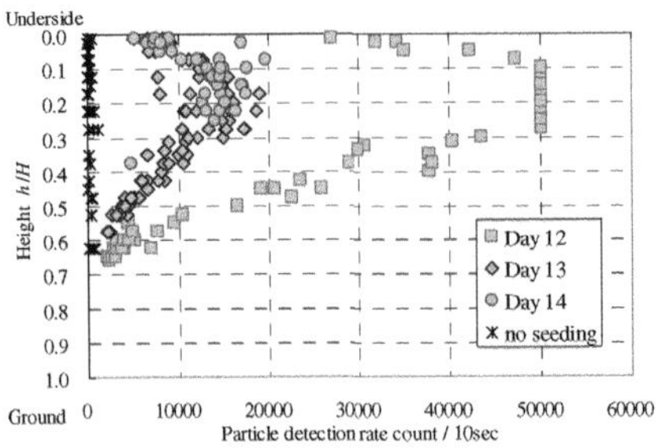

(a) Middle car

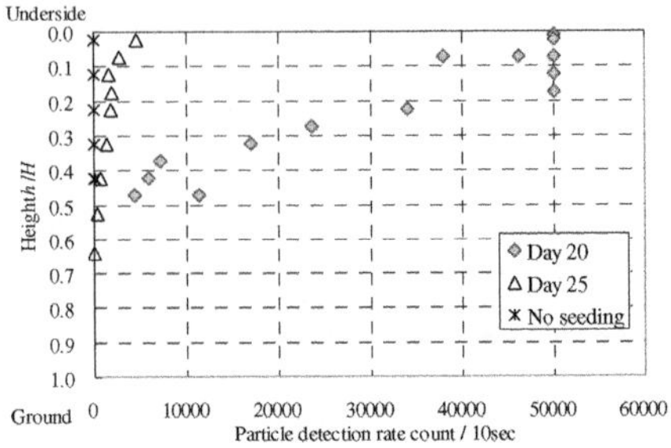

(b) Leading car

Fig. 8. Particle detection rate (color figure online)

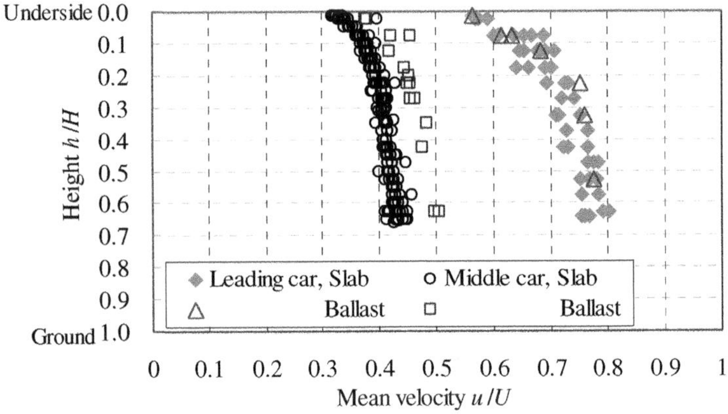

Fig. 9. Mean velocity profile under train cars

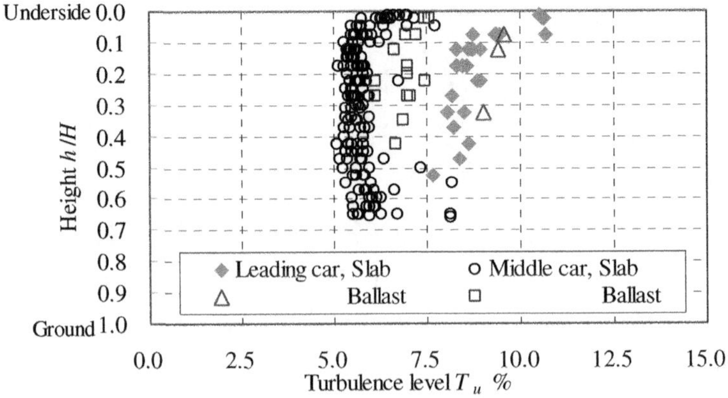

Fig. 10. Turbulence level profile under train cars

3.3.2 Velocity Distribution Underside of the Train Cars

Mean flow velocity u in the direction of the train is normalized by the train speed U. Turbulence level T_u is also defined as follows for the data in which the particle detection rate is more than 1000 counts / 10 sec.

$$T_u\,[\,\%\,] = u' \,/\, U \times 100$$

where u' is the root-mean square of the velocity fluctuation. The train speed U ranged from 194 km/h to 319 km/h.

Figures 9 and 10 show mean velocity and turbulence level distribution. While mean velocity around the point halfway between the underside of the floor and the ground is approximately 0.78 at the leading car, the velocity at the middle car decreases to 0.42 at the same point. Mean velocity distribution near the underside is also different between the measurement cars. The ratio of velocity change to the measurement point at the middle car is smaller than that of the leading car. This implies the existence of fully developed turbulent flow field around the middle cars. In the entire measured region, T_u at the leading car exceeds that of the middle car especially near the underside plane. This result also implies that the effect of separation at the leading edge of the train car remains at the rear bogie in the leading car.

Slab and ballasted tracks have been compared to investigate the influence of ground condition on the flow field near the underside plane. Different tendencies can be observed between the leading and the middle car in both the mean velocity and the turbulence level distribution. Both of them at the middle car on the ballast track are larger than those on the slab track. This result means that, in the ground coordinate, the flow induced by the train decreases due to the rough surface. Consequently, in the train system coordinate, the flow velocity increases on the rough ballast track compared with the smooth slab track. On the other hand, no difference between each ground condition can be seen at the leading car. This shows that the effect of the ground condition does not reach near the underside of the train.

4 Conclusion

Detailed vertical velocity profiles at the inlet of the bogie section were measured at a leading car and a middle (5th) car by using a non-intrusive LDV system to investigate flow field. This is the essential in understanding the aerodynamic noise generation mechanisms. Mean velocity at the point halfway between the underside of the car and the ground decreases to approximately 42% of the train speed at the middle car, while the velocity is 78% at the leading car. Since aerodynamic noise is closely related to flow velocity, aerodynamic noise from the bogie section should be considered individually. It was also found that the influence of ground condition appears at the rear part of the train. In the middle car, the mean velocity on rough ballasted track is larger than that on smooth slab track. On the other hand, ground conditions have little influence on the flow field at the leading car because the effect near the ground does not reach near the underside on the car.

References

[1] Goldstein, M. E.: Aeroacoustics, p.135, Kyoritsu Shuppan (1991)
[2] Nagakura, K.: Localization of aerodynamic noise sources of Shinkansen trains. Journal of Sound and Vibration 293(3), 547–556 (2006)
[3] Yamazaki, N., et al.: Wind Tunnel Tests on Reduction of Aeroacoustic Noise from Car Gaps and Bogie Sections. Quarterly Report of RTRI 48(4), 229–235 (2007)
[4] Morita, K., et al.: Visualization of Aeroacoustic Sound Sources of Bogie Section for High Speed Trains. In: Proceedings of 2009 Symposium on Environmental Engineering, No. 127 (2009) (in Japanese)
[5] Crespi, P., et al.: LASER Doppler velocimetry measurements and boundary layer survey on-board the TGV high speed train. In: Proceedings of the World Congress on Railway Research (1994)
[6] Ido, A., et al.: Study on under-floor flow to reduce ballast flying phenomena. In: Proceedings of World Congress on Railway Research (1999)

Experimental Study of Noise Barriers for High-Speed Trains

P. Belingard[1], F. Poisson[2], and S. Bellaj[3]

[1] SNCF, Infrastructure Engineering Department, Eurostade, 6 avenue François Mitterand 93574 La Plaine Saint-Denis Cedex
e-mail: pascal.belingard@sncf.fr
[2] SNCF, Innovation and Research Department, 45 rue de Londres, 75379 Paris cedex 08
e-mail: franck.poisson@sncf.fr
[3] SNCF, Test Department, 21 Avenue du Président Salvador Allende, 94407 Vitry-sur-Seine Cedex
e-mail: selim.bellaj@sncf.fr

Summary

Noise reduction in the environment of the railway system can be achieved by acting on the source and the transfer path. The opening of the transport market leads to different operators running on the same track with inhomogeneous rolling stock. In this context, a noise barrier is still a relevant mitigation measure.

An experimental study has been carried out to evaluate the relevancy of using noise barriers to reduce the noise exposure caused by high-speed trains running at 380 km/h. First, a bibliography study summarizes the main results of previous studies and available products on the market. Some improvements seem to be possible to increase the noise reduction of classical noise barriers. Consequently, a modular noise barrier has been defined following the most promising results of the bibliography to test several configurations of a "Y" shape. A prototype has been built along a TGV line in France. Six configurations of the "Y"-shaped noise barrier have been tested, with or without absorbent materials, for a TGV running below 380 km/h.

Absorbent material and multiple diffracting edges confirm their potential to increase the efficiency of the noise barrier. For the current commercial speed (320km/h), the noise reduction provided by the noise barrier compared to the free-field conditions, for a receiver located 25m from the track and 3.5m high, increases from 4 to 5 dB(A) by using absorbent materials. The multiple diffracting edges provide a noise barrier efficiency equal to or even better than a "classical" noise barrier for the current commercial running speed.

At higher running speeds, the energy of aerodynamic noise sources located on the roof of the train increases and the barrier height is not sufficient.

T. Maeda et al. (Eds.): Noise and Vibration Mitigation for Rail Trans. Sys., NNFM 118, pp. 495–503.
springerlink.com

1 Introduction

Environmental noise is an important target for railway transportation. In Europe, the technical specifications for interoperability (TSI) limit the noise at source, which is experimentally assessed during the acceptance process of each new train type. For residents, noise at the receiver is mapped as required in the European Directive 2002/49CE. In parallel, limit values at the receiver are defined by some member states. In France, the indicators are the sound pressure levels L_{DEN} and L_{NIGHT} that correspond to a mean level integrated over a long period. In a context where the commercial speed and the traffic increase, limit values in L_{DEN} and L_{NIGHT} can be exceeded. Then, the source level or the number of pass-bys has to be reduced. Noise reduction at source is the most promising solution. However, due to the opening of the transportation market, several companies are running on the same lines. In this context, mitigation measures at source are much more difficult to impose, apart from a revision of the TSI with stricter limit values. In this context, sound barriers remain one way of achieving a noise reduction in the short to medium term.

Therefore, an experimental study has been carried out to evaluate the ability of noise barriers to reduce the noise exposure caused by high-speed trains running at 380 km/h. First, a bibliography study summarizes the main results of previous studies and available products on the market. The modular noise barrier is described in section 3 and the field test presented.

2 Bibliography Study

Several years ago, a collaborative project called "Euroecran" [1] was carried out to design more efficient noise barriers, validate numerical models, and test some improved noise barriers at full scale. The main results can be summarized as follows:

- The distance to the track, the height of the noise barrier, and the shape of the top of screen are the main parameters.
- The thickness of the noise barrier is not an important parameter.
- A minimum surface of absorbent material is needed to achieve an effect (third of the total surface).
- A maximum noise reduction is reached with less than the total surface covered with absorbent material.
- Helmholtz resonators located on the top of the noise barriers are not efficient.

Ray-tracing simulations confirm the results:

- Reflections between the train and the noise barrier must be limited using absorbent material or tilted noise barriers.
- If the distance to the track is more than 5m, a tilted noise barrier or a noise barrier with absorbent material provides the same improvement.

Thirty-four scale models of noise barriers have been tested by the RTRI in 2002 [2].The most efficient one was a Y-shaped noise barrier, 3m high, with 45° angle.

JR East has also carried out an experiment with scale models to validate solutions to improve an L-shaped noise barrier. An optimal noise barrier has been designed with a large plate on the top. Half of the vertical surface covered by absorbent material provides a noise reduction increase of 1.5 dB(A) for a Shinkansen train running at 110 km/h and a measurement point located at 25m from the track at the rail height. With the same top covered with absorbent material, 1 dB(A) more can be obtained.

At the same time, new configurations of noise barriers are proposed by the manufacturers. "Curved" noise barriers using PVC and vegetation offer a great design to be installed in the city with efficiency close to a tilted noise barrier. The most innovative products are:

- The "fractal" noise barrier promoting the fact that the effective surface is 100 times the surface of a flat panel and composed of numerous small acoustic cavities that provide absorbent.
- The combination of passive and active solutions onto a noise barrier with active control devices inserted on the top [3].

In all the cases, the location of the barrier is constrained by safety criteria. The height, the shape, and the acoustic properties of the panels are mainly linked to the cost of the noise barrier. In this context, an increase in the number of diffracting edges ("Y" shape is promising) is a way of improvement in accordance with the constraints of the railway system.

3 Field Test

3.1 Definition of the Modular Noise Barrier

In France, the majority of the acoustic barriers used for high-speed trains are reflective, made of concrete, and 2.2 m high above the head of the rail. According to the results of the bibliography study, it seems to be important to test several shapes of noise barriers and to validate the effect of absorbent material. To make possible the comparisons (with the same test train) and preserve the experimental cost, a modular noise barrier has been designed (see figure 1).

Fig. 1. Pictures of the modular noise barrier (color figure online)

The lower part of the barrier consists of a fix concrete wall 2m high and 20cm thick. The top is made with a “Y” shape to integrate metallic panels fitted with absorbent materials in three positions. The panels are metallic boxes 7.5cm (3 inches) thick with microperforated plate on one side and metal plate on the other side. Panels are fitted with glass wool inside (100kg/m^3). Acoustics characteristics of the panels according to NF EN ISO 354 are the following:

Table 1. Sound absorption coefficient of the panels (NF EN ISO 354)

Frequency (Hz)	Sound absorption coefficient	Frequency (Hz)	Sound absorption coefficient
100	0.38	800	0.96
125	0.41	1000	0.98
160	0.41	1250	0.85
200	0.69	1600	0.81
250	0.94	2000	0.85
315	0.99	2500	0.79
400	1.16	3150	0.77
500	1.03	4000	0.79
630	1.01	5000	0.88

The panels are used on one side as absorbent materials and as reflective panels on the other side.

By changing the location of the panels, 6 noise barriers are tested (see figure 2):

- NB1 is the reference noise barrier, completely reflective, 3.0m high (2.1m above the head of the rail), located 5.05m from the center of the nearest track.
- NB2 is an improved reference noise barrier equipped with absorbent material on the top 100cm of the noise barrier.
- NB3 is like the reference NB1 with the diffracting edge moved closer to the track.
- NB4 is an improved NB3 with absorbent material on the top.
- NB5 is a double-diffracting edge noise barrier (Y shape).
- NB6 is an improved NB5 with absorbent material on the top.

For each noise barrier, the location of the concrete wall is fixed for practical reasons. Then, the comparison between the noise barriers is not completely fair because the relative position to the source of the diffracting edge of the barriers NB1 and NB2 compared with the barriers NB3, NB4, NB5, and NB6 is different. The diffracting edge position remains the same for NB3, NB4, NB5, and NB6. Nevertheless, all the barriers fulfill the safety requirements.

Due to the constraints of the experiment, a configuration NB7 with absorbent materials inside the Y shape has not been tested.

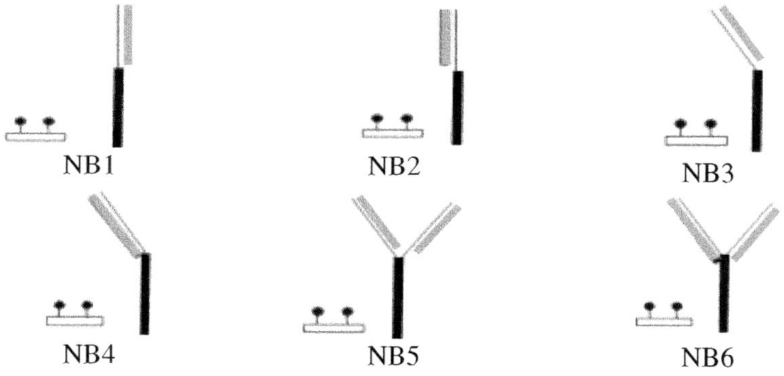

Fig. 2. Configurations of the modular noise barrier (NB) (absorbent materials are represented by the thick gray lines)

Fig. 3. NB6 in place for measurement (color figure online)

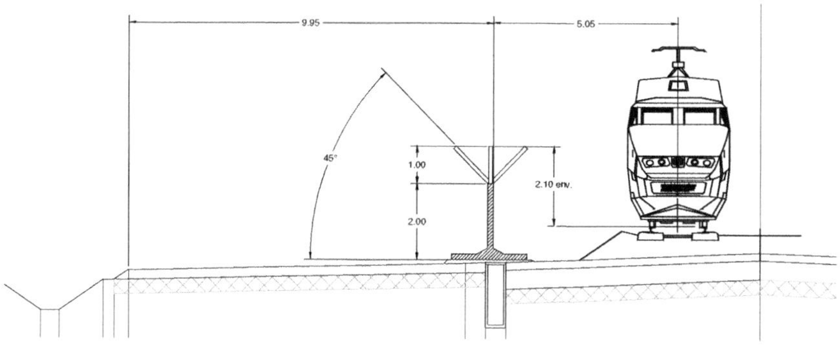

Fig. 4. Location of the noise barrier (color figure online)

3.2 Measurement Campaign

The prototype was installed in France along the east high-speed line just before its commercial opening. Two hundred meters of noise barrier have been built and removed after the test (see figure 3). The ballast is 0.90m high; thus, for NB1 the top of the noise barrier is 2.1 m above the head of the rail (see figure 4). The rail is a UIC60 fixed with Pandrol 8494 fastening system onto bibloc concrete sleepers.

The rolling stock is a test train TGV-POS without any modification running below 380 km/h.

Measurements are carried out in free field and behind the noise barrier, at 25m from the center of the track and 3.5m above the head of the rail. $L_{pAeq,tp}$ and TEL are measured for each of the six configurations. Each train pass-by is measured behind the noise barrier and in free-field conditions, 500m far from the end of the noise barrier. All the results correspond to a mean value of 2 or 3 pass-bys of the same train running at the same speed (+/-5 km/h) with the same time integration for TEL assessment with and without barriers.

4 Results

4.1 Pass-by Noise in Free-Field Conditions

First, the regression law of the $L_{pAeq,tp}$ in free field has been identified :

$$L_{pAeq,tp}(V) = l(V_0) + k\log_{10}\left(\frac{V}{V_0}\right).$$

Measurement results are presented figure 5.

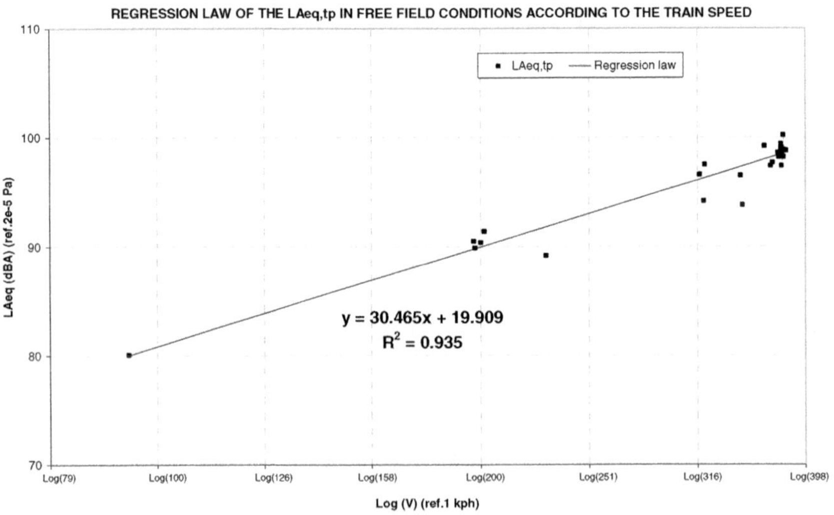

Fig. 5. Regression law of the $L_{pAeq,tp}$ in free-field conditions according to the train speed (color figure online)

The value of the k factor is around 30 that is representative of the rolling noise increase factor. For commercial high-speed trains in France, the rolling noise remains the main source below 320 km/h [4]. In this particular case, the wheels of the test train are in very bad conditions probably due to numerous braking tests and the track is TSI compliant in terms of roughness and track decay rates.

4.2 Noise Reduction of the Reference Noise Barrier (NB1)

The noise reduction provided by the reference noise barrier is 6dB(A) at 320 km/h and 5dB(A) at 375 km/h. The figure 6 shows the noise reduction according to the frequency in third octave band for the two running speeds.

The noise reduction becomes significant over 300Hz and reaches 7dB(A) around 3KHz. The increase in the noise reduction around 100Hz may be explained by an interaction phenomenon between the train and the barrier. The distance between the train and the barrier is around the wavelength of the noise source at this frequency.

The efficiency of the noise barrier decreases when the train speed increases. This result is validated up to 380km/h. It is probably due to the increased contribution of aerodynamic noise with the train speed, especially around the pantograph and its cavity that are higher than the noise barrier.

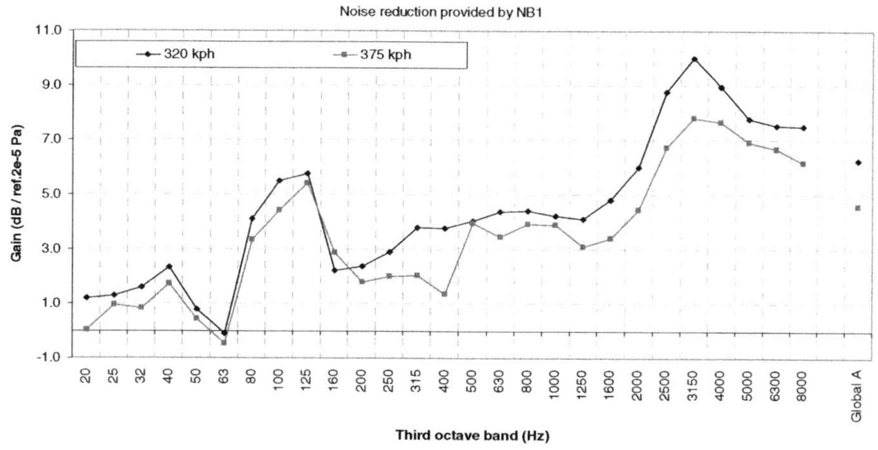

Fig. 6. Noise reduction provided by NB1 according to the frequency (in third octave band) (color figure online)

4.3 Noise Reduction Provided by the Absorbent Material

Configurations NB2, NB4, and NB6 correspond to the configurations NB1, NB3, and NB5 plus absorbent materials onto the top of the noise barrier (path differences remain the same). The table 2 gives a comparison of the noise reduction provided by each noise barrier.

Table 2. Noise reduction provided by NB1 to NB6

Noise barrier	Train speed (km/h)	Noise reduction (dB(A))	Absorbent effect (dB(A))
NB1	320	6.5	+5
NB2	320	11.5	
NB3	285	8	+3
NB4	285	11	
NB5	375	8.5	0
NB6	375	8.5	

The increase in the noise reduction provided by the absorbent material is higher on a classical vertical noise barrier. On a "Y"-shaped noise barrier, the effect of adding absorbent material can be neglected. The diffraction effect is higher than the absorbent one.

4.4 Noise Reduction Provided by the Shape of the Noise Barrier

The improvement in the noise reduction provided by each noise barrier compared with the reference one (NB1) cannot be measured at each running speed. Results are presented in table 2 for a train running at 375 km/h for barriers NB1, NB3, and NB5 and a train running at 320 km/h for barriers NB1 and NB5. For these noise barriers, absorbent material is not on the side of the barrier facing track.

Table 3. Noise reduction provided by NB1, NB3, and NB5

Noise barrier	Train speed (km/h)	Noise reduction(dB(A))	Shape effect (dB(A))
NB1	375	4.5	Reference
NB3	375	6	+1.5
NB5	375	8.5	+4
NB1	320	6.5	Reference
NB5	320	10	+3.5

In table 3, the effect of a single diffracting edge closer to the source (NB3 versus NB1) is not very high, but the "Y"-shaped noise barrier leads to an increase in the noise reduction from 3.5 to 4dB(A). In this case, path differences are higher for the Y shape.

5 Conclusion

A dedicated measurement campaign has been carried out to characterize the noise reduction provided by optimized noise barriers along a TGV high-speed line. The vertical reflective noise barrier has been taken as a reference to compare the measurement results. Absorbent material and multiple diffracting edges confirm their own potential to increase the efficiency of the noise barrier. Indeed, the effect of the absorbent material increases the noise barrier efficiency by an additional 4 to 5 dB for current commercial running speed (320 km/h). Moreover, at higher commercial speeds (for example 375 km/h), multiple diffracting edges provide a noise barrier efficiency equal to or even better than a "classical" noise barrier for current commercial running speed.

Nevertheless, these solutions are limited at higher running speed when aerodynamic noise sources located on the roof of the train cannot be neglected.

On the one hand, a solution consists in increasing the height of the noise barriers for high-speed line if the commercial speed of trains increases. On the other hand, aerodynamic noise can be tackled at source by optimizing the train shape, by designing skirts, and by using active or semi-active solutions to constrain the air flow.

References

[1] Collaborative project "Euroécran", final report (1998)
[2] Nagakura, K., Kitagawa, T.: Study on effective shapes of sound barriers for Shinkansen. RTRI report 16(12) (2002)
[3] Berkhoff, A.P.: Control strategies for active noise barriers using near-field error sensing. JASA 118(3) (September 2005)
[4] Gautier, P.-E., Poisson, F., Letourneaux, F.: High Speed Trains external noise: a review of measurements and source models for the TGV case up to 360 km/h. In: Proceedings of the WCRR 2008, Seoul (2008)

‘Noise Mapping’ of a Railway Network: Validation and Use of a System Based on Measurement of Axlebox Vibration

E. Bongini[1], S.L. Grassie[2], and M.J. Saxon[3]

[1] SNCF, I&R Department, 45 rue de Londres, 75379 Paris, France
Tel.: +33 1 5342 2356
e-mail: estelle.bongini@sncf.fr
[2] RailMeasurement Ltd, Abbauernring 1, 30900 Wedemark, Germany
Tel.: +49 5130 375775
e-mail: stuart.grassie@railmeasurement.com
[3] RailMeasurement Ltd, Barrington Road, Orwell, Royston, UK SG8 5QP
Tel.: +44 1223 208791
e-mail: martin.saxon@railmeasurement.com

Summary

A series of tests have been undertaken on the French railway system to examine whether a commercially available system based on axlebox accelerometers offers a viable method of measuring amplitudes of railhead roughness that are significant for wheel/rail rolling noise. The HSRCA system that was tested gives repeatable measurements and a good estimate of acoustic roughness provided the measuring speed is similar to that at which the system is calibrated and the dynamic behaviour of the trackform is similar to that on which the system is calibrated.

The tests described below will demonstrate tentatively that the HSRCA offers a promising means of undertaking noise monitoring of a railway network.

1 Introduction

The issue of environmental noise in Europe is addressed specifically by the European Commission in Directive 2002/49/EC [1], which states amongst other things, the desirability of determining ‘exposure to environmental noise through noise mapping, by methods of assessment common to the Member States’. To this end, the Research and Innovation Department of SNCF, the French national railway operator, is undertaking the LECAV project, which has the following objectives:

- to provide track roughness mapping of the entire French railway network as a critical step towards the ‘noise mapping’ anticipated by ref [1]
- to monitor acoustic performance of the track for a noise grinding policy.

T. Maeda et al. (Eds.): Noise and Vibration Mitigation for Rail Trans. Sys., NNFM 118, pp. 505–513.
springerlink.com

Longitudinal irregularities can be measured directly over short lengths of track, but this is impractical for measurement of a network of even modest size, let alone a system with thousands of kilometres of track. There are essentially two methods of indirect measurement of irregularities that offer the potential of undertaking such a survey of track roughness:

- measurement of noise on-board a vehicle (an example of which is the Schallmesswagen, or SMW, used by the German railway system, DB [2])
- indirect measurement of railhead irregularities using axlebox accelerometers, as proposed originally by Lewis and Richards [3] and used by several others since. Such a system is, for example, now used routinely by Banverket to provide corrugation information that is used amongst other things to plan rail grinding [4].

The principles underlying several techniques of measuring longitudinal irregularities, including axlebox accelerometers, are presented in ref [5], which discusses also the advantages and disadvantages of the different techniques.

The purpose of the first phase of the LECAV project was to examine the potential of measuring systems that are commercially available and that can be fitted to a conventional vehicle. To this end, a textbook example of a validation and testing exercise was undertaken on the French railway system in November 2009 using two sets of equipment: one based on noise measurement that is similar to the SMW system [2] and the HSRCA (High Speed Rail Corrugation Analyser), produced by RailMeasurement Ltd, which is based on axlebox accelerometers. Some results from an early version of the HSRCA, which was used for routine measurement of corrugation and welds on the Australian National railway system, are presented in ref [6]. This paper presents results that were obtained from the HSRCA system and discusses the potential of this system for routine measurement of longitudinal irregularities and acoustic roughness for the purpose of 'noise mapping' of a railway network.

2 HSRCA within LECAV Test Campaign

The HSRCA was initially developed to provide routine measurements of the severity of corrugation and discrete railhead irregularities (in particular, welds and joints), and thereby provide the information required to plan a campaign of weld straightening and rail grinding. The HSRCA supplied to Australian National was used for this purpose for many years [6].

The HSRCA samples axlebox accelerations on both rails simultaneously at a rate of 8000 samples per second, giving a measurement interval of approximately 4mm at the maximum recommended running speed of 120km/h. Vehicle speed and relative position are also measured using a digital tachometer signal supplied from the vehicle. The HSRCA software was designed to accommodate a longest wavelength component corresponding to a frequency of 5Hz and a shortest wavelength component corresponding to a frequency of 4000Hz.

The typical output of the system to date has been the severity of rail corrugation, shown as RMS amplitudes in 5 wavelength ranges as functions of distance, and 'percentage exceedences' of these RMS amplitudes, which are based on a value for each wavelength range that is selected by the user. The severity of discrete irregularities, if desired, is related to the maximum wheel/rail contact force, normalised to a specified speed (typically 80km/h).

For the LECAV measurement campaign, the HSRCA system was modified to provide one-third octave spectra of railhead roughness, which could be compared with limits such as that proposed in ISO 3095:2005 [7]. These spectra were calculated every 10m for both rails for 100m of track. The first phase of the test, described here, was designed to test, as simply as possible, the repeatability of the measuring system, the effect of measuring speed on reproducibility and the accuracy in order to assess whether the equipment was viable for noise mapping of the network. For the test, accelerometers were simply attached to studs that were fixed with epoxy adhesive to the axleboxes, with cables into the coach. All of the test equipment was transported to site by two people using public transport, fitted in less than a day and removed in only a few hours. The test coach formed part of a consist that was used for other test purposes (Figure 1).

Fig. 1. Consist for LECAV tests (colour figure online)

In order to assess repeatability, reproducibility and accuracy, the test consist ran from Paris to Orange in the south of France (day 1), three times around a test loop passing through Avignon (day 2), and then back to Paris (day 3). The consist ran over six sites, each of 100m length, which had been marked out in advance with transponders, to give an accurate indication on board the train of both ends of each site. A direct measurement of the roughness at all six sites had been made in advance using the Müller-BBM straight-edge instrument following the protocol in ref [7]. Two reference sites, on the northbound and southbound lines between Paris and Orange, had also been measured using the CAT (Corrugation Analysis Trolley). SNCF provided spectra for the two reference sites to calibrate the HSRCA. With the calibration thus made, the HSRCA was used to provide one-third octave spectra for the other test sites and test conditions. The test consist ran over reference sites 1 and 2 at 120km/h and 100km/h, respectively.

There were two other test sites on the northbound and southbound lines near Pierrelatte. On one of these sites, there were rail dampers, while there were no dampers on the rail at the adjacent site. For reasons of space, no results are presented for these sites.

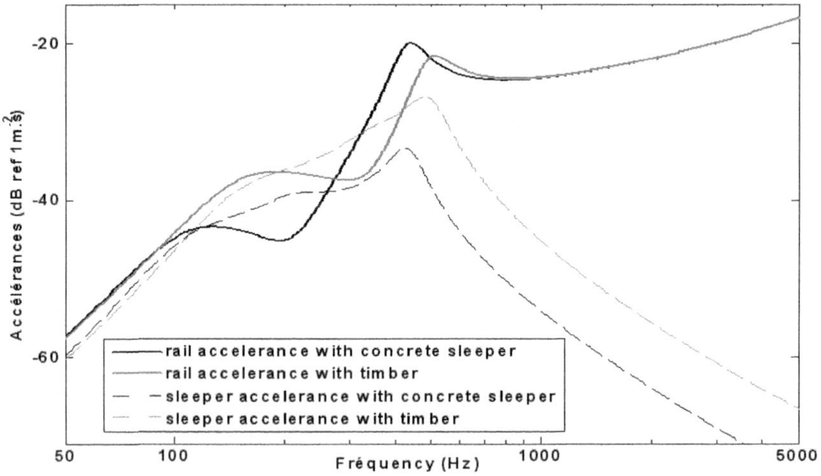

Fig. 2. Rail and sleeper accelerances, for concrete and wooden sleeper track. Track modelled with a Timoshenko beam continuously supported with a double stiffness layers: rail UIC60,ballast stiffness equal to 100MN/m^2, railpad stiffness equal to 300 MN/m^2, sleeper masses: concrete: 245kg timber: 96kg

The principle underlying equipment such as the HSRCA is that a transfer function relating axlebox acceleration to railhead irregularities is used to estimate irregularities from a direct measurement of axlebox accelerations. The accuracy of the estimate therefore depends critically upon the transfer function. The LECAV test was a particularly severe test of the influence of the transfer function, since the two reference sites and four test sites gave four significantly different sets of dynamic behaviour (even without the inevitable variation in ballast properties):

- timber sleepers, RN rail fastenings, 4.5mm railpad: reference site 1
- concrete sleepers, Nabla fastenings, 9mm railpad, no rail dampers: reference site 2, test sites 1 and 2
- concrete sleepers, Nabla fastenings, 9mm railpad, rail dampers: test site1
- concrete sleepers, RN fastenings, 4.5mm railpad:, test site 3

The sleeper alone significantly influences the transfer function for most of the frequency range of interest, i.e., 50-1500Hz, as illustrated in Figure 2. This shows the calculated transfer function between acceleration and contact force for a continuous track model in which the rail is modelled as a Timoshenko beam, the ballast and railpads as damped elastic layers and the sleepers as rigid bodies.

3 Test Results

3.1 Calibration of Equipment

Although in principle, the HSRCA can be calibrated using a single site for which 'reference' measurements are available, it is usually desirable to have at least two sites with extremes of corrugation, i.e., pronounced corrugation and very little

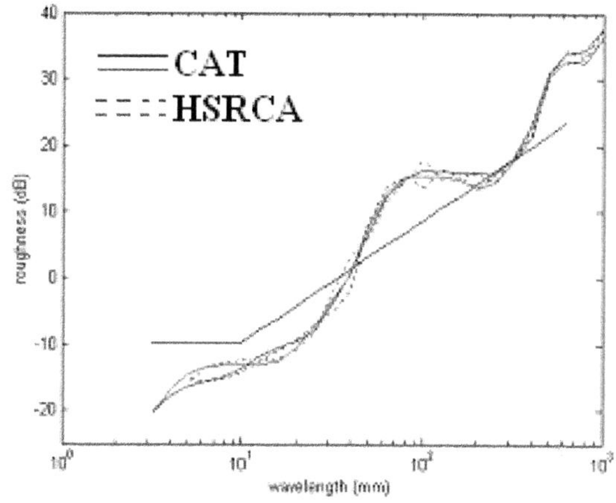

(a)

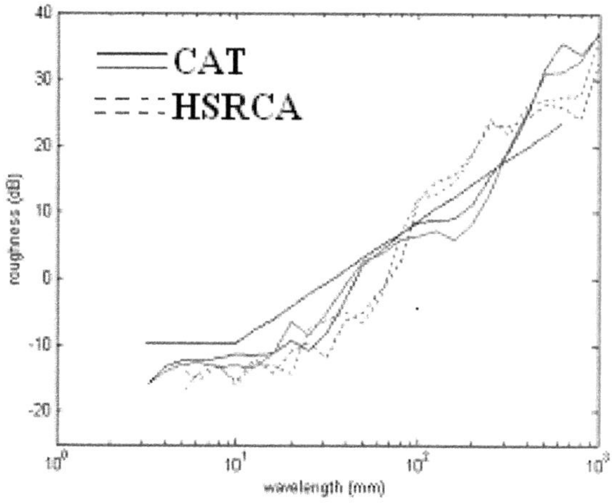

(b)

Fig. 3. Correlation of HSRCA and direct measurements of railhead roughness at reference sites (colour figure online)

corrugation. This helps to validate the HSRCA for measuring both large and small irregularities. In the present case, reference site 1 had moderate corrugation while reference site 2 was extremely smooth. However, since there was also a different trackform at the two sites (Section 2), the calibration was undertaken only for reference site 1 with reference site 2 used to check the results. The correlation of HSRCA and direct measurements for the two reference sites is shown in Figure 3. Since most of the measurements were made on test sites with concrete sleepers, it may have been better to use reference site 2 for calibration.

The principal difference between HSRCA and direct measurements of roughness at reference site 2 is for wavelengths of about 35-300mm. For the speed of 102km/h at this site, this corresponds to a frequency range of 100-800Hz. It is apparent from Figure 2 that the most likely cause of this difference is the influence on the transfer function of the different sleepers at the two reference sites.

3.2 Repeatability; Correlation of Direct and HSRCA Measurements

To assess repeatability of the system, three pass-bys were made at nominally the same speed of 120km/h over test site 3, between Miramas and Cavaillon (Figure 2). HSRCA and direct measurements of railhead roughness are shown in Figure 4.

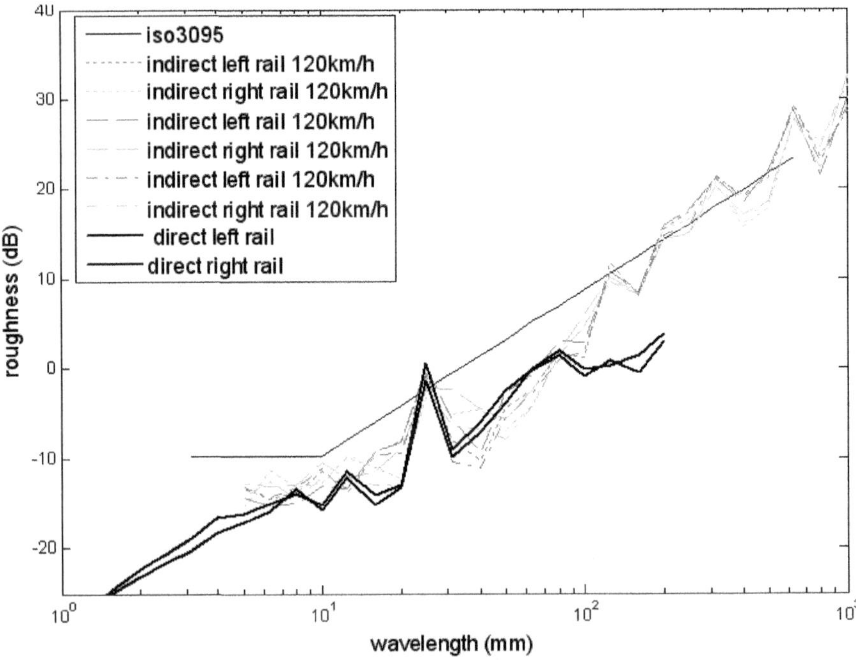

Fig. 4. Repeatability of measurements at 120km/h and comparison of HSRCA and direct measurements of railhead roughness, test site 3

Repeatability of the HSRCA measurements is very good throughout the wavelength range shown. The difference is typically no more than 3dB, with a maximum of 7dB in a few frequency bands around 30mm. The direct measurements (made with the MBBM instrument) are limited to 200mm wavelength. Correlation between HSRCA and the direct measurements is generally good at the shorter wavelengths, with the periodicity from grinding marks at 25mm shown particularly well. However, the HSRCA overestimates roughness at longer wavelengths. This is consistent with the different sleepers at test site 3 compared with those at the reference site used for calibration (Section 3.1).

3.3 Influence of Speed

To examine the effect of speed on the measurements, test runs were made at speeds of about 60km/h, 120km/h and 160km/h over test site 2, between Avignon and Miramas (Figure 2). HSRCA and direct measurements of railhead roughness are superposed in Figure 5.

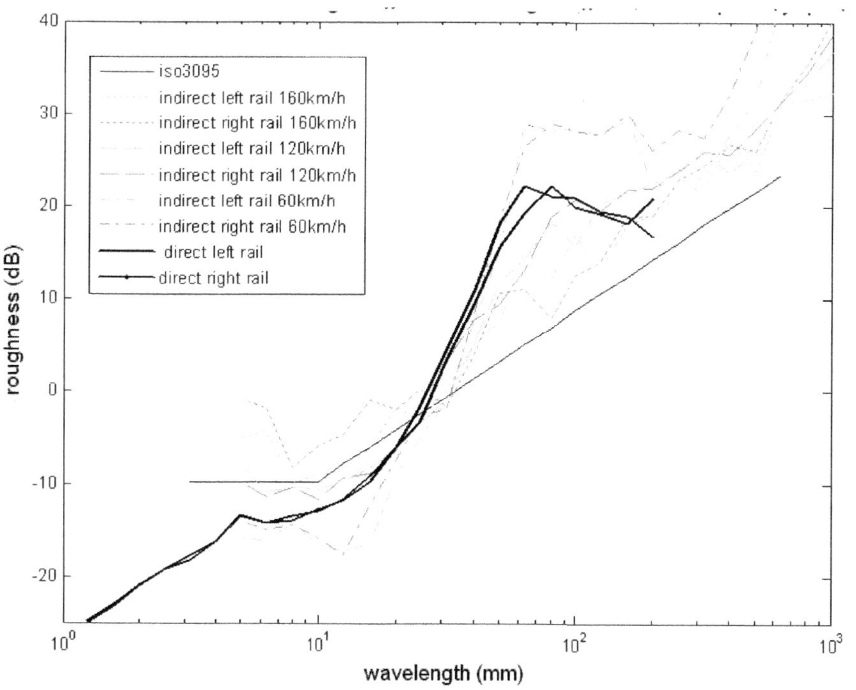

Fig. 5. Effect of speed on HSRCA measurements and comparison of HSRCA and direct measurements of railhead roughness, test site 2

There is reasonable correlation of measurements at different speeds in the 20-60mm wavelength range, with an average deviation of about 3dB. At other wavelengths, the measurements differ by as much as 20dB from one another depending on running speed. The difference between measurements made at

60km/h and 120km/h is greater than that between those made at 120km/h and 160km/h.

The HSRCA measurements at 120km/h again tend to underestimate roughness at wavelengths of 30-100mm and overestimate it at longer wavelengths, which is consistent with the influence of the different sleepers (Section 3.1). This site also had more resilient, 9mm railpads, which would be a further influence on this mid-frequency range.

4 Conclusions

The HSRCA corrugation-measuring system produced by RailMeasurement Ltd gives repeatable measurements of rail corrugation and acoustic roughness in the wavelength range 10-1000mm. The accuracy appears also to be satisfactory provided measurements are made at a similar speed to that at which the equipment is calibrated and that the dynamic behaviour of the trackform does not vary greatly from that on which the system is calibrated. Although the system is somewhat sensitive to both speed and dynamic behaviour of the trackform, there is nevertheless impressive correlation of direct and indirect measurements at a level of microns. This would be extremely difficult to obtain with any other technology that could measure at line speeds.

The effect of sensitivity to speed and dynamics of trackform could be alleviated in practice by attempting to run at a more constant speed than the almost threefold variation examined here and by calibration of the system for the predominant trackform on which it would be used. The possibility of having different calibrations for different trackforms could be examined.

The tests described here demonstrate tentatively that the HSRCA can be used to measure irregularities of an amplitude that is significant for acoustic work, and accordingly that it offers a promising means of undertaking noise monitoring of a railway network.

References

[1] Directive 2002/49/EC of the European Parliament and of the Council of 25 June 2002 relating to the assessment and management of environmental noise, OJEC 189/12, July 18 (2002)

[2] Asmussen, B., Onnich, H., Strube, R., Greven, L.M., Schröder, S., Jäger, K., Degen, K.G.: Status and perspectives of the 'Specially Monitored Track. In: Procs of 8th International Workshop on Railway Noise, Buxton, UK, pp. 575–583 (2004)

[3] Lewis, R.B., Richards, A.N.: A new method for the routine measurement of railhead corrugation. Rail International, 37–41 (February 1986)

[4] Berggren, E.G., Li, M.X.D., Spännar, J.: A new approach to the analysis and presentation of vertical track geometry quality and rail roughness. Wear 265, 1488–1496 (2008)

[5] Grassie, S.L.: Measurement of railhead longitudinal profiles: a comparison of different techniques. Wear 191, 245–251 (1996)
[6] Grassie, S.L.: Corrugation on Australian National: cause, measurement and rectification. In: Procs. of 4th International Heavy Haul Railway Conference, Brisbane, The Institution of Engineers, Australia, National Conference Publication No. 89/12, pp. 188–192 (1989)
[7] EN ISO 3095: 2005. Railway applications – acoustics – measurement of noise emitted by railbound vehicles. CEN, European Committee for Standardisation, Brussels (August 2005)

sonRAIL – The New Swiss Calculation Model for Railway Noise

M. Hecht[1], J.M. Wunderli[2], T. Thron[3], and D. Sehu[4]

[1] TU Berlin, Department of Rail-vehicles,
Office SG14, Salzufer 17-19, D-10587 Berlin, Germany
Tel.: +49 314 25150; Fax: +49 314 22529,
e-mail: markus.hecht@tu-berlin.de
[2] EMPA, Swiss Federal Laboratories for Materials Science and Technology,
Laboratory for Acoustics, Ueberlandstrasse 129
CH-8600 Duebendorf, Switzerland
Tel.: +41 44 823 47 48; Fax: +41 44 823 47 93,
e-mail: jean-marc.wunderli@empa.ch
[3] PROSE GmbH,
Agricolastraße 28, D-10555 Berlin, Germany
Tel.: +49 30 3974 4975; Fax: +49 30 39063972,
e-mail: Thomas.thron@prose-deutschland.de
[4] Federal Office for the Environment FOEN, Noise Abatement Division
CH-3003, Bern, Switzerland
Tel.: +41 31 322 68 93; Fax: +41 31 323 03 72,
e-mail: fredy.fischer@bafu.admin.ch

Summary

The new Swiss calculation model for railway noise called sonRAIL is presented. The sound propagation model as well as the emission model is designed to simulate fundamental physical processes. Therefore, they allow a reproduction of current situations and prediction of the effects of different mitigation measures. The model precisely calculates emission levels, taking into account all relevant variables. The emission levels are determined as sound power spectra for five source heights. The rolling noise is calculated with wheel and rail roughness spectra. The influence of different superstructure types is taken into account. The propagation model considers the direct sound propagation, meteorological effects, and reflections at buildings, walls, and other rigid surfaces, and diffuse reflections at forest edges and cliffs are described. As a first broad application of the model, the results of the sound exposure along the Gotthard railway line in Switzerland are shown.

1 Introduction

sonRAIL is the new Swiss calculation model for railway noise, designed to set new standards in the accuracy of sound emission prediction [0]. The sound propagation

T. Maeda et al. (Eds.): Noise and Vibration Mitigation for Rail Trans. Sys., NNFM 118, pp. 515–522.
springerlink.com

model as well as the emission model is designed to simulate the fundamental physical processes, including reproduction of current situations, prediction of the effects of different mitigation measures. sonRAIL was funded by the Swiss federal office for the environment FOEN and was completed in three years, from 2007 until autumn 2009. It was developed by an interdisciplinary project team composed of several enterprises and institutions: Empa, Swiss Federal Laboratories for Material Testing and Research, Laboratory of Acoustics; Technical University of Berlin, department of rail vehicles; PROSE, measuring and testing; LCC Consulting.

2 Emission Model

The calculation model of sonRAIL is composed of an emission model that describes the generation of railway sound and a propagation model. Both employ one-third-octave bands from 100 Hz to 8 kHz. The emission model yields sound power levels for five predefined source heights along the vehicle surface for each vehicle, depending on infrastructure and operation conditions [1]. Track sections with constant properties, i.e., track superstructure, traveling speed, traffic volume, and composition, are combined into line sources that can be described by their total sound power. The structure of the emission model is shown in Fig. 1. The emission model describes rolling noise as the primary sound source based on roughness spectra of the wheel and rail. First, an effective roughness spectrum is derived using a contact patch filter. In combination with train speed and type-dependent transfer functions for wheel and track, the total rolling noise can be calculated. Special solutions are implemented for switches, several types of bridges and curved tracks.

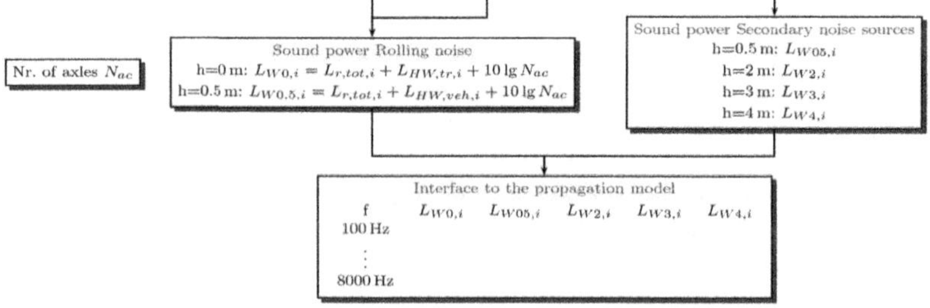

Fig. 1. Diagram showing the procedure for calculating emissions for primary and secondary noise sources [1]. (color figure online)

For secondary sound sources such as traction noise, gear noise, or aerodynamic noise, sound power spectra have been gathered and stored in a database together with a speed dependency factor.

3 Emission Measurements

Extensive measurements have been performed to collect model data for the majority of vehicles and track systems that are currently in use in Switzerland [2]. Pass-by-measurements carried out according to EN ISO 3095 of totally 15'000 vehicles with train speeds ranging from 40 to 200 km/h were performed at 18 measurement sites. The attenuation rate at each measurement site was described according to EN 15461, the local rail roughness according to TSI Noise, and the local sound propagation conditions, determined with loudspeaker experiments. Pass-by noise levels of trains were measured at sound pressure level measurement sites. Additionally, a line array was used to gather information on secondary sound sources. Additionally, two test trains were assembled for the purpose of this measurement campaign with known wheel roughness. In Fig. 2, the test train composed of 3 D-braked, 4 K-braked passenger cars, and 3 4-axle, 3 2-axle freight wagons is shown. For every single vehicle pass-by the pass-by level, the total roughness of the rail-wheel contact patches, the indirect wheel roughness, and the vehicle transfer functions were measured.

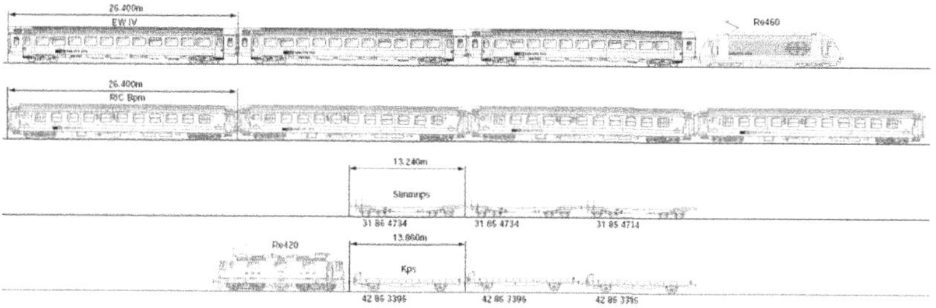

Fig. 2. The sonRAIL test train was measured simultaneously at 12 measurement sites

4 Propagation Model

The propagation model is organized in four different modules that are run independently. The module 'Basic' is mandatory for each calculation, because it performs a calculation of direct sound propagation under the assumption of a homogenous atmosphere. The model is basically an implementation of the ISO Standard 9613 [3], [4]. The most important deviation from the standard concerns the ground effect calculation. Ground reflections are calculated for spherical waves over flat and homogenous ground [5], [6]. The solution is extended to uneven terrain and varying ground properties using a Fresnel-zone-approach and accounts for the coherence loss between direct and reflected sound as a function of frequency and propagation distance. In the second module, meteorological effects on sound propagation are accounted for. Additional input data include vertical profiles of wind speed, temperature, and humidity. While temperature and humidity affect atmospheric absorption, wind and temperature gradients with height in combination with a wind direction are used to derive effective sound speed profiles. Based on these sound speed profiles, a ray tracing algorithm is applied to estimate changes in shielding effect and the evolution of acoustical shadow zones as a consequence of sound propagation along curved rays [7].

The third and fourth module yields independent contributions, one for reflections at buildings, walls, and other rigid surfaces, and one for diffuse reflections at forest edges and cliffs. The model for reflections at buildings, walls, and other rigid

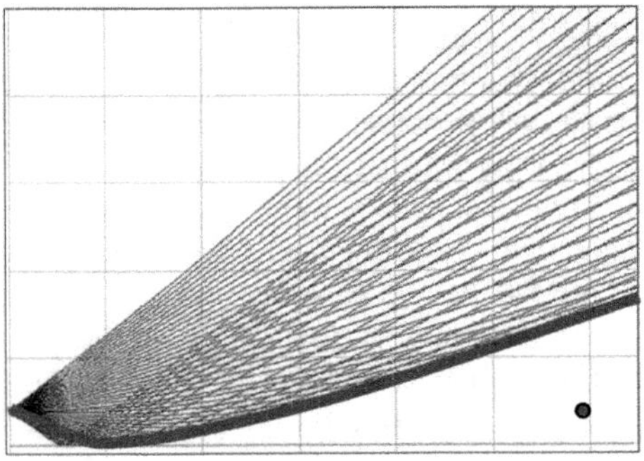

Fig. 3. Meteorological effect in shadow zones. The ray that comes closest to the receiver is highlighted. Together with few geometry parameters, it determines the attenuation to the receiver. (color figure online).

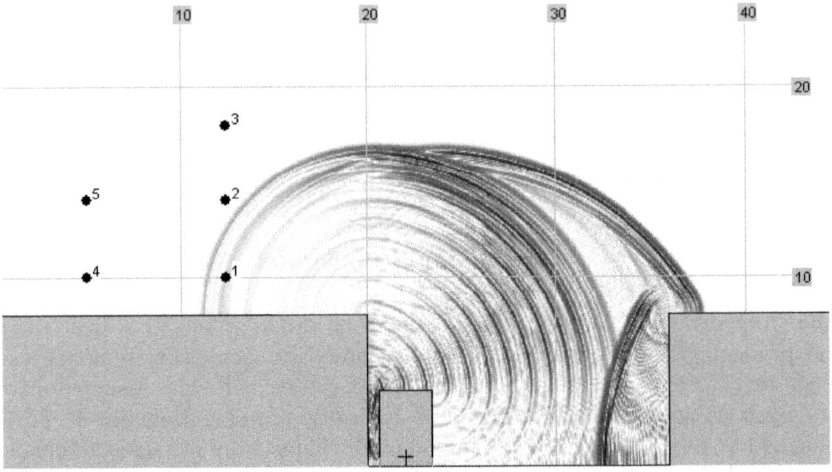

Fig. 4. Sound propagation in a railway line cutting showing multiple reflections between lateral walls and the vehicle body, reflections on the opposite wall and diffractions on the edges. (color figure online).

surfaces is designed for sound fields in urban environments. The calculation procedure is based on two analytical solutions of the reflection problem, one for coherent reflections (mirror-reflections) and one for scattering [8]. For railway noise, the ground effect close to the source shows a distinct behavior as a consequence of the specific properties of the ballast bed. Therefore, an extended ground effect model is implemented for this situation [9]. Also typical for railway noise are situations with hard surfaces in close vicinity of the vehicles, leading to additional reflections. This is, for example, the case for tunnel openings or railway line cuttings. To reduce the calculation effort, these multiple reflections are not dealt with within the reflection module but are treated with empirical formulas [10], [11].

5 Assessment of the Model Accuracy

The accuracy of the emission and propagation models was assessed separately. The analysis has been performed on the assumption that all relevant input parameters are known with an accuracy that is representative for real life conditions. The accuracy of the emission model has been determined on the basis of data collected during the measurement campaign described in section 2. Comparisons have been performed based on pass-by measurements at 7.5 m distance according to ISO 3095.2 [12]. For the emission model, an average mean quadratic error of 0.4 dB(A) was derived. The validation of the propagation model was performed on the basis of 119 propagation situations with either measurements, scale-model experiments, or

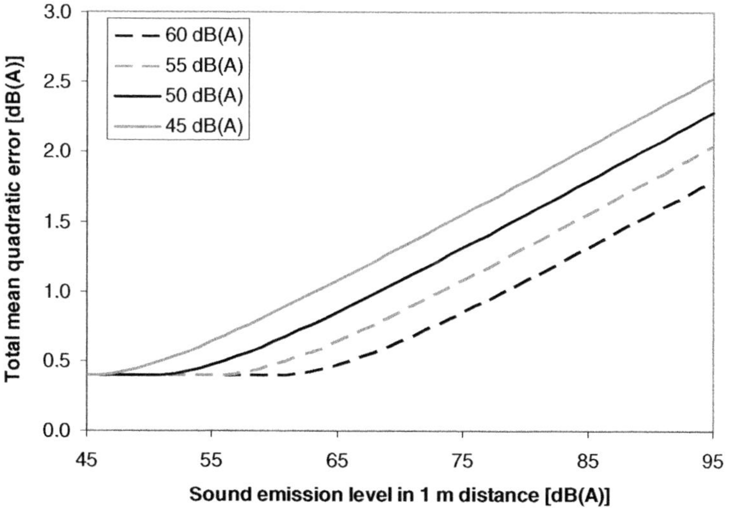

Fig. 5. Overall model uncertainty for the prognosis of A-weighted equivalent continuous sound level Leq in dependence of the sound emission level at a reference distance of 1 m. Depicted are four emission levels that correspond to typical noise limits for areas with only housing.

calculations with a FDTD-Model as a reference. A representative mean quadratic error of 5% of the propagation attenuation in dB was derived. The reflection models for cliffs and buildings thereby yielded a slightly smaller uncertainty than the direct sound calculation. This greater accuracy is due to the various contributions that are incorporated in the reflected sound exposure level, and that these numerous contributions average out uncertainties of single sound propagation paths. This conclusion also holds for reflections from the forest. Nevertheless, the uncertainty is significantly greater for these calculations as the relevant properties of individual forests vary much more than those of buildings or cliffs. The uncertainty of the resulting sound exposure calculations is composed of the uncertainty of the emission model and the uncertainty of the sound propagation calculation. Fig. 5 gives the resulting overall model uncertainties for four A-weighted equivalent continuous sound level *Leq* that correspond to typical noise limit values in housing areas. The model uncertainty remains below 2.5 dB(A) even in the vicinity of railway lines with high sound emissions.

6 Implementation and Application

As the accuracy of the calculation not only depends on the correctness of the applied algorithms but also on the accuracy of the input data, great attention was also paid to the latter. Most of the necessary data are geo-referenced. Geographical information systems yield potent tools to prepare the data and to present and analyze the results. Therefore, the calculation model was attached to a GIS-platform

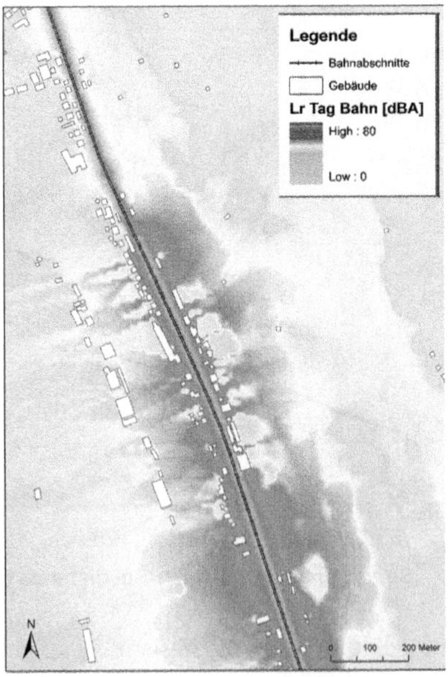

Fig. 6. Noise mapping example along the north-south-corridor (Gotthard line) through Switzerland

called EnvGDB (Environmental GeoDataBase), which is working on the basis of ESRI-products and an Oracle database. While the emission model was directly implemented in the GIS-system, the sound propagation model was designed as a separate application that is started and controlled by the main program but works independently from the rest. The multi-processor-structure splits big projects up into several calculation tasks and permits mapping noise levels over greater areas within reasonable time. The propagation calculations are performed on a Linux-Cluster at Empa called Ipazia. Ipazia currently consists of 46 compute nodes and a total of 248 cores and will be enlarged further in the future. The model is designed to give calculation results with high accuracy for open-track railway situations. Fig. 6 shows the calculation results for a section of the main Gotthard line. The model can be used to assess existing situations or to evaluate different mitigation measures. Projects from single receiver positions up to noise mapping for greater areas can be handled. As a first broad application of the model, the sound exposure along the north-south-corridor through Switzerland will be calculated.

Acknowledgements

The investigations presented were carried out within the framework of the development of the new Swiss sonRail rail noise model and were financed by the Swiss Federal Office for the Environment.

References

[1] Thron, T., Hecht, M.: The sonRAIL emission model for railway noise in Switzerland. Acta Acustica united with Acustica. Accepted by Acta Acoustica, April 20 (2010)
[2] Rohrbeck, A., Thron, T.: sonRAIL – Die Erweiterung des schweizerischen Lärmmodells für den Schienenverkehr, ZEVrail Gl. Ann. 132, 95–105 (2008)
[3] Standardization, I.O. o., ISO 9613-1: Acoustics - Attenuation of sound during propagation outdoors, Part 1: Calculation of the absorption of sound by the atmosphere (1993)
[4] Standardization, I.O.o., ISO 9613-2: Acoustics - Attenuation of sound during propagation outdoors, Part 2: General method of calculation (1996)
[5] Chessell, C.I.: Propagation Of Noise Along A Finite Impedance Boundary. Journal of the Acoustical Society of America 62(4), 825–834 (1977)
[6] Delany, M.E., Bazley, E.N.: Acoustical properties of fibrous absorbent materials. Applied Acoustics 3, 105–116 (1970)
[7] Hofmann, J., Heutschi, K.: An engineering model for sound pressure in shadow zones based on numerical simulations. Acta Acustica United with Acustica 91(4), 661–670 (2005)
[8] Heutschi, K.: Calculation of Reflections in an Urban Environment. Acta Acustica United with Acustica 95(4), 644–652 (2009)

[9] Heutschi, K.: Sound Propagation over Ballast Surfaces. Acta Acustica United with Acustica 95, 1006–1012 (2009)
[10] Heutschi, K., Bayer, R.: Sound radiation from railway tunnel openings. Acta Acustica United with Acustica 92(4), 567–573 (2006)
[11] Heutschi, K.: Sound propagation in railway line cuttings. Applied Acoustics 69(12), 1189–1194 (2008)
[12] Standardization, I.O.o., ISO 3095: Railway applications – Acoustics – Measurement of noise emitted by railbound vehicles (2005)

Small Deviations Procedure: A New Way to Introduce Flexibility in the Conformity Assessment of Reference Tracks for Pass-by Acoustic Tests

F. Létourneaux[1], N. Meunier[2], and P. Fodiman[3]

[1] SNCF Agence d'Essai Ferroviaire,
21 avenue du Président Allende, F94407 Vitry-sur-Seine, France
Tel.: +33 1 47 18 82 32; Fax: +33 1 47 18 82 30
e-mail: fabien.letourneaux@sncf.fr
[2] Deutsche Bahn AG, DB Systemtechnik, Völckerstraße 5, D80939 Munich, Germany
Tel.: +49 89 1308 7617; Fax: +49 89 1308 7311
e-mail: nicolas.meunier@deutschebahn.com
[3] SNCF Infrastructure - Department of European Affairs & Sustainable development
34 rue du Commandant Mouchotte, F75699 Paris, France
Tél.: +33 1 53 25 32 69; Fax: +33 1 53 25 32 26
e-mail: pascal.fodiman@sncf.fr

Summary

When measuring the noise emitted by a train passing at constant speed, a significant contribution may come from the track. Hence, the accurate assessment of the noise emission of railway vehicles requires a tight control of the track acoustic parameters. The concept of reference track was developed to meet this requirement. However, the need to carry out measurements on tracks which strictly comply with the limits of track decay rates and rail acoustic roughness may lead to undue stringent requirements: small deviations from these limits may not have a significant effect on the measured noise value. The purpose of this paper is to propose a method which introduces some flexibility in the assessment of conformity of track sections with regard to these specifications.

1 Introduction

When measuring the noise emitted by a train passing at constant speed, a significant contribution may come from the track. Hence, the accurate assessment of the noise emission of railway vehicles requires a tight control of the track acoustic parameters. The concept of reference track was developed to meet this requirement. The purpose of the reference track is to improve (guarantee) the reproducibility of pass-by measurements and limit the acoustic contribution of the track in the total pass-by noise.

T. Maeda et al. (Eds.): Noise and Vibration Mitigation for Rail Trans. Sys., NNFM 118, pp. 523–531.
springerlink.com

The reference track specification includes:

- a list of specifications regarding the acoustic environment of the test section and the track geometry,
- an upper limit for the rail acoustic roughness wavelength spectrum to control the excitation at the wheel-rail contact by the track (see [1]),
- a lower limit for the vertical and transverse track decay rates frequency spectra to control the vibration attenuation and acoustic radiation of the track (see [2]).

The reference track principle was implemented in Europe in the most recent High Speed and Conventional Rail Technical Specifications for Interoperability (TSI see [3]). This principle is not a new concept: it was already introduced, although in a simplified way, in standards related to noise emission and comfort for railway vehicles (see [4] & [5]) since 2001, and it was also used for type testing of road vehicles. However, the need to carry out measurements on tracks which strictly comply with the limits of track decay rates and rail acoustic roughness may lead to undue stringent requirements: small deviations from these limits may not have a significant effect on the measured noise value.

This paper proposes a method which introduces some flexibility in the assessment of conformity of track sections with regard to these specifications.

2 Procedure

2.1 Principle

The proposed method relies on a calculation of a correction to the measured level based on the effect of any deviation of a measured spectrum of acoustic rail roughness and/or decay rates from the specified limit curves. The difference between the corrected pass-by noise level and the measured one is then compared to an acceptance criterion. If the criterion is fulfilled, the acoustic impact of the rail roughness and decay rate deviations is deemed 'small' and the measured pass-by noise level is considered to be valid. This method is train speed dependent.

2.2 Theoretical Approach

All the spectra considered hereafter are one-third octave band spectra, either in the frequency or in the wavelength domain. The rolling noise, which is part of the total noise emitted during passage of a train, can be divided into terms related to the acoustic radiation efficiency and terms related to the excitation at the wheel/rail interface.

$$L_{pAeq,Tp}^{rolling} = \left(L_H^{vehicle} \oplus L_H^{track}\right) + \left(L_r^{rail} \oplus L_r^{wheel}\right) + C_{filter} \tag{2.1}$$

- $L_{pAeq,Tp}^{rolling}$ is the pass-by noise level frequency spectrum,
- $L_H^{vehicle}$ and L_H^{track} are the A-weighted sound pressure level to roughness level transfer functions corresponding, respectively, to the vehicle and the track. L_H characterizes the acoustic radiation of the subsystem,

- L_r^{rail} and L_r^{wheel} are the acoustic roughness frequency spectra of the rail and the wheel,
- C_{filter} is the frequency dependant contact filter due to the wheel and rail profiles and
- $\oplus$ stands for the dB summation ($A \oplus B = 10^{A/10} + 10^{B/10}$)

In this equation, the effect of the rail roughness on the total noise is obvious, whereas the effect of the track decay rate is hidden in the track transfer function.

Assuming a simplified model for the track vibration behaviour where the rail response is the superposition of only one vertical bending wave and one transverse bending wave (see annex A of [2]), the track transfer function may be written as follows:

$$L_H^{track} = L_H^{track,vertical} \oplus L_H^{track,transverse} \tag{2.2}$$

Assuming also that the deviations of the track decay rates do not significantly change the receptance of the railhead vibration, the following relationship of the variations of the track transfer function induced by the track decay rate deviations may be derived:

$$\Delta L_H^{track,i} = 10 \cdot \log_{10}\left(\frac{DR_i^{measured}}{DR_i^{corrected}}\right) \tag{2.3}$$

- $\Delta L_H^{track,i}$ is the correcting spectrum of the track radiation in direction i,
- $DR_i^{measured}$ is the measured track decay rate in direction i, and
- $DR_i^{corrected}$ is the corrected track decay rate in direction i.

Any deviation in the track decay rate may now be related to a rolling noise level change. The three following sections detail the process which leads to an upper bound of the noise level change due to the track decay rates and/or the acoustic rail roughness deviations.

2.3 Data Processing for the Acoustic Rail Roughness Deviation

The processing comprises 2 steps:

- _Step 1_: generate a 'just compliant' corrected spectrum from the measured acoustic rail roughness wavelength spectrum and from the limit spectrum. The corrected acoustic rail roughness spectrum is equivalent to the measured one except in the wavelength bands where the measured spectrum exceeds the limits. It complies with the limit spectrum:

$$\tilde{L}_{r,rail}^{corrected} = \min\left[\tilde{L}_{r,rail}^{measured}, \tilde{L}_{r,rail}^{\lim it}\right] \tag{2.4}$$

$\tilde{L}_{r,rail}^{measured}$ is the wavelength spectrum of the measured acoustic rail roughness;

$\tilde{L}_{r,rail}^{\lim it}$ is the wavelength limit spectrum; and

$\tilde{L}_{r,rail}^{corrected}$ is the wavelength spectrum of the corrected acoustic rail roughness.

- *Step 2*: quantify the deviations in the rail roughness frequency spectrum. First, produce normalized roughness frequency spectra from wavelength spectra and then derive a correcting spectrum corresponding to the difference of the measured and the corrected acoustic roughness spectra:

$$\Delta L_{r,rail} = L_{r,rail}^{measured} - L_{r,rail}^{corrected} \tag{2.5}$$

$L_{r,rail}^{measured}$ is the frequency spectrum of the measured rail acoustic roughness;

$L_{r,rail}^{corrected}$ is the frequency spectrum of the corrected rail acoustic roughness; and

$\Delta L_{r,rail}$ is the frequency correcting spectrum.

2.4 Data Processing for the Track Decay Rate Deviation

The process includes calculating a correcting track transfer function for each direction i (vertical and transverse).

$$\Delta L_H^{track,i} = \min\left[0,\ 10 \cdot \log_{10}\left(\frac{DR_i^{measured}}{DR_i^{\lim it}}\right)\right] \tag{2.6}$$

- $\Delta L_H^{track,i} = 0$ for frequency bands where the decay rate in direction i complies with the corresponding limit curve,
- $\Delta L_H^{track,i} < 0$ for frequency bands where the decay rate in direction i is below the corresponding limit curve.

As the relative weight of the vertical and transverse vibration waves on the total track transfer function is not known, an upper bound of the impact of the decay rate deviations is obtained by:

$$\Delta L_H^{track} = -\min\left(\Delta L_H^{track,vertical}, \Delta L_H^{track,transverse}\right) \tag{2.7}$$

2.5 Assessment of the Noise Level Changes Due to the Deviations

A revised noise spectrum is calculated from the measured noise spectrum, the correcting roughness spectrum and the correcting track transfer function spectrum as follows:

$$L_{pAeq,Tp}^{revised} = L_{pAeq,Tp}^{measured} - \Delta L_H^{track} - \Delta L_{r,rail} \tag{2.8}$$

The rail roughness influences the wheel-rail contact force, whereas the track decay rates have affected the track response. The equation (2.8) shows that these effects are cumulative.

The revised noise spectrum is derived from a simplified process, which assumes that the excess rail roughness and the track decay rate weakness directly affect the total noise. This procedure cannot then be considered to be a prediction method. The revised noise spectrum is the minimum that might have been measured.

An upper bound of the noise impact of the rail roughness and track decay rate deviations can then be derived from the measured and revised noise spectra.

$$\Delta L_{pAeq,Tp} = \bigoplus_i \left\{ L_{pAeq,Tp}^{measured}(f_i) \right\} - \bigoplus_i \left\{ L_{pAeq,Tp}^{revised}(f_i) \right\} \quad (2.9)$$

where $\bigoplus_i \{\ \}$ stands for the dB sum of all the one-third octave frequency bands i.

2.6 Acceptance Criterion

The acoustic rail roughness spectrum of the track shall be considered compliant if the noise level change $\Delta L_{pAeq,Tp}$ due to rail roughness and/or track decay rate deviations is less than or equal to 1 dB. This compliance shall be examined for each train speed.

3 Application to Real Data

The small deviation procedure has been applied on measured data from nearly compliant test track sections. Two different cases are presented below.

3.1 Case of a Conventional Rail Field Test Campaign

The first application concerns the pass-by measurements of wagons equipped with composite brake blocks. The test campaign was carried out in 2004 in CLARO

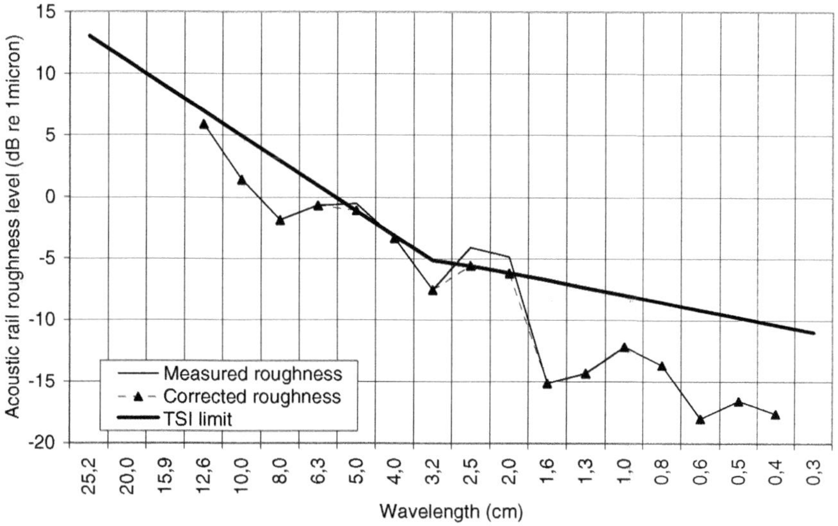

Fig. 1. Acoustic rail roughness of the conventional rail test site

(Switzerland) as part of the UIC project ERS (see [6]). The characteristics of the test site met the requirements of the measurement standard that was in force at that time but did not fully comply with the specifications of the reference track, defined one year later. The deviations concerned both the acoustic rail roughness as shown in Fig. 1 as well as the track decay rates in the vertical direction as shown in Fig. 2.

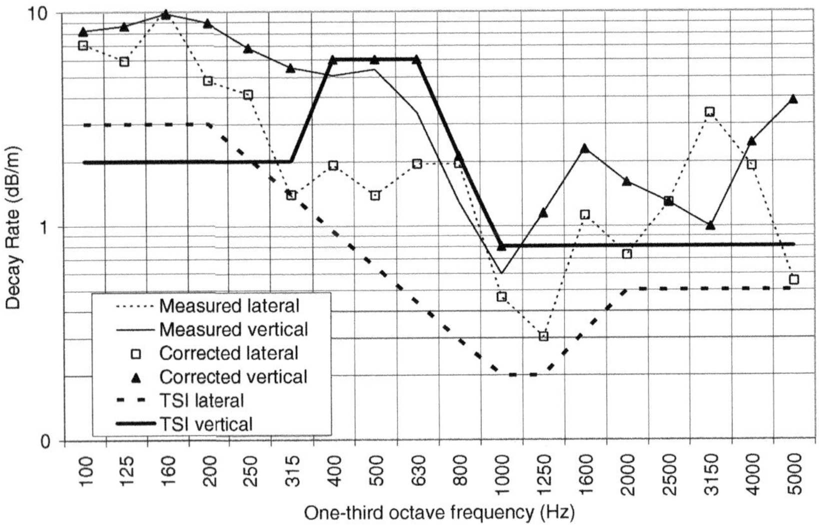

Fig. 2. Track decay rates of the conventional rail test site

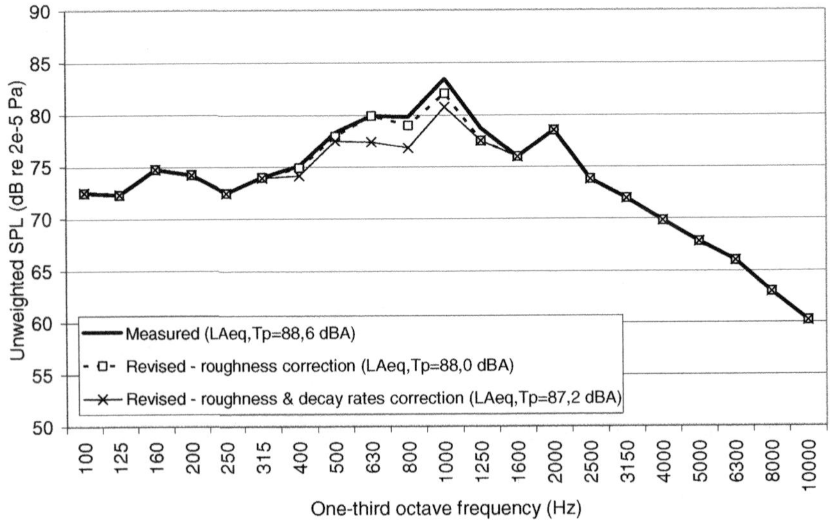

Fig. 3. Pass-by noise spectra of the wagon equipped with LL-blocks

We have applied the small deviations procedure as presented in Section 2 to assess whether the pass-by noise results are strongly affected by the nonconformity of the track against the reference track requirements. The pass-by noise results at 80 km/h supplied in Fig. 3 prove that the deviations in roughness lead to an increase in the overall noise level of up to 0,6 dB, which would be considered as acceptable according to the criterion set in section 2.6. However, when considered in conjunction with the track decay rate deviations, the calculated upper bound of the noise level change grows to a value of 1,4 dB. According to the acceptance criterion, the deviation is not small enough for the track to be considered as compliant with reference track requirements.

3.2 Case of a High Speed Field Test Campaign

The second application concerns the noise measurement during passage of a high speed train at 320km/h on a track with recently ground rail.

The measured track decay rates fulfilled the limit curves, whereas the acoustic rail roughness included a large deviating peak in the 5cm one-third octave wavelength band (see Fig. 4). This peak was due to the presence of periodic grooves on the railhead surface produced by the rotating stones during the rail grinding operation. When applying the small deviation procedure, Fig. 5 shows that such a large nonconformity of the rail roughness does not have a significant effect on the pass-by noise level (less than or equal to 0,4 dB). The calculated noise level change is an upper bound of the actual noise level change of the rail roughness deviation, because it assumed that, in that one-third octave frequency band, the rail roughness dominates the wheel roughness and that the aerodynamic and traction noise do not influence the emitted noise in that particular band. The procedure proposed in section 2 is then a general procedure which can be used at all speeds, even if acoustic sources other than rolling noise are present during the pass-by.

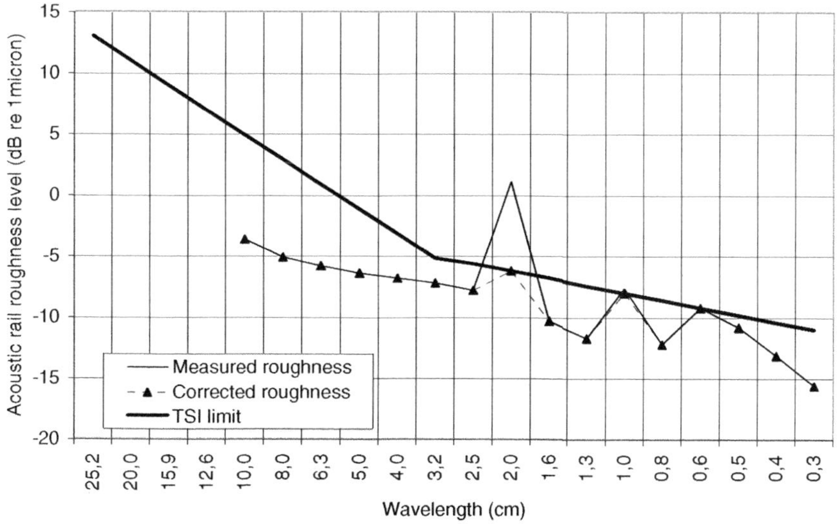

Fig. 4. Acoustic rail roughness of the high speed test site

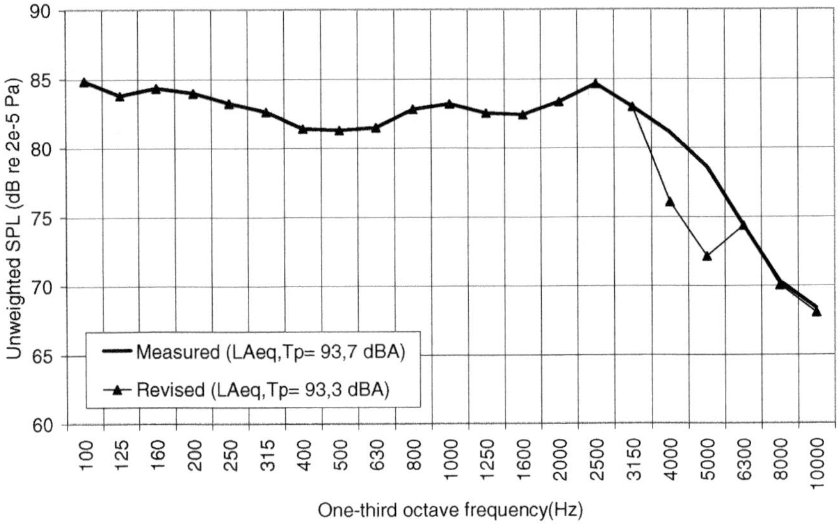

Fig. 5. Pass-by noise spectra of the high speed train

4 Conclusion

A general method to provide flexibility in the conformity assessment of reference tracks was developed and named 'small deviation procedure'. The different tests performed on real data demonstrate its usefulness. The method was proposed, for the rail roughness deviations only, to be introduced both in the TSI noise and the standard under revision ENISO3095 dedicated to the measurement of the noise emitted by railway vehicles.

Considering the limit value of the acceptance criterion, the method may be considered as a tool to adjust the stringency of the reference track requirement (from $\Delta L_{pAeq,Tp}=0$, which means 'no flexibility at all', to $\Delta L_{pAeq,Tp}=\infty$, which means 'complete flexibility', allowing measurement on any track). Conversely, it may also be used to show the potential variations in noise emission due to noisy track, although it should not be considered as a prediction method.

References

[1] EN15610, Railway applications – Noise emission – Rail roughness measurement related to rolling noise generation (2009)

[2] EN15461, Railway applications – Noise emission – Characterization of the dynamic properties of track sections for pass by noise measurements (2008)

[3] Commission decision 2006/66/EC of December 23, 2005 concerning the technical specification for interoperability relating to the subsystem 'rolling stock — noise' of the trans-European conventional rail system. Official Journal of the European Union L37 (49) (2005)

[4] ENISO3095, Railway applications – Acoustics – Measurement of noise emitted by railbound vehicles (2005)
[5] ENISO3381, Railway applications – Acoustics – Measurement of noise inside railbound vehicles (2005)
[6] SNCF test report, UIC project ERS, Noise emitted in the environment during the pass-bys of wagons equipped with composite blocks (LL type), n°AEF-D R 04048 / 04 D-054 (2004) (in French)

Numerical Simulations of Propagation of Bogie Noise in 3D Field by Duhamel's Transformation Using Transient Solutions Calculated by 2D-FDTD Method

M. Hiroe[1,3], S. Ishikawa[2], R. Shiraga[2], and T. Iwase[3]

[1] Kobayasi Institute of Physical Research,
3-20-41 Higashi-motomachi, Kokubunji, Japan
Tel.: +81-42-321-2841; Fax: +81-42-322-4698
e-mail: hiroe@kobayasi-riken.or.jp
[2] Research and Development Center of JR East Group,
[3] Niigata University, Faculty of Engineering, Department of Architecture

Summary

We applied a numerical technique to transform a two-dimensional (2D) solution to a three-dimensional (3D) solution based on Duhamel's transformation method using the transient solution obtained by finite-difference time-domain (FDTD) simulation to calculate the propagation of bogie noise, which is the predominant noise from the lower part of a train. Different shapes of reflective barriers were used in this investigation, and the source model of bogie noise was treated as an array of incoherent line sources. To verify the 3D transformation results obtained using Duhamel's method, we carried out an experiment using a 1/25 3D scale model of a reflective inverted-L barrier. The good agreement between the numerical and experimental results indicates that the 3D transformation technique is a useful tool for precisely predicting the sounds that propagate from incoherent line sources such as bogie noise sources. We also investigated several reflective barriers of different edge shapes and clarified the differences between the shielding effects of the barriers in both 2D and 3D transformed using Duhamel's method.

1 Introduction

In Japan, noise barriers are widely used to reduce noises radiated from the Shinkansen (Bullet train) and conventional trains. Various noise barriers have been developed to control increasing radiated noise caused by the recent and rapid increase in train speeds. These barriers have complicated top shapes and boundary conditions such as absorbing materials or an interference device for achieving greater attenuation of noise than that achieved using rigid straight barriers. Therefore, it is quite difficult to accurately predict the shielding effect of such noise barriers against railway noise, especially the noise from the lower part of trains.

T. Maeda et al. (Eds.): Noise and Vibration Mitigation for Rail Trans. Sys., NNFM 118, pp. 533–540.
springerlink.com

Experiments using scaled models have been conducted for estimating the shielding effect of noise barriers against railway noise [1, 2].

In our previous study [3], we showed that it was possible to approximately calculate the improved effect against bogie noise due to changes of absorbing conditions or barrier shapes using the two-dimensional finite-difference time-domain (2D-FDTD) method. For this research, we applied to a numerical technique to convert from a 2D solution to a 3D solution [4] to calculate the propagation of bogie noise, such as sounds radiated from incoherent line sources of finite length. We compared the numerical results with a scale model experiment to verify the precision of the 3D transformation technique [5]. Moreover, we calculated the shielding effects of several barrier shapes against bogie noise using the 2D-FDTD method and clarified differences between the shielding effects of barriers in 2D and in 3D transformed with Duhamel's method.

2 Numerical Analysis

A distribution of sound sources, an elevated structure with a slab track, train cars, and a reflective barrier were used in this investigation, as shown in Fig. 1. Both experiment and simulation were carried out in a free sound field condition, and the bogie noise sources were treated as an array of incoherent line sources of finite length in this investigation.

2.1 2D-FDTD Simulations

As shown in Fig. 1, the cross-section of the railway structure with the barrier was modeled in a 2D sound field. The sound source was set under the train body at the center of the nearest track (x_s, y_s) and at the height of the railhead, and the initial sound pressure distribution with a Gaussian profile was used. Around the calculation area, a perfectly matched layer was set to produce a free sound field. In this investigation, we used a fourth-order spatial finite-difference scheme [6], to reduce dispersion error, and discrete spatial intervals of 0.008 m and a time step of around 1/128,000 sec in the

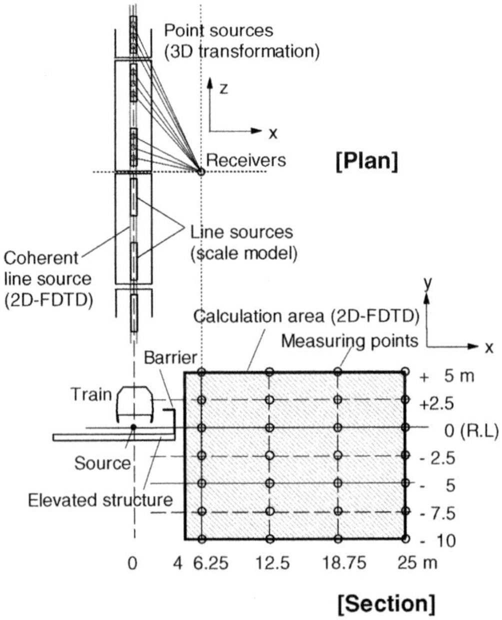

Fig. 1. Geometrical setting of source, receivers, train, and railway structure

2D-FDTD simulation. We obtained the transient solutions $p_{2D}(x, y, t)$ in the calculation area.

2.2 Transformation Technique from 2D to 3D Using 2D-FDTD Results

By including both Fourier transform $\Phi_{2D}(x,y,k)$ and Laplace transform results $\Phi_{2D}(x,y,jk)$ of integrated functions $\varphi_{2D}(x,y,t)$ to Duhamel's transformation, as shown in Eqs. (1) to (4), we obtain the 3D solution $\Phi_{3D}(x, y, z, k)$ for a point source at positions (x, y, z) in Cartesian coordinates with k as wave number ($=\omega/c$, ω is the angular frequency and c is the speed of sound) [4].

$$\Phi_{3D}(x,y,z,k)=\frac{1}{2\pi}\int_{-\infty}^{+\infty}\Phi_{2D}\left(x,y,\sqrt{k^2-\alpha^2}\right)e^{-j\alpha z}d\alpha \tag{1}$$

$$=\frac{1}{\pi}\int_{0}^{k}\Phi_{2D}\left(x,y,\sqrt{k^2-\alpha^2}\right)\cos(\alpha z)d\alpha+\frac{1}{\pi}\int_{k}^{+\infty}\Phi_{2D}\left(x,y,j\sqrt{\alpha^2-k^2}\right)\cos(\alpha z)d\alpha$$

$$\Phi_{2D}(x,y,k)=\int_{-\infty}^{+\infty}\phi(x,y,t)e^{j\omega t}dt \tag{2}$$

$$\Phi_{2D}(x,y,jk)=\int_{-\infty}^{+\infty}\phi(x,y,t)e^{-\omega t}dt \tag{3}$$

$$\phi(x,y,t)=\int_{0}^{t}p_{2D}(x,y,\tau)d\tau \tag{4}$$

Using the reciprocity theorem, we calculated a set of the 3D solutions at source positions (x_s, y_s, z_s) for z_s=0 to 200 m at 1-m intervals at each point in the calculation area. By summing the squared values of sound pressure at the positions (x_s, y_s, z_s) corresponding to each incoherent line source, we obtained 3D results for the bogie noise sources as an array of incoherent line sources [5].

2.3 Scale Model Experiment

The 1/25-scaled models, consisting of a train, an elevated structure and a barrier, were arranged in a hemi-anechoic room. All scale models were constructed with chloroethylene boards to perfectly reflect the sound off the surface. The model represented a 150-m-long train, 5-m-long incoherent line sources, which were positioned at each bogie engine on the nearest track, and 200-m-long barrier, which was set at one end of an elevated railway. We spread absorptive mats on the floor of the experimental room and kept both the temperature and humidity in the room constant to produce an ideal 3D free sound field. The arrangement of the measuring points in the scale model experiments is shown in Fig. 1. We measured 1/3 octave band levels averaged over 10 sec at an actual frequency range from 200 Hz to 2.5 kHz at each measuring point [3].

2.4 Comparisons between Numerical and Experimental Results

The overall A-weighted sound pressure level (SPL) was calculated at frequencies between 200 Hz and 2.5 kHz. The sound source used, which was measured at the vicinity of a passing train, was assumed to have the characteristics of an A-weighted sound power spectrum as shown in Fig. 2.

Figure 3 shows a comparison of spatial distribution of A-weighted SPLs of the 2D-FDTD simulation, 3D transformation, and experimental results. The averaged differences between the numerical and experimental results were 0.2 dB for 3D transformation and 2.9 dB for 2D-FDTD. The 3D transformation results showed good agreement with those of the scale model experiment in the spatial distribution of the overall A-weighted SPL of the bogie noise.

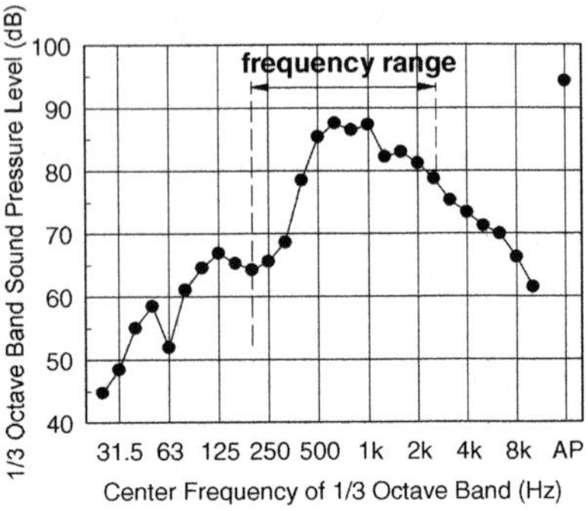

Fig. 2. Frequency characteristics of bogie noise measured at R=1 m from train

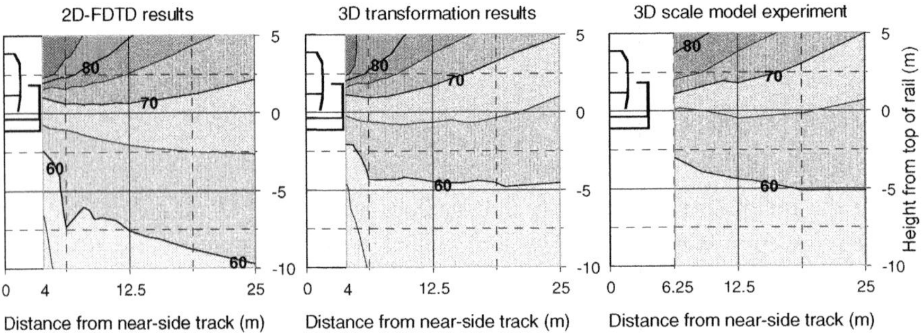

Fig. 3. Comparison of spatial distributions of A-weighted sound pressure level of 2D-FDTD, 3D transformation, and experimental results

Figure 4 shows the frequency characteristics of A-weighted SPLs obtained using the numerical simulations and the scale model experiment at a position R=25 m and H=-10 m. In these frequency characteristics, the 3D transformation results agreed well with the experimental ones. This good agreement indicates that the 3D transformation technique is a useful tool for precisely predicting the sounds that propagate from incoherent line sources, such as bogie noise sources. In the next section, we show applications of the 3D transformation technique for estimating the shielding effects of several different barrier shapes in 3D.

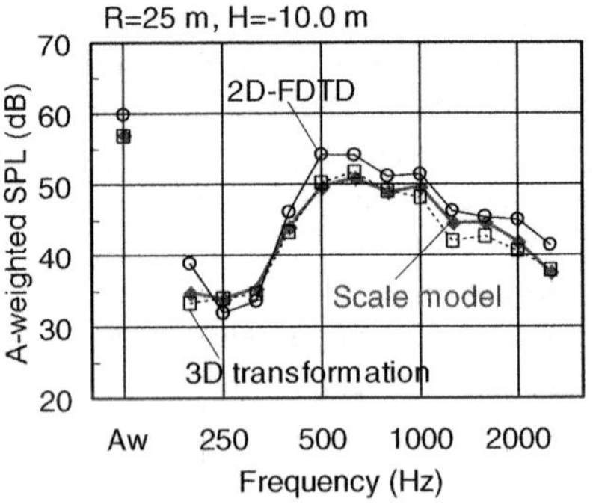

Fig. 4. Frequency characteristics of A-weighted SPL among numerical and experimental results.

3 Applications of 3D Transformation Technique

We show applications of the 3D transformation technique for estimating improved effects, due to changes in the edge shapes of reflective barriers, against bogie noise radiated from a 400-m-long Shinkansen train.

3.1 Four Reflective Noise Barriers with Different Edge Shapes

The cross-section of four reflective noise barriers with different edge shapes (Case (a): straight, Case (b): round, Case (c): T shaped, and Case (d): Y shaped) are shown in Fig. 5. Each noise barrier was arranged at one end of the elevated structure with a slab track, as shown in Fig. 1, and we calculated the transient solutions $p_{2D}(x, y, t)$ in the calculation area by 2D-FDTD simulation under the same conditions of the simulations discussed in Section 2.1.

By including the calculated results of Eqs. (2), (3), and (4) into Eq. (1), we transformed the 2D-FDTD results to 3D solutions for a point source. In a similar way, discussed in Section 2.2, we obtained 3D results for the bogie noise radiated from a 400-m-long Shinkansen train.

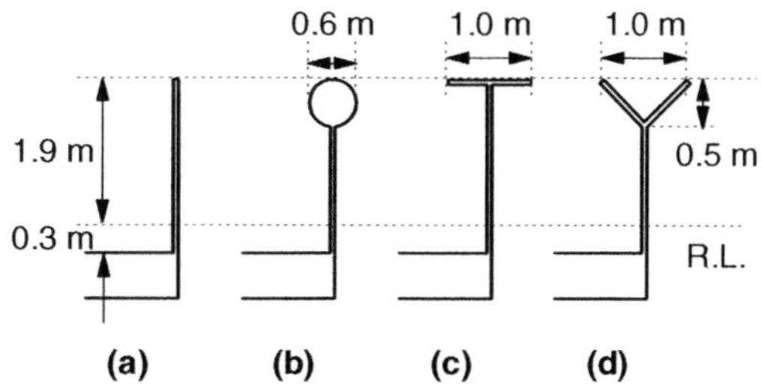

Fig. 5. Four shapes of reflective noise barriers

3.2 Improvement in Shielding Effect

For the four barrier types shown in Fig. 5, the frequency components included in the 1/3 octave band of each center frequency between 200 Hz and 2.5 kHz were calculated from the 2D-FDTD and 3D results. The improved shielding effect resulted from the level differences between the straight barrier, Case (a), and the other barriers, Cases (b) to (d). The overall A-weighted values of the improved effect against bogie noise were calculated at these center frequencies, and the noise source used in this calculation was assumed to have characteristics of an A-weighted sound power spectrum, as shown in Fig. 2.

Figs. 6 and 7 show spatial distributions of the A-weighted SPLs of the bogie noise for Cases (a) and (d), and Fig. 8 shows the improved shielding effect for Case (d), which is difference of level between Cases (d) and (a). In Figs. 6, 7, and 8, the left figure shows the results of the 2D-FDTD simulation, and the right one shows the results of 3D transformation. In Figs. 6 and 7, although spatial distributions of the A-weighted SPLs in 3D are similar to those of the A-weighted SPLs in 2D, the A-weighted values in 2D seem to become larger than those in 3D at the areas of 20 m further away from the center of the nearest track. The level differences of the A-weighted SPLs in 2D and in 3D are less than 3 dB in the calculation area. As shown in Fig. 8, the improved shielding effects in 3D are also similar to those in 2D

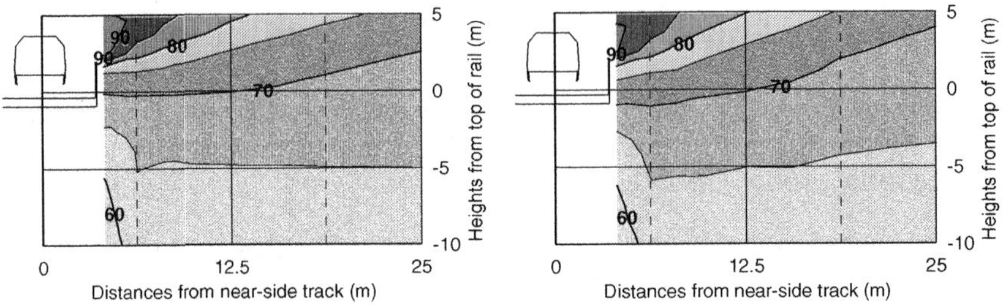

2D-FDTD results 3D transformation results

Fig. 6. Spatial distribution of A-weighted SPLs of bogie noise for Case (a)

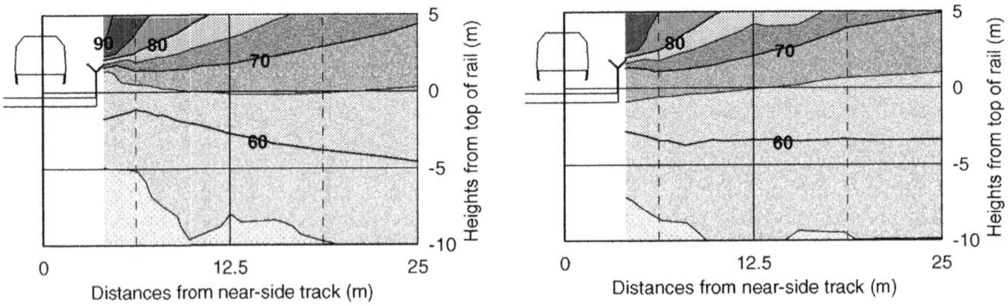

2D-FDTD results 3D transformation results

Fig. 7. Spatial distribution of A-weighted SPLs of bogie noise for Case (d)

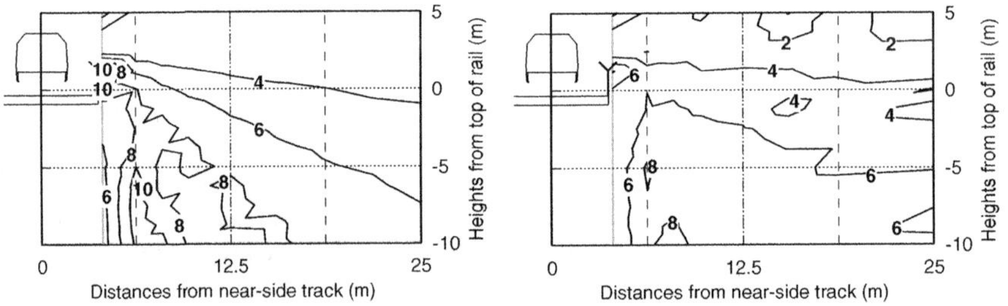

2D-FDTD results 3D transformation results

Fig. 8. Spatial distribution of improved shielding effect for Case (d)

as well as the relationship of the spatial distributions of the A-weighted SPLs in 2D and in 3D.

Figure 9 shows the improved shielding effects at points of 10 m away from the diffracted edge of Case (a) as a function of the angle θ, which is defined in the left side of Fig. 10, for Cases (b), (c), and (d). Figure 11 shows the frequency characteristics of the improved shielding effect both in 2D and in 3D at θ=0 and θ=70 degrees for Cases (b), (c), and (d).

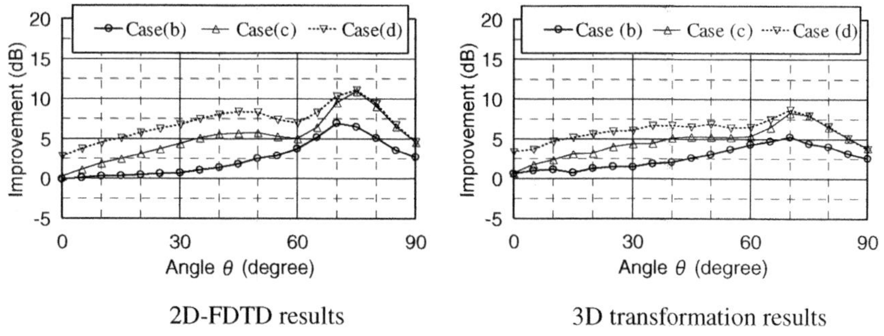

Fig. 9. Improved shielding effect as function of angle θ in Cases (b), (c), and (d)

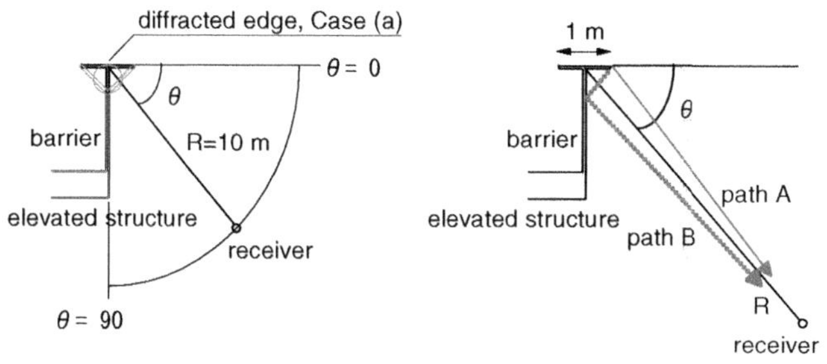

Fig. 10. Definition of angle θ and interference of sounds between A and B

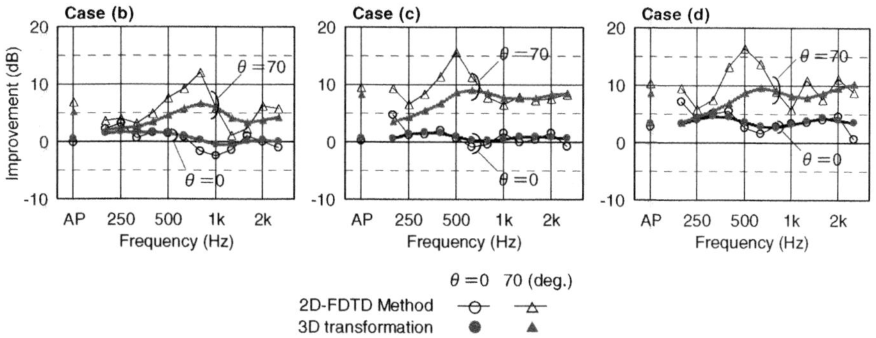

Fig. 11. Frequency characteristics of improved shielding effects in Cases (b), (c), and (d)

From the results in Figs. 9 and 11, we can see the main factors are attributed to the improved shielding effects for Cases (b), (c), and (d). The improved shielding effect for Case (d) at θ=0 degrees, which is around 3 dB, depends on the depth of the Y-shaped edge rather than the width of such an edge, because Case (c) with the same width of Case (d) has a little improvement at the same angle θ. The improved shielding effects in 2D at θ=70 degrees had peak frequencies of 500 Hz for Cases (c), (d) and 800 Hz for Case (b). These peak frequencies are dependent on the destructive interference of sounds propagated through paths between A and B in the right side of Fig. 10. However, as shown in Fig. 11, the frequency characteristics in 3D that are smoothed by summing the squared values of sound pressures propagated from bogie source positions differ greatly from those in 2D, and the difference in the overall values in 2D and in 3D is within 3 dB. These facts indicate that we should apply carefully the calculated results in 2D, such as the improved shielding effect, to prediction of railway noise in 3D.

4 Conclusions

We applied a numerical technique to transform a 2D solution to a 3D solution based on Duhamel's method using a transient solution obtained by 2D-FDTD simulation to calculate the propagation of bogie noise behind a reflective inverted-L barrier, and we confirmed the 3D transformation results with a scale model experiment. From the comparison between numerical and experimental results, the 3D transformation results agreed with the scale model experiment very well. This good agreement indicates that the 3D transformation technique is a useful tool for precisely predicting the sounds that propagate from incoherent line sources, such as bogie noise. We also calculated the shielding effects of several noise barrier shapes against bogie noise using the 2D-FDTD method and clarified the differences between the shielding effects of barriers in 2D and in 3D transformed using Duhamel's method.

References

[1] Hothersall, D.C., Horoshenkov, K.V., Morgan, P.A., Swift, M.J.: Scale modeling of railway noise barriers. Journal of Sound and Vibration 234(2), 207–223 (2000)

[2] Abe, Y., Nagakura, K., Kitagawa, T., Murata, K.: A study on effective configurations of sound barriers for Shinkansen track. In: Proceedings of inter-noise 2005, pp. 1868–1876 (2005)

[3] Hiroe, M., et al.: Numerical analysis on the shielding effects of noise barriers against bogie noise by Finite-Difference Time-Domain method. In: Proceedings of Inter-Noise 2007, in07_326, Istanbul, Turkey (2007)

[4] Sakamoto, S.: Calculation of sound propagation in three-dimensional field with constant cross section by Duhamel's efficient method using transient solutions obtained by finite-difference time-domain method. Acoust. Sci. & Tech. 30, 72–82 (2009)

[5] Hiroe, M., et al.: Numerical simulations of propagation of bogie noise using 2D-FDTD method, - Treatment of boundary conditions of railroad tracks, and applications of transformation results calculated by 2D-FDTD method to 3D results -. Technical Committee Meeting on Noise and Vibration, N-2010-02 (January 2010) (in Japanese)

[6] Sakamoto, S.: Phase-error analysis of high-order finite difference time domain scheme and its influence on calculation results of impulse response in closed sound field. Acoust. Sci. & Tech. 28, 295–309 (2007)

Difference between Using 2 and 4 Meter Receiver Height in Railway Noise Prediction

M. Ögren[1], T. Jerson[2], E. Öhrström[3], and A. Gidlöf Gunnarsson[3]

[1] VTI the Swedish National Road and Transport Research Institute,
Box 8077, SE-402 78 Gothenburg, Sweden
Tel.: +46 31750 26 04
e-mail: mikael.ogren@vti.se
2 WSP Acoustics
[2] Rullagerg. 6, SE-415 26 Gothenburg, Sweden
Tel.: +46 31 7272 647
e-mail: tomas.jerson@wspgroup.
[3] Sahlgrenska Academy at the University of Gothenburg
Box 414, SE-405 30 Gothenburg, Sweden
Tel.: +46 31 786 36 10
e-mail: evy.ohrstrom@amm.gu.se, anita.gidlof@amm.gu.se

Summary

In the Nordic countries, a receiver height of 2 m has often been used when calculating noise levels over large areas for socio-acoustic surveys, but within the EU 4 m is used. Here, results are presented for railway noise calculations at both heights in 1459 points across several areas in Sweden. The average difference in equivalent level is 2.5 dB higher at 4 m height than at 2 m, which in turn leads to 10 % – 40 % less predicted annoyance if the exposure is calculated at 4 m.

1 Introduction

Within the Swedish research program "Train Vibration and Noise Effects" (TVANE) [1, 2], a large number of noise calculations have been performed both at a height of 2 m (Nordic standard) and 4 m (EU Environmental Noise Directive standard) above the ground. This paper describes the differences between the two approaches, both in th statistical terms and in the theoretical explanations in a few example cases. The impact on estimated annoyance is also briefly investigated.

In this paper, the equivalent A-weighted sound pressure level over 24 hours is denoted $L_{\mathrm{AEq,24h}}$, and the maximum A-weighted level with time Weighting FAST (125 ms) is denoted L_{AFmax}. The special noise indicator used in the EU "Level day-evening-night" is denoted L_{DEN}, which is an equivalent level where events that occur at night are given a 10 dB penalty and events in the evening a 5 dB penalty when forming the equivalent during 24 hours.

T. Maeda et al. (Eds.): Noise and Vibration Mitigation for Rail Trans. Sys., NNFM 118, pp. 541–546.
springerlink.com

2 Calculated Equivalent and Maximum Levels

When calculating the noise level from railway traffic, the receiver height is of great importance. During detailed investigations, it is possible to take the different heights of buildings with several floors into account, but in large noise mapping endeavors, it is more efficient to use a standard height. For Swedish conditions, 2 m is often used, but for the EU Environmental Noise Directive (END) noise mappings calculations at a height of 4 m is advised. Therefore, it was decided to work with both heights in the TVANE project.

Within TVANE, five different areas were selected for calculations and questionnaire surveys: Falköping, Töreboda, Kungsbacka, Alingsås and Sollentuna. The calculations were performed using the Nordic method revised 1996 [3] in a total of 1479 points up to 350 m[1] from the railway and were based on digital maps including height information of the terrain, buildings and the railway embankment. Specular reflections of order 1 were included, which was thought to be enough since the areas mostly consisted of detached houses and no street canyons. The traffic flow varied from 10 to 34 trains per hour at peak traffic for the different areas.

As expected, the calculated noise levels are slightly higher at 4 m receiver height when compared to 2 m. The histogram of the differences in equivalent level can be

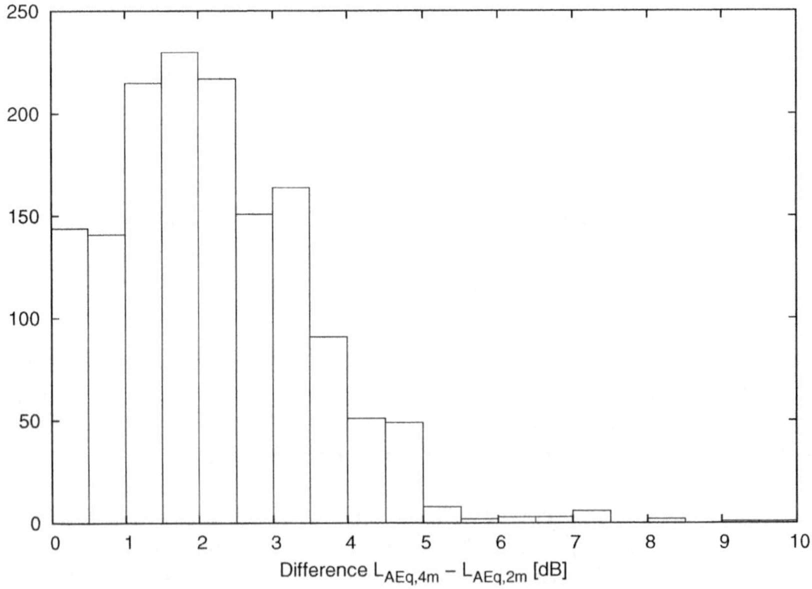

Fig. 1. Histogram of differences in equivalent level between 2 m and 4 m receiver height

[1] A few receiver positions were more distant, up to 450 m from the track.

found in Fig. 1, and the difference versus distance to the railway can be found in Fig. 2. The overall mean difference is 2.2 dB (95 % confidence interval -0.4 – 4.8 dB). There are differences up to 11 dB, and as can be seen in Fig. 2, those extreme points occur at all distances from the railway. There is a slight trend of decreasing differences at longer ranges, which is visible in Fig. 2 where the dotted line is the least-squares linear regression line, i.e., the best-fit linear function of distance that predicts the difference.

After investigating the receiver points where the differences are large and making calculations in a few test cases, it was found that the large differences occur almost exclusively at receiver points where screening is important, either from noise barriers or other buildings.

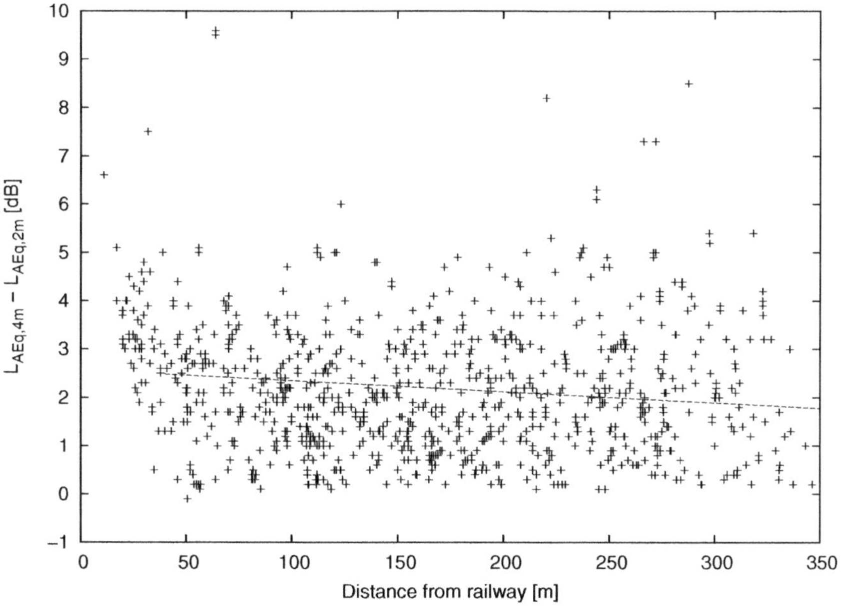

Fig. 2. Difference in equivalent level between 2 m and 4 m receiver height versus distance from source

The differences in the maximum level (time weighting F 125 ms) are more pronounced than that for the equivalent level, which is partly due to the method that specifies an incremental search for the position of the noisiest train passage which gives the highest noise level. This method needs to be simplified by the program manufacturer to reduce the necessary calculation time, and different programs use different approaches. Part of the difference may also be attributed to numerical problems related to finding the relevant rays in a 3D environment, and possibly to program errors. The difference in maximum level is plotted versus the difference in equivalent level in Fig. 3. At receiver positions where the difference is large for the equivalent level, the same is true for the maximum level. The line shows the

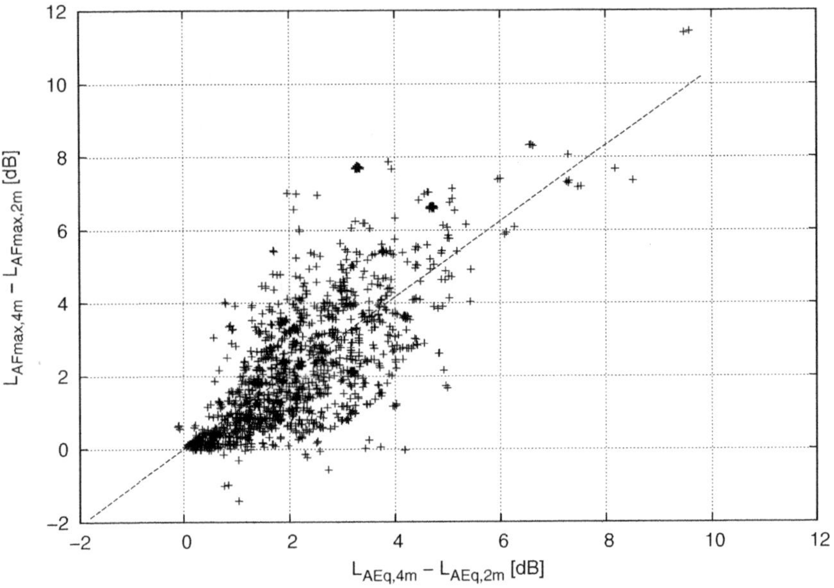

Fig. 3. Difference in equivalent and maximum level between 2 m and 4 m receiver height

least-squares linear regression, and the 95 % confidence interval is +/- 2.5 dB from this line.

3 Screening and Ground Effect

By making a few example calculations for flat ground and a low source (no embankment) both with the Nordic method from 1996 [3] and the newer Nord 2000 [4], it was clear that the difference in ground effect between 2 and 4 m height was 1 – 3 dB (more in individual frequency bands). Thus, differences in ground effect could not explain the higher differences in the calculations even at long ranges. In other words, screening by terrain, buildings or sound barriers is what differs between 2 m and 4 m receiver height.

Therefore, a set of calculations were performed using the Nord 2000 method [4] with a screen and two receivers at 2 and 4 m height, respectively. The resulting difference is plotted versus distance from the screen in Fig. 4, where the screen height is 4 m above the top of the rail and the distance from the track to the barrier is 15 m. Note that very close to the barrier, the difference is as high as 10 dB in this example. This is reasonable from a theoretical perspective since the closer the receivers are to the barrier the larger the difference in diffraction angle becomes. It should be noted though that if reflections between the barrier and the receiving façade are included the difference decreases. It decreases even more if scattering and higher-order diffraction are taken into account. Therefore, it is reasonable to expect differences between 2 and 4 m height from field measurements to be less than the calculated values in Fig. 2.

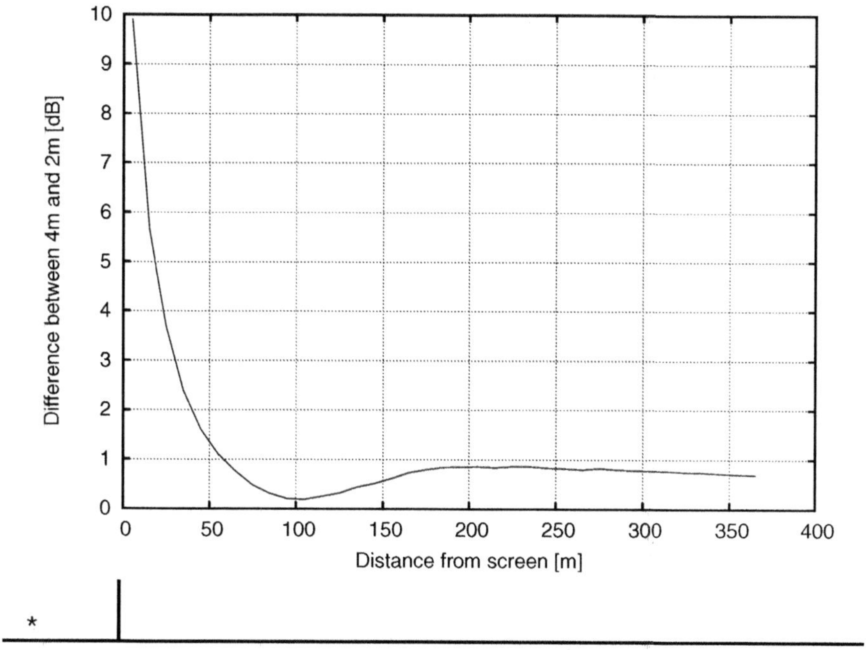

Fig. 4. Difference in equivalent level at 2 and 4 m height behind a screen versus distance behind the screen using the Nord 2000 method. (color figure online)

4 Discussion

Since the calculated noise levels are higher at a receiver height of 4 m above ground than at 2 m, the annoyance measured at a certain equivalent or maximum interval such as 55 – 60 dB will be reduced if 4 m receiver height is used. In the TVANE study, one of the questions in our questionnaire was the ISO standardized annoyance question [5]. Evaluating the proportion of annoyed respondents for different receiver heights in the area, Sollentuna gives approximately 40 % less annoyance at low levels which decreases to about 10 % less annoyance for high levels. In other words, the effect would make comparisons between surveys based on 2 m receiver height noise calculations difficult to compare with those based on 4 m.

This difference is important to keep track of when comparing different studies, or when performing meta-analysis of many studies. The effect may be as large as more well-known problems in comparing studies; such as different annoyance scales and difference between noise indicators such as $L_{AEq,24h}$ vs L_{DEN}.

References

[1] Jerson, T., Ögren, M., Öhrström, E.: Combined effects of noise and vibration from train and road traffic, Paper P08 - IWRN 9, Munich (2007)

[2] http://www.tvane.se/ (accessed June 1, 2010)

[3] Naturvårdsverket. Buller från spårburen trafik – Nordisk beräkningsmodell. Rapport 4935, Naturvårdsverket (Swedish Environmental Protection Agency), Stockholm

[4] Jonasson, H., Storeheier, S.: Nord2000. New nordic prediction method for rail traffic noise. SP Report 2001:11, SP the Swedish Technical Research Institute, Borås (2001)

[5] ISO/TS 15666:2003. Assessment of noise annoyance by means of social and socio-acoustic surveys

Stationary Noise Monitoring for High-Speed Trains with Single or Multiple Microphones

Koji Watanabe[1], Shigeru Naito[1], Ken-ichi Tagawa[1], and Toshikazu Sato[2]

[1] Research and Development div. Central Japan Railway Company
1-45-33 Ohyama, Komaki, Aichi, Japan
Tel.: +81-568-47-5394; Fax: +81-568-47-5364
e-mail: koji-wata@jr-central.co.jp
[2] Brüel & Kjær div. Spectris, Co. Ltd., Japan

Summary

In Japan, high-speed train (*"Shinkansen"*) noise is strictly regulated by the government due to dense population. However, there is also a strong need for journey or travel time reduction. We have been developing high-speed, low noise, and low-energy consumption Shinkansen trains as well as noise barriers with high noise reduction efficiency for over 40 years. To evaluate the train noise and the performance of the noise barriers, we have been measuring the noise with portable noise equipment, which is a quite troublesome job and costly. Recently, we developed automatic noise measuring systems with combined Internet technology and noise measurement technologies. We discuss single microphone and multiple microphone noise monitoring systems in this paper.

1 System Outline

1.1 Single Microphone System: Noise Monitoring Terminal

First, we use the Brüel & Kjær noise monitoring terminal (NMT) that is used in airport noise monitoring systems. This is a powerful noise monitoring terminal that is optimized for outdoor use, in all climatic environments, with a heat- and humidity-resistant enclosure. This system is especially designed for permanent and semi-permanent environmental noise monitoring.

We usually use a pistonphone or a calibrator periodically (once a day, etc.) for calibration, but the Brüel & Kjær patented Charge Injection Calibration (CIC) allows us to avoid calibrating microphones with such calibrators.

The system calculates the peak noise level L_p, the energy mean or equivalent level, L_{eq} and other noise metrics. Noise events can be recorded for later analysis or noise source identification with MP3 files. Also, the system produces hourly and daily reports of the noise events and weather data.

T. Maeda et al. (Eds.): Noise and Vibration Mitigation for Rail Trans. Sys., NNFM 118, pp. 547–554.
springerlink.com

However, the NMT system is designed for airport noise monitoring, so the system has several problems when we try to use it in for railway noise monitoring.

1. Although the system is basically a local area network (LAN), which is intended to be connected to a local network, not Internet connections. The system did have a modem for the Internet connections, but it was intended for sending reports only. We had to record WAV files for each noise event, the size of which is much larger than that of the MP3 file. Thus, we had to transmit the broadband data over the Internet.
2. We developed a train detection system that triggers the noise analysis and data recording and records train speed as well.
3. We also have a need for measuring the wheel wheel/track noise by setting the microphone in the vicinity of the track and structure borne noise by setting the microphones under the elevated track. So, we need at least three microphones for measuring such noise. The multiple microphone system will be described later.
4. Train information is needed, but the NMT data do not include any train information. After downloading the data from the NMT, we combine the data with COMTRAC train operating system data used in our Tokaido Shinkansen line.
5. In the NMT, the noise of each event is recorded in MP3 format. However, the MP3 compression process may affect the frequency of the noise, so we need to record noise with a non-compressed linear WAV format.

Regarding No.3., we have to develop another noise measurement system (multiple microphone system), but in some locations, only one microphone is needed. In such a case, the NMT system can be used, if we make some modifications. Thus, we modified the software of the NMT to apply train detection and train speed calculation and to record analog noise data in WAV files. We also developed special browser software, which is described later, to combine the noise data with train information for both NMT and multiple microphone systems.

1.2 Multiple Microphone System

To measure the noise of the track or wheels of the trains, we must install a microphone close to the track. Also, we installed a microphone under the elevated track to measure structure-radiated noise. In this case, we measured the at frequencies extending below 10Hz with one low-frequency microphone. A fourth microphone was installed at 25m from the center of the track, giving a total of four microphones for one noise monitoring site.

We used the Brüel & Kjær PULSE noise and vibration analysis system with Brüel & Kjær outdoor microphones that are supplied with the NMT.

However, the PULSE system is not designed as a train noise monitoring system, and we had to develop custom software, especially for noise analysis and recording, to work with PULSE system.

We developed three software programs:

(1) PULSE noise monitoring software

The software performs 1/3 octave band frequency analyses, narrow band FFT analyses (low-frequency noise), and peak noise detection, similar to conventional sound level meter (SLM). It also calculates the equivalent sound level (L_{eq}) and the sound exposure level (L_{AE}) during the train passing event. Finally, the system records all of the four microphone signals output in WAV format for the later detailed analyses.

PULSE itself has VBA macro ability, so we programmed the PULSE system with VBA macros.

(2) Train detection software

The trains are detected with an ultrasonic sensor. The signal starts the PULSE noise monitoring software, calculates the train speed, and records the output of the analyzers in a text file. The raw analog data of each noise event are recorded in WAV files. The trains of both directions (up and down) are detected individually, and if both up and down trains pass simultaneously, the software records such events by setting the pass bit in the file.

(3) Weather monitoring software

As with the NMT, we developed Vaisala weather transmitter monitoring software. The software records the output of the Vaisala transmitter: wind speed, wind direction, temperature, humidity, rain intensity, barometric pressure, etc., each second, in a text file.

1.3 Data Acquisition and Transmission

When a train passes in front of the laser or ultrasonic sensors, a trigger signal is generated. The PC in the NMT or PULSE calculates the train speed from the output of the sensors and records the noise from the microphones. It also starts the noise analyses, sound level measurement, and 1/3 octave bandwidth analyses and measures the L_{eq}, L_{AE}, L_{P}. The system summarizes the data hourly and daily, and the noise event report is transmitted through the Internet. Noise data are recorded in WAV format, which is downloaded at night for later analyses by the Brüel & Kjær PULSE TIME.

The data, such as FFT and 1/3 octave band analysers, and WAV files of each event, and train data, are transmitted through the Internet to the center located in Komaki City, Aichi Prefecture. For security reasons, we use VPN (virtual private network), which is the business network of NTT. However, the telecommunication protocol is based on the same TCP/IP protocol, so the system described in this paper can be used with the Internet.

1.4 Installation

The NMT is installed at Kikugawa City, Shizuoka prefecture, and the multiple microphone system based on PULSE is installed at Tarui town, Gifu prefecture. These sites are selected to test noise barriers and the future high-speed experimental trains. As for the transmitting system, we chose fiber optic cable with virtual private network (VPN).

Fig. 1. Kikugawa City noise monitoring site

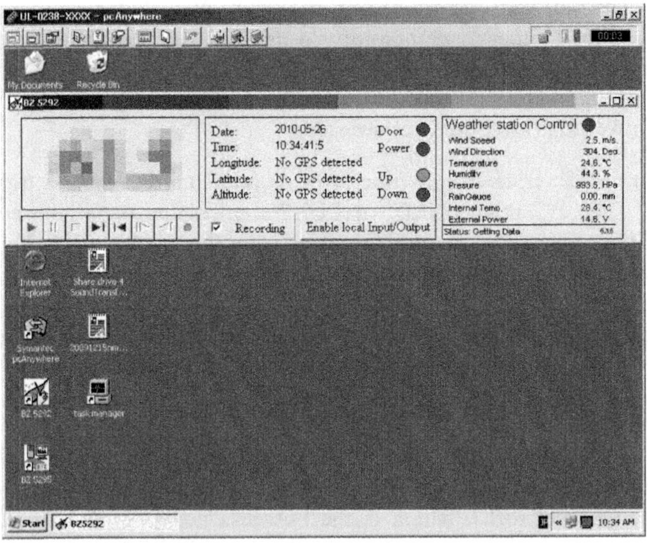

Fig. 2. GUI of the NMT in Kikugawa site

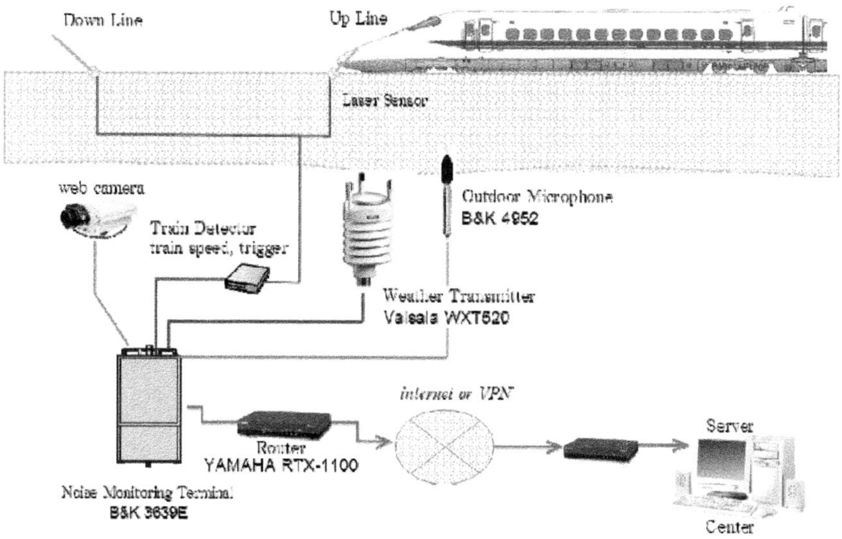

Fig. 3. Noise monitoring system for Shinkansen (sited at Kikugawa City, Shizuoka pref.) (color figure online)

Fig. 4. Tarui Town noise monitoring site and 25-m microphone

Fig. 5. Structure noise microphones (wide-range and low-frequency microphones)

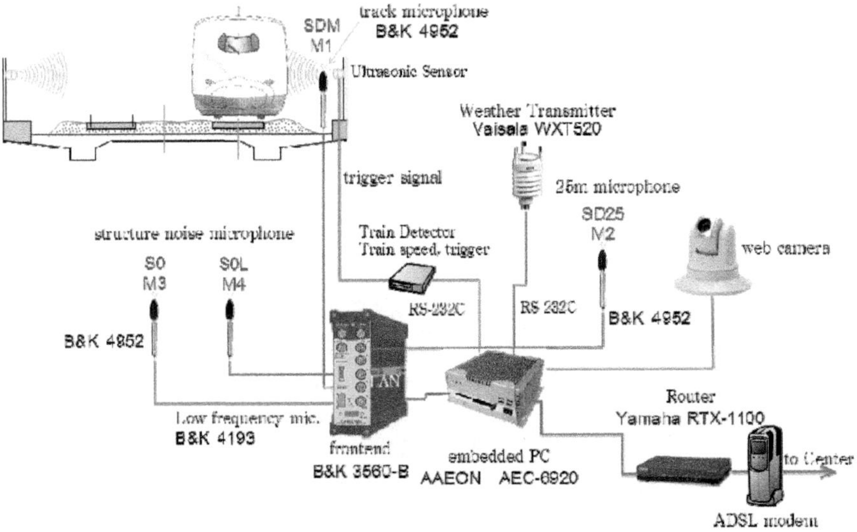

Fig. 6. Multiple microphone system installed at Tarui Town, Gifu pref. (color figure online)

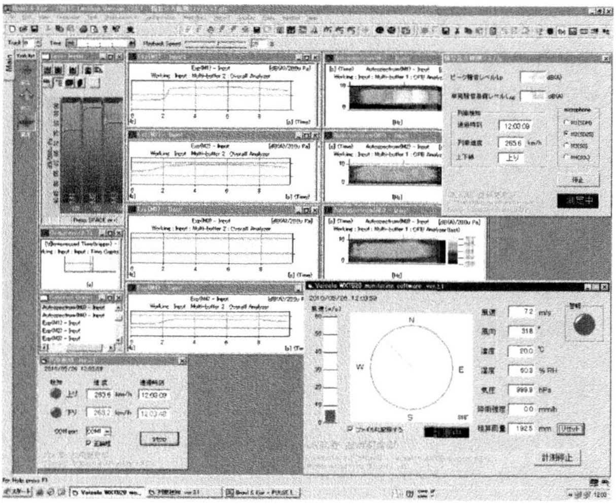

Fig. 7. GUI of multiple microphone system in Tarui site

2 Post-processing

2.1 Data Browsing

The collected data are combined with the output data of COMTRAC train information system for each train. This post-processing uses a web based browser. We developed the browser with PERL (Practical Extraction and Report Language). By combining the noise event data with COMTRAC train information,

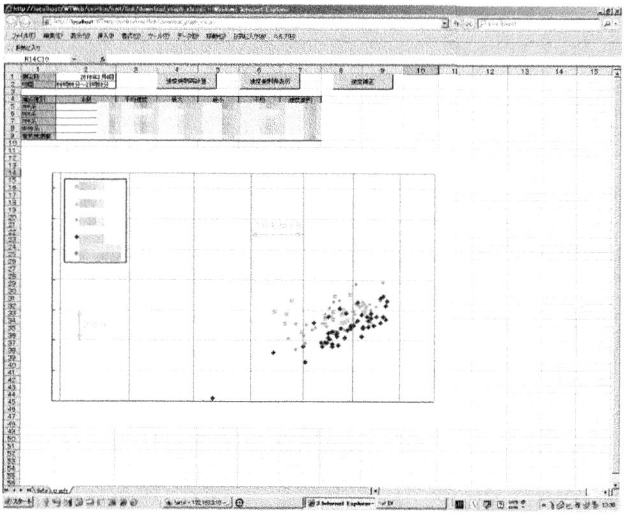

Fig. 8. Train speed v.s. peak noise level L_p

we obtain the noise data of individual trains in a table and noise level L_P and L_{AE} v.s. velocity graphs shown in Fig.8.

The browser shows the following data for each train:

(1) time, train speed, train number, train type, rolling stock, L_P, L_{AE}, L_{eq} of determined time (15h, 24h, or other averaging time period).
(2) Graph of the L_P, L_{AE} v.s. train speed
(3) Users can hear the train noise by clicking the train number. We link the noise data with each WAV file automatically.
(4) Calculation result of the environmental quality standards value in Japanese noise regulations.

2.2 Database Function

The browser software has a database function as well. When developing a new high-speed train, we must track the noise data of the new train. Also, we periodically grind the surface of the rail to reduce wheel noise. However, the effect of the rail grinding is not confirmed by tracking the noise data for a long period, such as one or half year. Temperature, barometric pressure, wind speed and wind direction may affect the noise, but we do not have any weather data. By selecting rolling stock, specific day, period, and weather conditions, we can select the noise data matching these conditions.

3 Conclusions

The stationary noise monitoring system for Tokaido Shinkansen will greatly assist development of future low noise, high-speed, Shinkansen trains, to reduce noise and meet environmental conditions.

References

[1] Brüel, Kjær: Noise Monitoring Terminal Plus Types 3639 E and 3639G User Manual (2005)
[2] Brüel, Kjær: PRODUCT DATA Outdoor Microphone Type 4952 (2006)
[3] Manvel, D., et al.: SADMAM – Combining Measurements and Calculations to Map Noise in Madrid. In: Proceedings of InterNoise (2004)
[4] Ministry of the Environment, Japanese Government, Environmental Quality Standards for Shinkansen Superexpress Railway noise, `http://www.env.go.jp/en/air/noise/railway.html`

Rail Vehicle Noise Source Identification Using Moving Frame Acoustical Holography

H.I. Koh and W.H. You

Korea Railroad Research Institute,
#360-1, Woramdong Uiwang City Gyonggido, Korea
Tel.: +82 31 460 5207; Fax: +82 31 460 5279
e-mail: hikoh@krri.re.kr, whyou@krri.re.kr

Summary

In this paper, the principal theories of the moving frame acoustical holography and main processes are presented. To build databases of noise maps of noise sources under moving vehicles, several measurements were taken with EMU vehicles using microphone arrays. Moving frame acoustical holography noise maps were obtained in overlap with each component of the under part of the EMU vehicles. Some results are shown and discussed.

1 Introduction

Recently, a need for identification of unusual noise sources has been arising in urban railway lines. The structure of the under part of the rolling stock, which is a complex assortment of various equipment and their interactions, makes it hard to define the sources of an unusual noise and its exact location. Unusual noise often has a dominant sound level, a characteristic frequency, or is impulsive in operating condition. This study is primarily aimed at establishing a technology for identifying noise sources and their locations on moving rail vehicles from the wayside. If we are not aware of the occurrence of unusual or abnormal noise sources, which can lead to a critical situation for the vehicle, an overall monitoring and diagnosis method is needed for the whole vehicle. For that case, databases of sound maps of passing vehicles for various conditions are essential.

A moving frame acoustical holographic method is developed, which uses sound signals measured with microphone arrays installed under the vehicle near the ground or alongside the track. Possible applications cases are holograms of under car equipment noise in each frequency band, identification of the most dominant source location at certain operating conditions, and comparison of holograms with maintenance conditions according to vehicle type. Various databases are constructed through field experiments, and algorithms will be enhanced for special target applications.

T. Maeda et al. (Eds.): Noise and Vibration Mitigation for Rail Trans. Sys., NNFM 118, pp. 555–562.
springerlink.com

2 Principles

2.1 Measurement of the Pass-by Acoustical Signal

The sound of a moving source is measured with a line microphone array of microphones spaced at regular intervals. The size of the interval is related to the wavelength of the observing signal and determines the spatial sampling parameters of the acoustical map. To avoid spatial aliasing, at least 4 microphones per wavelength at the shortest wavelength are used. The length of the array should be longer than the width of the passing sound source. These conditions determine the number of the microphones. Two arrays can be used for the purpose of figuring out the effect of the ground reflection using the phase information of the direction of propagation. In this case, the distance between two measurement planes should be less than the half wavelength of the measuring signal but should large enough to measure phase differences to distinguish the incident wave from the reflected wave. The horizontal distance between two arrays is related to the sizes of the acoustical hologram.

For the purpose of the mapping of the sound source with the holograms, two pairs of the photo sensors are used to record the positions of the sound source and match the locations of the moving vehicle and the measured sound signal. The moving speed of the sound source is calculated from the photo-sensor signals.

2.2 Moving Acoustical Holography

2.2.1 Theory

In the moving frame, acoustical holograms of moving pure tones or sounds with narrow band frequencies are measured by a fixed microphone array, but the Doppler effect is eliminated and the sound pressure at the source plane is imaged. Signal processing is carried out in the spatial domain by means of the spatial Fourier transform. Two coordinates are considered, namely a fixed coordinate in the measurement position and a hologram coordinate moving with the source plane with a constant speed (Fig.1) [1]. First, the sound pressure at the hologram coordinate frame can be calculated in the spatial domain from the measured sound pressure of a moving source moving at speed u, and the sound pressure at the sound source plane can be also obtained. Since the sound level is pronounced at the sound source position, the source can be localized and imaged.

Moving periodic impulsive signals can be expressed as a combination of moving pure tones. Band pass filters are used to image the sound field for each center frequency. The frequency bandwidth of the moving impulse signal depends on the frequency and the translation speed of the source. So there exists frequency range, in which the moving acoustical hologram can be obtained without the error caused by the overlapping bands of two adjacent frequency components.

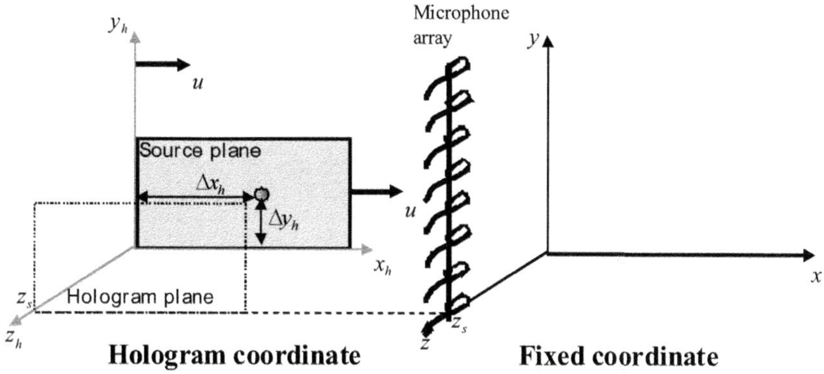

Fig. 1. Coordinates in the moving acoustical holography (color figure online)

2.2.2 Background Noise

To enhance the resolution in the source image and to reduce the effect of the background noise, such as the ordinary rolling noise and other stochastic broad band noise, the holograms can be averaged, assuming that these holograms are initiated from the same sound source. Fig.2 shows a result of the moving periodical impulsive source localization.

A periodical impulsive noise is embedded in the measured pass-by noise from an EMU train moving at speed u=20km/h. The upper result of the two acoustical maps in Fig.2 shows the effect on the impulsive noise source location image due to the other uncorrelated noise during the EMU pass-by at the frequency of 240Hz. The lower result shows the enhanced impulsive source localization after the holograms were averaging in the frequency range of 120Hz~240Hz. The effect of the background noise of the vehicle during the pass-by is nearly eliminated in the impulsive noise localization.

2.2.3 Boundary Conditions

The bandwidths of each center frequencies f_i and the distinguishable intervals between two adjacent spectral components are related to the vehicle speed with the following condition.

$$f_i < \frac{1-M}{2M}\frac{1}{\Delta T} \tag{2.1}$$

With (2.1), several boundary conditions are derived, which determine other parameters for the measurement setting analysis of the holograms. The lowest frequency limit for the hologram is related to the resolution in the frequency domain and is determined by the vehicle speed of the source. The number of the hologram that are needed for the averaging process is related to the required background noise

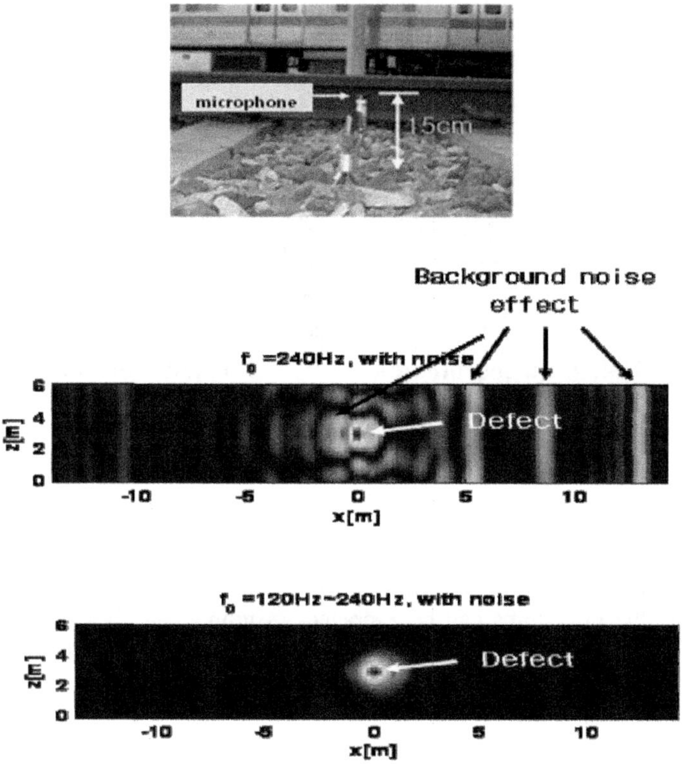

Fig. 2. Moving periodic impulsive signal embedded in EMU vehicle noise of the under part (color figure online)

reduction. This number is determined from the periodicity of the impulse and then the upper frequency is determined [2,3].

3 Field Test

To implement the theory and algorithm into the railway field, databases for different conditions have to be established and database classified for several levels of the conditions if we want to get even a diagnostic result. Since the moving acoustical hologram method yields acoustical maps of the source location for pure tones or for tones or for individual spectral components, the application target has to be chosen and the background noise type has to be investigated. Fig.3 shows an acoustical hologram of noise from the under part of a train vehicle that was passing by a roller rig with rail joints. The target impulsive noise was located at the end of the vehicle beginning at this begins at 12.3 sec. In the acoustical hologram, the location of the target noise is hard to find. Since the impulsive characteristic of the

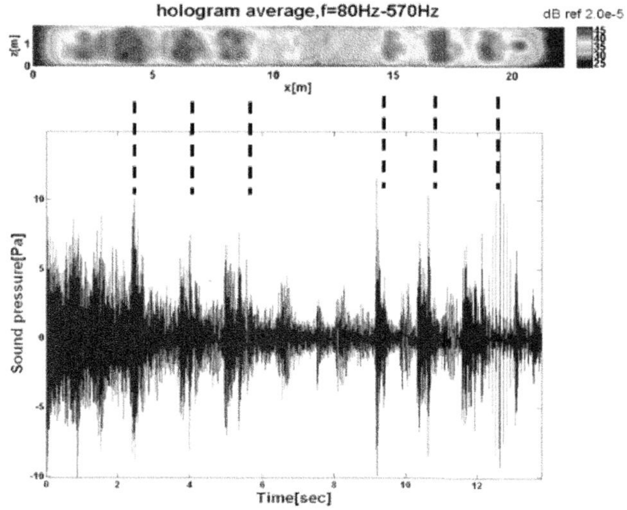

Fig. 3. Impulsive noise signal radiated from the under part of a vehicle with impulsive noise from the rail joints (color figure online)

rolling noise at the rail joints was known and could be measured beforehand, its effect could be reduced or eliminated before generating the acoustical hologram.

In Fig.4, the effects of the rail joint noises are reduced and the hologram shows the location of the target source, which is in red. To detect or identify any extraordinary noise source or to compare different measurement results, a reference acoustical map of the inspection vehicle is essential. To reduce the background noise effects, we choose a test track in a maintenance center as a test site. Noise data before and after the regular vehicle maintenance process were obtained.

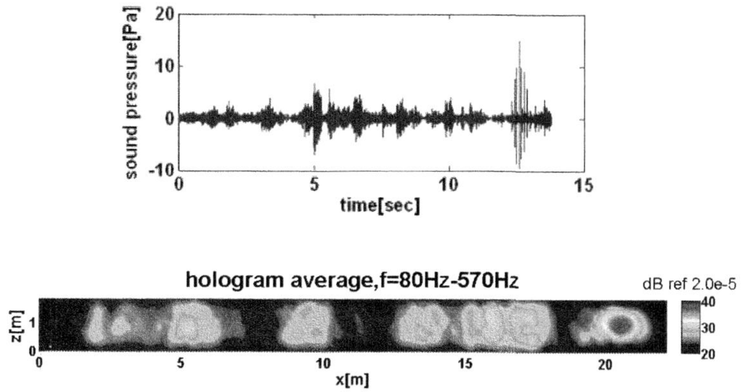

Fig. 4. Reduced impulsive signal of the rail with the impulsive noise signal radiated from the under part of a train vehicle (top), acoustical hologram (bottom) (color figure online)

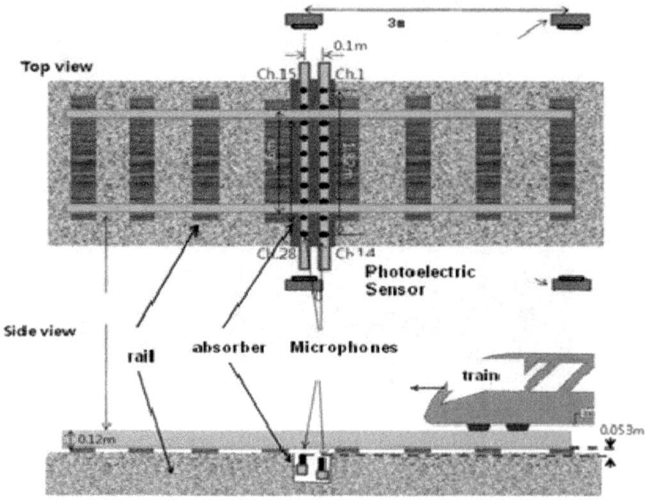

Fig. 5. Field experiment setup for the EMU pass-by noise measurement with a microphone array and photoelectric sensors (color figure online)

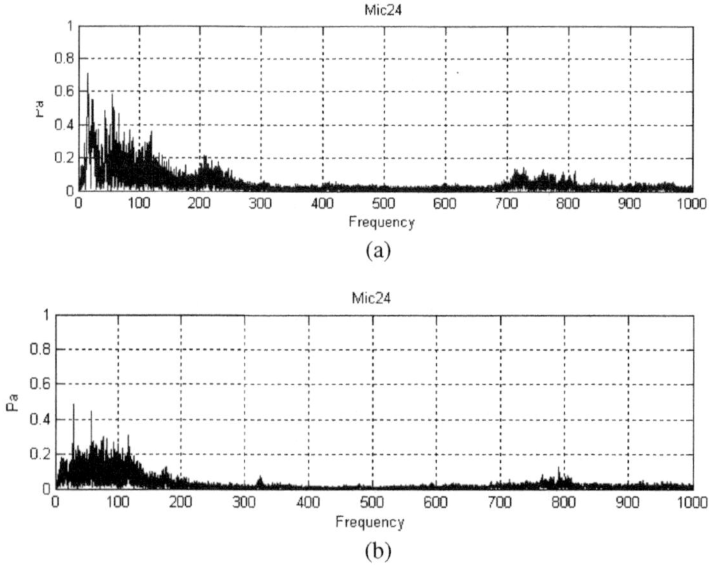

Fig. 6. Sound spectrums from mic24 before (top) and after (bottom) the 4Y maintenance (color figure online)

The field test setup is shown in Fig.5. A total of 28 Microphones are used, 14microphnes in each of two arrays. The array is fixed under the rails between two sleepers. The train passed by constant speed of u=20km/h. The measured sound is analyzed for the whole vehicle length. The pass-by noise is measured before and after the 4Y maintenance of the same vehicle. The driving direction of the train during measurement after the maintenance was reversed from the direction during the measurement before the maintenance. Fig.6 shows the measured sound pressure radiated from the under part of the EMU vehicle. Overall, the sound pressure was reduced after the 4Y maintenance and the dominant frequency range is also changed.

The sound pressure map is shown for several frequency ranges in Fig.7. The dominant noise characteristics and the differences before and after the 4Y maintenance can be compared and analyzed. The acoustical holograms resulted from the averaging process are shown in Fig.8. The location of the most dominant noise source can be observed.

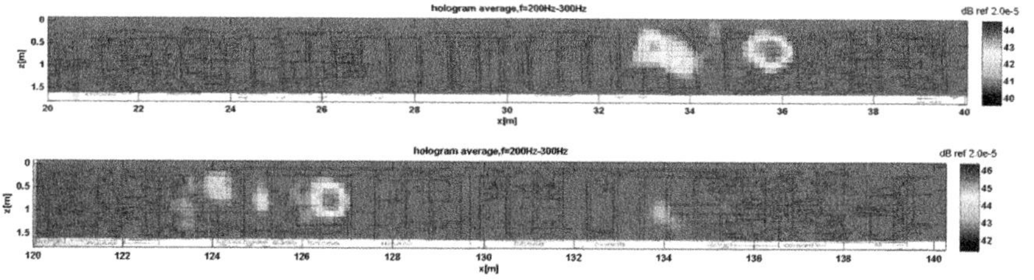

Fig. 7. Comparison of the sound source localization from 200Hz to 300Hz for the EMU under car noise (color figure online)

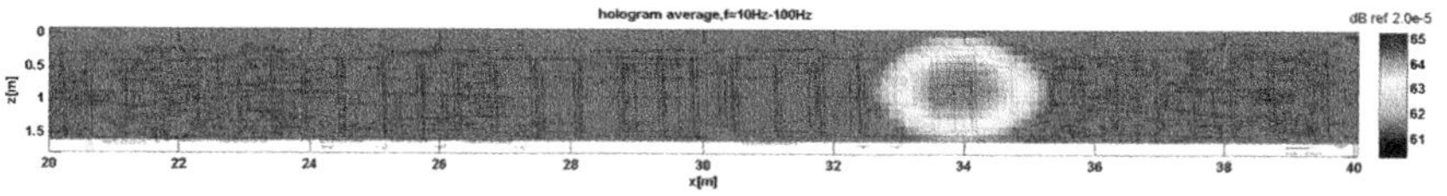

Fig. 8. Sound source localization from averaged hologram over the range of 10Hz to 100Hz of the EMU under car noise.

4 Concluding Remarks

A moving frame acoustical holography algorithm and testing technology have been developed to localize and image noise sources of the under car equipment of moving railway vehicles. This technology is being test because the acoustical hologram can yield more physical information about the noise source than the beam forming method where frequently occurring extraordinary noise events are one of the types of noise, such as tonal noise or periodic impulsive noise. A database of the reference condition is essential and the acoustical information of the background noise has to be investigated.

Acknowledgements

The study presented is carried out within the framework of the research project "Railroad safety technology R&D" funded by the ministry of land, transport and maritime Affairs, Korea. The authors gratefully acknowledge the support of the Seoul Metropolitan Rapid Transit Corporation for the field test. The contributions by Korea Advanced Institute of Science and Technology for the development of the algorithm are acknowledged.

References

[1] Kwon, H.S., Kim, Y.H.: Moving frame technique for planar acoustic holography. J. Acoust. Soc. Am. 103(4), 1734–1741 (1998)

[2] Jeon, J.H., Kim, Y.H.: Fault localization method of a train in cruise. In: Spring conference of the Korean Society for Railway (2007)

[3] Jeon, J.H., Park, C.S., Kim, Y.H., Koh, H.I., You, W.H.: Bearing fault localization of a moving vehicle by using a moving frame acoustic holography. Transactions of the Korean Society for Noise and Vibration Engineering 19(8), 816–828 (2009)

Indirect Rail Roughness Measurement: The ARRoW System within the LECAV Project

A.H.W.M. Kuijpers[1], W. Schwanen[1], and E. Bongini[2]

[1] M+P - consulting engineers
P.O. Box 2094 5260 CB Vught, The Netherlands
Tel.: +31 73 6589050; Fax: +31 73 6589051
e-mail: ArdKuijpers@mp.ml
[2] SNCF Innovation and Research Department
45 rue de Londres, 75379 Paris cedex 8 France
Tel.: +33 153422356; Fax: +33 153429784
e-mail: estelle.bongini@sncf.fr

Summary

ARRoW has been evaluated in the framework of SNCF's LECAV project. The aim of the LECAV is to characterize the acoustic condition of the railway network in different ways: to provide a map of the track roughness for the entire railway network; to monitor the acoustic performance of the track for the grinding policy; and to easily identify and locate possible TSI-compliant measurement locations. For this task, ARRoW was tested along other onboard measurement systems. More specifically, ARRoW was used to provide rail roughness levels and spectra during a 2,000 km measurement campaign. A global network map of the roughness was delivered, and, in addition, more detailed results were obtained at some test and reference sites that were measured multiple times to check the repeatability and representativeness of the ARRoW results compared to direct roughness measurement results. ARRoW proved to live up to the task. However, to get the most out of the system, attention should be paid to the selection of calibration sites and maintaining a low wheel roughness.

1 Introduction

The Innovation and Research Department of SNCF, the French national railway operator, has started a project dedicated to the acoustic characterization of the track (LECAV project). This project includes three main objectives:

- To provide track roughness mapping of the entire railway network in order to anticipate the requirements of the future European Directive EN 2002/49 [1];
- To easily identify sites that comply with the TSI specifications for rolling stock homologation;
- To monitor the acoustic performance of the track for grinding policy.

T. Maeda et al. (Eds.): Noise and Vibration Mitigation for Rail Trans. Sys., NNFM 118, pp. 563–570.
springerlink.com

These three objectives can be reached by onboard track-monitoring systems that provide track roughness levels and, when required, roughness spectra according to ISO 3095 [2] specifications.

The first step of LECAV project is testing of existing onboard measurement systems: selected systems have been evaluated in their ability to provide rail roughness on a global level as well as third-octave band spectra according to the requirements of SNCF. So, the systems have been used for acoustic mapping of a large network section, and the indirectly measured roughness has been compared to the directly measured roughness.

One of these measurement systems is the ARRoW system developed by M+P. To evaluate ARRoW, the system was installed on rolling stock and used on over 2,000 km of track. During this measurement campaign, two reference sites and four test sites were visited. The reference sites were used to calibrate the system. This was necessary because ARRoW was used on this rolling stock for the first time. The test sites were used to compare directly measured and indirectly measured rail roughness data. Some of these test sites were tested several times at the same speed, and other sites were tested at different speeds to check the repeatability and representativeness of the measurement results obtained with the ARRoW system.

2 Direct and Indirect Rail Roughness Measurement

Rail roughness of a track can be measured directly and accurately with a profilometer. The results can directly be transformed into a noise reduction, wheel roughness notwithstanding. However, the rail roughness is only measured over a short section of the track. Secondly, the measurements need to be performed when the track is out of service, and the method is rather labor intensive.

To overcome the disadvantage of the direct method for long track lengths, ARRoW (acoustic rail-influence recording on wheels) can be used. ARRoW measures the roughness indirectly with rolling noise measurements and translating these to roughness levels, using direct rail roughness measurements to calibrate the system. ARRoW measures sound levels at close proximity to the wheels. This is the basic principle: for a given (sufficiently low) wheel roughness and a certain track system, a direct relationship exists between the change in rail roughness and the change in rolling noise sound levels. All noise variations are assumed to come from variations in rail roughness (for a given dynamic behavior of the track). Thus, when we measure the sound variation at a certain track, we know the variation in rail roughness.

The indirect method yields relative changes in noise level close to the wheel. These results need to be translated into (changes of the) absolute rail roughness level. We use the direct measurement results at a couple of reference sections to calibrate the results of the indirect measurements. Combining the results of both measurements yields an absolute rail roughness level of the whole track part.

2.1 ARRoW

ARRoW is a multichannel system that simultaneously measures sound levels, position and speed. ARRoW was mounted on a SNCF measurement train. The sound levels are measured at close proximity to the wheels rolling over the rails with four microphones, one microphone near each wheel of the measurement bogie. This has the benefit that the left and right side rail can be distinguished. Using multiple microphones also introduces redundancy in the measurement chain that increases the robustness of the system. The position and the speed are recorded with a GPS receiver and used in the postprocessing.

Fig. 1. ARRoW microphone in close proximity to the wheel (color figure online)

2.2 Direct Rail Roughness Measurement

SNCF directly measured rail roughness at two reference sites with the Müller-BBM m|rail rail roughness measurement device. The direct measurements were done at different positions over 100 m of track at different lateral positions on the rail, all according to the specifications from ISO 3095:2005. All measurements were analyzed according to the EN 15160 standard. The final result was an average rail roughness spectrum per rail for each reference site.

2.3 Disturbances

In the postprocessing phase, the rail roughness level is determined from the sound measurements using the technique described in [3]. In this phase, we have to eliminate a few factors that, besides rail roughness, directly influence the rolling noise level. The factors that we have to take into account are the following:

1. vehicle speed during the measurements
2. wheel roughness and wheel dynamics of the measurement vehicle
3. track dynamics

This section addresses each of these parameters and how we deal with these disturbances in the postprocessing.

2.3.1 Vehicle Speed

The rolling noise from a wheel on a certain track is influenced by vehicle speed. Generally, the noise will change with increasing speed, both in level as in spectral distribution. We have to remove the speed influence from the sound measurements, because changes in the noise due to the vehicle speed are considered to be caused by rail roughness. The solution is to evaluate all sound measurements at a certain chosen speed that we call the nominal speed. Measurement results obtained at speeds that are different from the nominal speed are 'corrected' to eliminate the speed effect.

To eliminate the speed effect, a simple overall level correction is not sufficient. We have to consider the noise spectrum. The total rolling noise consists of the contributions of the sleepers, the rail and the wheels. The sleepers are the most important source of noise at low frequencies, the rail is the dominant noise source in the mid frequency range, and the wheels are the dominant noise source at high frequencies. At higher vehicle speeds, the contribution of the wheel relative to that of the rail increases. Thus, the sound level increases differently in the various frequency bands. Secondly, the effect of roughness at a certain wavelength differs at different frequencies for different speeds: doubling the speed implies that the frequencies shift one octave ($f=v/\lambda$). Because of the dependency of the rolling noise spectrum on speed, we need to have a correction algorithm that takes all this into account.

2.3.2 Wheel Roughness

The wheel roughness was measured before and after the measurement campaign. The wheel roughness and wheel dynamics are assumed to be constant for the whole measurement campaign, but different for each wheel on the measurement bogie. The recorded noise level is proportional to the combined (wheel and rail) roughness. We translate the measured noise levels into combined roughness levels. Since we are interested in the rail roughness and not the combined roughness, we subtract the wheel roughness to obtain the rail roughness. This is done for each measurement wheel separately.

2.3.3 Dynamic Track Properties

The last parameter that influences the rolling noise and needs to be considered is the track construction and the resulting dynamic track properties. The sleeper type and the rail pad in particular determine the rail damping and sound radiating from the surface area and thus the rolling noise. There are two ways to compensate for this effect: 1) correct the measured noise spectrum for this effect, similar to the effect of vehicle speed, and 2) or determine the noise-to-roughness transformation separately for each track type.

We have chosen not to compensate for the track dynamics since we had only results for two reference sites, and the track dynamic properties were unknown to us at the time of the analysis.

3 Measurement Campaign

The measurement campaign consisted of three measurement days. On the first day, the track between Paris and Orange was measured. On this track, one reference site and two test sites were visited. The second day we measured the rail track in the triangle between Avignon, Miramas and Cavaillon multiple times. One test site was measured three times at the same speed to check the repeatability of the ARRoW system. Another test site was measured three times at different speeds. Hence, we could check the speed dependency of the ARRoW system. At the final day, the track between Orange and Paris was measured. On this track, we visited one reference site.

4 Results

4.1 Global Roughness Levels

We determined the global roughness level over the measured track. For this purpose, we calculated the roughness indicator, $L_{\lambda CA}$, per 100 m of track. The single value indicator $L_{\lambda CA}$ quantifies the roughness of wheel and rail in a number that is proportional to the noise level [4]. Figure 2 shows the results over a small part of the track. We see that a range of more than 15 dB is covered in a short part of the total network. For low values of this indicator, we are likely to comply with the TSI limits regarding rail roughness and track decay rate. For TSI rail roughness and a defined speed, $L_{\lambda CA}$ equals a certain value (e.g., at 120 km/h, $L_{\lambda CA}$ for TSI rail roughness equals 8.9 dB(A)). If at the same speed the measured track has a value for $L_{\lambda CA}$ equal or lower than TSI rail roughness, the track is likely to be TSI compliant.

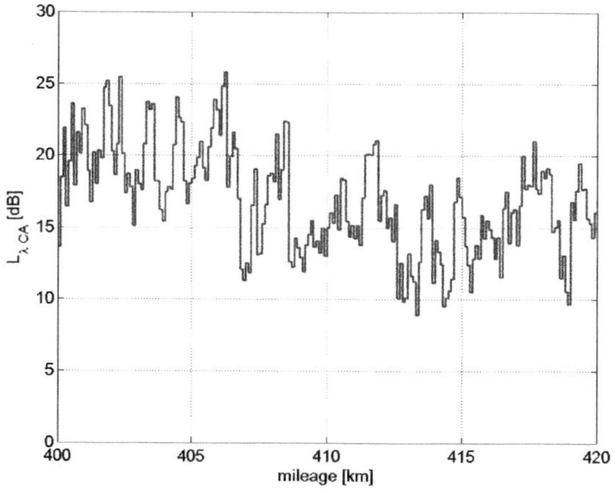

Fig. 2. Value for $L_{\lambda CA}$ (averaged over 100-m segments) for 20 km of the track. (color figure online)

4.2 Comparison with the Direct Measurement Results

Direct measurements were carried out at two reference sites to calibrate the ARRoW system. The two reference sites are situated between Paris and Orange at km 402.5 and km 414.5. The sites are indicated with the mileage and the track side (e: east, w: west). Fig. 3 shows the average rail roughness spectra from ARRoW and direct measurements.

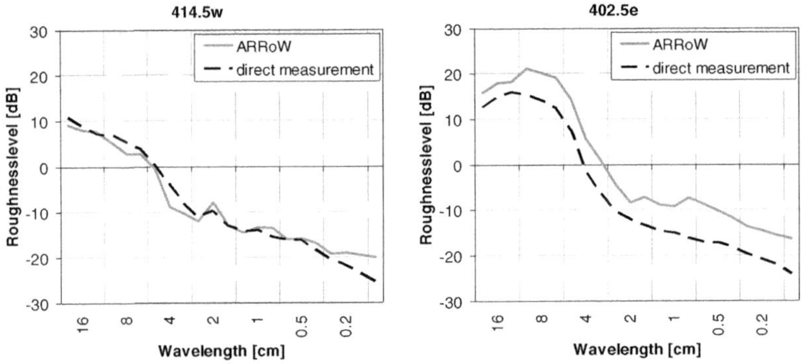

Fig. 3. Rail roughness spectra at two reference sites. We distinguish between the results of the direct measurement (black dashed line) and of ARRoW (gray solid line). The site on the left has concrete sleepers, and the site on the right has wooden sleepers.

We notice that for reference site 414.5w, the results of the direct and ARRoW measurements are very close. For the reference site 402.5e, the results differ: ARRoW overestimates the roughness. This difference can be explained by the differences in track dynamics. The second test site has wooden sleepers and a different fastening system that changes the track dynamics and decay rate. This results in a larger sound emission than we would expect at the reference site with the same roughness but with concrete sleepers and different decay rate. In the ARRoW postprocessing, we make no correction for the influence of the track dynamics, so the higher sound emission is interpreted as a higher rail roughness. This would explain the differences we find in the long/medium-wavelength range.

For short wavelengths (< 0.5 cm) and very low roughness (< -20 dB), we observed that the directly measured roughness was lower than the roughness measured with ARRoW. This difference is due to the wheel condition at the time of measurement. The wheel roughness was much higher than the rail roughness and hence dominates the combined roughness. The wheel roughness determines the lower sensitivity limit of any indirect rail roughness measurement system.

Whether or not to compensate for the 'track dynamics' effect depends on the purpose of the measurement. If ARRoW is used for track quality monitoring (e.g., for EU noise mapping), the measured sound levels itself indicate whether the track is in good or bad condition (either due to high roughness or due to low decay rate) and there is no need for compensation. For grinding monitoring, however, the track dynamics should be taken into account.

4.3 Repeatability

To assess the repeatability of ARRoW results, the test site between Miramas and Avignon at km 29.3 was measured three times at the same speed. The results of ARRoW and the direct rail roughness measurement are shown in figure 4. From these results, we conclude that ARRoW delivers repeatable results throughout the whole wavelength range.

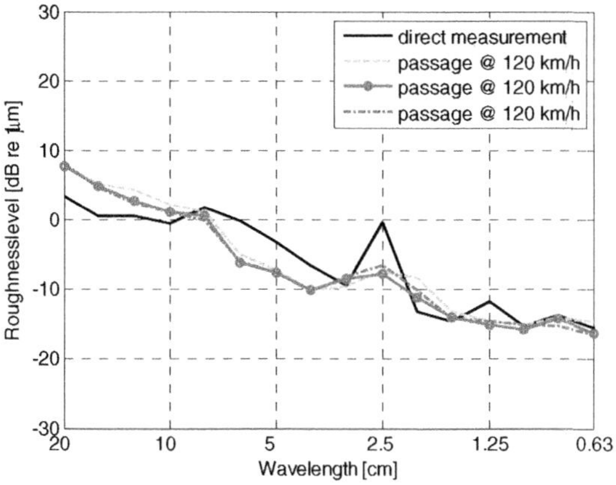

Fig 4. Repeatability of ARRoW measurements at 120 km/h

4.4 The Influence of Speed

To investigate the effect of driving speed, we measured one test site three times at different speeds: 60, 120 and 160 km/h. Fig 5. shows the results of ARRoW and direct measurements. For longer wavelengths, we find almost no influence of the speed. For wavelengths shorter than 6.3 cm, we see that the ARRoW spectra are

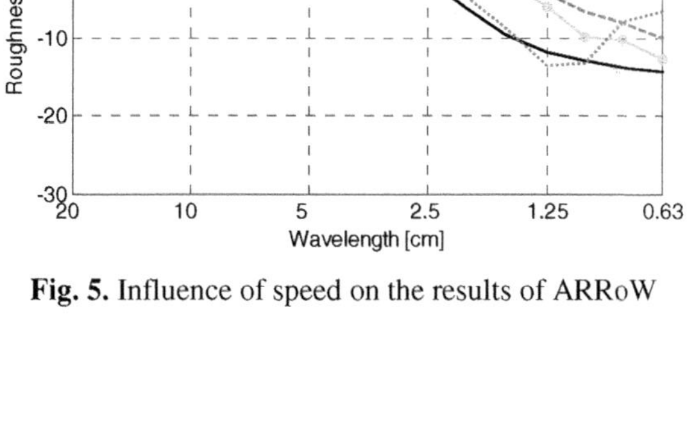

Fig. 5. Influence of speed on the results of ARRoW

close to the direct measurement result at different wavelengths. At lower speeds, we are closer to the direct measurement spectrum at shorter wavelengths. From this, we conclude that the calibration at low speeds differs from the calibration at moderate and high speeds. This completely agrees with earlier work [4].

The comparison between direct and indirect measurement shows that even if the calibration has been carried out at 120km/h, it is not really the speed at which we get the best agreement. Moreover, at longer wavelengths, corresponding to lower frequencies, the agreement deteriorates, probably because of the dynamic behavior of the track that differs from reference site to this test site.

5 Conclusions

ARRoW has been tested in the LECAV campaign to provide roughness spectra and an acoustic quality indicator for a significant part of the French railway network. ARRoW produced results that were in good agreement with the roughness spectra obtained with direct measurements. It was also shown that the repeatability of the results is good when measuring at the same speed.

The campaign also showed the importance of wheel roughness and track dynamics for the results. The lower sensitivity limit of the system is determined by the wheel roughness: the weakest (i.e., roughest) component determines the total accuracy, so it is important to keep the wheel in good condition. If the effect of track dynamics is not taken into account, then noise level changes due to lower damping or wooden sleepers might be erroneously attributed to rail roughness. For roughness monitoring, this is unwanted, but, for EU noise mapping, this is a unique opportunity: ARRoW measures *directly* the acoustic quality of the track, albeit influenced by rail roughness or other track properties. This is the reason we call ARRoW a "rail influence" measurement system rather than strictly a "rail roughness" measurement system.

References

[1] Directive 2002/49/EC of the European Parliament and of the Council of June 25, 2002 relating to the assessment and management of environmental noise. OJEC 189(12), July 18 (2002)

[2] ISO 3095: 2005: Railway Applications - Acoustics - Measurement of noise emitted by railbound vehicle, August 15 (2005)

[3] Kuijpers, A.H.W.M.: Rail roughness monitoring in the Netherlands. NNFM, vol. 99, pp. 348–354 (2008)

[4] Verheijen, E., van Beek, A.: Definition of track influence: roughness in rolling noise, Harmonoise deliverable D12 part 1, appendix 7, doc. id. HAR12TR-020813-AEA10, July 17 (2003)

[5] Kuijpers, A.H.W.M., Schwanen, W., van Dijck, B.H.C.A.: Controlling rail roughness as a noise mitigation measure: from pilot project to implementation (in Dutch), M+P report M+P.RAIL.06.15.11 rev 4, February 2 (2009)

Estimation of Rail Irregularity by Axle-Box Accelerometers on a High-Speed Train

J.S. Lee[1], S. Choi[2], S.S. Kim[2], and C. Park[2]

[1] University of Science and Technology/Korea Railroad Research Institute, 360-1, Woram-dong, Uiwang-si, Gyeonggi-do, 437-757, Republic of Korea
Tel.: +82 31 460 5272; Fax: +82 31 460 5649,
e-mail: junseok@krri.re.kr

[2] Korea Railroad Research Institute, Division of High Speed Rail System Research, 360-1, Woram-dong, Uiwang-si, Gyeonggi-do, 437-757, Republic of Korea

Summary

This paper is focused on the rail irregularity estimation from axle-box vibration of an in-service high-speed train. Usually, the rail irregularity has been measured by a special railway inspection system during maintenance work. However, the procedure requires some mechanical or optical sensors that are sensitive to harsh railway environments, and it is too slow for in-service trains. Therefore, the procedure has to be performed in limited time and track sections. As a solution to this problem, measuring axle-box acceleration is proposed to estimate the lateral and vertical rail irregularities. For reliable estimation of displacement from the measurement, a method based on a Kalman filter and wavelength band-pass filters is proposed. The estimated results are compared with a railway inspection system. It is confirmed that the proposed method can provide useful rail estimates without any mechanical or optical sensors.

1 Introduction

Vibration of the railway vehicles is mostly caused by an interaction between the wheel and rail and is transmitted to the car-body via the bogie. Dynamic characteristics of the vehicle and the rail irregularity play important roles in determining passenger comfort. In particular, the rail irregularity needs to be monitored and managed continuously with appropriate methods.

The rail irregularity is composed of various wavelengths, and each of them affects the safety and ride quality of the railway vehicles. For conventional speeds below 150 km/h, the waveband from 3 to 25 m contributes to running safety of vehicles and failure of rail and the waveband from 25 to 70 m contributes to ride comfort of passengers. In addition, for high speeds above 200 km/h, rail irregularity

T. Maeda et al. (Eds.): Noise and Vibration Mitigation for Rail Trans. Sys., NNFM 118, pp. 571–578.
springerlink.com

with the waveband from 3 to 70 m affects running safety of vehicles and failure of rail and the waveband from 70 to 150 m significantly affects ride comfort of passengers [1].

Rail irregularity has been measured by a special inspection system using the versine with 10 m of symmetrical chord length during maintenance work. However, it needed some mechanical or optical sensors that were sensitive to a harsh railway environment. With those sensors, the inspection system was very expensive; hence, only a small number of special inspection vehicles could be implemented. Moreover, the special inspection vehicles were slower than the in-service trains because of using contact probes. Thus, the measurement could be taken only for limited sections at night. For example, the railway inspection vehicle, EM-120, which has been widely used in Korea for rail irregularity measurements of the conventional and high-speed railway track with the versine method, is limited to 120 km/h [2]. As a solution of the problem, measuring the acceleration of an in-service high-speed train has been proposed to monitor the rail irregularity repeatedly in the revenue service operations [3].

In this research, an estimation method is proposed to measure the rail irregularity from the axle-box vibration data. To estimate the rail irregularity, a combined method using a Kalman filter and several sets of the wavelength band-pass filters are proposed. To verify the estimation method, the results from a commercial railway inspection system are used as reference data. Finally, performances of the proposed estimation method are discussed.

2 Data Processing Procedure

In the measuring rail irregularity procedure, both acceleration and displacement data have been required. Theoretically, the displacement data, x, can be simply calculated by double integrating the acceleration data, a, as following:

$$x = x_0 + v_0 \cdot t + \int_0^t \left[\int_0^\tau a(\tau')d\tau' \right] d\tau \tag{2.1}$$

where x_0 and v_0 are initial condition of the integration. In practice, the double integration causes unrealistic drifts in the displacement result. The drifts have been identified with nonzero initial condition of the data and the direct current (DC) offset or measurement noise of mechanical or electrical hysteresis in the sensor or cable [4]. In this research, some signal processing techniques, based on statistical methods, are adopted to resolve the unrealistic drift problem.

2.1 State-Space Model

To establish a statistical model for the estimation, two assumptions are introduced as follows: The model follows a random process whose future probabilities are only determined by present values, called a Markov process, and it can be represented by process and measurement models.

Based on the previous assumptions, a statistical model based on a random walk model is established in a discrete-time state-space model to estimate displacement from acceleration data as following [5]:

$$\boldsymbol{x}_{n+1} = \boldsymbol{F}\boldsymbol{x}_n + \boldsymbol{G}\boldsymbol{u}_n \tag{2.2}$$

$$\boldsymbol{y}_n = \boldsymbol{H}\boldsymbol{x}_n + \boldsymbol{v}_n \tag{2.3}$$

where

$$\boldsymbol{F} = \begin{bmatrix} \alpha & 0 & 0 \\ 1 & 0 & 0 \\ 0 & 1 & 0 \end{bmatrix}, \boldsymbol{G} = \begin{bmatrix} 1 \\ 0 \\ 0 \end{bmatrix}, \boldsymbol{H} = \frac{1}{T_s^2}[1 \quad -2 \quad 1] \tag{2.4}$$

T_s: sampling interval, α: model parameter ($0 << \alpha < 1$)

In the state-space form, $\boldsymbol{x}_n$ and $\boldsymbol{y}_n$ are the estimated displacement and the measured acceleration, respectively. $\boldsymbol{F}$ is the state transition matrix that is applied to the previous state, and $\boldsymbol{H}$ is the measurement matrix that maps estimated displacement into measured acceleration. The process noise and measurement noise, $\boldsymbol{u}_n$ and $\boldsymbol{v}_n$, are assumed to be zero-mean white Gaussian and mutually independent process. The initial displacement x_0 is assumed to have zero. Their auto-covariance and cross-covariance are compactly denoted as following:

$$E\left[\begin{bmatrix} \boldsymbol{u}_n \\ \boldsymbol{v}_n \\ \boldsymbol{x}_0 \\ 1 \end{bmatrix} \begin{bmatrix} \boldsymbol{u}_m \\ \boldsymbol{v}_m \\ \boldsymbol{x}_0 \end{bmatrix}^*\right] = \begin{bmatrix} \begin{bmatrix} \boldsymbol{Q}_n & \boldsymbol{S}_n \\ \boldsymbol{S}_n^* & \boldsymbol{R}_n \end{bmatrix} \delta_{nm} & \begin{matrix} 0 \\ 0 \end{matrix} \\ \begin{matrix} 0 & 0 \end{matrix} & \boldsymbol{\Pi}_0 \\ \begin{matrix} 0 & 0 \end{matrix} & 0 \end{bmatrix} \tag{2.5}$$

In this equation, $\boldsymbol{Q}_n$ and $\boldsymbol{R}_n$ are the auto-covariance of $\boldsymbol{u}_n$ and $\boldsymbol{v}_n$, and $\boldsymbol{S}_n$ is the corresponding cross-covariance of the noise, and $\boldsymbol{\Pi}_0$ is an auto-covariance of the initial displacement x_0.

2.2 Kalman Filter

To estimate the displacement from the measured noisy axle-box acceleration data, a Kalman filter is introduced as an optimal state estimation method. The optimal state estimation minimizes the expected value of the squared error between the measured noisy acceleration data and the computed acceleration data from the estimated displacement.

The discrete-time Kalman filter algorithm for the previous state-space model can be summarized as follows [6]:

(1) Initial condition

$$\hat{\boldsymbol{x}}_{0|-1} = 0 \qquad , \boldsymbol{P}_{0|-1} = \boldsymbol{\Pi}_0 \tag{2.6}$$

(2) Recursion relations

- Innovation: $$\boldsymbol{e}_n = \boldsymbol{y}_n - \boldsymbol{H}\hat{\boldsymbol{x}}_{n|n-1} \tag{2.7}$$

- Innovation covariance: $$\boldsymbol{R}_{e,n} = \boldsymbol{H}\boldsymbol{P}_{n|n-1}\boldsymbol{H}^* + \boldsymbol{R}_n \tag{2.8}$$

- Kalman prediction gain: $$\boldsymbol{K}_{p,n} = (\boldsymbol{F}\boldsymbol{P}_{n|n-1}\boldsymbol{H}^* + \boldsymbol{G}\boldsymbol{S}_n)\boldsymbol{R}_{e,n}^{-1} \tag{2.9}$$

- State estimation: $$\hat{\boldsymbol{x}}_{n+1|n} = \boldsymbol{F}\hat{\boldsymbol{x}}_{n|n-1} + \boldsymbol{K}_{p,n}\boldsymbol{e}_n \tag{2.10}$$

- State error covariance:

$$\boldsymbol{P}_{n+1|n} = \boldsymbol{F}\boldsymbol{P}_{n|n-1}\boldsymbol{F}^* + \boldsymbol{G}\boldsymbol{Q}_n\boldsymbol{G}^* - \boldsymbol{K}_{p,n}\boldsymbol{R}_{e,n}\boldsymbol{K}_{p,n}^* \tag{2.11}$$

2.3 Band-Pass Filter

As mentioned in the introduction, the rail irregularity is composed of various wavelengths. Therefore, the estimated displacement should be classified into each waveband using some wavelength band-pass filters.

The band-pass filter is realized by the linear difference equation [7]:

$$w_n = -\sum_{k=1}^{N} a_k w_{n-k} + \sum_{k=0}^{M} b_k \hat{x}_{n-k} \tag{2.12}$$

This difference equation sets up a frequency selective filtering operation on the raw displacement x_n to produce the filtered displacement w_n with the two parameters, b_k and a_k, which determine the frequency response characteristics of the filter. To realize the wavelength band-pass filter, frequency should be transformed to the wavelength by following relationship:

$$\lambda = \frac{v}{f} \tag{2.13}$$

(v: train speed, λ: wavelength)

For example, the train speed at 300 km/h, the wavelengths 3, 25, 70, and 150 m correspond to the frequencies 27.8, 3.3, 1.2, and 0.6 Hz, respectively.

Finally, a Kalman filter and the wavelength band-pass filters are consecutively realized in a recursive structure, and they repeat until the end of data. Those frequencies are updated along with the train speed.

3 Measurement Setup

Sensors and a data acquisition system as well as a railway inspection system were mounted on a bogie of the Korean High-Speed Train (Hanvit 350). The train is a result of national research projects for the development of the high-speed train technology.

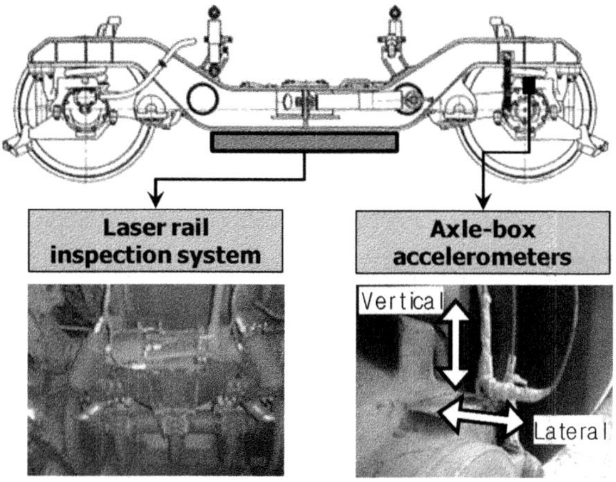

Fig. 1. Measurement setup

3.1 Railway Inspection System

For verification of the proposed method, the measurement result from a laser railway inspection system on the same bogie, shown in Fig. 1, is used as a reference data. The inspection system can measure the profile and geometry of rail up to 320 km/h. The measured parameters are track gauge, horizontal alignment, surface, super-elevation, and twist [8].

3.2 Sensors and Data Processing

Two integral electronics piezo electric (IEPE) accelerometers were mounted on the axle-box in both vertical and lateral directions near the railway inspection system to minimize relative delay in the measurement. The reasons for using IEPE-type accelerometers were its small size, moderate price, and broadband characteristics, without any additional signal conditioner. The measureable frequency range of the accelerometers was from 1 to 10,000 Hz. Each acceleration signal was conditioned and converted from analog to digital at 20,480 Hz sampling frequency with a data acquisition board.

4 Results

From Fig. 2 to Fig. 3, the train speed and the estimated rail irregularities are presented with the results from the railway inspection system. The focused section is the slab track where the train passed some notable rail irregularity at a speed of nearly 300 km/h.

4.1 Estimated Lateral Rail Irregularity and Horizontal Alignment

For the waveband from 3 to 25 m, the result from a railway inspection system is similar to the estimated result from the lateral acceleration. However, for the

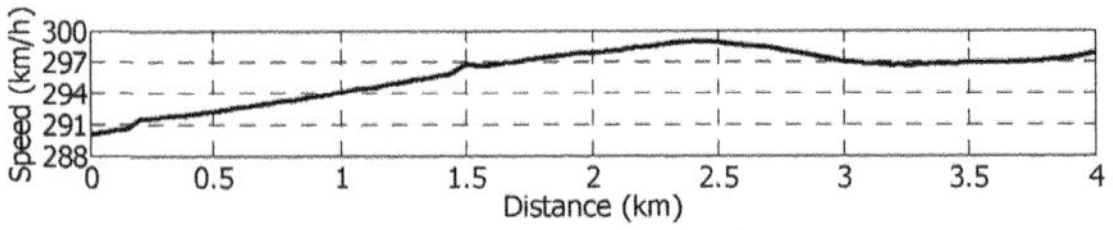

Fig. 2 Train speed

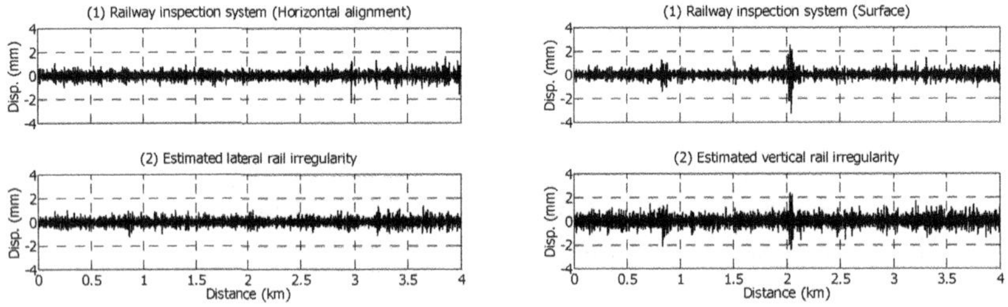

(a) Waveband: 3 ~ 25 m

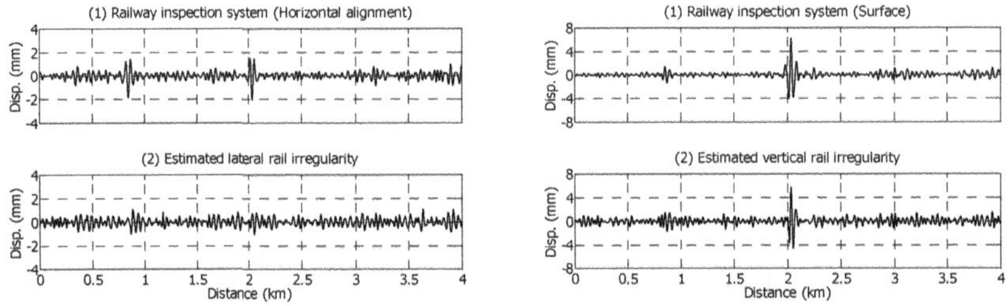

(b) Waveband: 25 ~ 70 m

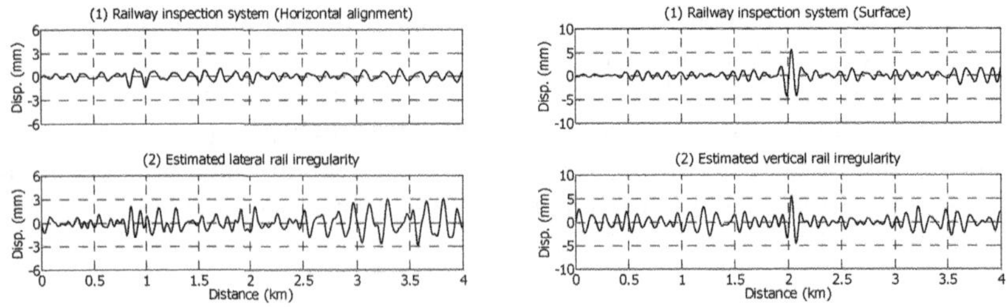

(c) Waveband: 70 ~ 150 m

Fig. 3. Comparison of Rail irregularities: Railway inspection system and estimated from axle-box mounted accelerometers

waveband from 25 to 70 m and from 70 to 150 m, the estimated result is larger than the result obtained from the railway inspection system, except some peaks in the horizontal alignment for the waveband from 25 to 70 m.

For the waveband from 25 to 70 m, some additional minor waves are observed in the estimated lateral rail irregularity. The additional waves are assumed to be caused by relative lateral motion of the wheel and rail. For the waveband from 70 to 150 m, amplitude in the estimated results is largely different from the result obtained from the railway inspection system. This is probably caused by the frequency characteristics of the IEPE accelerometers. The accelerometers are able to measure from 1 to 10,000 Hz. However, the waveband from 70 to 150 m is equivalent to the frequency band from 1.2 to 0.6 Hz for a train speed of 300 km/h, and this is within or beyond the accelerometer's capability, and leads to the large error.

From the result, the lateral acceleration includes not only the horizontal alignment of rail, but also other lateral dynamic motions, which are presumably caused by the wheel and rail interaction. Therefore, some additional processing methods should be adopted for reliable measurement of the lateral rail irregularity.

4.2 Estimated Vertical Rail Irregularity and Surface

In contrast to the lateral rail irregularity, the vertical rail irregularity shows good agreement with the result from the railway inspection system.

For the waveband from 3 to 25 m, one major irregularity is clearly observed and its position and amplitude are consistent with each other. For the waveband from 25 to 70 m, the major irregularity as well as some minor irregularities is also clearly observed and they are very consistent. For the waveband from 70 to 150 m, the major irregularity in the result obtained from the railway inspection system agrees with the estimated result, but other irregularities are different. This is probably caused by the low-frequency characteristics of IEPE accelerometers.

These results indicate that vertical motion of the axle-box closely follows the surface. Hence, displacement estimation is sufficient for the estimation of the vertical irregularity without additional processing.

5 Conclusion

This research examined a method to measure the lateral and vertical rail irregularities of high-speed railway track from the axle-box vibration of the high-speed railway vehicles. The lateral and vertical axle-box vibration is measured by IEPE accelerometers. A Kalman filter and wavelength band-pass filters are used to estimate the displacement from the acceleration. These analyses lead to several concluding remarks as follows:

(1) The estimation method can estimate displacement stably from the noisy axle-box acceleration data in both the lateral and the vertical directions. Therefore, the method could be a useful tool to estimate the rail irregularity from axle-box acceleration of the high-speed trains.

(2) Comparisons with the results from the railway inspection system suggest that the estimated displacement results from accelerometers provide information about the rail irregularities. The estimated lateral irregularity is fairly consistent with the horizontal alignment results of the railway inspection system. The estimated vertical irregularity is good agreement with the surface results of the inspection system, especially the wavebands from 3 to 25 m and from 25 to 70 m. However, the waveband from 70 to 150 m, both irregularities show relatively large difference due to the low-frequency characteristics of the IEPE accelerometers. Presumably, extending the low-frequency response of the accelerometers might improve the agreement.

Capacitive accelerometers with DC coupling will be applied to resolve the problem of low-frequency characteristics, and some additional processing methods will be applied to improve the result of lateral rail irregularity.

Acknowledgment

This research was supported by the Next-generation High-speed Train Project (grant number A01) from the Development of Future Rail Technology Program funded by the Ministry of Land, Transport and Maritime Affairs of the Korean government. The authors thank the anonymous reviewers for a lot of helpful comments.

References

[1] Kang, K.: A study of the long wave track defect analysis for high speed railways. Journal of the Korean Society for Railway 8(2), 111–115 (2005)

[2] Kang, K.: A study of analysis method for the track geometry measuring data on high speed railway. Journal of the Korean Society for Railway 7(4), 291–295 (2004)

[3] Takeshita, K.: Track irregularity inspection method by revenue-earning train. RTRI report 9(2), 29–34 (1995)

[4] Yang, J., Li, J.B., Lin, G.: A simple approach to integration of acceleration data for dynamic soil-structure interaction analysis. Soil Dynamics and Earthquake Engineering 26(8), 725–734 (2006)

[5] Ogata, K.: Discrete-time control systems. Prentice-Hall Inc., Englewood Cliffs (1995)

[6] Sayed, A.H.: Adaptive filters. Wiley Inter-science, Chichester (2008)

[7] Proakis, J.G.: Digital signal processing. Pearson Education International, London (2007)

[8] Kim, S.S., Park, C.: High speed rail measurement system of HSR-350x. Journal of the Korean Society for Railway 11(2), 115–119 (2008)

Author Index

Notes on Numerical Fluid Mechanics and Multidisciplinary Design

Available Volumes

Volume 118: Tatsuo Maeda, Pierre-Etienne Gautier, Carl E. Hanson, Brian Hemsworth, James Tuman Nelson, Burkhard Schulte-Werning, David Thompson, and Paul de Vos (eds.): Noise and Vibration Mitigation for Rail Transportation Systems. ISBN 978-4-431-53926-1

Volume 116: Michael Klaas, Edmund Koch, and Wolfgang Schröder (eds.): Fundamental Medical and Engineering Investigations on Protective Artificial Respiration. ISBN 978-3-642-20325-1

Volume 115: Egon Krause, Yurii Shokin, Michael Resch, Dietmar Kröner, Nina Shokina (eds.): Computational Science and High Performance Computing IV. ISBN 978-3-642-17769-9

Volume 114: Piotr Doerffer, Charles Hirsch, Jean-Paul Dussauge, Holger Babinsky, and George N. Barakos (eds.): Unsteady Effects of Shock Wave Induced Separation. ISBN 978-3-642-03003-1

Volume 112: Andreas Dillmann, Gerd Heller, Michael Klaas, Hans-Peter Kreplin, Wolfgang Nitsche, and Wolfgang Schröder (eds.): New Results in Numerical and Experimental Fluid Mechanics VII – Contributions to the 16th STAB/DGLR Symposium Aachen, Germany 2008. ISBN 978-3-642-14242-0

Volume 111: Shia-Hui Peng, Piotr Doerffer, and Werner Haase (eds.): Progress in Hybrid RANS-LES Modelling – Papers Contributed to the 3rd Symposium on Hybrid RANS-LES Methods, Gdansk, Poland, June 2009. ISBN 978-3-642-14167-6

Volume 110: Michel Deville, Thien-Hiep Lê, and Pierre Sagaut (eds.): Turbulence and Interactions – Proceedings the TI 2009 Conference. ISBN 978-3-642-14138-6

Volume 109: Wolfgang Schröder (ed.): Summary of Flow Modulation and Fluid-Structure Interaction Findings – Results of the Collaborative Research Center SFB 401 at the RWTH Aachen University, Aachen, Germany, 1997-2008. ISBN 978-3-642-04087-0

Volume 108: Rudibert King (ed.): Active Flow Control II – Papers Contributed to the Conference "Active Flow Control II 2010", Berlin, Germany, May 26–28, 2010. ISBN 978-3-642-11734-3

Volume 107: Norbert Kroll, Dieter Schwamborn, Klaus Becker, Herbert Rieger, Frank Thiele (eds.): MEGADESIGN and MegaOpt – German Initiatives for Aerodynamic Simulation and Optimization in Aircraft Design. ISBN 978-3-642-04092-4

Volume 106: Wolfgang Nitsche, Christoph Dobriloff (eds.): Imaging Measurement Methods for Flow Analysis - Results of the DFG Priority Programme 1147 "Imaging Measurement Methods for Flow Analysis" 2003–2009. ISBN 978-3-642-01105-4

Volume 105: Michel Deville, Thien-Hiep Lê, Pierre Sagaut (eds.): Turbulence and Interactions - Keynote Lectures of the TI 2006 Conference. ISBN 978-3-642-00261-8

Volume 104: Christophe Brun, Daniel Juvé, Michael Manhart, Claus-Dieter Munz: Numerical Simulation of Turbulent Flows and Noise Generation - Results of the DFG/CNRS Research Groups FOR 507 and FOR 508. ISBN 978-3-540-89955-6

Volume 103: Werner Haase, Marianna Braza, Alistair Revell (eds.): DESider – A European Effort on Hybrid RANS-LES Modelling - Results of the European-Union Funded Project, 2004–2007. ISBN 978-3-540-92772-3

Volume 102: Rolf Radespiel, Cord-Christian Rossow, Benjamin Winfried Brinkmann (eds.): Hermann Schlichting – 100 Years - Scientific Colloquium Celebrating the Anniversary of His Birthday, Braunschweig, Germany 2007. ISBN 978-3-540-95997-7

Volume 101: Egon Krause, Yurii I. Shokin, Michael Resch, Nina Shokina (eds.): Computational Science and High Performance Computing III - The 3rd Russian-German Advanced Research Workshop, Novosibirsk, Russia, 23–27 July 2007. ISBN 978-3-540-69008-5

Volume 100: Ernst Heinrich Hirschel, Egon Krause (eds.): 100 Volumes of 'Notes on Numerical Fluid Mechanics' - 40 Years of Numerical Fluid Mechanics and Aerodynamics in Retrospect. ISBN 978-3-540-70804-9

Volume 99: Burkhard Schulte-Werning, David Thompson, Pierre-Etienne Gautier, Carl Hanson, Brian Hemsworth, James Nelson, Tatsuo Maeda, Paul de Vos (eds.): Noise and Vibration Mitigation for Rail Transportation Systems - Proceedings of the 9th International Workshop on Railway Noise, Munich, Germany, 4–8 September 2007. ISBN 978-3-540-74892-2

Volume 98: Ali Gülhan (ed.): RESPACE – Key Technologies for Reusable Space Systems - Results of a Virtual Institute Programme of the German Helmholtz-Association, 2003–2007. ISBN 978-3-540-77818-9

Volume 97: Shia-Hui Peng, Werner Haase (eds.): Advances in Hybrid RANS-LES Modelling - Papers contributed to the 2007 Symposium of Hybrid RANS-LES Methods, Corfu, Greece, 17–18 June 2007. ISBN 978-3-540-77813-4

Volume 96: C. Tropea, S. Jakirlic, H.-J. Heinemann, R. Henke, H. Hönlinger (eds.): New Results in Numerical and Experimental Fluid Mechanics VI - Contributions to the 15th STAB/DGLR Symposium Darmstadt, Germany, 2006. ISBN 978-3-540-74458-0

Volume 95: R. King (ed.): Active Flow Control - Papers contributed to the Conference "Active Flow Control 2006", Berlin, Germany, September 27 to 29, 2006. ISBN 978-3-540-71438-5

Volume 94: W. Haase, B. Aupoix, U. Bunge, D. Schwamborn (eds.): FLOMANIA - A European Initiative on Flow Physics Modelling - Results of the European-Union funded project 2002 - 2004. ISBN 978-3-540-28786-5

Volume 93: Yu. Shokin, M. Resch, N. Danaev, M. Orunkhanov, N. Shokina (eds.): Advances in High Performance Computing and Computational Sciences - The Ith Khazakh-German Advanced Research Workshop, Almaty, Kazakhstan, September 25 to October 1, 2005. ISBN 978-3-540-33864-2

Volume 92: H.J. Rath, C. Holze, H.-J. Heinemann, R. Henke, H. Hönlinger (eds.): New Results in Numerical and Experimental Fluid Mechanics V - Contributions to the 14th STAB/DGLR Symposium Bremen, Germany 2004. ISBN 978-3-540-33286-2

Volume 91: E. Krause, Yu. Shokin, M. Resch, N. Shokina (eds.): Computational Science and High Performance Computing II - The 2nd Russian-German Advanced Research Workshop, Stuttgart, Germany, March 14 to 16, 2005. ISBN 978-3-540-31767-8

Volume 87: Ch. Breitsamter, B. Laschka, H.-J. Heinemann, R. Hilbig (eds.): New Results in Numerical and Experimental Fluid Mechanics IV. ISBN 978-3-540-20258-5

Volume 86: S. Wagner, M. Kloker, U. Rist (eds.): Recent Results in Laminar-Turbulent Transition - Selected numerical and experimental contributions from the DFG priority programme 'Transition' in Germany. ISBN 978-3-540-40490-3

Volume 85: N.G. Barton, J. Periaux (eds.): Coupling of Fluids, Structures and Waves in Aeronautics - Proceedings of a French-Australian Workshop in Melbourne, Australia 3-6 December 2001. ISBN 978-3-540-40222-0

Volume 83: L. Davidson, D. Cokljat, J. Fröhlich, M.A. Leschziner, C. Mellen, W. Rodi (eds.): LESFOIL: Large Eddy Simulation of Flow around a High Lift Airfoil - Results of the Project LESFOIL supported by the European Union 1998 - 2001. ISBN 978-3-540-00533-9

Volume 82: E.H. Hirschel (ed.): Numerical Flow Simulation III - CNRS-DFG Collaborative Research Programme, Results 2000-2002. ISBN 978-3-540-44130-4

Volume 81: W. Haase, V. Selmin, B. Winzell (eds.): Progress in Computational Flow Structure Interaction - Results of the Project UNSI, supported by the European Union 1998-2000. ISBN 978-3-540-43902-8

GPSR Compliance

The European Union's (EU) General Product Safety Regulation (GPSR) is a set of rules that requires consumer products to be safe and our obligations to ensure this.

If you have any concerns about our products, you can contact us on ProductSafety@springernature.com

In case Publisher is established outside the EU, the EU authorized representative is:

Springer Nature Customer Service Center GmbH
Europaplatz 3
69115 Heidelberg, Germany

Batch number: 10755248

Printed by Printforce, the Netherlands